T0224546

High-Temperature Ordered Intermetallic Alloys II

MATERIALS RESEARCH SOCIETY SYMPOSIA PROCEEDINGS

ISSN 0272 - 9172

MATERIALS RESEARCH SOCIETY SYMPOSIA PROCEEDINGS

MATERIALS RESEARCH SOCIETY SYMPOSIA PROCEEDINGS

MATERIALS RESEARCH SOCIETY CONFERENCE PROCEEDINGS

VLSI-I—Tungsten and Other Refractory Metals for VLSI Applications, R. S. Blewer, 1986; ISSN: 0886-7860; ISBN: 0-931837-32-4

VLSI-II—Tungsten and Other Refractory Metals for VLSI Applications II, E.K. Broadbent, 1987; ISSN: 0886-7860; ISBN: 0-931837-66-9

TMC—Ternary and Multinary Compounds, S. Deb, A. Zunger, 1987; ISBN:0-931837-57-x

MATERIALS RESEARCH SOCIETY SYMPOSIA PROCEEDINGS VOLUME 81

High-Temperature Ordered Intermetallic Alloys II

Symposium held December 2-4, 1986, Boston, Massachusetts, U.S.

EDITORS:

N. S. Stoloff
Rensselaer Polytechnic Institute, Troy, New York, U.S.A.

C. C. Koch
North Carolina State University, Raleigh, North Carolina, U.S.A.

C. T. Liu
Oak Ridge National Laboratory, Oak Ridge, Tennessee, U.S.A.

O. Izumi
Tohuku University, Sendai, Japan

MRS MATERIALS RESEARCH SOCIETY
Pittsburgh, Pennsylvania

CAMBRIDGE UNIVERSITY PRESS
Cambridge, New York, Melbourne, Madrid, Cape Town,
Singapore, São Paulo, Delhi, Mexico City

Cambridge University Press
32 Avenue of the Americas, New York NY 10013-2473, USA

Published in the United States of America by Cambridge University Press, New York

www.cambridge.org
Information on this title: www.cambridge.org/9781107405615

Materials Research Society
506 Keystone Drive, Warrendale, PA 15086
http://www.mrs.org

First published 1987
First paperback edition 2012

Single article reprints from this publication are available through
University Microfilms Inc., 300 North Zeeb Road, Ann Arbor, MI 48106

CODEN: MRSPDH

ISBN 978-0-931-83746-3 Hardback
ISBN 978-1-107-40561-5 Paperback

Contents

*Invited Paper

*Invited Paper

*Invited Paper

*Invited Paper

Preface

Many advances in research have been made since the first Materials
Research Society Symposium on High-Temperature Ordered Intermetallic
Alloys, held in Boston, November 26-28, 1984. That conference demonstrated
a resurgence of interest in potential uses of intermetallic alloys for
structural applications. The second conference, held in Boston,
December 2-4, 1986, underscored the proliferation of research activities
in this field. These proceedings of the 1986 Conference contain most
of the 18 invited and 42 contributed papers presented at the meeting. A
significant fraction of the papers were presented by overseas authors.
All manuscripts were peer-reviewed, with the invaluable assistance of
session co-chairmen as coordinators of the reviews. Thanks are due to
the individual reviewers, many of whom are also contributors to this
volume.

These proceedings demonstrate recent advances in determining the
mechanical properties of aluminides and other intermetallics, with
special emphasis upon the beneficial effects of boron on grain boundary
ductility of Ll_2 superlattices. Progress has also been made in
evaluating new (as well as more conventional) processing techniques,
especially in the powder metallurgical area. Two sessions devoted to
structure, thermodynamics and kinetics provided the major theoretical
underpinnings to the Conference. More such work is needed to help
establish a foundation for further alloying efforts. The symposium
also suggested the potential for development of multi-phase alloys
(preciptiation hardened, oxide dispersion strengthened and fiber
reinforced intermetallics), although work in this area is still in a

very early stage. Interest in multi-phase intermetallics is expected to grow rapidly.

Several organizations co-sponsored this conference: Dept. of Energy, Division of Energy Conversion and Utilization Technologies (E-Cut); NASA-Lewis Research Center; Dept. of Energy, Division of Materials Science, Office of Basic Energy Science; and the General Electric Corporation. We are grateful for their financial support and participation in the activities of the conference. Finally, we look forward to continued international cooperation in advancing this fascinating and important branch of materials science and engineering.

N.A. Stoloff
C.C. Koch
C.T. Liu
O. Izumi

March 1987

INTRODUCTION

Keynote Lecture

HIGH TEMPERATURE ORDERED INTERMETALLIC ALLOYS

D. P. Pope, University of Pennsylvania, Department of Materials Science and Engineering, Philadelphia, PA 19104-6391

ABSTRACT

This paper is intended to be a general introduction to this conference and is therefore not a review of the state of our current knowledge. Instead, it will address questions like the following: Why are intermetallic compounds interesting? What alloys are being studied, and which are being ignored? Since most research work is now being performed on Ll_2 alloys, with by far the greatest emphasis on Ni_3Al, the balance of the paper will concentrate on strengthening mechanisms and mechanisms of ductility control in Ni_3Al, pointing out the interesting questions and controversies which arose during this conference.

The conclusions to be drawn from this paper are that ordered intermetallic alloys are very valuable materials for high temperature use, but engineers probably must become more sophisticated in the use of materials with limited ductilities at low temperatures before intermetallics will gain widespread usage. Furthermore, additional research needs to be performed on more complex intermetallic compounds than Ll_2 since Ll_2 compounds, as a group, do not have particularly high melting temperatures. However, since alloys with complex structures, e.g. Laves phases, are well known for their brittleness at low temperatures, it is all the more important that the properties of such alloys be studied and methods found to improve their ductilities.

WHY ARE ORDERED INTERMETALLIC ALLOYS INTERESTING?

Intermetallic compounds occupy an intermediate position between metallic alloys based on solid solutions or solid solutions with second phase strengtheners, on the one extreme, and ceramics on the other. In this section, the properties of intermetallic compounds will be contrasted with those of disordered metallic alloys to show that it is possible to produce higher strength, higher use temperature alloys by basing them on intermetallic compounds rather than on disordered solid solutions.

Intermetallic compounds have a number of properties that make them intrinsically more appealing than disordered alloys for high temperature use. First, and perhaps most important, intermetallic compounds tend to be intrinsically very strong (high yield or fracture stress) and the strength tends to be maintained up to high temperatures. For example, the strength of TiAl is 450 MPa up to about 600°C, and only then begins to drop with increasing temperature [1]. Other intermetallic compounds, many with the $L1_2$ structure, actually show an increasing yield strength with increasing temperature, e.g. Ni_3Al [2]. Not only is the strength of intermetallic compounds maintained to high temperatures, the modulus also tends to be high and tends to decrease more slowly with increasing temperature than does that of disordered alloys. For example, Ray et al [3] have compared the temperature dependence of Young's modulus of an alloy based on Fe_3Al with that of an alloy based on a disordered Fe-base solid solution and shown that the modulus of the Fe_3Al-based alloy is the larger by a substantial margin between room temperature and about 800°C. Since stiffness is the controlling property for many applications, e.g. in systems subjected to vibrations, a high modulus is an important benefit.

In addition to having high strength and high stiffness, those compounds based on light elements, such as Ti_3Al, can have extremely low densities. The low density combined with the high strength and modulus give rise to very attractive specific properties (property divided by density), which are especially important for rotating machinery and aerospace applications. These kinds of considerations are discussed at greater length in chapter 5 of ref. [4].

Because of the ordered structure, intermetallic compounds tend to have much lower self diffusion coefficient than do disordered alloys. This can be demonstrated most clearly in those systems which show an order-disorder transition, e.g. CuZn, which has the bcc-based B2 structure [5]. The slope of the log D vs 1/T plot for both Cu and Zn shows a discontinuous increase as the temperature is reduced below Tc, indicating that the activation energies for self diffusion are much higher for the ordered than for the disordered alloy. As a result, the self diffusion coefficients of an ordered alloy can be several orders of magnitude smaller than that of the disordered alloy at a given temperature. Slow rates of diffusion bring with them the attendant advantages at elevated temperatures of improved microstructural stability and, since the creep rate is proportional to the diffusion coefficient, improved creep strength. This was shown very dramatically by Liu and coworkers [6] on an Fe-Ni-V-Ti $L1_2$ alloy where they showed that there is a discontinuous drop in the creep rate of over two orders of magnitude when the material is cooled through T_c.

Of course the advantages of improved strength, modulus, density, microstructural stability and creep rate which intermetallic compounds offer do not come without attendant disadvantages. The single largest disadvantage is low ductility - most intermetallic compounds tend to be quite brittle, especially at low temperatures. For example, the TiAl [1] referred to above as having a high strength up to 600°C is also very brittle up to that temperature and the ductility begins to increase only in the temperature range where the strength is rapidly decreasing. Also, the tensile ductility of Fe-Al alloys at room temperature goes almost to zero as the aluminum content is increased to 25 at.%, producing Fe_3Al [7]. In fact a loss of ductility is commonly the first indication of the occurrence of an intermetallic compound in an alloy. For example, my students and I once prepared an Fe-3 wt% Si alloy, which is normally very ductile, but because we did not allow adequate mixing of the liquid, the top of the melt contained most of the silicon. As a result, the first ingot poured from the crucible had sufficiently high Si content to produce Fe_3Si and the ingot shattered into small pieces when it was swaged. Also, small single crystals of Ni_3Al have been produced by several investigators simply by cooling a large-grained polycrystalline sample to a low temperature and breaking the sample along the grain boundaries with a sharp blow. As will be discussed later in this paper, there are many reasons for brittleness in intermetallic compounds, intergranular brittleness is only one of them.

WHICH INTERMETALLIC COMPOUNDS ARE CURRENTLY BEING STUDIED?

Most disordered alloys used for structural applications have either the fcc or bcc structure and only a very few have the hcp or other structures. In fact it is generally found that metals having the fcc and bcc structure are more ductile than those with more complex structures. So by analogy, it is perhaps not surprising that similar results have been found for intermetallic compounds. As a result, most studies have been performed on intermetallic compounds having fcc- or bcc- derivative structures. By far the most widely studied are those having the fcc-derivative $L1_2$ structure with composition A_3B, in which A atoms occupy the face centers and B occupy the corners. Examples are Ni_3Al and Cu_3Au. A brief perusal of the papers presented in this conference will convince the reader that current research is very heavily concentrated on these alloys. The next most widely studied intermetallic compounds are those having the bcc-derivative B2 structure with composition AB. A great deal is known about the mechanical properties of $L1_2$ and B2 alloys, although there are some surprisingly large gaps in our knowledge of creep and fatigue properties which are only now being filled, e.g., the papers in this volume by Stoloff, Krueger et al, Sauthoff et al, Anton et al, Schneibel et al, Chang et al, Vedula et al, Mendiratta and Ehlers, Shah and Duhl, and Hsu et al. Studies of oxidation and sulfidation are very scarce, and therefore the work of Meier on metal silicides and that of Natesan on nickel aluminides presented in this volume are particularly welcome.

The amount of research currently devoted to intermetallic compounds having structures other than $L1_2$ and B2 is exceedingly small. For example,

the work on hcp-derivative DO_{19} intermetallics (composition is A_3B) is
almost entirely devoted to Ti_3Al and Mn_3Sn, see the work by Lee et al and
Rowe et al in this volume. We know very little about the deformation and
strength-controlling mechanisms in these materials. Some materials having
more complex bcc-derivative structures have also been studied, e.g. those
with the DO_3 structure, such as Fe_3Al and Fe_3Si, see the papers by McKamey
et al, Diehm and Mikkola, and Mendiratta and Ehlers. If one now excludes
all intermetallics with fcc- bcc- and hcp-derivative structures, the number
of studies is vanishingly small. In this category, and reported in this
volume, are the studies of Yamaguchi et al on DO_{22} Al_3Ti, Kao et al on
Ni_4Mo, Huang et al on TiAl and then the two papers by Fleischer and by Shah
and Duhl in which they investigate possibly useful intermetallic compounds
for high temperature use, with special emphasis on materials of low density.
Separate quotes from these two papers are informative. R. Fleischer (from
his abstract): "A subtitle of this abstract might be, Why are Ll_2 structures unlikely to be the best intermetallics for high temperature use?" D.
Shah (from his oral presentation): "Give me an intermetallic compound based
on Ni_3Al and I will give you a better Ni-base superalloy". Both authors are
saying that future development of superalloys having better properties than
Ni-base superalloys will be based neither on Ni_3Al nor on Ll_2 materials in
general, but on other intermetallic compounds which have higher melting
temperatures and lower densities than Ni_3Al. This does not mean that alloys
based on Ni_3Al will have no utility, it only means that they will probably
not be used as substitutes for Ni-base superalloys, in heat engines for
example. If Fleischer and Shah are correct, and I believe they are, then
much more work needs to be performed on intermetallic compounds with high
melting temperatures. In general, these intermetallics do not have Ll_2, B2
or DO_{19} structures, they have many atoms per unit cell, and are likely to be
quite brittle at low temperatures. Our challenge is to find ways to impart
acceptable ductilities to alloys based on these compounds. This also implies that researchers must begin investigating these more complex
intermetallic compounds so that a data base can be built up on which to base
future alloy development. I end this discussion with a note of caution
supplied by N. Stoloff during the discussion of R. Fleischer's paper.
Professor Stoloff pointed out that if 30 years ago those trying to develop
better superalloys had considered melting temperature as the main criterion,
they would have abandoned Ni in favor of the refractory metals, and as a
result, the current generation of superalloys would not have been developed.
Consequently, as we screen the list of possibly useful intermetallics we
must be careful not to overemphasize one property.

Having discussed which intermetallic compounds are currently being
studied and having shown that the emphasis of current research is, for
better or worse, on Ll_2 materials, I will now discuss several of the major

themes of this conference, highlighting the questions and controversies
involved. The emphasis will necessarily be on Ll_2 materials, in general,
and on Ni_3Al in particular.

STRENGTHENING MECHANISMS IN INTERMETALLIC COMPOUNDS

Intermetallic compounds can be strengthened by all the usual methods
used to strengthen disordered alloys, but there are some mechanisms which
are peculiar to intermetallics. Those will be discussed in this section.

The ordered structure of intermetallic compounds gives rise to the fact
that the Burgers vector of a dislocation will, in general, be quite long
compared to that of disordered materials because the Burgers vector is not a
nearest neighbor distance. For example, in fcc materials it is $1/2[110]$,
but in Ll_2 materials it is $[110]$, i.e. it is twice as long in Ll_2 materials.
The increased length of the Burger's vector in intermetallic compounds gives
rise to two additional effects: (1) Complex dissociations of the total
dislocation into several partial dislocations can occur, each separated from
the other by a fault, such as an antiphase boundary (APB), superlattice
intrinsic stacking fault (SISF), or a complex stacking fault (CSF). These
faults will have low energies on only certain planes, and hence the number
of slip systems may be limited, especially at low temperatures, resulting in
brittleness at low temperatures, as discussed earlier for TiAl. In addition, anisotropy of fault energies can give rise to complex cross-slip or
dislocation core transformation phenomena, as is observed in Ni_3Al and
(probably) in CuZn. In addition, these complex dissociations may also mean
that the dislocation core is non-planar and hence lies on no slip plane.
When this occurs dislocations can move only when the temperature is sufficiently high to provide the necessary thermal energy to nucleate double
kinks on the dislocation [8]. As a result, the strength will be high (and
commonly the ductility will be low) at low temperatures.

Perhaps the best known example of an intermetallic compound whose
strength is apparently controlled by dislocation core transformations is
Ni_3Al [8]. In this material the APB energy on (111) planes is expected to
be lower than on (010) planes and therefore screw dislocations are expected
to cross-slip from (111) planes where they are mobile to (010) planes where
they are sessile. Since this process is thermally driven, the flow stress
increases with increasing temperature, as was first observed by Flinn [2].
Since the rate of cross-slip is also affected by different components of the
stress tensor, the CRSS for $(111)[\bar{1}01]$ slip depends on the orientation and
sense of the applied uniaxial stress, i.e. Schmid's law does not hold [9].
A similar orientation and temperature dependence of the CRSS for $(110)[111]$

slip is also observed in CuZn [10], although no detailed analysis of this
process has been performed for B2 materials. This unusual temperature
dependence of the CRSS observed in Ni_3Al is the subject of papers in this
volume by Tien, Yoo, Pope and Heredia, Miura et al, and Lin et al. Yoo
shows in his paper, for example, that cross-slip of screw dislocations from
(111) to (010) planes in Ll_2 materials can be driven primarily by elastic
interactions between 1/2[101] superpartial dislocations and that anisotropy
of the APB energy is not necessary. This is due to the fact that the elastic anisotropy in Ni_3Al is large, giving rise to a non-central force
interaction between the superpartials, which, in turn, produces a torque on
the leading dislocation, pushing it from the (111) towards the (010) plane.
The torque can result in a larger driving force for cross-slip than does the
APB energy anisotropy.

The questions of solid solution strengthening of Ni_3Al is addressed
directly and indirectly by a number of papers in this volume: Brenner and
Burke, Miller and Horton, Bohn and Viander, Williams et al, Lin et al and
Pope and Heredia. Two separate questions are addressed in these papers:
(1) Given that certain ternary elements, such as Hf, greatly increase the
strength of Ni_3Al, is the increased strength due to increased cross-slip,
and if it is, is the increased cross-slip due to increased anisotropy of the
APB energy (or increased torque), or is some other mechanism involved? The
answers to these questions are, yes, the strengthening is due to increased
cross-slip, but increased anisotropies seem not to be the reason. (2) The
strength increase of Ni_3Al which results from the addition of a given ternary element depends strongly on which site the element apparently occupies.
If the direction of the Ni_3Al solid solubility lobe in the ternary phase
diagram is taken as an indication of the site occupied by the ternary element [11], then elements which substitute for Al increase the CRSS of Ni_3Al
much more than those which substitute for Ni. Question: Do these ternary
elements really occupy those sites indicated by the ternary phase diagrams?
Several different techniques have been used to answer this question, including Rutherford backscattering, field ion microscopy and the perturbed
angular correlations method, and based on the results of papers presented at
this conference, the answer to this question appears to be a resounding
"maybe"! It can be safely concluded, however, that the determination of
site occupancy from the direction of the solubility lobes is, at best, a
questionable practice, but more experimental work is clearly required.

Finally, a number of papers in this volume relate strength to other
aspects of microstructure. The paper by Hazzledine and Hirsch show
unequivocally that APB tubes do, indeed, exist in Ni_3Al and Fe_3Al, and that
these tubes can provide substantial dislocation drag. Khadkikar et al
investigate the possibility of producing high strength two phase Ni_3Al +
NiAl alloys, Koch reviews the literature on the strengthening of intermetal-

lic compounds by dispersoids, and Noebe et al describe the results of their unique surface film-softening experiments performed on both NiAl and Ni_3Al.

DUCTILITY OF INTERMETALLIC COMPOUNDS WITH EMPHASIS ON Ni_3Al

As was demonstrated by Takasugi and Izumi [12], some Ll_2 alloys are brittle at room temperature due to grain boundary failure, e.g. Ni_3Al and Ni_3Si, while others are ductile at room temperature, e.g. Co_3Ti and Ni_3Mn. The brittleness of Ni_3Al at room temperature can be removed, however, by the addition of a few hundred wt. ppm of B [13] and making the alloy slightly Ni-rich [14]. This gives rise to two questions: (1) Why does B improve the grain boundary cohesion and (2) Why is stoichiometry so important?

Consider first the question of B: A small amount of B is sufficient to improve the ductility because it preferentially segregates to the grain boundaries [15] but the segregation is not uniform and there may be a B-rich phase in some boundaries [16]. B greatly improves the grain boundary cohesion of Ni_3Al, but it has a much smaller effect on Ni_3Si, as shown by Taub and Briant in this volume, and of FeAl, as shown by Vedula, also in this volume, and B seems to have *little* effect on grain boundary cohesion in any other intermetallic compound. The reasons for this are not currently understood. However, a great deal is now known about the kinetics and the homogeneity of B segregation in Ni_3Al, see the paper in this volume by White. The papers of Izumi and Liu also discuss the role of B in improving the ductility. One novel approach to this question is that of Schulson et al in this volume. They have shown that the slope of the yield strength vs. $(d)^{-0.8}$ curve, a modified Hall-Petch relation, for Ni_3Al decreases when B is added. They conclude from this that B promotes dislocation generation in the grain boundaries by increasing the mobility of grain boundary dislocations, thereby facilitating the propagation of slip through grain boundaries and relieving local stresses at the grain boundaries.

Now consider the question of stoichiometry: There are at least two approaches to this question. Consider first the approach taken by Chen et al in this volume. In single crystalline Ni_3Al the Al atoms are completely surrounded by 12 nearest neighbor Ni atoms, but Ni atoms are surrounded by 4 nearest neighbor Al atoms and 8 Ni atoms, i.e., there are no Al-Al nearest neighbors but many Al-Ni and Ni-Ni nearest neighbors. This means that Al-Al bonds are of higher energy than the others. In a random grain boundary there will, of necessity, be some of these high energy Al-Al bonds, but their number will decrease as the alloy is made increasingly Ni-rich. Consequently the trend of increased cohesion with increased Ni content is at least qualitatively explained. A second approach is that of Takasugi and Izumi [12] in which they argue that the average valence difference between the A and B atoms in an A_3B Ll_2 alloy determines the cohesion, a large difference implies good cohesion and a small difference implies poor cohesion. This idea is discussed in this volume by Izumi. The paper of Taub and Briant in this volume further explores this concept, and they suggest that average electronegativity difference should be used instead of valence difference. Using various electronegativity scales they show that the effects of many ternary additions on ductility, including B, can be rationalized using the modified Takasugi and Izumi approach. There is one problem with this approach, as shown in the paper in this volume by Dimiduk et al. They test the Takasugi and Izumi hypothesis by measuring the ductility of Ni_3Al as a function of the amount of Fe or Mn added. They found that the critical Fe or Mn necessary to improve the ductility is precisely the same amount necessary to cause the disordered fcc phase to appear in the microstructure. They then question the valence (or electronegativity) difference criterion since microstructure, not valence, appears to control the ductility. This is still an open question and requires more research.

Finally, we consider the ductility of Ni_3Al at elevated temperatures, as discussed in this volume by Liu. The ductility of polycrystalline Ni_3Al at room temperature can be greatly improved by control of the stoichiometry and by the addition of B, as discussed above. However, the ductility of polycrystalline Ni_3Al when tested in air shows a deep minimum at about $760^\circ C$, and neither B nor control of the stoichiometry has any beneficial effect. This ductility minimum is shown to be the result of oxygen penetration along grain boundaries during the test, since samples tested in a hard vacuum are ductile, samples exposed to air at temperature and then tested at temperature in vacuum are ductile, and samples deformed at temperature in a hard vacuum immediately fail intergranularly when air is introduced. The ductility at elevated temperatures can be dramatically improved through the addition of Cr and by producing an elongated grain structure in the sample.

SUMMARY AND CONCLUSIONS

Intermetallic compounds have properties which make them extremely interesting from both a scientific and technological viewpoint. They are scientifically interesting because they provide a whole range of phenomena which are not seen in disordered alloys, but against which we can test the applicability of classical theories of strengthening, deformation and fracture. Great progress has been made and continues to be made in this area. Intermetallic compounds are interesting from a technological point of view because they tend to be strong and rigid, they can have low densities, and they have low self diffusivities, but unfortunately they tend to be brittle, especially at low temperatures. The ductility problem is a particularly important one, since the development of future high temperature alloys with better high temperature properties than modern nickel base superalloys almost certainly will involve the use of intermetallics with complex atomic structures. These intermetallics will probably have very low ductilities at low temperatures. Consequently, it is important that the horizons of our research be expanded from our current overwhelming emphasis on $L1_2$ and B2-derivative alloys to include studies of more complex materials with higher melting temperatures.

ACKNOWLEDGEMENTS

The author gratefully acknowledges the support of the National Science Foundation under grant No. DMR85-01974 and the Office of Naval Research under grant No. 5-21233.

REFERENCES

1. H.A. Lipsitt, D. Shechtman and R.E. Shechtman and R.E. Schafrik, Met. Trans. A, 6A, 1991 (1975).
2. P.A. Flinn, Trans TMS-AIME, 218, 145 (1960).
3. R. Ray, V. Panchanathan and S. Isserow, J. of Metals 35(6), 30 (1983).
4. "Structural Uses for Ductile Ordered Alloys", National Materials Advisory Board Rpt. #NMAB-419, National Academy Press, Washington, D.C. (1984).
5. L.A. Girifalco in Diffusion, ASM, Metals Park, Ohio, p. 185 (1973).
6. C.T. Liu, Int. Met. Rev. 29, 168 (1984).
7. M.J. Morcinkowski, M.E. Taylor and F.X. Kayser, J. Mat. Sci. 10, 406 (1975).
8. D.P. Pope and S.S. Ezz, Int. Met. Rev. 29, 136 (1984).
9. V. Paidar, D.P. Pope, and V. Vitek, Acta Met. 32, 435 (1984).
10. Y. Umakoshi, M. Yamaguchi, Y. Namba and K. Murakami, Acta Met. 24, 89 (1976).
11. S. Ochiai, Y. Oya and T. Suzuki, Acta Met. 32, 289 (1984).
12. T. Takasugi and O. Izumi, Acta Met. 33, 1247 (1985).
13. K. Aoki and O. Izumi, Nippon Kinzoku Gakkaishi 43, 1190 (1979).
14. C.T. Liu, C.L. White and J.A. Horton, Acta Met. 33, 213 (1985).
15. C.L. White, R.A. Padgett, C.T. Liu and S.M. Yalisove, Scr. Met. 18, 1417 (1984).
16. J.A. Horton and M.K. Miller, Acta Met. 35, 133 (1987).

PART I

Ordering Behavior and Theory

AB INITIO THEORY OF THE GROUND STATE PROPERTIES OF ORDERED AND DISORDERED ALLOYS AND THE THEORY OF ORDERING PROCESSES IN ALLOYS

G. MALCOLM STOCKS,* D.M. NICHOLSON,* F.J. PINSKI,* W.H. BUTLER,*
P. STERNE,† W.M. TEMMERMAN,† B.L. GYORFFY,§ D.D. JOHNSON,§ A. GONIS,**
X.-G. ZHANG,** AND P.E.A. TURCHI††
*Oak Ridge National Laboratory, P.O. Box X, Oak Ridge, TN 37831
†Science and Engineering Research Council, Daresbury Laboratory, Daresbury, Warrington, United Kingdom
§H. H. Wills Physics Laboratory, University of Bristol, Bristol, United Kingdom
**Department of Physics, Northwestern University, Evanston, IL 60201
††University of California, Berkeley, CA, and Lawrence Livermore National Laboratory, Livermore, CA 94550

ABSTRACT

We review some of the advances in the calculation of the electronic structure and energetics of ordered and disordered alloys that hold out the possibility of obtaining, in the not-too-distant future, an ab initio theory of ordering and phase stability in alloys. In particular, we focus on the calculation of the ground state properties of Ni_3Al and discuss the competition between the $L1_2$ and DO_{22} ordered structures. We review the ab initio concentration functional theory of ordering developed by Gyorffy and Stocks and its application to the short-range-ordered solid-solution state in Cu_cPd_{1-c} alloys. Finally, we review the generalized perturbation method approach to calculation of multisite interchange potentials in Ni_3Al, Pd_3V, and Al_3Ti and again discuss $L1_2/DO_{22}$ competition as well as antiphase boundary energies in Ni_3Al.

INTRODUCTION

Much of the modern theory of the electronic structure and energetics of metals and alloys is based on the density functional theory (DFT) of Kohn, Hohenberg, and Sham.[1,2] Within this theory the complicated many-body problem of interacting nuclei and electrons is mapped onto a set of single particle-like equations that describe the interaction of a single electron or quasi-particle with the assembly of nuclei through an effective electron ion potential $V_{eff}(\vec{r})$. The complex exchange and correlation effects are included in $V_{eff}(\vec{r})$ through an exchange correlation potential that adds to the usual Hartree potential of the nuclei and electron cloud. Though the DFT formulation is, in principle, exact, approximations must be made since the precise form of the exchange-correlation potential is not known. Over the last few years, it has become clear that the local-spin-density approximation (LSDA) is of sufficient accuracy to allow the accurate calculation of the electronic structure and ground state (T ≈ 0 K) properties[3] (crystal structures, equilibrium lattice spacing, total energy, heats of formation and mixing, bulk modulus, etc.). Thus, the calculation of these quantities is reduced to self-consistently solving a set of coupled Hartree-like equations. In this self-consistency process the electronic charge density distribution $\rho(\vec{r})$ is made self-consistent with V_{eff}.

At the heart of self-consistency equations is a single-particle-like Schrödinger equation involving $V_{eff}(\vec{r})$. Since $V_{eff}(\vec{r})$ depends on all of the positions and types of nuclei, the solution of the Schrödinger equation for a macroscopic condensed system is still a complex problem. However, over the last two decades a great deal of progress has been made in solving the equations of DFT for systems of increasing complexity. It is now a routine matter to calculate the ground state properties of pure metals and ordered alloys having a few atoms per unit cell. Here the underlying lattice periodicity allows the use of any one of a number of band theory methods of varying degrees of sophistication and accuracy. Recently, the machinery of DFT has been implemented for random substitutional alloys.[4,5] Here, solution of the Schrödinger equation is a much more complex task since the underlying lattice periodicity is lost as a result of the occupancy of any site by the species comprising the alloy being (ideally) random. However, a solution to this problem that is of sufficient accuracy to allow extension of LSDA-DFT to random alloys is provided by the Korringa-Kohn-Rostoker coherent-potential approximation (KKR-CPA).[6,7] In this method the underlying Hamiltonian is a generalization of the muffin-tin potential model used in many of the standard band theory methods.[3] The disorder is treated within mean-field theory by use of the coherent-potential approximation. In the limit of an ordered compound or pure metal, the KKR-CPA method is rigorously the KKR band theory method.[6] With this availability of ab initio methods for treating pure metals, ordered intermetallic alloys, and disordered substitutional alloys, it now becomes possible to contemplate constructing a general theory of order-disorder processes and of alloy phase stability within the context of DFT. In the remainder of this paper, some attempts in this direction are briefly described. The results and methods are, in general, those of the authors and no attempt is made, within the strict length constraints of this paper, to make a general survey of the field and to compare and contrast approaches to understanding phase stability (see, for example, the papers of Carlsson,[8] of Chen, Voter, and Srolovitz,[9] and of Foiles[10] at this meeting).

GROUND STATE PROPERTIES OF ORDERED COMPOUNDS

Structural Energies

It is now 15 years since the first total energy calculations of metallic systems were performed by Frank Averill.[11] However, it was not until the publication of the results of calculations by the IBM group (see ref. 3) of the ground state properties (cohesive energy, equilibrium lattice spacing, and bulk modulus) of the metallic elements having $Z < 40$ that the power of DFT and band theory methods was fully demonstrated. Since this work, many successful applications of DFT including calculation of the heats of formation of ordered intermetallic compounds[12] have further demonstrated the utility of the method.

Within the context of calculating the energetics of ordered compounds, the task of calculating the small energy differences $E^{\alpha/\beta}$ associated with different structural arrangements of α and β phases of a given compound is amongst the most problematical. Typically, $E^{\alpha/\beta}$ is the order of 1 to 5 x 10^{-3} Ry/atom (1 to 5 mRy) whilst the total energies, E_{tot}, of even the 3d-transition metals are greater than 10^3 Ry/atom. In general, such studies require the use of the most sophisticated band structure methods available, the so-called full potential methods, e.g., FLAPW[13] and QKKR.[14] These methods do not make the muffin-tin potential approximation to V_{eff}, rather the full non-spherical shape dependence of V_{eff} imposed by the crystal symmetry is accounted for. In studies of $E^{\alpha/\beta}$ between structures where the volume fraction excluded by touching muffin-tin spheres is different (e.g., fcc and bcc) use of full potential methods is of paramount importance. However, studies of $E^{\alpha/\beta}$ in situations where excluded volume effects are negligible require much less sophisticated (and therefore computationally faster) methods, e.g., ASA-LMTO[15] and ASW[12] methods. This is the situation for A_3B compounds in the $L1_2$ and DO_{22} structures. Consider both $L1_2$ and DO_{22} as tetragonal structures with $c/a = 2$ that have eight atoms per unit cell. Tetragonal $L1_2$ is no more than a double unit cell of $L1_2$. Both crystal structures can be constructed from eight interpenetrating simple tetragonal sublattices, six A sublattices and two B sublattices. For the $L1_2$ structure, the A sublattice origins are at (1/2, 1/2, 0), (0, 1/2, 1/2), (1/2, 0, 1/2), (1/2, 1/2, 1), (1/2, 0, 3/2), and (0, 1/2, 3/2); the B sublattice origins are at (0, 0, 0) and (0, 0, 1) in units of the lattice spacing a. For the DO_{22} structure the sublattices have the same origins as those of $L1_2$ excepting that the species occupying the (0, 0, 1) and (1/2, 1/2, 1) sublattices are interchanged. Pictured in this way, the structures are identical in terms of the numbers and kinds of atoms, excluded volume, and the numbers of like and unlike nearest-neighbors. It therefore becomes possible to calculate the small energy differences between the $L1_2$ and DO_{22} using fast band theory methods.

Relative Stabilities of $L1_2$ and DO_{22} Structures in Ni_3Al

In Fig. 1 we show results of ASA-LMTO[15] calculations of the energies (per formula unit) of Ni_3Al in both the $L1_2$ and DO_{22} structures at several lattice spacings close to that observed experimentally. The particular self-consistent total energy code used was that developed at the Daresbury Laboratory by two of the authors (PS and WMT). With this code, energies can be converged to a numerical precision of approximately 1 μRy. The calculations are spin polarized and are based on the spin-dependent exchange correlation potential of von Barth and Hedin.[16]

The calculated equilibrium lattice spacing is essentially the same for both structures and is within 1% of that measured for the $L1_2$ structure. For the DO_{22} structure, no attempt was made to minimize the total energy with respect to the c/a ratio. On the basis of this calculation the $L1_2$ structure is stable with respect to DO_{22} (in obvious accord with experiment) and $E^{L1_2/DO_{22}} = 13.6$ mRy/formula unit (0.34 mRy/atom). Unfortunately, we have not yet analyzed these calculations sufficiently to see if we can find an explanation of the relative stability of the $L1_2$ structure in terms of the underlying energy band structure and density of states. However, the calculation serves to highlight the fact that modern electronic structure methods, when carefully applied, are capable of describing the energetics of structural rearrangements.

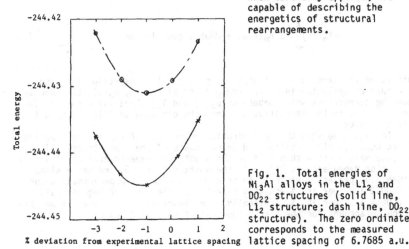

Fig. 1. Total energies of Ni_3Al alloys in the $L1_2$ and DO_{22} structures (solid line, $L1_2$ structure; dash line, DO_{22} structure). The zero ordinate corresponds to the measured lattice spacing of 6.7685 a.u.

TOWARDS A THEORY OF ORDERING IN ALLOYS

KKR-CPA Theory of the Disordered State

If we are to contemplate constructing an ab initio theory of alloy formation and stability, it is necessary to be able to treat the electronic structure and energetics of pure metals, ordered alloys, and disordered alloys on an equal footing. The KKR-CPA method provides a way of solving the Schrödinger equation for a random alloy that is of sufficient accuracy to allow its use in such a global theory. For a review of the development of the KKR-CPA see the reviews of Faulkner[17] and of Stocks and Winter.[4] The KKR-CPA method does not provide an exact solution of the DFT equations for a random alloy in the same sense as band theory methods do for ordered compounds. Rather, it provides a mean-field description of the effects of disorder on the electronic structure. This is accomplished by defining some effective ordered system, the scattering properties of which are chosen to best model the real disordered system. The KKR-CPA method involves a particular single-site prescription for choosing the effective ordered system and from this for calculating such quantities as the electronic density of states, charge density, and Bloch spectral function. Of course, being able to solve the Schrödinger equation for a random alloy is only one part of the full density functional theory. It is still necessary to obtain charge self-consistency and to have a method for calculating the total energy. Charge self-consistency is obtained by requiring the single-site charge densities $\bar{\rho}_\alpha(\vec{r})$ and effective muffin-tin potentials $V_\alpha(\vec{r})$ associated with the αth species to be self-consistent.[18] Once charge self-consistency has been achieved, the configurationally averaged total energy $\bar{E}$ is given by[5]

$$\bar{E} = \sum_\alpha c_\alpha \bar{E}_\alpha (\bar{\rho}_\alpha, Z_\alpha) , \tag{1}$$

where c_α is the concentration of the αth species and $\bar{E}_\alpha$ can be thought of as the average energy associated with α^{th} species. The expression for $\bar{E}_\alpha$ is of the form

$$\bar{E}_\alpha = \int_{-\infty}^{\varepsilon_F} \varepsilon \, n_\alpha(\varepsilon) d\varepsilon + \text{double counting terms} . \tag{2}$$

This is of the same form as that for a pure metal, consisting of a band structure contribution involving the density of states, $n_\alpha(\varepsilon)$, plus double counting corrections which depend on $\bar{\rho}_\alpha(r)$ and $V_\alpha(r)$ and which correct for the fact that in the band structure term the electron self-interaction is included twice.

In Fig. 2 we show the concentration dependence of the lattice spacing of fcc $Cu_c Zn_{1-c}$ alloys calculated on the basis of the above theory. The absolute values are within 1.5% of those obtained experimentally. The concentration dependence is linear in the α-phase ($c < 0.3$) and has a slope of 0.0045 a.u. per at. % which is to be compared with the experimental value of 0.0042 a.u. per at. % and that expected on the basis of Vegards rule[19] of 0.0054 a.u. per at. %. Evidently, the theory is of sufficient sophistication to account for the slight departure from Vegards rule.

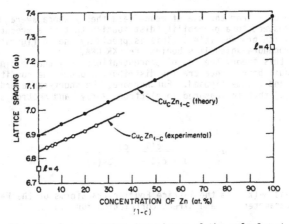

Fig. 2. Concentration dependence of the calculated and measured lattice spacing in Cu_cZn_{1-c} alloys. The experimental results are plotted for the α-phase (fcc) alloys only and are taken from Pearson.[20]

Concentration Functional Approach to Alloy Phase Stability

In the previous sections we have reviewed some of the progress that has been made in calculating the electronic structure and energetics of systems where the arrangement of the atoms is known a priori (either they are arranged in some periodic fashion or were assumed to be distributed randomly amongst the lattice sites). However, the scope of a full theory of ordering phenomena and of phase stability should be much broader than this. It should allow us to predict the arrangement of atoms expected at particular temperatures and pressures, be it random, short-range ordered, clustered, ordered, etc. After all, the electron glue that holds the atoms together, the nature of which we have been calculating, is what determines how the atoms are arranged. At zero temperature the methods outlined previously are, in principle, sufficient since from total energy calculations of the various ordered and disordered phases alone we could obtain the ground state phase diagrams. However, at finite temperatures the problem is much more complex since the statistical mechanics of alloy configurations must be treated along with the energetics associated with the electronic arrangement.

A proper ab initio electronic mean-field theory of the free energy of substitutional alloys has been given by Gyorffy and Stocks.[21] For a two-component A_cB_{1-c} alloy the mean-field free energy takes the form

$$F^{MF} = k_BT \sum_i \left[c_i \ln c_i + (1 - c_i)\ln(1 - c_i) \right] + \Omega^{KKR-CPA}(\{c_i\}) - \nu \sum_i c_i \ . \qquad (3)$$

The first term in this expression is the mean-field entropy associated with the compositional arrangements (c_i is the local concentration of the A species at the site i), the second term is the mean-field grand potential, and the final term involves the chemical potential difference $\nu = \nu^A - \nu^B$ which governs the fraction of sites occupied by the A and B species. That the grand potential, Ω, appearing in the theory is the KKR-CPA grand potential results from the use of mean-field theory where averages are taken with respect to a probability distribution that is a product of independent-site probabilities. This is precisely the configurational averaging procedure which lies behind the KKR-CPA.

Within this theory the set of concentrations $\{c_i\}$ that minimizes F^{MF} gives the equilibrium concentration distribution under the conditions of temperature and volume imposed. Furthermore, the short-range order parameter $\alpha(k)$ that is measured in electron, X-ray, and neutron scattering is given by

$$\alpha(k) = \frac{c(1-c)}{1 + c(1-c)\beta\, S^2(\vec{k})} , \tag{4}$$

where $\alpha(k)$ and $S^2(k)$ are the lattice Fourier transforms of the Warren-Cowley SRO-parameter α_{ij} and direct correlation function

$$S^2_{ij} = \frac{\partial\Omega^{KKR-CPA}}{\partial c_i \partial c_j} , \tag{5}$$

respectively.

Because we were able to derive an expression for the total energy within the KKR-CPA, it is a straightforward matter to generalize this to finite temperatures and to obtain an expression for $S^2(\vec{k})$ that involves only quantities available at the end of KKR-CPA calculations.

Thus, we now have a fully ab initio theory of alloy phase stability and of concentration fluctuations in the disordered state. The only inputs to the theory are the atomic numbers and concentrations of the alloying species. The great merit of the theory is that on the one hand it offers specific predictions regarding alloy phase stability that do not depend on adjustable parameters, while on the other offering a picture of the electronic structure (based on the KKR-CPA) to which the origins of specific ordering processes can be traced. Our strategy is to study the disordered state and examine its stability with respect to the formation of ordered states. Since the theory is fully ab initio, we cannot build in our prejudices regarding the expected nature of the ordered phases. If the disordered state is unstable to the formation of the correct ordered state, we can then trace the underlying reasons to the electronic structure. Even in the case where we do not predict the correct phase, we have a much better idea of the ingredients that are missing from theory than we would if it were based on adjustable parameters. In the next subsection, we show results that are based on this theory. In the actual calculations, only the band structure contributions to the total energy were included in the derivations of the equations for $S^2(\vec{k})$. "Charge transfer" corrections which result from the double-counting terms in Eq. (2) are not included, as a result the theory in its present form is best suited to alloys whose charge transfer is very small. This is the case for the $Cu_c Pd_{1-c}$ alloy system.

The first problem the concentration functional approach was applied to was that of understanding the origin of the concentration waves[22] seen in copper-rich $Cu_c Pd_{1-c}$ alloys.[23] In Fig. 3, we show the electron diffuse scattering patterns of Ohshima and Watanabe.[23] The sharp diffuse maxima that occur as a result of scattering from the concentration waves are clearly seen, a four-fold pattern about (011) and a two-fold pattern about (001). Furthermore, a pronounced concentration dependence of the

separation between the diffuse maxima is seen. As was pointed out by
Moss,[24] this rapid concentration dependence poses several problems for the
usual pairwise interchange description of alloy phase stability since it
implies a very concentration-dependent interchange potential. If this is
so, what is the origin of this concentration dependence? Moss suggested
that it comes from a singularity in the screening of the copper and
palladium ions resulting from there being flat, parallel sheets of Fermi
surface of some characteristic dimension k_F. The diffuse maxima then arise
at scattering vectors $\vec{k}_c = 2\vec{k}_F$ connecting these flat sheets. As was shown
by Gyorffy and Stocks,[21] the concentration functional theory gives a solid
theoretical foundation for these ideas and specific predictions about the
concentration dependence of $\vec{k}_c$ based on calculations of the alloy Fermi
surface. These predictions are in excellent agreement with experiment.

In Fig. 4 we show calculated SRO diffuse scattering patterns for three
$Cu_c Pd_{1-c}$ alloys having c = 0.93, 0.75, and 0.60 at temperatures just above
the phase transition. The region of $\vec{k}$-space shown corresponds to one
quadrant of the experimental diffuse patterns shown in the previous figure.
In the calculated patterns there is no contribution from the Bragg scattering which gives rise to the "over exposed" regions in the electron
diffraction patterns. Close examination of the theoretical and experimental SRO diffuse scattering reveals a close correspondence between the two.
As the concentration increases, the single (011) maximum which is seen at
c = 0.93 in the theory and c = 0.12 in the experiment splits into a four-
fold pattern of increasing separation as the palladium concentration is
increased. These calculations confirm the suggestion of Gyorffy and Stocks
that the diffuse scattering maxima are at positions corresponding to Fermi
surface spanning vectors.

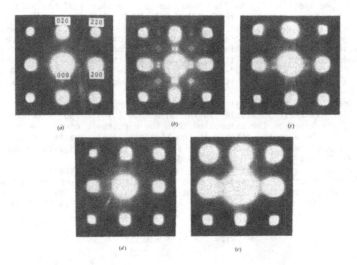

Fig. 3. Experimental electron diffraction patterns for five
$Cu_c Pd_{1-c}$ alloys after Ohshima and Watanabe. (a) $Cu_{0.87}Pd_{0.13}$,
(b) $Cu_{0.75}Pd_{0.25}$, (c) $Cu_{0.67}Pd_{0.33}$, (d) $Cu_{0.59}Pd_{0.41}$,
(e) $Cu_{0.52}Pd_{0.48}$.

22

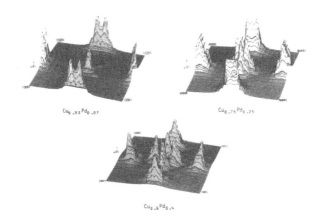

Fig. 4. Calculated short-range order diffuse scattering patterns for three Cu_cPd_{1-c} alloys. (a) $Cu_{0.93}Pd_{0.07}$, (b) $Cu_{0.75}Pd_{0.25}$, (c) $Cu_{0.60}Pd_{0.40}$.

Interchange Potentials

In the concentration functional theory described in the previous section, we specifically avoided expansion of the energy of the alloy in terms of a power series in fluctuations of the local site concentrations, eliminating assumptions that the energy can be so factored. Consequently, the theory looks very different from the normal development in terms of pairwise interchange potentials. Thus, an internally consistent theory is obtained in which the statistical mechanics and the electronic structure are both treated within mean-field theory. While retaining this kind of internal consistency, this is probably as far as one can go and still base the theory on the KKR-CPA.

Traditional statistical mechanical approaches to alloy phase stability (for example, the cluster variation method (CVM)[25] and Monte Carlo[26] method) are based on the assumption that the electronic energy can be apportioned between the sites. At the simplest level, the interactions are supposedly pairwise and short-ranged and are treated as adjustable parameters which are, at best, empirically determined.

In the generalized perturbation method (GPM) of Ducastelle and Gautier,[27] an attempt is made to obtain the multisite interchange potentials from electronic structure calculations based on the CPA. Here it is argued that the electronic grand potential can be written in the form

$$\Omega(\{\varsigma_i\}) = \overline{\Omega}(c) + \Omega'(\{\varsigma_i\}) , \qquad (6)$$

where $\overline{\Omega}(c)$ is a concentration dependent but configuration independent energy associated with the homogeneously random alloy and $\Omega'(\{\varsigma_i\})$ is the part of the energy that depends on the particular configuration defined by the set of occupation variables $\{\varsigma_i\}$. In the GPM the homogeneously disordered alloy is described by the CPA, and $\Omega'(\{\zeta_i\})$ is expanded about this effective medium

$$\Omega'(\{\zeta_i\}) = \frac{1}{2} \sum_{i \neq j} v_{ij}^{(2)} \delta c_i \delta c_j + \frac{1}{3} \sum_{\substack{i \neq j \\ j \neq k \\ k \neq i}} v_{ijk}^{(3)} \delta c_i \delta c_j \delta c_k + \cdots . \qquad (7)$$

The multisite expansion coefficients can then be obtained from the results of CPA calculations. The GPM has been widely applied to transition metal/ transition metal alloys where it is possible, at least for qualitative studies, to base the theory on a tight-binding Hamiltonian. Within this theory, it turns out that the expansion is rapidly convergent in terms of the separation between sites and in terms of the order of the interaction.[28]

It is fairly straightforward[29] to generalize the GPM to the KKR-CPA, thereby ridding the theory of adjustable parameters and extending it to alloys involving nontransition metals. By doing this, we are able to investigate in a systematic way the range and complexity (order) of the interactions that would have to be included in CVM and Monte Carlo calculations. Furthermore, the concentration and volume dependence of the parameters can also be studied.

In the calculations presented in the following section, only the band structure term in the total energy [Eq. (2)] has been considered.

Interchange Potentials and the Competition Between $L1_2$ and DO_{22} Ground States

Table I lists the calculated values of the first four near-neighbor pairwise interchange potentials for three different alloys, $Pd_{0.75}V_{0.25}$, $Al_{0.75}Ti_{0.25}$, and $Ni_{0.75}Al_{0.25}$ in the fcc phase. Perhaps the most striking feature of this table is that the interchange potentials converge slowly and are oscillatory. That they converge slowly for alloys containing aluminum, which is a simple metal, is expected. That they do for the transition metal/transition metal alloy $Pd_{0.75}V_{0.25}$ is surprising. Evidently, even in transition metal alloys, one cannot just assume, because the overall disorder scattering is large, that it is large for all bands.

Once the pairwise interchange potentials are known, it is a simple matter to study the competition between the $L1_2$ and DO_{22} phases since the ordering energies E^{ord} of each of the phases can be expressed in terms of these parameters. For $L1_2$ and DO_{22}, the energy difference $\Delta E^{L1_2/DO_{22}}$ ($\equiv E^{ord}_{L1_2} - E^{ord}_{DO_{22}}$) is given by[28]

Table I. Pairwise interchange energies in $A_{0.75}B_{0.25}$ alloys.

| System | Interchange energy (mRy/at.) | | | | ΔE | Stable |
	V_1	V_2	V_3	V_4		
$Pd_{0.75}V_{0.25}$	4.1	−3.9	0.8	2.0	0.3	DO_{22}
$Ni_{0.75}Al_{0.25}$	14.2	−1.0	1.7	−1.0	−2.9	$L1_2$
$Al_{0.75}Ti_{0.25}$	7.5	−6.0	1.9	−0.8	−4.2	$L1_2$

$$\Delta E^{L1_2/D0_{22}} = \frac{1}{4} v_2 - v_3 + v_4 \ . \tag{8}$$

Because the numbers and types of nearest neighbors are the same in the $L1_2$ and $D0_{22}$ structures, v_1 does not enter the question of the relative stability of these phases. The calculated values of $\Delta E^{L1_2/D0_{22}}$ are shown in Table I. For Pd_3V and Ni_3Al, the theory predicts the correct ground state; for Al_3Ti, $D0_{22}$ is stable at low temperatures. Interestingly, for Ni_3Al $\Delta E^{L1_2/D0_{22}}$ is very close to that calculated previously on the basis of band theory: −2.9 mRy/at. compared with −3.4 mRy/at.

While for Pd_3V and Ni_3Al the results are encouraging, there are a number of caveats. Firstly, only the band structure term was included in the calculation of the v_{ij}'s. Thus for systems in which charge transfer is large, these results must be regarded as being suspect. For the systems considered, charge transfer (defined in terms of the departure from neutrality of the charge inside equal volume Wigner-Seitz cells) increases in the order Al_3Ti, Pd_3V, and Ni_3Al. For the latter the charge transfers associated with Ni and Al Wigner-Seitz cells are −0.089 and 0.267 electrons, respectively. From this point of view, it becomes even more surprising that it is for Al_3Ti we obtain the incorrect ground state. It should be stressed that even for quite small charge transfers, the Madelung energy is of the same order as the ordering energy. For Ni_3Al the Madelung energy is −8.0 mRy/at. and −8.1 mRy/at. for the $L1_2$ and $D0_{22}$, respectively. Interestingly, the Madelung energy slightly favors the $D0_{22}$ ground state.

A second important caveat regards the interactions of higher order than pairwise. For $Pd_{0.75}V_{0.25}$ they are small. However, for the other two systems this is not the case. Indeed, for both Ni_3Al and Al_3Ti some of the triplet and quadruplet interactions are as large as the pairwise potentials. However, they involve only lines, triangles, and squares involving near-neighbor chains and do not affect the question of stability.

Antiphase Boundaries in Ni_3Al

If we consider the intersite interactions out to only fourth neighbor, then the energy for creation of a (100) antiphase boundary is simply related to $\Delta E^{L1_2/D0_{22}}$. In Table II we have tabulated the antiphase boundary energies calculated from the v_{ij} and band structure energies of this work along with those obtained experimentally and those from two different implementations of the embedded atom method (EAM).[9,10] The ab initio values we calculate by two vastly different techniques agree rather closely and are within a factor of two of the difficult to obtain experimental value.

Table II. $Ni_3Al(100)$ antiphase boundary energy.

γ_{100} (mJ/m^2)	Method	Reference
90	Weak-beam TEM	Dorin, Veyssiere, and Beaucamp[30]
83	EAM method	Chen, Voter, and Srolovitz[31]
28	EAM method	Foiles and Daw[32]
197	v_{ij}	Present work
230	LMTO	Present work

CONCLUSIONS

The advances outlined in this paper give some idea of the progress that is being made towards an ab initio theory of alloy phase stability. For the ground state, the methods are very well developed and, as a consequence, it is becoming possible to obtain theoretically the nature of the ground state of increasingly complex systems. These methods can be extended to treat the energetics of antiphase boundaries, in general, as well as low-angle grain boundaries. Thus, much more can be expected in the future. So far as theories of the finite temperature states are concerned, while implementation of such methods as the concentration functional technique are not yet complete, these hold out the possibility of obtaining a complete theory of alloy phase stability.

ACKNOWLEDGMENT

Research sponsored by the Division of Materials Sciences, U.S. Department of Energy, under contract DE-AC05-840R21400 with the Martin Marietta Energy Systems, Inc. We also acknowledge a number of very helpful discussions with M. H. Yoo.

REFERENCES

1. P. Hohenberg and W. Kohn, Phys. Rev. 136, B864 (1964).

2. W. Kohn and L.J. Sham, Phys. Rev. 140, A1133 (1965).

3. For a succinct presentation of the LSDA and of the results of its application to metals see Calculated Electronic Properties of Metals, by V.L. Moruzzi, J.F. Janak, and A.R. Williams (Pergamon Press, New York, 1978).

4. G.M. Stocks and H. Winter, in The Electronic Structure of Complex Systems, edited by P. Phariseau and W.M. Temmerman, NATO ASI Series B, Vol. 113 (Plenum Press, New York, 1984) p. 463.

5. D.D. Johnson, D.M. Nicholson, F.J. Pinski, B.L. Gyorffy, and G.M. Stocks, Phys. Rev. Lett. 56, 2088 (1986).

6. B.L. Gyorffy, Phys. Rev. B 5, 2382 (1972).

7. G.M. Stocks, W.M. Temmerman, and B.L. Gyorffy, Phys. Rev. Lett. 41, 339 (1978).

8. A.E. Carlsson, presented at the 1986 MRS Fall Meeting, Boston, MA, 1986 (unpublished).

9. S.-P. Chen, A.F. Voter, and D.J. Srolovitz, presented at the 1986 MRS Fall Meeting, Boston, MA, 1986 (unpublished).

10. S.M. Foiles, presented at the 1986 MRS Fall Meeting, Boston, MA, 1986 (unpublished).

11. F.W. Averill, Phys. Rev. B 6, 3637 (1972).

12. A.R. Williams, J. Kübler, and C.D. Gelatt, Jr., Phys. Rev. B 19, 6094 (1979).

13. E. Wimmer, H. Krakauer, M. Weinert, and A.J. Freeman, Phys. Rev. B 24, 864, (1981).

14. J.S. Faulkner, Phys. Rev. B 26, 1597 (1982); D.M. Nicholson and J.S. Faulkner (unpublished).

15. H.L. Skriver, "The LMTO Method," Solid State Sciences, Vol. 41 (Springer-Verlag, New York, 1984).

16. U. von Barth and L. Hedin, J. Phys. C 5, 1029 (1972).

17. J.S. Faulkner, Prog. Mater. Sci. 27, 1 (1982).

18. G.M. Stocks and H. Winter, Phys. Rev. B 27, 882 (1983).

19. L. Vegard, Z. Phys. 5, 17 (1921); Z. Kristallogr. Kristallgeom., Kristallphys., Kristallchem 67, 239 (1928).

20. W. B. Pearson, A Handbook of Lattice Spacings and Structures of Metals and Alloys (Pergamon, New York, 1958).

21. B.L. Gyorffy and G.M. Stocks, Phys. Rev. Lett. 50, 374 (1983).

22. A.G. Khachaturyan, Theory of Structural Transformations in Solids (John Wiley and Sons, New York, 1983).

23. K. Ohshima and D. Watanabe, Acta Crystallogr. A 33, 520 (1977).

24. S.C. Moss, Phys. Rev. Lett. 22, 1108 (1969).

25. R. Kikuchi, Phys. Rev. 81, 988 (1951).

26. "Monte Carlo Methods in Statistical Physics," in Topics in Current Physics, Vol. 7, 2nd edition, edited by K. Binder (Springer-Verlag, New York, 1986).

27. F. Ducastelle and F. Gautier, J. Phys. F 6, 2039 (1976).

28. P. Turchi, F. Ducastelle, and G. Treglia, J. Phys. C 15, 2891 (1982); P. Turchi, Thèse de Doctarat d'Etat es Sciences, Univ. Pierre et Marie Curie, Paris VI (1984).

29. A. Gonis, P.E. Turchi, W.H. Butler, and G.M. Stocks (unpublished).

30. J. Dorin, P. Veyssiere, and P. Beaucamp, Philos. Mag. 54, 375 (1986).

31. S.-P. Chen, A.F. Voter, and D.J. Srolovitz, Sci. Metallogr. 20, 1389 (1986).

32. S.M. Foiles and M.S. Daw, J. Mater. Res. (in press).

ANTIPHASE DOMAINS, DISORDERED FILMS AND THE DUCTILITY OF ORDERED ALLOYS BASED ON Ni3Al

R. W. CAHN
Cambridge University, Dept. of Materials Science & Metallurgy, Pembroke Street, Cambridge CB2 3QZ, England

ABSTRACT

In elements and intermetallic compounds alike, tensile ductility is favoured by a small spacing of obstacles to dislocation movement, whether these obstacles be grain boundaries or other structural features. The evidence for this generalisation is reviewed.

Antiphase domains (APD) form only when T_t is below the freezing temperature, T_f: this is a rule which has not hitherto been recognized. In Ni-Al, $T_t < T_f$ when the Al content is less than ≈ 23.5at.%. APD's can serve as heterogeneous nuclei where films of disordered (γ) phase deposit. Domains, especially if decorated by γ films, are associated with enhanced ductility, and there is circumstantial evidence that the ductility is indeed caused by these microstructural features.

INTRODUCTION

1. Models for ductility/brittleness of binary and B-doped Ni3Al

Binary nickel aluminides based on the $L1_2$ superlattice – the γ' phase – are highly ductile in the form of single crystals but brittle in polycrystalline form. Boron additions of 0.05-0.2at.% to alloys containing 24at.% or less Al renders them ductile, though too much boron restores brittleness[1,2]. Melt-quenching tends to enhance ductility of the B-doped alloys. This much is now generally accepted. When it comes to interpreting the facts, opinions differ: a number of papers at the present Symposium are concerned with the mechanisms governing ductility or brittleness; since the brittleness problem arises from the presence of grain boundaries, much attention is currently being devoted to the structure and behaviour of grain boundaries in the presence of an ordered arrangement of atoms.

The common denominator of several recent discussions of the ductilisation of $Ni_{3+x}Al_{1-x}$ by boron-doping has been that the dopant — which is known to segregate strongly to the grain boundaries — enhances the resistance of the boundary to crack propagation and also the ability of dislocation arrays impinging on a grain boundary to activate a new dislocation source in the adjoining grain: that is, the grain boundary is strengthened with respect to fracture propagation and weakened with respect to slip propagation. In an alternative formulation of the same ideas, boron-doping reduces the grain-boundary energy and thus increases the grain-boundary cohesion[1,2]. This approach has been varied by Hack et al.[3], who analysed the fracture behaviour of Ni_3Al in terms of the stability or otherwise of a sharp crack up to a stress intensity which will enable such a crack (if not previously blunted by localised plasticity) to spread brittly along a grain boundary. The stability of a sharp crack is determined by the yield stress of the matrix which fixes the resistance to crack-blunting by localized plastic flow. The level of yield stress, according to this analysis, thus plays a crucial role in determining brittle or ductile behaviour: in this way, Hack et al. claim to rationalize the intrinsic grain-boundary weakness of pure (strong) Ni_3Al, in spite of the similarity of its grain-boundary energy to that of pure nickel.

When Hack et al. wrote their paper a few months ago, they had available no clear observations concerning the effect of grain size on ductility. What little experimental evidence there was showed no systematic effect of grain size on ductility of Ni_3Al, which was consistent with Hack's scepticism concerning the relevance, to ductile/brittle behaviour of this phase, of the ease/difficulty of slip transmission across grain boundaries. Very recently, however, Weihs et al.[4] completed a very comprehensive study of grain-size effects in stoichiometric Ni_3Al: they studied a wide range of grain sizes and enough duplicate specimens to be able to draw for the first time convincing conclusions concerning the influence of grain size on ductility. Briefly, at room temperature a fine grain size ($\leq 10\mu m$) in the presence of B converts a ductile into a more ductile material while not creating any ductility in the B-free alloy. At higher temperatures ($\geq 673K$), however, a grain size of $10\mu m$ or less greatly enhances ductility even in the absence of boron. While in no way invalidating the ideas of Hack et al., these important findings make it clear that the slip-transmitting role of dislocation pile-ups at grain boundaries (the efficacy of which depends on grain size) cannot be neglected in rationalizing the ductility or otherwise of Ni_3Al polycrystals.

2. Ternary phases based on Ni_3Al

In an important study, Takasugi and Izumi[5] have sought to interpret the very widely varying ductility of different $L1_2$ intermetallic phases — Cu_3Au, Ni_3Fe, Co_3Ti, Ni_3Al, etc. — on the basis of the influence of the type and strength of the interatomic bonds on grain-boundary cohesion. The more different the electronic structure of the constituent metals, the more does the bonding across a (necessarily) somewhat disordered grain boundary weaken its resistance to fracture.

However, these two papers make no reference to another, equally important study[6,7] from the same Japanese laboratory, which casts a quite different light on the varying ductility of melt-quenched single-phase $L1_2$ ternary alloys, not doped with B, of Ni-Al-X type, where X_1 =

Cr, Mn, Fe, Co or Si; X_2 = Ti, Zr, V, Nb or Cu. When X_1 solutes were added at levels of 4at.% or more, the alloy became both stronger and more ductile; when X_2 solutes were added, the alloys became even more brittle than the binary precursor. The two families of ternary alloys did not differ significantly in grain size (always in the range 1.4-2.6μm) but did differ in respect of their antiphase domain (APD) structure. The X_1 group alloys had fine APD's (30-55nm diameter), the X_2 group alloys had very much coarser APD's. When the X_1 group alloys were annealed at 600-700°C, the domains grew up to several μm diameter, and this growth was accompanied by a fall of yield stress and a sharp reduction in plastic elongation.

It would appear from Inoue's study that the ductility of Ni+Al+X γ' alloys is correlated with the domain structure rather than with particular bonding at grain boundaries, if only because the latter would not be affected by heat-treatment and therefore ductility could not be altered by heat-treatment either if that hypothesis were applicable.

A very recent study of $L1_2$-type Ni-Al-Cr alloys[8] has confirmed the above study, in that the presence of APD's (decorated by thin γ films) conferred enhanced ductility on the alloys.

3. Effect of grain and subgrain size on ductility: elements

The observations of Inoue et al. on the influence of APD size on ductility raise the more general question: how does the spacing of obstacles to free dislocation motion, such as grain boundaries and subgrain boundaries, affect ductility?

Leaving aside the superplastic domain, which is not linked primarily with dislocation motion, very little research has been done on the variation of plastic ductility with grain size. Armstrong[9] has assembled and analysed several earlier studies of magnesium: in the range 20-1.5μm grain diameter, d, the ductility increases from ≈2% to ≈55% as the grain size is reduced. Armstrong analyses this in terms of the Hall-Petch relationship for yield ($\sigma_y = \sigma_{oy} + k_y d^{-1/2}$) and for fracture ($\sigma_f = \sigma_{of} + k_f d^{-1/2}$) and a mean work-hardening coefficient, $\bar{h}$; one finds that quite generally, the tensile elongation, ε_f, is given by

$$\varepsilon_f = \varepsilon_o + (\Delta k/\bar{h})d^{-1/2}$$

where $\Delta k = (k_f - k_y)$; ε_o is a term involving the σ_o's and $\bar{h}$, and represents the fraction of strain due to the frictional stress increase within the grains, while the second righthand term is attributed to grain-boundary strengthening. − A related analysis has been published by Schulson[10]: his paper deals specifically with ordered intermetallic phases, and aims to calculate a critical grain diameter at which slip propagation across grain boundaries gives way to crack propagation − in other words, a brittle/ductile transition takes place at the critical grain size.

One of the few other relevant studies I have found refers to the ductility and strength of polycrystalline chromium[11,12]. This metal is highly subject to brittle fracture and at normal grain sizes shows very limited tensile ductility. It is possible, however, at slow strain rates to draw a rod down to wire, to very high strains (up to 99.7% R.A.). Under these circumstances, one finds very sharply defined subgrains or cells, down to $0.1\mu m$ diameter, virtually free of interior dislocations; grain boundaries can then no longer be identified by TEM. If the drawn wires are tested in slow tension, the elongation is now several per cent, which for chromium represents considerable ductility. The authors point out that the cell wall misorientation is quite large enough for the walls to be effective obstacles to the passage of dislocations, so that dislocation pile-ups are restricted to single cells. σ_y, σ_f and ε_f are then determined by the cells and not by the grains (as has also been concluded for cold-worked and recovered aluminium, see Cahn[13]). The heavily drawn wire has almost run out of work-hardening potential, so that $\bar{h}$ in Armstrong's formula is small, and this also favours enhanced ductility. – In their detailed analysis of their Cr results, Ball et al. conclude that their results are consistent with crack initiation as the crucial stage of fracture, but not consistent with the idea that crack propagation is the critical stage. Schulson's analysis[10], however, is based on the idea of crack propagation as the critical stage.

4. Other intermetallic compounds

The A-15 compounds, as a class, are extremely brittle and this can cause serious problems in their application as superconducting windings. Recently, Clapp and Shi[14], examining a $Ti_3Nb_6Mo_3Si_4$ compound, found it possible to turn it into a glass by melt-spinning, and then to crystallize the glass into an ultra fine-grained microstructure. They found that the tensile ductility ranged from <0.05% for grain sizes >0.6μm to 1.6% for a grain size of 0.05μm. The strength also rose with decreasing grain size.

5. General comments

The foregoing remarks, taken together, make it very probable that the scale either of grain structure or of internal "subgrains" must affect the ductility of polycrystalline γ' phases; however, the "subgrains" in these materials are actually APD's and not polygonized cells. As we shall see, the dislocation stopping power of APD boundaries is further enhanced under some circumstances by the presence at them of films of disordered material.

The objective of the remainder of this paper is to analyse, in the light of recent experimental findings, the role of APD's in enhancing ductility of γ' alloys based on Ni_3Al.

ANTIPHASE DOMAINS AND DISORDERED FILMS IN ORDERED Ni₃Al-BASE ALLOYS

1. Structural observations

Cahn, Siemers, Geiger and Bardhan[15,16] have recently completed a detailed study of the order-disorder transformation in several γ' alloys, both binary Ni-Al and ternary Ni-Al-Fe compositions. The study was based on dilatometric curves such as Fig. 1, which show a distinct kink at the O-D transition: high-temperature diffractometry confirmed this interpretation.

Fig. 2 shows the modified binary Ni-Al phase diagram which emerged from this work. The dashed curve labelled "locus of T_{tm}" represents the concentration dependence of the completion of disordering of the γ' phase; $T_{tm} \equiv$ "transition, metastable". For details of the observations and arguments which led to this locus, the original papers should be consulted. The essential requirement for γ' to remain metastably single-phase in the hatched region is the absence of APD's: if such domains are present, they serve as heterogeneous nucleation sites for disordered γ to form. This kind of γ nucleation can be seen clearly in Figs. 3 and 4 (which actually refer to a ternary Ni-Al-Fe alloy, in which T_{tm} is much lower, at 1160°C). Again, the detailed evidence and theoretical background will be found in ref. 16.

It is important for our present purposes to be quite clear about the circumstances under which a network of antiphase domains can form in an ordered alloy. Such APD networks are formed when a disordered crystal cools through the transition temperature for ordering: each APD corresponds to a distinct nucleus of order. (Homogeneous or dering of a disordered crystal likewise generates an APD network). This process is sequential ordering. If, on the other hand, an alloy freezes directly into the ordered state, then a single nucleus grows into a complete grain. This process is direct ordering. The notion, repeatedly expressed, that stoichiometric Ni₃Al is domain-free because domains form on freezing but then grow exceedingly fast (so that they disappear even during melt–spinning) entirely misses the point. The only exception to this hitherto unrecognized distinction would be if a directly ordered grain suffers superlattice stacking faults during growth from the melt; under these circumstances, a very coarse APD structure (but scarcely a network) can form.

The validity of the distinction between sequentially and directly ordered alloys can be seen by reference to Fig. 5, from Cahn et al.[16]. This is an electron micrograph of a meltspun Ni₇₈Al₂₂ ribbon and shows, within a single grain, a duplex structure of circular regions containing fine APD's complete with associated γ films, surrounded by regions containing only a few coarse APD's without γ. A microanalysis profile of a similar structure (Fig. 6) shows that the duplex structure is due to microsegregation: the fine-domained ($\gamma' + \gamma$) isfound when the Al content

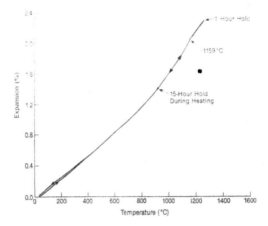

Fig. 1. Dilatometric cycle through order-disorder transition, for alloy $(Ni_{77.9}Al_{20.1})_{0.87}Fe_{0.13}$.

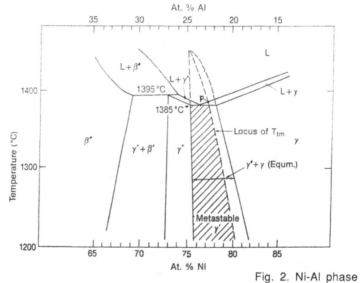

Fig. 2. Ni-Al phase diagram, showing metastable order-disorder transition.

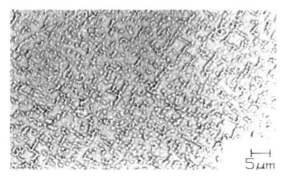

Fig. 3. Optical (oil immersion) micrograph of sample of Fig. 1, after dilatometric cycle.

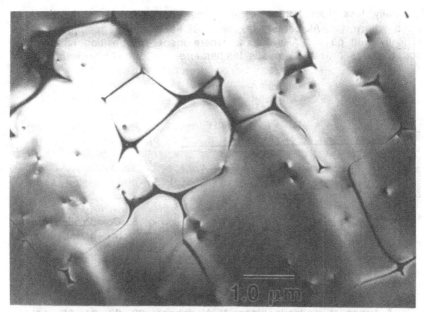

Fig. 4. As Fig. 3, dark-field electron micrograph: γ' with γ films.

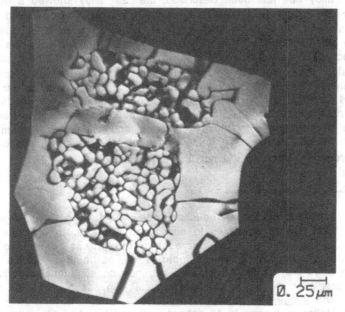

Fig. 5. Meltspun Ni₇₈Al₂₂ ribbbon, dark-field electron micrograph: γ'. The fine domain walls are decorated with γ films, the coarse ones are not.

is locally less than 23.5at.%, while the coarse-domain γ' structure is found where the Al content exceeds 23.5at.%. This changeover fits perfectly with point "P" in Fig. 2, where the O-D transition temperature locus crosses over the liquidus temperature.

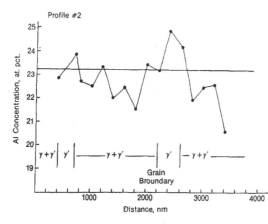

Fig. 6. Composition profile across two grains in melt-spun $Ni_{78}Al_{22}$.

A series of meltspun binary Ni-Al ribbons (22, 23, 24, 25 at.%Al): The "22" alloy has just been described; the "23" alloy showed a duplex structure as in Fig. 5 but apprently without γ films, the "24" and "25" ribbons showed no domains at all. (This last observation is at variance with the observations of Koch et al.[17], who examined arc-hammer-quenched stoichiometric Ni_3Al and found a duplex domain structure.) Presumably the extent of microsegregation in these alloys depends sensitively on the exact quenching conditions, and the local appearance of APD"s for local compositions to the left of "P" in Fig. 2 is contingent on the presence of sufficiently severe microsegregation. - Our results are also at variance with those of Horton and Liu[18] who examined meltspun $Ni_{76}Al_{24}$: they also found a duplex APD structure, associated with segregation, but they did not find any γ films.

2. APD's, γ films and ductility

As explained in the Introduction, one would expect APD's and γ films, in their role of obstacles to dislocation passage, to enhance the ductility of polycrystals, by decreasing the stress exercised by dislocation pile-ups at grain boundaries in particular. Since the formation of APD's, as we have seen, depends on sequential ordering, I call the prediction of greatly enhanced ductility in association with such networks, the sequential ordering criterion.

(a) Meltspun binary Ni-Al ribbons without B. – $Ni_{78}Al_{22}$ was very ductile - the ribbon could be bent back sharply on itself. $Ni_{77}Al_{23}$ was brittle (though slight plastic bending was sometimes possible), and

$Ni_{76}Al_{24}$ and $Ni_{75}Al_{25}$ very brittle. The "22" alloy, as we have seen, contained (locally) a fine APD network and γ films at the APDB's; the "23" alloys had local APD networks only, the other two had neither feature. These observations fit the sequential ordering criterion, except that more ductility was to be expected for the "23" alloy. (It is true that the sequential ordering criterion is only ambivalently satisfied by a microsegregated specimen which has sequentially ordered in some regions but not in others).

(b) Ni-Al-Fe Alloys. – The ternary alloy, of composition $(Ni_{79.9}Al_{20.1})_{0.87}Fe_{0.13}$ (no boron), to which Figs. 1, 3 and 4 refer, proved highly deformable by rolling, and accepted >70% R.A. before some edge cracking was detected. This alloy, which was coarse-grained, contained both APD networks and γ films. Another Ni-Al-Fe alloy, containing more Al and B-doped, accepted less rolling deformation (≈50% R.A.) because of the presence of embrittling β' phase[16].

(c) Role of Ni-Al ratio. – It is well established that, even with benefit of B doping, the ductility of binary Ni_3Al drops sharply on going from 24 to 25at.% Al. Liu et al.[19] found by Auger spectroscopy that the efficacy of B segregation to the grain boundaries decreases as the Al content is raised through this range, and they hold this variation of segregation responsible for the dramatic change in ductility. A contributory factor might be that (depending on freezing rate) some APD's can form in the "24" alloy (the composition is quite close to "P" in Fig. 2), while this is less likely to happen in the "25" alloy - or where it does, the volume fraction of APD's will be small. Since no systematic study of microstructure in "24" and "25" binary alloys has been made as a function of solidification conditions, this can only be a guess.

(d) Role of excess B. – Measurements reported by Cahn et al.[15] indicate that substantial B in solution lowers the freezing temperature of Ni-Al-Fe alloys faster than the T_{tm} (ordering transition) is lowered. It is possible that a contributory factor in the re-embrittlement of binary γ'-phase alloys as the B content is raised beyond 0.7at.% is a reconversion from sequential to direct ordering, so that excessive boron excludes APD formation. Confirmation of this hypothesis awaits TEM examination of heavily B-doped alloys. The usual explanation of the embrittling effect of excess boron is the appearance of boride particles at grain boundaries.

(e) Ti-bearing alloys. – Experiments were performed[15] with two meltspun Ti-bearing alloys. $Ni_{76}Al_{19}Ti_5$ + 0.25at.%B proved to be entirely brittle, whereas $(Ni_{76}Al_{19}Ti_5)_{0.90}Fe_{0.10}$ + 0.25at.%B was highly ductile. The former had an ordering transition temperature far above the freezing temperature and thus ordered directly (for details of how this was deduced, see ref. 15), while the Fe-bearing alloy was found to order sequentially, with an ordering transition below the freezing temperature. TEM showed that the brittle alloy contained no APD's whereas the ductile one contained a fine APD network. No γ films were noted.

(f) Interpretation of Inoue's observations. – The observations, outlined above, on a range of Ni-Al-X ternaries[6,7] fall into place when it is recognized that the X_1 solutes (which ductilise the alloys) lower the order-disorder transition temperature to below the freezing temperature (we have confirmed this experimentally for Fe and Cr[15,16]); such alloys therefore order sequentially and thus form APD networks. The X_2 solutes (e.g., Ti) raise the transition temperature[15] and this is why few or no APD's form and the alloys are brittle.

CONCLUSIONS

It is clear that fine APD networks are apt to improve ductility and that this tendency is enhanced if γ films are present at APDB's. Plainly, γ films render APDB's more opaque to the passage of dislocations, but it is impossible to judge on the evidence at present available how substantial an effect γ films have, or how that effect depends on the thickness of the γ films. Neither is it clear at present how small the grain size has to be before grain size takes over from domain size as the ductility-defining feature (probably below 1 μm). More systematic experiments are needed.

The sequential ordering criterion is by no means an absolute criterion. This emerges clearly from Liu's[20] comparative study of cubic (e.g., $Fe_{48}Co_{27}Ni_{25})_3V$ and hexagonal (e.g., Co_3V) ordering alloys: the former (which ordered sequentially and were also apt to develop γ films at APDB's[21]) were highly ductile, the latter (which also ordered sequentially) nevertheless had very little ductility. Likewise, α_2-Ti_3Al, a hexagonal superlattice which orders sequentially and has a fine APD network[22,23,24], nevertheless has very limited ductility because its structure is hcp and it thus possesses insufficient independent slip systems to permit each grain to follow the overall plastic strain[25]. (A recent publication[24] on the anneal-induced coarsening of APD networks in Ti_3Al is, however, silent on the effect of such coarsening on the mechanical properties).

At the other extreme from Ti_3Al, there is at least one directly ordered alloy, Co_3Ti, which according to the sequential ordering criterion should be brittle but in fact shows considerable polycrystalline ductility. There is reason to believe (e.g., because of the ready recrystallization of this alloy[26], in contrast to other ordered alloys) that its grain-boundary structure is in some way anomalous for an ordered alloy, and this suggests that Takasugi and Izumi's ductility criterion based on the nature of the interatomic bonds at grain boundaries operates in parallel with the sequential ordering criterion: i.e., appropriate bonding at grain boundaries of an ordered alloy will ensure ductility, but if the bonding is inappropriate then a fine APD structure (especially if abetted by disordered skins) can still confer ductility - either by itself in favourable cases (such as the Ni-Al-Fe alloy mentioned above) or else with the help of boron doping which modifies the grain-boundary structure. All the foregoing is contingent on the availability of a sufficient number of independent slip systems.

In predicting ductility for intermetallic phases one thus has to proceed down a kind of "decision tree" of linked criteria.

This decision tree of criteria should also include reference to the grain size. The recent work of Weihs et al.[4] (reported on by Schulson elsewhere in this Symposium) proves unambiguously that grain size, in B-doped Ni_3Al, plays a major role in determining ductility, in the absence of APD's. It would now be interesting to have the results of a similar study of, say, $(Ni_{80}Al_{20})_{0.87}Fe_{0.17}$, where both copious APD's and γ films are present: here, I would predict, grain size will have much less effect, or perhaps none at all, on ductility. - Where grain size does have a major effect, the nature of the average misorientation across individual boundaries (and therefore the preferred orientation, if any) appears to influence ductility[27] and this factor deserves further investigation.

Finally, it is not yet clear what sort of APD structure, if any, a worked and recrystallized Ni_3Al sample will have. Calvayrac and Fayard[28] found that whereas cast stoichiometric Ni_3Al contained no APD's, some coarse APD's were present in rolled and recrystallized material. My own unpublished results for a Ni-Al-Fe alloy recrystallised below the order-disorder transition temperature suggests a substantial, but incomplete, network of APD's. (Recent work on a model system, Cu_3Au, heavily rolled and recrystallised below the order-disorder transition, shows copious APD formation, with some anomalous features[29]). Since, increasingly, mechanical properties of γ' Ni-base alloys are being measured with samples made by deforming single crystals or columnar polycrystals and then recrystallizing these, (e.g., Hanada et al.[30]), it is plainly desirable that the APD's and γ films in recrystallized ordered alloys should be systematically explored.

The present Symposium has been called in order to explore the present status of single-phase ordered intermetallic compounds as possible high-temperature constructional materials, to replace - in some applications - the current generation of multiphase superalloys. The ductilising role of γ films at APD's suggests that perhaps γ' phases may after all have the best properties if they contain a fine γ network: if so, then we shall be moving from a family of duplex alloys containing γ' dispersed in γ to another family containing γ dispersed in γ'; plus ça change, plus c'est la même chose.

ACKNOWLEDGMENTS

I am deeply grateful to Drs. E.L. Hall and K.N. Westmacott for preparing the electron micrographs presented here, and other micrographs to which reference is made. Prof. R.W. Armstrong, Dr. J.E. Hack, Dr. C.T. Liu and Prof. E.M. Schulson kindly provided preprints of their latest papers. I am also grateful to the authorities of GE Corporate R&D, Schenectady, for the opportunity of carrying out the research which furnished the basis for part of this paper.

38

REFERENCES

1. A. Aoki, O. Izumi, Nippon Kinzoku Gakkaishi 43, 1190 (1979).
2. C.T. Liu, C.L. White, J.A. Horton, Acta Metall. 33, 213 (1985).
3. J.E. Hack, D.J. Srolovitz, S.P. Chen, Scripta Metall., in press.
4. T.P. Weihs, V. Zinoviev, D.V. Viens, E.M. Schulson, Acta Metall. 34, in press.
5. T. Takasugi, O. Izumi, Acta Metall. 33, 1247, 1259 (1985).
6. A. Inoue, H. Tomioka, T. Masumoto, Met. Trans. 14A, 1367 (1983).
7. A. Inoue, T. Masumoto, H. Tomioka, N. Yano, Int. J. Rapid Solidif. 1, 115 (1985).
8. S.C. Huang, E.L. Hall, K.-M. Chang, R.P. LaForce, Met. Trans. 17A, 1685 (1986).
9. R.W. Armstrong, in Strength of Metals and Alloys (ICSMA-7-CIRMA) (Pergamon Press, 1985).
10. E.M. Schulson, Res Mechanica Lett. 1, 111 (1981).
11. A. Ball, F.P. Bullen, F. Henderson, H.L. Wain, in Fracture (Proc. 2nd Int. Conf. on Fracture, Brighton, 1969) (Chapman & Hall, London) p. 327.
12. A. Ball, F.P. Bullen, F. Henderson, H.L. Wain, Phil. Mag. 21, 701 (1970).
13. R.W. Cahn, in Physical Metallurgy, edited by R.W. Cahn & P. Haasen, (North-Holland Physics Publishers, Amsterdam, 1983), p. 1611.
14. M.T. Clapp, D. Shi, J. Appl. Phys. 57, 4672 (1985).
15. R.W. Cahn, P.A. Siemers, J.E. Geiger, P. Bardhan, submitted to Acta Metall.
16. R.W. Cahn, P.A. Siemers, E.L. Hall, submitted to Acta Metall.
17. C.C. Koch, J.A. Horton, C.T. Liu, O.B. Gavin, J.O. Scarbrough, in: Proc. 3rd Conf. on Rapid Solidification Processing, Principles and Technologies, III, 1982, edited by R. Mehrabian (National Bureau of Standards, Gaithersburg, MD, 1983), p. 264.
18. J.A. Horton, C.T. Liu, Acta Metall., in press.
19. C.T. Liu, C.L. White, J.A. Horton, Acta Metall. 33, 213 (1985).
20. C.T. Liu, in High-temperature Alloys: Theory and Design, edited by J.O. Stiegler (The Metallurgical Society of AIME, Warrendale, PA, 1984), p. 289.
21. J.A. Horton, A. Dasgupta, C.T. Liu, in High-Temperature Ordered Intermetallic Alloys, (Mat. Res. Soc. Symp. Proc., Vol. 39, 1985), p. 109.
22. S.M.L. Sastry, H.A. Lipsitt, Met. Trans. 8A, 1543 (1977).
23. H.A. Lipsitt, as ref. 21, p. 351.
24. A.G. Jackson, K.R. Teal, D. Eylon, F.H. Froes, S.J. Savage, in Rapidly Solidified Alloys and Their Mechanical and Magnetic Properties, (Mat. Res. Soc. Symp. Proc., Vol. 58, 1986), p. 365.
25. D. Banerjee, J.C. Williams, Defence Sci. J. (India) 36, 191 (1986).
26. T. Takasugi, O. Izumi, Acta Metall. 33, 39 (1985).
27. S. Hanada, T. Ogura, S. Watanabe, O. Izumi, T. Masumoto, Acta Metall. 34, 13 (1986).
28. Y. Calvayrac, M. Fayard, Acta Metall. 14, 783 (1966).
29. R.W. Cahn, K.N. Westmacott, to be published.
30. S. Hanada, S. Watanabe, O. Izumi, J. Mat. Sci. 21, 203 (1986).

EFFECTIVE-PAIR-INTERACTIONS FROM SUPERCELL TOTAL ENERGY CALCULATIONS: Al-TRANSITION METAL ALLOYS

Anders E. Carlsson, Department of Physics, Washington University, St. Louis, Missouri 63130

ABSTRACT

Effective-pair-interactions (EPI) are computed for alloys of Al with transition metals, Li, and Zn, using a method in which concentration-*independent* cluster interactions are resummed to obtain the concentration-*dependent* EPI. The method includes alloy fluctuations in the interatomic charge transfer, enables one to transcend the muffin-tin approximation and thus treat surfaces and layered structures, and allows the inclusion of lattice strain effects. The calculated EPI have a large magnitude when d-bonding effects are important. For transition metals the EPI are strongly concentration-dependent. In Ni-Al, results for bcc and fcc lattices are similar and exhibit a quick decay of the EPI with interatomic separations. The concentration dependence of the transition metal EPI exhibits rapid oscillations with the number of valence electrons. The concentration-*averaged* EPI varies less dramatically. The oscillations in the concentration dependence of the EPI are interpreted in terms of the position of the Fermi level relative to peaks and valleys in the one-electron density of states.

INTRODUCTION

Modeling both bulk phase diagrams and defect properties in intermetallic alloys requires a detailed understanding of the relevant interatomic force laws. A feature of the interatomic force law particularly relevant for alloy ordering is the effective-pair-interaction (EPI), which is the energy parameter that specifies whether like or unlike neighbors are preferred; it is directly analogous to the exchange coupling strength in an Ising model calculation. It is not *a priori* obvious that the energies associated with atomic rearrangements, even as a fixed lattice, can be described by pair interactions. However, a large body of analysis [1-5] has shown that by using a completely random alloy as a starting point for a perturbation expansion (the "generalized perturbation theory", or GPT) one can obtain a concentration-dependent EPI which describes the ordering energy fairly accurately [6]. The methods used to obtain these EPI have described the random alloy with the coherent potential approximation, and, with the exception of a few tight-binding model calculations [4], have included only the one-electron contribution to the EPI.

In the present work we use a new approach based on the resummation of cluster potentials [7] obtained from supercell total energy calculations. Since the alloy averaging in this approach is performed *after* the total energies are calculated, the method has several unique features. The most important of these are the following: First, the effects of alloy fluctuations on the electrostatic energy are included through the incorporation of several charge states for each atom, with relative probabilities weighted according to the concentration and degree of order in the alloy. This enables one to treat alloy systems with large atomic size mismatches, such as some of the more important high-temperature ordered alloys. Second, one can use total energies obtained from schemes considerably more sophisticated than the muffin-tin approximation; this feature will be particularly useful in calculations for surfaces and layered structures, but is not implemented in our present results. Finally, by relaxing the atomic positions in the supercell calculations, one can include lattice strain effects. The primary disadvantage of the method is that one must use very large supercells to obtain the large-separation behavior of the EPI, with correspondingly rapid increases in computer time.

Mat. Res. Soc. Symp. Proc. Vol. 81. ᶜ 1987 Materials Research Society

METHOD

Since the method is described in detail elsewhere [8], we only summarize it here. The total energy of an A-B alloy is first expressed [9] in terms of concentration-*independent* cluster interactions:

$$E = \sum_{\mathcal{C}} V_{\mathcal{C}} \, \xi_{\mathcal{C}} \quad , \tag{1}$$

where $\mathcal{C}$ denotes a type of cluster, $V_{\mathcal{C}}$ is the potential associated with this cluster type, and $\xi_{\mathcal{C}} = <\sigma_1 \cdots \sigma_n>$ is a cluster correlation function. Here the σ_i are ± 1 according to whether site i is an A or a B atom and n is the order of the cluster. Given energies from M supercell calculations one can invert (1) to obtain the M most important $V_{\mathcal{C}}$, if one assumes that the remaining cluster interactions vanish [7]. For fcc calculations we take M = 5 and use energies for pure A and B, A_3B and AB_3 in the Cu_3Au structure, and AB in the CuAu structure (ideal c/a ratio), as input, as in Ref. [7]. For bcc calculations we take M = 6 and use energies for pure A and B, A_3B and AB_3 in the Li_3B structure, and AB in the CsCl and NaTl structures [10]. As discussed in Ref. [8], truncation of the cluster expansion at this stage provide a reasonably good level of convergence.

While the $V_{\mathcal{C}}$ obtained in this fashion are useful in their own right, it is useful to convert them into concentration-*dependent* EPI, since the series of concentration-dependent interactions provides much more rapid convergence of the ordering energy with cluster size. For example, ordering energies in simple d-band models are obtained within 10-20% already at the pair level, provided the concentration-dependence is taken into account [3]. To obtain the EPI $\phi(R_i, R_j)$ we begin by making the truncated approximation

$$\xi_{\mathcal{C}} = <\sigma_1 \cdots \sigma_n>$$

$$= \left(<\sigma_1\sigma_2> + <\sigma_1\sigma_3> + \cdots <\sigma_{n-1}\sigma_n> \right)<\sigma>^{n-1} - [n(n-1) - 1] <\sigma>^n \tag{2}$$

for the cluster correlation functions, which is valid in the high-temperature limit. Using the relation

$$\phi(R_i, R_j) = 4 \frac{\partial E}{\partial <\sigma_i \sigma_j>} \quad ,$$

we then obtain $\phi(R_i, R_j)$ as a polynomial in $<\sigma>$ of order N-2, where N is the order of the largest cluster included, and i and j denote atomic sites. (This results in a quadratic polynomial in the present case.) The neglect of higher-order concentration fluctuations in (2) is parallel to the approximations made in deriving the EPI in the GPT; the primary difference is that only the short-ranged part of the EPI is obtained.

RESULTS

The calculated EPI for fcc and bcc Ni-Al alloys are shown in Figure 1. In the fcc case our choice of clusters provides information only about the nearest-neighbor interaction ϕ_1; in the bcc case we obtain the second-neighbor potential ϕ_2 as well. Here "unrelaxed" means that only one lattice constant is used for all of the calculations. "Globally relaxed" means that a different value of the lattice constant, obtained by linear interpolation from the pure element values, is used for each concentration. Finally, "locally relaxed" means that the energy of each cluster in the alloy is obtained at the lattice constant which minimizes the energy of the cluster, thus artificially ignoring the constraints on the bond length resulting from the rest of the lattice. The supercell total energies are obtained using the self-consistent augmented-spherical-wave method [11] with an exchange-correlation functional of the Hedin-Lundquist form [12]. Magnetic effects on the total energies are neglected; these are small even for pure Ni and drop off rapidly with increasing Al content.

Furthermore, they are weaker at the elevated temperatures where Ni-Al alloys display their most interesting properties. The salient features of the results are the following:

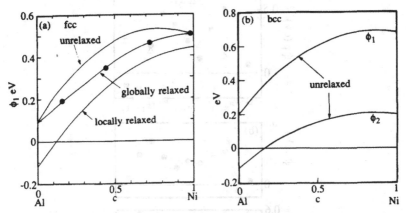

Figure 1: EPI for Ni-Al system as functions of concentration, for fcc and bcc structures. ϕ_1 and ϕ_2 denote nearest-neighbor and second-neighbor interactions, respectively. "Unrelaxed" calculations performed at volume of 73.2 (Bohr)^3 per atom.

1) Both of the nearest-neighbor interactions are strongly concentration-dependent, with the ordering tendency much stronger at the Ni-rich end. The second-neighbor interaction, in the bcc case, displays a similar concentration dependence.

2) The lattice strain effects contribute 15-20% of the potential at the Ni-rich end.

3) The magnitudes of the potentials decay with increasing bond length:

$$|\phi_1(bcc)| > |\phi_1(fcc)| > |\phi_2(bcc)| \quad .$$

The decay is quite rapid; the nearest-neighbor bcc interaction is several times stronger than the second-neighbor interaction.

4) The overall behavior of the interactions is consistent with the observed phase diagram. Their large magnitude would lead one to expect a strong ordering tendency; in fact, both on the fcc and bcc lattices, ordered compounds persist up to the melting point [13]. Furthermore, the sign of the concentration dependence is consistent with the much larger solid solubility at the Ni-rich end relative to the Al-rich end.

Results for Al-transition metal alloys, at the Al-rich and transition metal-rich ends of the phase diagram, are shown in Fig. 2a versus the number of valence electrons. Al-Zn and Al-Li results are included for comparison. Lattice strain effects are treated with the "locally relaxed" approximation described above. Only fcc results have so far been obtained for most of these; however, the Ni-Al results described above would suggest a close correspondence between the fcc and bcc results. We note the large magnitude (several tenths of an eV) of the characteristic values of ϕ_1 for Al-transition metal alloys, compared to those for Al-Zn and Al-Li (less than 0.1 eV). This is not surprising in view of the large energies generally associated with d-bonding. Furthermore, rapid oscillations are seen in both $\phi_1(Al)$ and $\phi_1(T)$ as functions of d-band filling. This behavior is clarified in Figs. 2b and 2c, which show the concentration-averaged EPI $\bar{\phi}$ and the difference $\Delta\phi = [\phi_1(T) - \phi_1(Al)]$ versus valence electron number. Here $\bar{\phi}$ is seen to have a constant sign across the transition metal row, while $\Delta\phi$ oscillates rapidly.

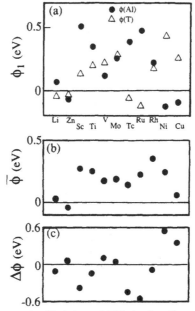

Figure 2: Variation of EPI ϕ_1 for alloys of Al with transition metals, Li and Zn, with number of valence electrons. a) $\phi_1(Al)$ and $\phi_1(T)$ denote values of ϕ_1 at Al end of phase diagram and opposite end, respectively. b) concentration-averaged EPI $\bar\phi$. c) $\Delta\phi = [\phi_1(T) - \phi_1(Al)]$

The oscillations in $\Delta\phi$ are closely analogous to those seen when the energy differences between fcc, bcc, and hcp structures in elemental transition metals are examined as functions of d-band filling [14-16]. The behavior of the structural energy differences can be explained [14] in terms of one-electron band effects involving the placement of the Fermi level relative to peaks and valleys in the electronic density of states (DOS). The necessity of oscillations in these energy differences can be understood via moment analyses [16] of the DOS, together with a theorem [17] placing a lower bound on the number of oscillations on the basis of the number of moments that are identical in different crystal structures. We do not have an interpretation of the oscillations in $\Delta\phi$ as elegant as these moment arguments. However, the calculated DOS [18] would suggest that one-electron band effects are very important. In the Ni-Al system, Fig. 3a, the Fermi level ε_F in pure Ni lies in a sharp peak in the DOS. As Al is added, the d-band fills and ε_F moves toward a minimum, with the DOS at ε_F falling rapidly as a function of band filling. As still more Al is added, the Fermi level moves above the Ni d complex and the value of the DOS at ε_F changes less dramatically with concentration. This behavior would suggest that the Ni-Al bond should be strongest in the Ni-rich part of the phase diagram. By contrast, in the Ru-Al system (Fig. 3b), ε_F for pure Ru lies in a dip in the DOS. As Al is added the Fermi level moves to a maximum in the DOS in Ru_3Al and RuAl. At Al_3Ru it has shifted close to a minimum, again consistent with the concentration dependence of $\Delta\phi$ seen in Fig. 2.

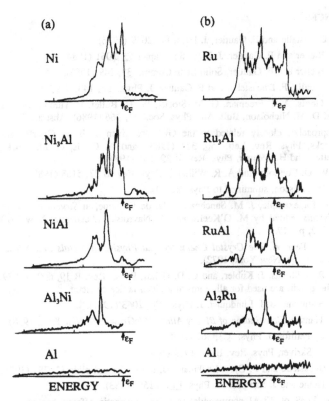

(a)
(b)

Ni
Ru

Ni₃Al
Ru₃Al

NiAl
RuAl

Al₃Ni
Al₃Ru

Al
Al

ENERGY
ENERGY

Figure 3: Electron densities of states (DOS) for supercell compounds used in fcc Ni-Al and Ru-Al calculations.

The relative constancy of $\bar\phi$ with changing band filling would suggest that this quantity is less sensitive than $\Delta\phi$ to the detailed shape of the DOS. Indeed, since $\bar\phi$ is an average, we would expect contributions from peaks and valleys in the DOS to be smoothed out to some extent. While we do not have a simple model for $\bar\phi$, it may be that more classical effects such as charge transfer play a significant role in determining its magnitude.

ACKNOWLEDGEMENTS

The author is grateful to Arthur Williams for an illuminating conversation and for the use of the augmented-spherical-wave code developed at IBM. This work was supported by the United State Department of Energy under Grant Number DE-FG02-84ER45130. The calculations were performed on the Production Supercomputer Facility of the Center for Theory and Simulation in Science and Engineering at Cornell University, which is funded, in part, by the National Science Foundation, New York State, and IBM Corporation.

44

REFERENCES

1. F. Ducastelle and F. Gautier, J. Phys. F6, 2039 (1976).
2. A. Bieber and F. Gautier, J. Phys. Soc. Japan 53, 2061 (1984).
3. A. Bieber and F. Gautier, Solid State Comm. 39, 149 (1983).
4. G. Treglia, F. Ducastelle, and F. Gautier, J. Phys. F8, 1437 (1978).
5. A. Gonis, A. J. Freeman, G. M. Stocks, W. H. Butler, P. Turchi, D. de Fontaine, and D. M. Nicholson, Bull. Am. Phys. Soc. 31, 366 (1986), Abstract EH1.
6. Approaches closely related to the GPT are given in B. L. Gyorffy and G. M. Stocks, Phys. Rev. Lett. 50, 374 (1983), and A. Gonis, G. M. Stocks, W. H. Butler, and H. Winter, Phys. Rev. B 29, 555 (1984).
7. J. W. D. Connolly and A. R. Williams, Phys. Rev. B 27, 5168 (1983).
8. A. E. Carlsson, submitted to Phys. Rev. B.
9. See, for example, J. M. Sanchez and D. de Fontaine, in *Structure and Bonding in Crystals*, edited by M. O'Keeffe and A. Navrotsky (Academic, New York, 1981), Vol. 2, p. 117.
10. W. B. Pearson, *The Crystal Chemistry and Physics of Metals and Alloys*, (Wiley-Interscience, New York, 1972).
11. A. R. Williams, J. Kübler, and C. D. Gelatt, Phys. Rev. B 19, 6094 (1979). Equal sphere radii are used for all atoms at a given lattice constant.
12. L. Hedin and B. I. Lundquist, J. Phys. C4, 2063 (1971).
13. M. Hansen, *Constitution of Binary Alloys (McGraw-Hill, New York, 1958)*.
14. D. G. Pettifor, J. Phys. C3, 367 (1970).
15. H. L. Skriver, Phys. Rev. B 31, 1909 (1985).
16. F. Ducastelle and F. Cyrot-Lackmann, J. Phys. Chem. Solids 32, 285 (1971).
17. V. Heine and J. Sampson, J. Phys. F13, 2155 (1983).
18. The DOS of Ni-Al compounds, including magnetic effects, has previously been given by D. Hackenbracht and J. Kübler, J. Phys. F10, 427 (1980).

ATOMISTIC SIMULATIONS OF [001] SYMMETRIC
TILT BOUNDARIES IN Ni$_3$Al

S. P. CHEN, A. F. VOTER, AND D. J. SROLOVITZ
Los Alamos National Laboratory, Los Alamos, NM 87545

ABSTRACT

We report a systematic atomistic simulation study of [001] symmetric tilt grain boundaries (GB) in Ni$_3$Al, Ni, and Al. We found that the grain boundary energies and cohesive energies of Ni$_3$Al and pure fcc Ni are approximately the same. Grain boundary energies and cohesive energies in Ni$_3$Al depends strongly on the grain boundary composition. The Al-rich boundaries have highest grain boundary energies and lowest cohesive energies. This offers an explanation for the stoichiometric effect on the boron ductilization.

INTRODUCTION

The intermetallic compound Ni$_3$Al, like many other L1$_2$ ordered alloys (Cu$_3$Au structure), exhibits increasing yield strength with increasing temperatures [1], a very desirable property in high temperature applications. Though Ni$_3$Al is ductile as a single crystal, it is intergranularly brittle in polycrystalline form [2]. This lack of ductility in the polycrystalline form prohibits the practical application of this material. Recently, Aoki and Izumi [3] were able to produce a substantial increase in the ductility of polycrystalline Ni$_3$Al by microalloying with small amounts of boron. Liu and coworkers [4] have found that boron increases the ductility only in Ni-rich Ni$_3$Al samples; the best ductility is achieved in samples with 24 atomic percent Al. Neither the boron effect nor the stoichiometry effect are understood. We report here the results of a systematic study of grain boundaries in Ni$_3$Al. A preliminary report has been published recently [5].

We have performed a series of computer simulations on symmetric tilt [001] grain boundaries in Ni$_3$Al using a high quality interatomic potential. The effect of grain boundary composition on the grain boundary energy and the cohesive strength of the grain boundary was studied for nine grain boundaries. The grain boundary composition was found to affect the grain boundary energy by approximately 10%. This finding may help shed some light on the stoichimetry effect mentioned above [4,6]. Also be comparing the grain boundary energies and cohesive energies the intrinsic brittleness of Ni$_3$Al polycrystal [6] can be understood.

COMPUTATIONAL PROCEDURE

Coincident site lattice (CSL) symmetric tilt [001] grain boundaries are generated for each of the nine tilt angles studied. For each of these boundaries, the Al can occupy one of four sublattices in each grain (top or bottom), leading to sixteen possible atomic arrangements for each boundary. However, for each boundary, there are only three unique atomic arrangements, as all others can be generated by reflection or by translations in the plane of the grain boundary. The composition of the grain boundary may be indicated by the composition of the first layer of each grain (percentage of atoms which are Ni: top grain/bottom grain): namely 100/100, 100/50, 50/50 grain boundaries. While the 100/50 grain boundary has the same stoichiometry as bulk Ni$_3$Al, the 100/100 boundary is Ni rich and the 50/50 grain boundary is Al rich. (Note that some grain boundaries, not studied here, have only one possible composition e.g. [011] tilt Σ3(111).)

The generated CSL grain boundaries were relaxed to their global energy minimum using a gradient technique. Periodic boundary conditions were employed in the directions parallel to the grain boundary. At least 160 atomic layers parallel to the grain boundary were employed and the top and bottom surfaces were free. The relaxation process allowed both rigid shifts of the two grains with respect to each other as well as individual atom motion. In order to guarantee that a global energy minimum was obtained, the minimization was started at more than 10 different rigid shifts of the two grains; local minima as high as 600 mJ/m^2 above the global minimum were found.

The interactions between Ni and Al were described using potentials [9] related to those of Daw and Baskes [7]. This approach allows for a simple description of atomic interactions in the vicinity of defects such as grain boundaries and free surfaces, and has proven quite successful in a variety of applications [8]. The method is inherently many bodied and involves two distinct terms: a local density or volume term and a pairwise term. These terms are determined by fitting empirical forms to experimental data. The details of fitting the potential are described elsewhere [9]; the following is a brief summary. The pairwise interaction is taken to be a Morse potential and the density function is of the form $r^6 e^{\xi r}$ (r is the radial distance and ξ is a parameter), leading to a total of 5 parameters using a variable distance for the (smooth) cut-off length. The shape of the embedding function [7] is chosen such that the crystal energy as a function of lattice constant matches the universal form given by Rose, et al. [10]. This leads to exact agreement with the experimental lattice constant (a_0), cohesive energy (E_{coh}), and bulk modulus. Concurrently, a best fit to the three elastic constants (two of which are independent), the vacancy formation energy (E_{vac}) and the diatomic bond length (R_e) and bond energy (D_e) is found by searching the 5 parameter space [while requiring E(fcc) < E(hcp), E(bcc)]. The RMS deviations between the calculated and experimental data in the fit are 0.8% for Ni and 3.9% for Al.

Without modifying the pure Ni and pure Al fits, a Ni-Al cross potential (Morse) is determined by optimizing a fit to the lattice parameter and cohesive energy of NiAl and Ni$_3$Al, as well as to the elastic constants, super intrinsic stacking fault energy (SISF), and anti-phase boundary energies of Ni$_3$Al and estimates of its ordering energy and vacancy formation energy. The resultant potential is capable of describing pure Ni, pure Al, diatomic Ni$_2$, diatomic Al$_2$, and Ni$_3$Al (L1$_2$). The energy of the (111) APB is higher than that of the (100) APB, as required for the validity of the Kear-Wilsdorf cross-slip mechanism [11] which explains the anomalous yield strength at higher temperatures.

RESULTS AND DISCUSSION

For pure Ni, pure Al and Ni$_3$Al, all three grain boundary types gave z-direction (perdendicular to the grain boundary) expansions of ~ 0.1 a_0 relative to the perfectly symmetric starting configuration. However, the x and y shifts are different for pure Ni, pure Al and for different Ni$_3$Al grain boundary compositions with the same Σ. Significant rigid shifts have been observed in earlier simulations of Al by Smith, et al. [12]. The grain boundary energy (γ_{GB}) is defined as

$$\gamma_{gb} = \frac{1}{A} \sum_{i=1}^{n} \Delta E_i(\alpha_i) \tag{1}$$

where A is the grain boundary area in our block of simulation, ΔE_i for an atom of type α ($\alpha = N_i$ or Al in the present study) is

$$\Delta E_i(\alpha_i) = E_\alpha - E_\alpha^{bulk} \tag{2}$$

where E_α is the energy of this atom in the presence of grain boundary and E_α^{bulk} is the energy of this atom in the perfect alloy, for Ni it is -4.509ev, for Al it is -4.821ev.

The sum in eq. (1) is over all atoms whose energy is perturbed by the proximity of the grain boundary. We find the sum converges to within 0.5 mJ/m^2 by including all atoms within 9 lattice parameters of the boundary. We also calculated the Griffith cohesive enery of grain boundary from

$$\gamma_{coh} = \gamma_{s1} + \gamma_{s2} - \gamma_{GB} \tag{3}$$

where γ_{s1} and γ_{s2} are the energies of the two surfaces created by pulling apart the boundary. Note that even for the symmetric grain boundaries studied here, γ_{s1} and γ_{s2} may be different due to the different possible surface compositions.

The grain boundary energies are show in Fig. 1 and Table I as a function of the misorientation angle (θ), measured from the (100) plane. Also shown are the average grain boundary energies ($\langle\gamma\rangle_{gb}$) for each composition (the averages do not include $\theta = 0°$ or $\theta = 90°$). Figure 2 and Table II show the cohesive energies.

The most striking feature of these results is that γ_{GB} varies greatly with the grain boundary composition. The energy difference corresponding to the variation in composition (due to the choice of the Al sub-lattice) is from 50 to 300 mJ/m^2 for the high angle cases presented here with an average value of 156 mJ/m^2. This variation is approximately 12% of the total grain boundary energy. It is noteworthy that the 50/50 grain boundary (Al rich) has the highest γ_{GB} for each Σ, with an average grain boundary energy higher than the stoichiometric or Ni-rich boundaries by approximately 10%. We can understand this trend by noting that γ_{GB} is not always zero at tilt angles of $0°$ and $90°$, since a perfect stoichiometry at the grain boundary is only obtained in the 100/50 case. For the 100/100 composition this leads to a higher grain boundary energy of 20 mJ/m^2 and 17 mJ/m^2 for $0°$ and $90°$ respectively. The 50/50 boundaries (which are Al-rich) lead to a 617 mJ/m^2 complex stacking fault at $0°$ corresponding to removal of a (100) layer of 100% Ni from the perfect crystal. This introduces Al-Al nearest neighbor interactions that do not exist in the perfect L1$_2$ crystal. As the tilt angle increases from $0°$, the atomic relaxations that occur can also act to lessen

Table I. Grain Boundary Energies (mJ/m^2)

$\theta(°)$	(index)	Ni	Al	Ni$_3$Al		
				100/100	100/50	50/50
0	$\Sigma 1(100)$	0	0	20	0	617
12.68	$\Sigma 41(540)$	866	278	904	938	1054
22.62	$\Sigma 13(320)$	1109	330	1101	1155	1351
28.07	$\Sigma 17(530)$	1198	338	1208	1303	1388
30.51	$\Sigma 65(740)$	1253	365	1305	1321	1393
36.87	$\Sigma 5(210)$	1278	351	1329	1213	1396
43.60	$\Sigma 29(730)$	1353	373	1417	1434	1467
46.40	$\Sigma 29(520)$	1347	370	1406	1397	1484
53.13	$\Sigma 5(310)$	1221	315	1166	1247	1468
61.93	$\Sigma 17(410)$	1261	349	1294	1334	1407
90.00	$\Sigma 1(110)$	0	0	17	0	441
$\langle\gamma\rangle_{gb}$		1210	341	1237	1260	1379

48

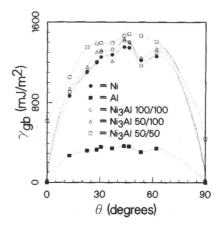

Fig. 1 [001] tilt boundary energies as a function of misorientation angle. The dashed lines are only guides for the eye.

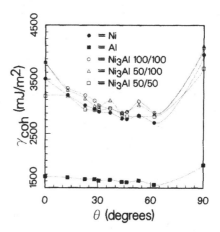

Fig. 2. Griffith cohesive energies as a function of misorientation angle. The dashed lines are only guides for the eye.

Table II. Griffith Cohesive Energy (mJ/m^2) as defined in eq. (3)

$\theta(^o)$	(Index)	Ni	Al	Ni_3Al		
				100/100	100/50	50/50
0	$\Sigma1(100)$	3510	1710	3814	3821	3191
12.68	$\Sigma41(540)$	3206	1690	3336	3302	3186
22.62	$\Sigma13(320)$	3023	1660	3201	3146	2949
28.07	$\Sigma17(530)$	2954	1658	3104	3014	2934
30.51	$\Sigma65(740)$	2905	1633	3019	3002	2929
36.87	$\Sigma5(210)$	2886	1647	2961	3114	2968
43.60	$\Sigma29(730)$	2777	1611	2891	2881	2855
46.40	$\Sigma29(520)$	2765	1604	2904	2909	2818
53.13	$\Sigma5(310)$	2835	1635	3092	3027	2822
61.93	$\Sigma17(410)$	2699	1557	2886	2867	2815
90.00	$\Sigma1(110)$	3954	1918	4077	4119	3703
$<\gamma>_{coh}$		2894	1663	3044	3029	2920

the severity of the Al-Al interactions. Indeed, the shift perpendicular to the grain boundary is found to be largest for the 50/50 composition, and the extra grain boundary energy due to these Al-Al interactions is reduced from 617 mJ/m^2 at 0° to an average of $<\gamma>_{GB}(50/50) - <\gamma>_{GB}(100/50) = 119$ mJ/m^2 for the other misorientations. The relaxed structures of $\Sigma5(210)$ grain boundaries are shown in Fig. 3.

The 100/100 (Ni rich) grain boundaries are seen to have the lowest average energy, in spite of the fact that at 0° and 90°, there is a very low stacking fault energy. Also note that cusp depth of γ_{GB} in Fig. 1 varies with grain boundary composition.

Tables I and II show that the grain boundary energies in Ni_3Al (1290 mJ/m^2) are much closer to those of pure Ni (1210 mJ/m^2) than to those of either pure Al (341 mJ/m^2) or a stoichiometrically weighted average of Ni and Al (993 mJ/m^2). Furthermore, the cohesive energies of the Ni_3Al grain boundaries (~ 3000 mJ/m^2) are comparable to or higher than those of pure Ni (2894 mJ/m^2). This raises a question: If the grain boundaries in Ni and Ni_3Al are of comparable strength, why is polycrystalline Ni_3Al intrinsically brittle while Ni is ductile? The answer may lie in he plastic response of the matrix. Hack et al. have proposed a simple model for the fracture behavior of Ni_3Al [13] which can be summarized as follows. If, in a simple model of material response, failure is controlled by either plastic flow or intergranular fracture, a decrease in grain boundary cohesion at fixed yield stress or an increase in yield stress for a fixed grain boundary cohesion leads to an increased propensity toward intergranular fracture. This latter case is pertinent in the comparison of Ni_3Al to pure Ni, suggesting that Ni_3Al should show a greater tendency for intergranular fracture than pure Ni.

CONCLUSIONS

We have performed atomistic simulations on nine [001] symmetric tilt grain boundaries. Depending on which sub-lattice in each of the two grains is occupied by Al, the grain boundary may have different stoichiometries. All of the simulations show that the Al-rich grain boundaries have the highest grain boundary energies. Thus Al-rich grain boundaries are more likely to fail than those which have the bulk stoichiometry or are Ni-rich. This conclusion is consistent with the observed stoichiometry dependence of the beneficial boron effect. The similarity between the grain boundary energies (cohesive energies) of Ni_3Al and Ni and the much higher yield stress of Ni_3Al provides a justification for the "inherent" brittleness of Ni_3Al grain boundaries.

50

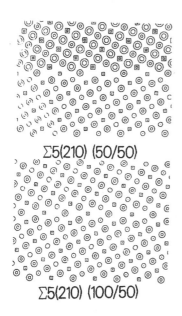

Σ5(210) (50/50)

Σ5(210) (100/50)

Σ5(210) (100/100)

Fig. 3 The relaxed structures for these Σ5(210) boundaries. The Al and Ni atoms are represented by squares and circles, respectively; the larger symbols are closer to the reader. Extra Al-Al nearest neighbor interactions are introduced in 50/50 grain boundary (Al-rich).

ACKNOWLEDGEMENTS

This work was supported by the Energy Conversion and Utilization Technologies (ECUT) Program of the U.S. Department of Energy.

REFERENCES

[1] J. H. Westbrook, Trans. A.I.M.E. 209, 898 (1957).
[2] J. H. Westbrook, in Mechanical Properties of Intermetallic Compounds, Ed. J. H. Westbrook, Wiley, New York (1960).
[3] K. Aoki and O. Izumi, Nippon Kinsoku Gakkaishi 43, 1190 (1979).
[4] C. T. Liu, C. L. White, and J. A. Horton, Acta Metall. 33, 213 (1985).
[5] S. P. Chen, A. F. Voter, and D. J. Srolovitz, Scripta, Metall. 20, 1389 (1986).
[6] T. Ogura, S. Havada, T. Masumoto, and O. Izumi, Metall. Trans. 16A, 441 (1985); T. Takasugi, E. P. George, D. P. Pope, and O. Izumi, Scripta Metall. 19, 551 (1985).
[7] M. S. Daw and M. I. Baskes, Phys. Rev. B 29, 6443 (1984).
[8] S. Foiles, M. I. Baskes, and M. S. Daw, Phys. Rev. B, in press (1986); M. S. Daw, Surf., Sci. Lett. 166, L161 (1986); S. P. Chen, A. F. Voter and D. J. Srolovitz, Phys. Rev. Lett. 57, 1308 (1986).
[9] A. F. Voter, to be published, A. F. Voter and S. P. Chen, MRS conference proceedings, Symposium I (1986).
[10] J. H. Rose, J. R. Smith, F. Guinea, and J. Ferrante, Phys. Rev. B 29, 2963 (1984).
[11] B. H. Kear and H. G. F. Wilsdorf, Trans. A.I.M.E. 224, 382 (1962); S. Takeuchi and E. Kuramoto, Acta Metall. 21, 415 (1973).
[12] D. A. Smith, V. Vitek, and R. C. Pond, Acta Metall. 25, 475 (1977); R. C. Pond, D. A. Smith, and V. Vitek, Acta Metall. 27, 235 (1979).
[13] J. E. Hack, D. J. Srolovitz, and S. P. Chen, Scripta, Metall. (1986) in press.
[14] M. K. Miller and J. A. Hoton, (1986), preprint.

CALCULATION OF THE DEFECT AND INTERFACE PROPERTIES OF Ni$_3$Al*

STEPHEN M. FOILES
Sandia National Laboratories, Livermore, CA 94550

ABSTRACT

The structure and energetics of point defects, surfaces and grain boundaries in Ni$_3$Al are investigated using the Embedded Atom Method. The approach is shown to reproduce the experimental phase diagram of the Ni-Al system and the elastic properties of Ni$_3$Al. The vacancy and anti-site defect energies are calculated and used to predict the vacancy concentration as a function of bulk composition. The preferred geometries and energies of the low index surfaces are also computed. The equilibrium structure of certain ideal grain boundaries are computed by Monte Carlo computer simulations as a function of bulk composition. It is found that the boundaries act as a sink for anti-site defects and the degree of ordering at the boundaries is strongly affected by the bulk composition. The cohesive energy of grain boundaries in Ni$_3$Al is computed and is found to be comparable to that for pure Ni.

INTRODUCTION

There is a great deal of current interest in nickel aluminide, Ni$_3$Al, as evidenced by the large number of papers concerning this material at this symposium. This interest stems in part from the unusual mechanical property that the yield strength increases with increasing temperature up to around 600 C.[1] Unfortunately, the pure alloy is brittle in polycrystalline form due to intergranular fracture. It has been discovered, however, that the addition of B to Ni-rich alloys produces a ductile material.[2] In this paper, the structure and energetics of point defects and interfaces, in particular grain boundaries, will be calculated using the Embedded Atom Method combined with energy minimization (lattice statics) techniques and with Monte Carlo computer simulations.

COMPUTATIONAL TECHNIQUE

To determine the atomic structure of defects and interfaces, a relatively simple model of the energetics is required so that a sufficiently large set of atoms can be considered to include the important relaxations. The Embedded Atom Method[3] (EAM) appears to be a good choice for such studies since previous work has shown that it provides a reasonable description of the energetics of metallic systems[4] yet does not require significantly more computational effort than the use of simple pair interaction models. In this model, much of the binding energy of the solid is attributed to the energy associated with embedding each atom into the local electron density provided by the remaining atoms of the system. This energy is assumed to depend only on the type of atom being embedded and the local electron density. Thus the same embedding energy applies in the pure metals as in an alloy. This embedding contribution is supplemented by a pair interaction term which accounts for core-core interactions. The embedding energies and pair interactions required by the model have been determined empirically for the Ni-Al system previously by fitting to the sublimation energy, lattice constant and elastic constants of the pure metals as well as the sublimation energy and lattice constant and preferred structures of the the ordered alloys NiAl and Ni$_3$Al.[5]

The thermal equilibrium structures are then determined by performing Monte Carlo computer simulations with the energies of the various atomic configurations computed by the EAM. This approach has been applied to the study of the equilibrium surface segregation in binary alloys and has been described in detail previously.[6] The simulations allow for both the compositional rearrangement of the system by the mathematical transmutation of the atoms (the simulations are performed in a fixed chemical potential ensemble) as well as the spatial displacement of the atoms. By adjusting the chemical potentials used for the simulations it is possible to study the variation in compositional order and structure as the composition of the bulk material is changed. The inclusion of local displacements allows for the relaxation of the atomic positions away from ideal lattice sites at defects and interfaces and so includes the local strain energies. In addition, since *all* of the atoms are free to move, larger scale relaxations, such as the increased interplanar spacing near a grain boundary, are also included.

BULK AND SURFACE RESULTS

This approach has been used previously to study some properties of both the bulk and the low index surfaces of Ni_3Al.[5] As a test of the approach the slice of the phase diagram at 1000 K predicted by the EAM energetics was computed. Monte Carlo simulations were performed for a variety of different chemical potentials corresponding to the Ni-rich half of the composition range to determine the predicted equilibrium phases. It was found that the method reproduced the three experimentally observed phases[7], namely the NiAl phase in the B_2 structure, the Ni_3Al phase in the $L1_2$ structure, and the solid solution phase. The range of homogeneity of the various phases was estimated and found to be in reasonable agreement with the experimental phase diagram. The largest discrepancy was that the solid solution phase was found to be stable to about 18 atomic percent Al while experimentally it is stable to about 11 atomic percent. As an additional test, the elastic constants were computed for the Ni_3Al system and found to be in good agreement with the experimental values. These results indicate that this approach provides a reasonable description of the Ni-Al system.

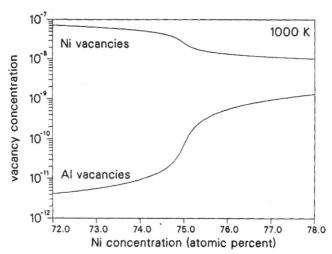

Figure 1. The calculated concentration of thermal vacancies on the Ni and Al sublattices of Ni_3Al at 1000 K.

The formation energy of vacancies on the two sub-lattices of Ni_3Al and the energy required to produce an anti-site defect, for example a Ni atom in an Al site, have also been computed in Ref. 5. The results show that the formation energy of anti-site defects is less than that for vacancies and so deviations from ideal stoichiometry should occur by the production of anti-site defects rather than by the production of constitutional vacancies. This is in accord with positron annihilation studies.[8] Further, the point defect energies can be used to estimate the vacancy concentration as a function of bulk composition. The results computed for a temperature of 1000 K are shown in Figure 1. Note that the majority of vacancies occur on the Ni sub-lattice. The vacancy concentration is also found to have a significant concentration dependence with the number of Ni vacancies enhanced for the Al-rich alloys. It should be noted that the overall concentration of thermal vacancies is predicted to be fairly small.

The structure of the low index surfaces of nickel aluminide was also investigated. For the (100) and (110) orientations there are two different planes that could be exposed on the equilibrium surface, one possible plane contains just Ni atoms and the other possible plane contains an equal mix of Ni and Al atoms. For these surfaces the energy of both possible surface geometries was computed and it was found that the surface containing an equal mix of the two elements is energetically preferred. This agrees with low energy electron diffraction (LEED) results for the (100) surface.[9] The surface relaxations were also computed for the (100), (110) and (111) surfaces. The relaxations were found to be small with the largest displacement being a 0.06 Å *outward* relaxation of the top layer Al atoms. The presence of a rippled surface with the Al atoms above the Ni atoms is the same as what is seen in LEED measurements of the structure of NiAl surfaces[10], in LEED measurements of the (100) surface of Ni_3Al[11], and in calculations for NiAl and Ni_3Al using similar techniques to those presented here by Chen, et al.[12] These functions predict the usual inward relaxations for the surfaces of both of the pure metals.

GRAIN BOUNDARY RESULTS

The grain boundaries of Ni_3Al are of interest due to its propensity for intergranular fracture. As a first step toward understanding the mechanical properties of the boundaries, the equilibrium structure of various ideal grain boundaries have been computed at a temperature of 1000 K. Two types of ideal boundaries will be considered here, symmetric twist boundaries and a symmetric tilt boundary. The two twist boundaries are the $\Sigma=5$ and the $\Sigma=13$ twist boundaries which are constructed by rotating one half of a crystal around the (100) axis normal to the boundary plane by 36.9° and 22.6° respectively. The tilt boundary is a $\Sigma=5$ symmetric tilt boundary. The two lattices on each side of this boundary are rotated around a cubic axis parallel to the boundary plane such that the boundary plane is parallel to a (210) plane in each lattice. These boundaries were chosen because they have relatively small unit cells and so are amenable to computer simulation. The simulations actually treat a bi-crystal slab which is periodically repeated in the plane of the slab. The linear dimensions of the periodic cell were different for each boundary but were about 15-20 Å in each direction and the slabs were about 35 Å thick. For each geometry, the simulations were performed with chemical potentials corresponding to a Ni-rich bulk, an Al-rich bulk and an ideal stoichiometric bulk. The initial condition for each simulation was an ideal coincidence site structure consisting entirely of Ni atoms.

The compositional structure determined by the simulations for the twist boundaries at ideal stoichiometry is very simple. The (100) planes of the Ni_3Al structure alternate between pure Ni and an equal mix of Ni and Al. At the grain boundary, this alternating pattern of the (100) planes parallel to the boundary continues uninterrupted through the boundary. Thus the compositional ordering is the same that would be obtained by taking an ideal Ni_3Al crystal and rotating the two halves to form the boundary. This result is not surprising since the interactions favor the presence of Ni-Al nearest neighbor pairs and this structure accomplishes that. The interplanar spacing normal to the boundary increases due to the lattice misorientation. For the $\Sigma=13$ boundary at ideal stoichiometry, the calculated net expansion normal to the boundary is 0.3 Å. The structure of the tilt boundary is more complicated. The atoms originally in the (210) plane on each side of the boundary combine to form one dense plane with little compositional order. The next (210) plane on each side of the boundary is close to its structure in the bulk crystal and is separated by 1.4 Å from the middle of the central plane. (The (210) interplanar spacing is 0.81 Å.) Thus there is a net expansion of about 0.4 Å normal to the boundary.

The local composition of the grain boundary need not be the same as the bulk composition. This deviation of the local composition near the interface can be described in terms of the interfacial excess of one of the two components. The interfacial excess of Ni is defined as the difference per unit area between the total number of Ni atoms in the sample (including the interface) and the number that would be present if the bulk composition is assumed throughout. The interfacial excess of Ni for the different geometries is plotted in Figure 2 as a function of the bulk composition. First note that the excess is essentially zero at the ideal bulk stoichiometry. This is reasonable for the twist boundaries since the alternation of the composition of the (100) planes is not disturbed by the boundary. Away from ideal stoichiometry, one sees that the grain boundary acts as a sink for the Ni in Ni-rich alloys and the Al in Al-rich alloys. This behavior can be qualitatively understood by considering the energy to create anti-site defects at various positions near the boundary. Figure 3 shows these energies for one of the atomic configurations generated for the tilt boundary. The anti-site formation energy is substantially reduced for sites within a couple Å of the boundary. Thus it is energetically favorable to create either anti-site defect at the boundary rather than in the bulk of the material. This is equivalent to a segregation of anti-site defects to the interface. An important consequence of this is that one would expect this segregation to be stronger at lower temperatures and so we would expect the interfacial excess to increase with decreasing temperature.

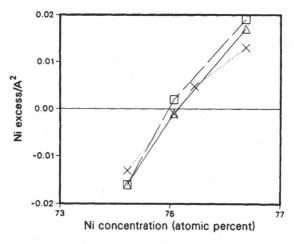

Figure 2. The grain boundary excess of Ni as a function of bulk composition at 1000 K calculated for the $\Sigma=5$ twist boundary (triangles), $\Sigma=13$ twist boundary (x), and the $\Sigma=5$ tilt boundary (squares).

In addition to the composition of the boundary region, the compositional order of the boundary is also affected by the bulk composition. This is most easily seen for the twist boundaries. For boundaries that are Al-rich, the excess Al could either be placed in the Ni plane or in the Ni sites of the mixed composition plane. However, the simulations show that the majority of the excess Al are placed on the Ni plane and that the mixed composition plane actually contains slightly less than 50% Al. For the Ni-rich case, one would simply expect to replace the Al atoms on the mixed composition plane with Ni atoms. The simulations show, though, that in addition there is a reduction in the ordering of the mixed plane. For the $\Sigma=13$ boundary at a bulk composition of 24.5% Al, there are three times more anti-site defects in the mixed compostion plane than are required by the reduced Al concentration in that plane. In addition, the Ni plane in this boundary is found to contain 4% Al atoms even though the system is deficient in Al. Thus away from ideal stoichiometry, there is a tendency to smear out the alternating composition of the (100) planes at the boundary. Also, for the Ni-rich alloy, the reduction of Al content in the mixed composition plane has lead to a significant reduction in the degree of order in that plane. The temperature dependence of this reduction of order is not clear. While lower temperatures reduce the entropic driving force for disorder, the Al content should be reduced more at lower temperatures making the disordering of the mixed plane easier.

The EAM can also be used to investigate the energetics of the grain boundary. This is of interest because the propensity of Ni_3Al to intergranular fracture could be attributed to a low cohesive energy of the grain boundaries. The energy of the tilt boundary has been computed at zero temperature for the $\Sigma=5$ grain boundary for both pure Ni as well as for Ni_3Al. In both cases the grain boundary energy is about 1250 ergs/cm^2. The cohesive energy represents the energy required to cleave the boundary and so is the sum of the surface energy of the two free surfaces that are created minus the energy of the grain boundary which is cleaved. For this tilt boundary, the surfaces created have a (210) orientation. The energy of this surface is 1830 ergs/cm^2 for pure Ni and is 1830 or 1940 ergs/cm^2 for Ni_3Al depending on whether the exposed surface plane is the

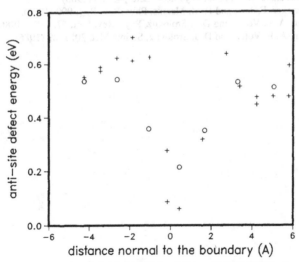

Figure 3. The energy to create anti-site defects in the vicinity of a $\Sigma=5$ tilt boundary as a function of position normal to the boundary. The + represent placing a Al atom in a Ni site (bulk enerrgy of 0.58 eV) and the circles represent placing a Ni atom in an Al site (bulk energy of 0.54 eV).

pure Ni plane or the mixed composition plane. Thus one would expect about the same cleavage energy for this boundary in both Ni and Ni_3Al, namely about 2400 ergs/cm^2. This value was compared for the case of Ni_3Al to the finite temperature simulation results by cleaving various of the configurations produced during the simulation of the boundary and comparing the initial and final energies. The cleavage energy computed in this manner was 2500-2600 ergs/cm^2. The difference between the energies computed by cleaving typical boundaries and that estimated from surface and grain boundary energies probably reflects the fact that the surfaces produced are not ideal. These results indicate that the cohesive energy of grain boundaries in nickel aluminide are not significantly different than for pure Ni which implies that the tendency to intergranular fracture is not simply due to poor cohesion of the boundaries. Similar results have been obtained from calculations performed by Chen, et al.[13]

REFERENCES

*This work supported by the United States Department of Energy, Office of Basic Energy Sciences.

1. D. P. Pope and S. S. Ezz, International Metals Review 29, 136 (1984).

2. C. T. Liu, C. L. White, and J. A. Horton, Acta Metall. 33, 213 (1984).

3. M. S. Daw and M. I. Baskes, Phys. Rev. B29, 6443 (1984).

4. S. M. Foiles, M. I. Baskes, and M. S. Daw, Phys. Rev. B33, 7983 (1986), and references therein.

5. S. M. Foiles and M. S. Daw, J. Mater. Res. (in press).

6. S. M. Foiles, Phys. Rev. B32, 7685 (1985).

7. M. Hansen, Constitution of Binary Alloys (McGraw-Hill, New York, 1958).

8. A. Dasgupta, L. C. Smedskjaer, D. G. Legnini, and R. W. Siegel, Mater. Lett. 3, 457 (1985).

9. D. Sondericker, F. Jona, and P. M. Marcus, Phys. Rev. B33, 900 (1986).

10. H. L. Davis and J. R. Noonan, Phys. Rev. Lett. 54, 566 (1985).

11. D. Sondericker, F. Jona and P. M. Marcus, Phys. Rev. B33, 900 (1986).

12. S. P. Chen, A. F. Voter, and D. J. Srolovitz, Phys. Rev. Lett. 57, 1308 (1986).

13. S. P. Chen, A. F. Voter, and D. J. Srolovitz, Scripta Met. 20, 1389 (1986).

MODELING OF ANTIPHASE BOUNDARIES IN L1$_2$ STRUCTURES

J.M. SANCHEZ, S. ENG, Y.P. WU and J.K. TIEN
Center for Strategic Materials and Henry Krumb School of Mines
Columbia University, New York, NY 10027

Abstract

The thermodynamic properties of conservative (111) antiphase boundaries in L1$_2$ ordered structures are modeled using the tetrahedron approximation of the cluster variation method. The concentration and long-range order parameter profiles are determined as a function of temperature and composition of the bulk alloy. Characteristic wetting transitions, with a macroscopic disordered layer growing from the antiphase boundary as the transition temperature is approached, are found for all cases investigated. The effect of antiphase boundaries on the disordering of ordered alloys and on the gliding of superdislocations are discussed.

Introduction

It has long been recognized that antiphase boundary (APB) energies play a significant role in the deformation behavior of superlattices [1-4]. In ordered alloys, such as Ni$_3$Al gamma prime, antiphase boundaries separate pair of dislocations, each with a Burgers vector less than the unit translation vector of the superlattice. Consequently, the slip of one of the dislocations will disturb the symmetry of the ordered structure by the shearing of two planes, creating an APB. Long range order is restored by the second dislocation of the pair. The pair of dislocations, whose sum of Burgers vectors equals a superlattice translation vector and combined movement restores order, is known as a superdislocation [5,6].

One of the earliest studies on the unusually high work hardening rate of Ni$_3$Al was conducted by Flinn [3]. Flinn attributed the increase in flow stress with temperature to the anisotropy of the APB energies, with a minimum occurring not on the {111} slip planes but rather on {010} sessile planes. Flinn's calculations involved the nearest neighbor interactions of a non-equilibrium APB created by the gliding of one of the components of the superdislocation. The non-equilibrium condition reflects an APB immediately after rapid shearing or at low temperatures where there is no diffusional atomic rearrangement.

Based on Flinn's model of APBs, a successful cross slip model, originally proposed by Kear and Wilsdorf [7] and subsequently refined [8-11], explains the anomalous yield behavior of Ni$_3$Al by the cross slipping of 1/2[101](111) screw dislocations from the (111) slip plane onto (001) planes on which the dislocations become immobilized. The driving force for cross slip is the reduction in APB energy. Paidar et al. [11] determined that in order for cross slip to take place, the anisotropy of APB energies between the (111) slip plane and the (001) plane should exceed $\sqrt{3}$.

Advances in high resolution microscopy have only recently made possible experimental verification of dislocation configurations and measurements of APB energies. Results reported by Veyssiere et al. [12,13] indicate that the anisotropy of APB energies in gamma prime is less than that required by the cross slip model. Consequently, an alternative model proposed by Yoo employs the concept of elastic anisotropy as the driving force for cross slip [14].

In view of the above, the actual role of APB in the deformation process of L1$_2$ has come into question. However, little is known about the actual or even the expected (111) APB configuration at high temperatures. For example, in addition to Flinn's non-equilibrium model of an APB generated by shear, other APB configurations may exist in the L1$_2$ superlattice. Metastable APBs in local equilibrium with the ordered structure may be generated by shear at high temperatures and/or slow deformation rates where there is sufficient time for diffusion to occur [15]. These "equilibrium" APBs will affect the degree of long-range order in the L1$_2$ phase. In addition, microsegregation of one of the alloy components to the APB, if present, could result in an additional drag force on the dislocation.

In this study, we characterize the thermodynamic properties of conservative APBs, such as those created by shear, which are in local equilibrium with the lattice. The calculations are done using the cluster variation method (CVM) in the tetrahedron approximation and assuming pair interactions between nearest-neighbors only [15]. This approximation of the CVM has been shown to reproduce accurately the bulk thermodynamic properties of Ni$_3$Al [16,17] and should provide reliable trends concerning the variation of long-range order (LRO) parameters with temperature and distance from the APB, degree of segregation to the APB, and temperature dependence of APB energy.

The Model

We describe the configuration of the alloy (binary or multicomponent) by introducing the probabilities z_{ijkl} of observing a given distribution of atomic species (labeled i, j, k, and l) on fcc tetrahedral clusters made of nearest neighbors. These tetrahedron probabilities uniquely define other subcluster probabilities such as the nearest-neighbor pairs (y_{ij}) and, of course, the atomic concentrations x_i. Thus, given the nearest-neighbor pair interactions V_{ij}, the configurational energy of the alloy can be written in terms of the z_{ijkl}. Furthermore, the tetrahedron probabilities z_{ijkl} can also be used to obtain an estimate of the configurational entropy by counting all possible alloy configurations with the same energy, i.e. all configurations with the same distribution of tetrahedron probabilities. This estimate of the configurational entropy is known to be a considerable improvement over that obtained using the Bragg-Williams approximation. The actual values of tetrahedron probabilities at a given temperature are obtained by minimizing the configurational free energy of the alloy.

Although the accuracy of the CVM increases with the size of the maximum cluster, so does the number of variables involved in the free energy minimization and, in practice, the calculations are limited to relatively small clusters. The usefulness of the technique stems from the fact that sufficient accuracy can be achieved with cluster of only four to six lattice points.

In the presence of a planar defect, such as an APB, the number of configurational variables z_{ijkl} becomes large even for small clusters since the translational symmetry of the lattice is lost. In addition, other point group symmetry elements of the parent structure may disappear, complicating even further the minimization of the free energy. Such is the case of the (111) conservative APB where, in addition to the translational symmetry, the three-fold axis perpendicular to the (111) plane is lost.

Bulk LRO in the L1$_2$ structure is described in the usual manner by defining sublattice concentrations x_i^α and x_i^β giving, respectively, the atomic occupancy at the face centers and corners in the fcc lattice.

With a conservative (111) APB present, LRO is characterized by probabilities $x_i^\alpha(n)$, $x_i^{\alpha'}(n)$ and $x_i^\beta(n)$ where n labels the plane away from the APB and α, α' and β stand for non-equivalent sites, or sublattices, on a given (111) plane (see Fig. 1). The configurational energy is given by:

$$E = (1/2) \sum_{n,n'}{}' \sum_{\nu,\nu'}{}' \sum_{i,j} V_{ij} \, y_{ij}^{\nu\nu'}(n,n') \qquad (1)$$

where V_{ij} is the pair interaction energy between species i and j and where $y_{ij}^{\nu\nu'}(n,n')$ stands for the probability of finding a nearest-neighbor pair of atoms i and j located, respectively, at sublattice ν on plane n and at sublattice ν' on plane n'.

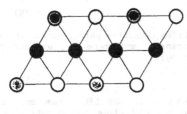

Figure 1:
Sublattices on a fcc (111) plane for the L1$_2$ structure with a conservative antiphase boundary.

⬒ = α   ● = α'   ○ = β

The expression for the configurational entropy follows from a straightforward generalization of the well known entropy for the bulk [18]. The somewhat cumbersome equation will not be reproduced here. It will be sufficient to point out that, as a consequence of the three sublattices present and the lack of translational symmetry, one must define, for each (111) plane, six different tetrahedron probability distributions. These, in turn, result in 11 possible pair probabilities and, as already mentioned, three point probabilities (or concentrations) per plane.

After minimization of the free energy, a very detailed picture of the state of short-and long-range order in the alloy is obtained from the full set of equilibrium cluster probabilities. In the next section we will concentrate our discussion on a few quantities of interest such as the average concentration, $x_i(n)$, and long-range order parameters, $\eta_1(n)$ and $\eta_2(n)$, for the nth (111) plane. In terms of the sublattice probabilities these quantities are given by:

$$x_i(n) = [\, x_i^\alpha(n) + 2\, x_i^{\alpha'}(n) + x_i^\beta(n) \,]/4 \qquad (2)$$

$$\eta_1(n) = x_A^\alpha(n) - x_A^\beta(n) \qquad (3)$$

$$\eta_2(n) = x_A^\alpha(n) - x_A^{\alpha'}(n) \qquad (4)$$

The LRO parameter η_2 is non-zero only in the neighborhood of the APB (due to the loss of the three-fold symmetry axis perpendicular to the (111) plane) and it vanishes in the bulk.

Also of interest are the different energies associated with either equilibrium or non-equilibrium APBs. The difference in configurational energies, ΔE, between the alloy with and without an APB gives a measure of the bond disorder associated with the defect. As we shall see, this energy increases with temperature as the APB becomes more diffuse, inducing disorder on several planes. The equilibrium energy γ_{APB} of the defect, i.e. the energy required to create an APB at constant tempera- ture and chemical potential, is given by the difference in grand potentials. Most likely, however, these energies are of little relevance to the formation of APBs by dislocation gliding since in order to attain the equilibrium APB configuration considerable atomic diffusion is required.

In considering the movement of a superdislocation, which is assumed to have reached thermal equilibrium with the lattice before the onset of deformation, two energies need to be considered [1-4]. These are the bond energy γ across a (111) plane involved in the creation of a non-equilibrium APB by the leading dislocation of the pair, and the bond energy γ^0 across the (111) plane of an equilibrium APB, which is gained by the movement of the trailing dislocation. For a binary alloy A-B, the energy (per atom) required to form a non-equilibrium APB by shear can be shown to be:

$$\gamma = V \eta^2 + V (\omega_{\alpha\alpha} + \omega_{\alpha\beta}) \tag{5}$$

where $V = (V_{AA} + V_{BB} - 2 V_{AB})/4$ is an effective pair interaction, where η is the bulk equilibrium LRO parameter and where $\omega_{\nu\nu'}$ are sublattice short-range order parameters given by:

$$\omega_{\nu\nu'} = 4 (x_A^\nu x_A^{\nu'} - y_{AA}^{\nu\nu'}) \tag{6}$$

The first term in Eq. (5), quadratic in the LRO parameter, is the usual expression obtained in the Bragg-Williams approximation and it vanishes in the disordered state [3]. The additional term in Eq. (5) is a short-range order contribution which is negligible for ordered phases. This term, however, gives rise to short-range order hardening in the high temperature disordered phase [19]. An expression similar to Eq. (5) is obtained for the energy γ^0 in the case of an equilibrium APB. Neglecting short-range order effects, the energy involved in the gliding of a superdislocation, proportional to the difference between γ and γ^0, is given by [1-4]:

$$\Delta\gamma = V (\eta^2 - \eta_1^2) \tag{7}$$

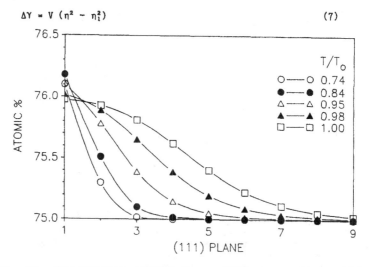

Fig. 2a: Concentration profile for a (111) conservative antiphase boundary located at plane n=1 in an alloy with bulk concentration equal to 0.75. The different temperatures are given in units of the congruent temperature T_0.

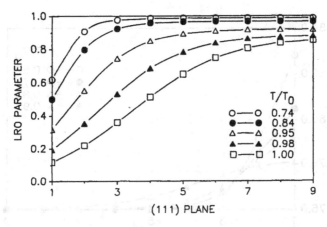

Figure 2b: Profile of long-range order parameter for a (111) conservative antiphase boundary located at plane n=1 in an alloy with bulk
concentration equal to 0.75. The different temperatures are given in
units of the congruent temperature T_0.

Results and Discussion

Figure 2 shows the concentration and LRO parameter profiles for a
conservative (111) APB in a stoichiometric A_3B (L1$_2$) binary alloy at
different temperatures, the latter given in units of the congruent
order-disorder transition temperature T_0. It is seen from Fig. 2a that,
at low temperatures, the effect of the APB is restricted to a few crystalline planes. As the temperature increases, however, a thick layer
rich in the majority component A develops in a fashion characteristic of
wetting transitions in systems with planar defects. This general behavior is a consequence of the disruption of translational symmetry in the
crystal and it also a characteristic phenomenon observed in alloy surfaces, grain boundaries, thin films, and other systems displaying first-
order transitions [20-22]. In Fig. 2b, we see that the A-rich layer is
characterized by a low value of the LRO parameter indicating that, at
the transition temperature, the APB is completely wet by the disordered
phase. The presence or absence of APBs should, therefore, play a key
role in the disordering kinetics of L1$_2$ structures. In particular, as
the temperature is increased into the two phase coexisting region of
such an alloy, the disordered phase will grow from the existing APB
without the need of a nucleation event.

The degree of segregation to the APB is strongly dependent on the
alloy stoichiometry. The concentration profile at different temperatures
for an alloy with bulk concentration of 0.76 is shown in Fig. 3a. In
this case, the enrichment of the APB in the majority component A is considerably larger than in the case of the stoichiometric A_3B alloy. The
LRO parameter profile, shown in Fig. 3b, is qualitatively similar to
that of the stoichiometric case indicating wetting of the APB by the
disorder phase. For both cases shown in Figures 2a and 3a, with bulk
concentrations of 0.75 and 0.76, respectively, the concentration of the
APB at the transition temperature approaches the value prescribed by the
two phase boundaries of the equilibrium phase diagram. This is expected
since the boundaries in question determine the concentrations of the
coexisting ordered (bulk) and disordered (APB) phases.

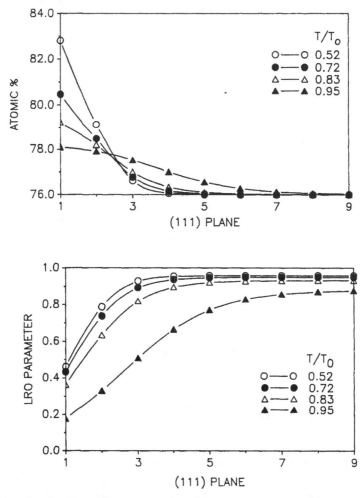

Fig. 3: Profiles of concentration (a) and long-range order parameter for a (111) conservative APB located at plane n=1 in an alloy with bulk concentration equal to 0.76. The different temperatures are given in units of the congruent temperature T_0.

Figure 4 depicts the equilibrium energy, defined as the change in grand potential (ΔG) relative to the bulk, and the internal energy (ΔE) of a conservative (111) APB for an A_3B stoichiometric alloy. Although the grand potential decreases monotonically with temperature, the internal energy increases due to the fact that the APB becomes more diffuse. In fact, the thickness of the wetting layer shown in Fig. 2a diverges logarithmically with temperature as the transition is approached. Consequently the energy of the equilibrium APB also diverges.

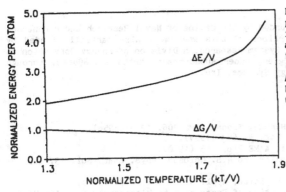

Fig. 4: Grand potential (ΔG) and internal energy (ΔE) as a function of temperature for a (111) conservative APB in an alloy with bulk concentration of 0.75.

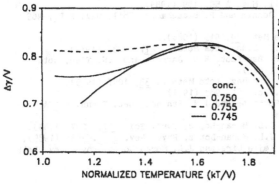

Fig. 5: Antiphase boundary drag on a superdislocation as a function of temperature for different bulk concentrations.

Finally, we have calculated the configurational energy involved in the gliding of a superdislocation as defined in the previous section (see Eq. 7). In order to calculate ΔY, the LRO parameter of the equilibrium APB was obtained by imposing bulk boundary conditions on the plane next to the defect. The variation of ΔY with temperature is shown in Fig. 5 for concentrations of component A equal to 0.745, 0.75 and 0.755. A rather complex behavior, which is largely controlled by segregation to the APB is observed as the temperature and bulk composition change. There is an expected decrease of ΔY at high temperatures due to the decrease of both η and η_1. At lower temperatures, however, the behavior is controlled by the temperature variation of η_1, with the equilibrium LRO parameter η being essentially constant. Figure 5 suggests that APB drag to dislocation motion, which could play a role in microyielding of $L1_2$ structures and/or superalloys, should be very sensitive to segregation to the APB of either alloying additions or point defects such as vacancies.

64

Acknowledgements

This work was funded by the Office of Naval Research under contract
N000114-83K-0223. Portions of this work were also partially funded by
the Office of Energy System Research, a Division of Energy Conversion an
Utilization Technologies, under subcontract Mrtta 19x-89664c, through
Martin Marietta Energy Systems, Inc.

References

1. N. Brown and M. Herman, Trans. AIME 206, 1353 (1956).
2. N. Brown , Phil. Mag. 4, 693 (1959).
3. P.A. Flinn, Trans. AIME 218, 145 (1960).
4. L.E. Popov, E. Kozlov and N.S. Golosov, Phys. Stat. Sol. 13, 569
   (1966).
5. J.S. Koehler and F. Seitz, Appl. Mech. 14A, 217 (1947).
6. A.H. Cottrell, Relation of Properties to Microstructures, ASM Mono-
   graph, p. 131 (1954).
7. B.H. Kear and H.G.F. Wilsdorf, Trans. AIME 224, 382 (1962).
8. P.H. Thornton, R.G. Davies and T.L. Johnston, Met. Trans. 1, 207
   (1970).
9. S. Takeuchi and E. Kuramoto, Acta Metall. 21, 415 (1973).
10. C. Lall, S. Chin and D.P. Pope, Met. Trans. A 10A, 1323 (1979).
11. V. Paidar, D.P. Pope and V. Vitek, Acta Metall. 32, 435 (1984).
12. P. Veyssiere, Phil. Mag. A 50, 189 (1984).
13. P. Veyssiere, J. Donnin and P. Beauchamps, Phil. Mag. A 51, 469
    (1985).
14. M.H. Yoo, Scripta Met. 20, 915 (1986).
15. R. Kikuchi and J.W. Cahn, Acta Metall. 27, 13237 (1979).
16. J.M. Sanchez, J.R. Barefoot, R.N. Jarrett and J.K. Tien, Acta
    Metall. 32, 1519 (1984).
17. C. Sigli and J.M. Sanchez, Acta Metall. 33, 1097 (1985).
18. R. Kikuchi, Phys. Rev. 81, 988 (1951).
19. T. Mohri, D. de Fontaine and J.M. Sanchez, Met. Trans. A 17A, 189
    (1986).
20. J.M. Sanchez and J.L. Moran-Lopez, Surf. Sci. 157, L297 (1985).
21. J.M. Sanchez and J.L. Moran-Lopez, Phys. Rev. B 32, 3534 (1985).
22. J.M. Sanchez, F. Mejia-Lira and J.L. Moran-Lopez, Phys. Rev. Lett.
    57, 360 (1986).

ORDER-DISORDER BEHAVIOR OF GRAIN BOUNDARIES
IN A TWO DIMENSIONAL MODEL ORDERED ALLOY

DIANA FARKAS AND HO JANG
Department of Materials Engineering
Virginia Polytechnic Institute
Blacksburg, VA 24061

ABSTRACT

The order-disorder behavior of a $\sum = 5$ grain boundary was
investigated using a two dimensional latice gas model and the
cluster variation method. It is found that a disordered layer
forms in the grain boundary region at temperatures significantly
below the order-disorder temperature for the bulk. Under certain
assumptions for the pair interaction energies the model predicts
grain boundary compositions different from the bulk.

INTRODUCTION

The present work is part of an effort to develop semiquantitative
understanding of the boron effect in Ni_3Al, partly described in
the preceding paper.

The potential applications of boron doped Ni_3Al are as a high
temperature structural material. Although the bulk material stays
ordered up to the melting point it is necessary to study the order
disorder behavior in the boundary region. Geometrical models and
computer relaxation studies yield zero temperature results, and
therefore cannot give adequate insight in the order-disorder behavior.
There are two known methods to study these finite temperature
effects. These are molecular dynamics and the cluster variation
method(CVM) (1).

In the present work we use the CVM formulation to study the
order-disorder behavior of a grain boundary. The calculation is
for a two dimensional lattice gas model alloy.

METHOD OF CALCULATION

We calculate the ordering state in the vicinity of a $\sum = 5$ grain
boundary in an ordered model alloy, as a function of temperature. It
is observed that the grain boundary region is disordered at
temperatures where the bulk material is still ordered.

We used the minimization of Gibbs free energy, where the energy is written in a pair approximation and the entropy is calculated using the cluster variation method (1). As the first calculation, we present a (21) $\sum$ = 5 boundary in two dimensions (Fig. 1). The underlying lattice is the the DSC lattice , such that atoms from both crystals fall on its nodes. In our model these sites can be occupied or empty.

Two types of ordering can be treated in this model. One is the ordering between empty and occupied sites, that is spacial ordering which corresponds to the melting transition of the material. The other is that of ordering between A and B atoms, that is chemical ordering. The model is also suitable to study the interplay between spatial and chemical ordering.

We used a nine point cluster, as shown in Fig. 1. It is essential to use this nine point cluster because otherwise the nearest neighbor atom interactions are not included in the cluster. Note that atoms cannot be allowed arbitrarily close within the cluster. The atoms cannot be allowed closer than the bulk nearest neighbor distance AC. When they are at distances larger than that (i.e. AE) the bonding energy is lower. There are 59 configurations possible for the nine-site basic cluster. Some cluster configurations are typical of crystal 1. Others are typical of crystal 2. Other types are only going to appear in the boundary region.

The Gibbs free energy will be minimized with respect to the concentration of each of these possible configurations at each atomic plane parallel to the boundary. The boundary conditions should be representative of the non-equilibrium nature of the grain boundary. The cluster concentrations will be fixed for the end planes at both sides far from the grain boundary, indicating the equilibrium bulk structure for the temperature of interest. This means that the bulk behavior has to be obtained from a separate bulk order-disorder calculation. This calculation (with the same nine point cluster) was done by using equilibrium boundary conditions (plane n = plane n-1). In this case the boundary does not appear and bulk transitions (order-disorder and melting) appear at T_c, T_m, respectively.

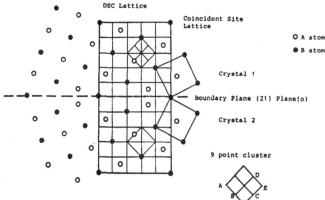

Fig. 1. The structure of two dimensional lattice and 9 point cluster used in our model.

RESULTS

A. High Temperature Grain Boundary Structure

For the present calculations we used the following assumption, as in reference 1 for a pure metal.

$$\varepsilon_{AC} = 1.2 \; \varepsilon_{AE}$$

where the subscripts refer to cluster points. This assumption is consistent with phase diagram simulations.(1).

In the binary alloy case we have taken the same assumption for AA and BB pairs. In addition we have assumed a stochiometric 50-50 alloy. In the first series of calculations we considered the the energy parameters corresponding to a symmetric phase diagram. We considered an asymmetric case as the second series of calculations. The energy parameters corresponding to both cases are as follows:

	energy parameter(Δ)	X= v/ε_0	$2\mu_A/(\varepsilon_{AA}+\varepsilon_{BB})$	$2\mu_B/(\varepsilon_{AA}+\varepsilon_{BB})$
case 1	0	0.45	−1.5	−1.5
case 2	1.33	0.45	−2.2	−0.78

where the energy parameter $(\Delta) = \dfrac{\varepsilon_{AA} - \varepsilon_{BB}}{V}$, , the ordering energy $V = \varepsilon_{AB} - \dfrac{\varepsilon_{AA} + \varepsilon_{BB}}{2}$, $\varepsilon_0 = \dfrac{\varepsilon_{AA} + \varepsilon_{BB}}{2}$, and $\mu_0 = \dfrac{\mu_A + \mu_B}{2}$

and the superscripts refer to atom types and x is the ratio of ordering energy to ε_0. Varying the value of x will vary T_c with respect to T_m.

All calculations are for an ordering energy of: $x = v/\varepsilon_0 = 0.45$ and a chemical potential of $\mu_0/\varepsilon_0 = -1.5$. A bulk calculation was performed to determine the melting temperature of the alloy, obtaining a value of $kT_m/\varepsilon_0 = 1.23$. This is a first order transition. Bulk calculations also give the order disorder temperature, which can be compared to well known results for a two-dimensional square lattice Ising Model. Our results for the order disorder transition indicate a second order transition at $KT/\varepsilon_0 = 0.78$, in agreement with the pair approximation solution to the above mentioned Ising problem.

Bulk calculations can also give the complete phase diagram for this model alloy, which for the parameters described above is completely symmetric as shown in Fig. 2

The grain boundary calculations were done using the values of the equilibrium cluster concentrations obtained from the bulk calculations as the boundary conditions far from the grain boundary plane. Results were obtained for a range of temperatures up to the order disorder temperature for the bulk.

Fig. 3 shows the variation of the long range order parameter across the grain boundary. Note that the effect of the grain boundary is extended to a larger region, as the order-disorder temperature is approached. That is, the grain boundary region disorders for a temperature significantly below the bulk order-disorder transition

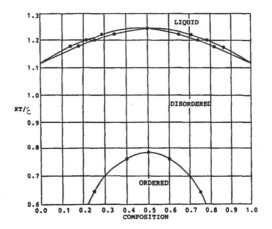

Fig. 2. The complete binary phase diagram corresponding to symmetric energy parameter (case 1).

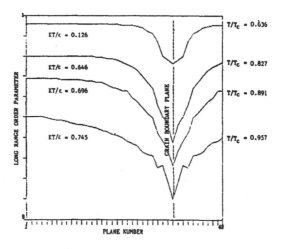

Fig. 3. Calculated long range order parameter in the vicinity of the grain boundary at various temperatures.

temperature.Fig. 4 shows the thickness of the disordere layer as a function of temperature. It is seen that it diverges logarithmically as the bulk critical temperature is approached.

B. The Grain Boundary composition

Our own work (2,3), as well as the work of other investigators (4,5) has shown that for ordered alloys the composition of the grain boundary is affected by the movement of the grain boundary plane parallel to itself and that this may lead to different grain boundary energies. In particular for Ni_3Al geometrical models based on size considerations suggested that the boundaries with high aluminum content would have higher energy (2). This has also been found in computer relaxation studies.

When energy parameters corresponding to an asymmetric phase diagram are introduced, the energy of the boundary varies with composition resulting in grain boundary segregation. The present calculations indicate that the asymmetry in the interaction energies also changes the order disorder behavior of the boundary. As pointed out previously the grain boundary region is more disordered than the bulk. However the transition temperature is defined as the temperature for which the order parameter goes to zero and in the energy parameter corresponding to symmetric system, it is the same for bulk and grain boundary. Similar results were observed for a free surface by other investigators (6), and this type of behavior is called an ordinary surface transition. Fig. 5 shows a comparison of the long range order parameter across the boundary for the energy parameters corresponding to symmetric and the asymmetric system.

For energy parameters corresponding to asymmetric phase diagrams the order-disorder temperature is different in the grain boundary and in the bulk. Fig. 6 shows the composition of the different atomic planes for the cases studied. In this figure the disordering of the boundary is seen in the decrease of the oscilation amplitude in the neighborhood of the boundary plane. The segregation effects are also clear in this figure. Fig. 7 shows the average composition profile across the boundary for $\Delta = 1.33$ case.

DISCUSSION

Although the calculations presented here are for a two dimensional model system the basic trends observed are expected to carry over to a three dimensional case with more realistic atom interactions. The present work shows that the vicinity of the grain boundary is significantly more disordered than the

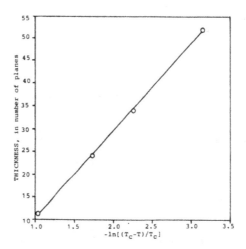

Fig. 4.   The thickness of disordered layer as a function of temperature.

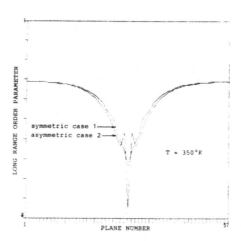

Fig. 5.   The long range order parameter across the boundary for the symmetric and asymmetric case.

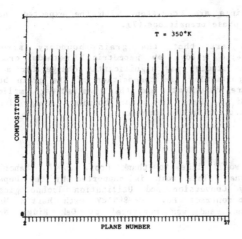

Fig. 6. The composition of different atomic planes for asymmetric case 2.

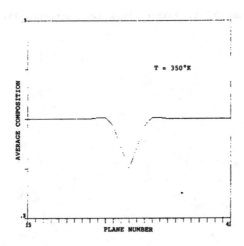

Fig. 7. The average composition profile across the boundary for asymmetric case 2.

bulk. These findings are consistent with the expected behavior from the theory of phase transitions (7).

The implications are that the grain boundary structures obtained for ordered alloys by geometrical methods or static computer relaxation studies may significantly change for a finite temperature. In particular the composition of the grain boundary will be temperature dependent, and this has strong implications for cases like Ni_3Al where the grain boundary energy increases with the Al concentration in the boundary.

ACKNOWLEDGEMENTS

The authers would like to thank Prof. J. M. Sanchez for valuable discussions. This work is supported by the Department of Energy, Energy Conversion and Utilization Technologies(ECUT) program under sub-contract No. 19x-89678V with Martin Marietta Energy Systems Inc. and was monitored by Oak Ridge National Laboratories.

REFERENCES

1. R. Kikuchi and J. W. Cahn, Phys Rev. 21,5,1893 (1980)
2. D. Farkas and A. Ran, Phys Status Solidi A. 93,45 (1986)
3. D. Farkas and V. RAngarajan,Acta Met. 35,2,353 (1987)
4. G. Kalonji, Ph.D thesis, MIT, (1982)
5. T. Takasugi and O. Izumi, Acta Met. 31,187 (1983)
6. J. L. Moran Lopez and J. M. Sanchez, Phys Rev. B 32,3534 (1985)
7. H. Nakanishi and M. E. Fisher, Phys Rev. Letters. 49,21,1565 (1982)

PART II

Microstructures

ANTIPHASE DOMAIN BOUNDARY TUBES IN ORDERED ALLOYS

P.M. HAZZLEDINE AND SIR PETER HIRSCH,
Department of Metallurgy and Science of Materials, Oxford University,
Parks Road, Oxford OX1 3PH, U.K.

ABSTRACT

APB tubes have been observed in both a B2 and an $L1_2$ ordered alloy by means of weak beam transmission electron microscopy. The tubes are attached to near-edge dislocations, either to single superdislocations or to superdipoles. The majority appear to have been formed by cross slip of screw dislocations. A computer model of the cross slip process in B2 alloys is described. The tubes formed by cross slip drag on edge dislocations and are capable in principle of explaining the extra work hardening shown by ordered over disordered crystals. The temperature and orientation dependence of the work hardening are similar to those of the proof stress which is also thought to be controlled by cross slip. This new mechanism of work hardening is shown to give order of magnitude agreement with experiment.

INTRODUCTION

Most alloys with long range order work harden more rapidly than those which are disordered [1,2] and many alloys have a work hardening rate which peaks at intermediate temperatures, broadly following the behaviour of the proof stress [3]. The work hardening rates are high in ordered alloy crystals despite the fact that slip is largely primary and the density of secondary dislocations is low. Vidoz and Brown [4] and Vasilyev and Orlov [5] suggested that the increase in hardening rate on ordering may be due to the generation of antiphase domain boundary (APB) tubes generated by the glide of superdislocation jogs which are not aligned along the Burgers vector direction. The explanation is attractive because it invokes a defect which is specific to ordered structures and which is capable of hardening the materials to the observed extent [6,7,8,9] but the temperature dependence of the work hardening [2,10] and, more particularly, the crystal orientation dependence [11,12] are not easily explained by this model.

Detailed observations of APB tubes in the electron microscope [13,14,15] have shown that many of them are probably not formed by the intersection jog mechanism but by a process of cross slip of screw dislocations. Such a process leaves mobile edge dislocations trailing APB tubes. The size and density of the tubes attached to edge dislocations depends on the ease of cross slip of screw dislocations; the resulting hardening will therefore be both temperature - and orientation-dependent, as observed.

In this paper the intersection-jog-tube work hardening mechanism is briefly reviewed and the new observations and cross slip mechanisms of tube formation and work hardening are described.

TUBE FORMATION AND GEOMETRY

In the Vidoz and Brown [4] mechanism of tube formation it is noted that if the leading superpartial of a superdislocation contains a jog it

must trail a stepped APB. If the corresponding jog on the trailing
superpartial is not perfectly aligned with the first jog, the APB is
imperfectly erased, leaving a tube of APB (Fig.1). As the
superdislocation moves the jogs glide in a direction parallel to their
Burgers vector b, so, to be 'aligned', the line joining the two jogs must
also be along b. An edge superdislocation will drag a tube which is
perpendicular to the dislocation line. As the character of the dislocation
θ decreases the tube becomes more nearly parallel to the dislocation until
the point is reached for screw dislocations when tubes cannot be formed by
conservatively moving jogs. It is generally assumed that the jogs whose
motion create an APB tube are formed at an intersection between two moving
dislocations, one primary and one secondary. In many intersections both
the primary and the secondary dislocation receive jogs and in this case
both dislocations will trail tubes. The two tubes will join at the point
of intersection P (Fig.1b).

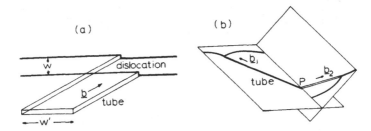

Fig.1 (a) Tube formation at misaligned jogs. (b) Two moving dislocations
cutting at P create a single bent tube.

All the material inside an APB tube has been displaced relative to
that outside by a vector which is parallel to the tube axis and which is a
translation of the disordered but not of the ordered lattice. In
principle, therefore, as Kear [16] pointed out, tubes can be formed in a
variety of ways, not just by the motion of jogged dislocations. For
example a prismatic loop whose glide cylinder is the cross-section of the
tube trails a tube as it moves along the cylinder. Alternative processes
giving the same translations involve screw dislocations moving by pencil
glide around the outside of the tube surface. A screw superdislocation
could form a tube by a simple process of double cross slip; a tube is
formed whenever the two superpartials do not cross slip on exactly the same
planes (Fig.2a-c). A variant of this process is the cross slip to
annihilation of a screw superdipole (Fig.2d,e). Again, a tube is formed if
the cross slip planes chosen by the superpartials differ from one another.
In both the cases involving screw dislocations, after cross slip, the edge
dislocations to which the screws are attached are left with high jogs
(Fig.3) and the process of tube formation may be continued by glide of the
jogged edges very much in the manner of the original Vidoz and Brown
mechanism.

Recently Chou et al [17] have considered the process of tube formation
by the annihilation of the screw superdipoles in some detail as it appears,
from direct observation of tubes in a B2 ordered alloy in the electron
microscope by Chou and Hirsch [13,14], that this particular process must be
a common one. The results of this calculation are described here at some
length because the details have not yet been published.

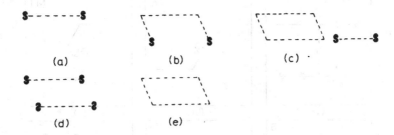

Fig.2 (a to c) Tube formation by double cross slip of a screw
superdislocation. (d,e) Tube formation by annihilation of a
superdipole (schematic).

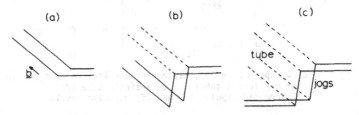

Fig.3 (a to c) When a screw superdislocation undergoes double cross slip
it may form a tube on an edge dislocation.

The four screw superpartial dislocations are set up in a starting
configuration on two primary slip planes separated by h. Each dislocation
is subject to a force from each other dislocation, from the applied stress
and from the antiphase boundary. The calculation refers to a B2 ordered
Fe 30% at Al alloy and elastic moduli, Burgers vectors etc. are chosen for
this alloy. The screw dislocations have a choice of primary and two cross
slip planes of type $\{110\}$. For any configuration of dislocations the
computer programme calculates the force on each dislocation on the three
planes, chooses the largest and moves the dislocations by a small step in
the appropriate direction, the calculation is then repeated. In this way
the full evolution of the cross slip process may be traced out.

Some examples of dislocation paths are given in Fig.4. In these
figures the tracks of the two superpartials in each superdislocation are
traced. Where they superimpose (Fig.4b) total annihilation has taken
place, leaving no trace of the original dipole. This occurs when h is
small. The remaining traces give examples of the behaviour as h is
increased successively. For 14 nm > h > 7 nm the trailing superpartials
follow exactly the leading superpartials at first, but after the leaders
have annihilated the followers take a different path, creating two APB
tubes before finally annihilating themselves. As h is increased again
there is a smaller range over which the four superpartials annihilate in
pairs but create a single tube with large cross-section (Fig.4d). At
higher still values of h the dislocations either cross slip partially and
form stable dipoles (Fig.4e) or else they remain on their original planes
and form stable dipoles (Fig.4f). In neither of these cases are tubes
formed. Finally, at very large values of h the screw dipoles pass one
another under the action of the applied stress. There are thus six
possible final configurations when two screw superdislocations interact,
only two of which give rise to APB tubes.

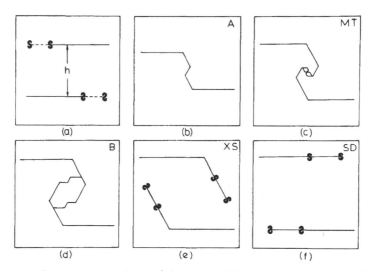

Fig.4 A screw superdipole (a), may annihilate leaving no trace (b),
or may form 2 tubes (c) or a single tube (d).
Rearranged dipoles (e) and (f) may also result.

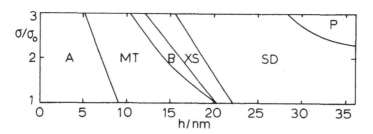

Fig.5 The relative importance of the configurations of Fig.4 shown as
functions of applied stress and slip plane separation h for
B2 ordered FeAl alloy.

The relative importance of the six configurations is indicated in
Fig.5 which shows the effects of the two parameters σ/σ_0 and h on the final
outcome. σ/σ_0 is the value of the applied stress relative to the yield
stress so the lower part of Fig.5 corresponds to lightly deformed material
and the upper part to heavily deformed, work-hardened material. Fig.5
shows the final configuration (A = annihilation with no tubes formed,
MT = multiple, usually double, tube formation with small cross section
tubes made, B = one big tube formed, XS = partial cross slip but no tube
formed, SD = stable dipoles and P = superdislocations pass without forming
dipoles) as functions of h and σ/σ_0 with σ_0 = 50 MPa. The effects of h on
the configurations have been described above; the main effect of
increasing σ/σ_0 is to reduce the value of h at which a boundary between two
fields occurs. This is especially true of the field labelled P. As the
stress rises the dislocations are more likely to pass one another and not
cross slip.

As can be seen in Fig.5 for a given value of σ/σ_0 it is only over a fairly narrow range of values of h that the screw dipoles form APB tubes as they annihilate and it is invariably true that the heights of the tubes h' are much less than h and the widths of the tube w' are considerably smaller than the equilibrium widths of the superdislocations w. From the model calculations both h' and w' are expected to be in the range 1 to 5nm, which broadly is what is found experimentally [14].

The model calculations are, of course, simplistic in that they consider long straight screw dislocations slipping with equal ease on three slip planes in an elastic continuum but the spiral dislocation paths generated in the calculation do give some insight into the fact that large dislocation dipoles should give rise to tubes with rather small cross sections. The cross sections of the tubes formed by cross slip are expected to be roughly equiaxed with linear dimensions equal to a few multiples of the elementary jog height.

OBSERVATIONS OF TUBES

In one of the first papers to describe the dislocation and APB structures in ordered alloys Kear [18] noted that in the $L1_2$ ordered Cu_3Au antiphase boundaries were generally found on cube planes but that, after deformation, new AP boundaries appeared along the primary Burgers vector direction <110>. The electron micrographs, taken in bright field using superlattice reflexions are probably the first direct observations of APB tubes. In his 1966 paper Kear [16] showed more evidence of antiphase defects lying along the primary slip direction. These defects were present in large numbers despite the fact that the density of secondary dislocations was small compared with primary dislocations.

Crawford [19] observed faint lines along b in electron micrographs of heavily deformed single crystals of B2 ordered Fe 30% at Al, taken in superlattice reflexions. These faint lines, together with streaks observed on diffraction patterns were considered to be consistent with the existence of APB tubes of varying heights and widths.

The most systematic observations of APB tubes in B2 alloys were described by Chou and Hirsch [13,14] who obtained images of linear defects parallel to the Burgers vector direction under weak-beam conditions in the transmission electron microscope. The contrast features which enabled the identification of these defects with APB tubes to be made were

(i) They were visible only in superlattice reflexions and so must have displacements of the disordered but not of the ordered lattice

(ii) They gave a uniform dark image when the crystal was at the reflecting position (w=0). This is consistent with the expected contrast from overlapping faults with a phase angle of π.

(iii) The oscillatory contrast observed when w≠0 had fringes which were symmetrical about the centre of the specimen under two beam conditions; the period of the oscillation was shorter than that of thickness fringes.

(iv) The contrast at the defects was high even though the intensity of both the tubes and the background was exceptionally low. This feature is consistent with the fact that both the diffusely scattered background intensity and the diffracted intensity under kinematic conditions are proportional to the square of the structure factor.

Fig.6

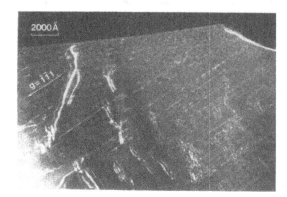

Weak beam (w=51)
image of (011) foil
of B2 ordered
Fe 30% at Al.
Tubes parallel to IĪ1
are attached to short
edge dipoles [14].

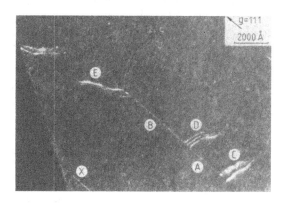

Weak beam image
of Fe 30% at Al.
Tubes A and B
are connected to
dipoles C, D and E.
Isolated tube at
X [14].

Weak beam image
of Fe 30% at Al.
In reflection g = 2Ī1
a dipolar cusp on a
screw is imaged. In
reflection g = IĪ1 tubes
are seen, one of which
is attached to A [14].

The measure of agreement between the observed and the calculated contrast leaves no doubt that the defects are APB tubes. Furthermore the cross sectional dimensions of the tubes may be estimated from the contrast: Since a tube has no strain field, the measured width of the line on the micrograph is a good estimate of the width of the tube. The height of the tube, h (dimension parallel to the electron beam) can be measured from w=0 micrographs because a slab of 'antiphase' material of thickness h has exactly the same effect as reducing the crystal thickness by 2h. The measured intensities inside and outside the tube may then be used to measure h. The tubes which Chou and Hirsch [14] measured had widths in the range 2 to 5nm and heights in the range 0.4 to 2.5nm.

Tubes in Fe 30% at Al were frequently attached to near-edge dislocations, either to cusps in single edge dislocations or to edge superdipoles, (Fig.6). In the former case the tubes could have been formed by double cross slip of a screw dislocation (Fig.3) or else by the Vidoz and Brown mechanism and in the latter case the configuration is consistent with screw superdipoles having annihilated leaving tubes behind (Fig.2). In one case a tube is shown to be attached to a screw superdislocation which is almost parallel to the tube. The cusp on the screw is so deep that it amounts almost to an edge dislocation dipole, (Fig.6).

Tubes showing the same contrast features as above have also been observed in $L1_2$ ordered Ni_3Al crystals [15]. In this material large densities of tubes lying along <110> directions were observed in a specimen which had been compressed by 4% at 400°C.

The electron microscope observations show that in both B2 and $L1_2$ ordered alloys large densities of tubes may be produced by deformation. The cusped shapes of dislocations to which the tubes are attached indicate that the tubes exert appreciable forces on the dislocations. The work hardening to be expected from this interaction is next assessed.

Work hardening from tubes formed by intersection processes

Vidoz and Brown [4] estimated the increment of flow stress $\Delta\tau$ caused by APB tubes: the force exerted by a tube of width w' and small height on a dislocation is $2\gamma w'$. If there are n tubes per unit length of dislocation then $\Delta\tau = \gamma w'n/b$. Now w' is related to the superdislocation width w (Fig.1a). Taking w' to be equal to w i.e. to $Gb^2/2\pi\gamma(1-\nu)$ for an edge dislocation, the flow stress increment is independent of APB energy γ

$$\Delta\tau = nGb/2\pi(1-\nu) \qquad (1)$$

As n increases with strain $\Delta\tau$ is a contribution to the work hardening.

Vidoz [6] visualized a rectangular loop of primary dislocation, in which edge dislocations slip by L_e and screws by L_c, spreading through a forest of single secondary dislocations with density ρ_f. If a fraction k of forest dislocations create a jog on the primary dislocations then the number of jogs per length on edges is $k\rho_f L_e$ and the density of tubes is one half of this. The work hardening stress increment for edge dislocations is

$$\Delta\tau_e = k\rho_f L_e Gb/4\pi(1-\nu) \qquad (2)$$

The number of jogs per length on screw dislocations is $k\rho_f L_s$ but these jogs cannot form tubes because, in order to do so, they must glide along the dislocation line; instead the jogs move non-conservatively and create point defects. Taking the energy of a vacancy or an interstitial to be αGb^3 (with $\alpha \simeq 0.2$ for vacancies and $\alpha \simeq 1$ for interstitials), the stress increment for screw superdislocations is

$$\Delta\tau_s = \alpha Gbk\rho_f L_s \qquad (3)$$

Vidoz equated $\Delta\tau_s$ to $\Delta\tau_e$ and obtained a ratio of slip lengths which is in the range 2 to 10

$$r = L_e/L_s = 4\pi\alpha(1-\nu) \qquad (4)$$

The dislocation structure on the primary plane would consist of elongated loops containing a large density of screw dislocations and a small density of edges. The edge dislocations would trail tubes but the screws would not. The screws would, however, be deeply cusped.

An estimate of the work hardening rate may be made by noting that if there are N primary loops per volume, then the strain $\varepsilon = 4NL_eL_sb$ and the primary dislocation density $\rho_p = 2N(L_e + L_s)$. Using equation (4) we may write ε and ρ_p in terms of the single slip length L_s and hence express equation (3) as

$$\theta \simeq \frac{\Delta\tau}{\varepsilon} = \left(\frac{\alpha k}{2}\right) G \left(\frac{\rho f}{\rho_p}\right) \left(\frac{1+r}{r}\right) \qquad (5)$$

Setting $k = 2/3$ and recalling that α lies in the range 0.2 to 1.0, the normalized work hardening rate is

$$\theta/G = \beta \left(\rho_f/\rho_p\right) \qquad (6)$$

with $0.1 < \beta < 0.4$

The two detailed intersection tube models [8,9] of work hardening give hardening rates of the same form as equation [6].

Even if ρ_f/ρ_p is as low as 0.01, the work hardening rate from this process is comparable with the work hardening rate of disordered crystals and is of the right order of magnitude to explain the extra work hardening specific to ordered crystals. However, an alternative explanation for L1$_2$ alloys, based on the formation of Kear-Wilsdorf locks has been proposed [20,21,3]. To decide which process controls the work hardening it is necessary to consider not just the magnitude of θ but its temperature and orientation dependence and the tension/compression asymmetry.

Work hardening controlled by cross slip processes

The Paidar, Pope and Vitek model [21] is a calculation of the critical resolved shear stress, or to be more accurate since some small plastic deformation is required, of the proof stress of L1$_2$ ordered alloys. In the model, screw superdislocations cross slip onto {001} planes and immediately redissociate onto {111} planes in a sessile configuration. The flow stress is proportional to the cube root of the rate of formation of sessile screw segments [20]. The factors which promote cross slip are (i) low APB energies on {001} planes relative to {111} planes, (ii) high Schmid factors on {001} planes, (iii) a large 'Escaig' stress acting on the edge components of the Shockley partials so as to constrict the partials on the primary plane; and (iv) the non-central force between screw dislocations caused by elastic anisotropy [22]. When these factors are all taken into account the model gives a remarkably good description of the proof stress of most L1$_2$ alloys and, in particular, of the rise in the proof stress with temperature at intermediate temperatures [3]. At higher temperatures the proof stress falls as slip on {001} becomes easy. In agreement with the model, most dislocations at and below the peak in the proof stress are screws [16,23,24] but, at variance with the model [25] is the fact that the dissociation of Kear-Wilsdorf locks on {001} planes is large enough to be seen clearly in electron micrographs [16,26,27] and is therefore much bigger than the distance b in the model. In this respect the original Kear-Wilsdorf model is a better description of the cross slip process than the computer simulations are [21].

The work hardening rate in L1$_2$ alloys (except Ni$_3$Fe [27]) behaves in a similar way to the proof stress [3,10,11] - it shows a peak (or usually a double peak) at intermediate temperatures and a similar orientation dependence to the proof stress [12]. This last point is crucial because it shows that cross slip onto {001} planes controls the work hardening to a greater extent than does conjugate slip (and hence the number of tubes formed from intersection jogs) which would be promoted by a high Schmid factor on the conjugate slip system. Clearly the proof stress, which requires microscopic strain, and the work hardening, which requires macroscopic strain, share a common mechanism, cross slip onto {001}.

An outline explanation of work hardening in L1$_2$ alloys which is consistent with the temperature - and orientation-dependence of θ and also with the experimental result that large numbers of tubes are formed and that many of them are created by cross slip of screw dislocations is as follows:

As a primary dislocation loop spreads from its source it picks up jogs from forest intersections and some hardening results from the Vidoz and Brown mechanism but, more significantly, as the primary screw dislocations advance they occasionally cross slip (for example when encountering opposite screws on a nearby slip plane) in such a way that the following superpartial does not trace exactly the same path as the leading superpartial. The resulting tubes (Figs.2,3) on edge dislocations cause further work hardening.

In the case of B2 alloys there is no exact equivalent of the Kear-Wilsdorf lock. However, the work hardening rate may be exceptionally high [9], up to G/50, and in some B2 alloys there is a peak in the yield stress at intermediate temperatures [28,29]. The peak in β brass and ternary brasses has been related from slip traces and from the orientation dependence of the yield stress to the cross slip of screw dislocations onto {112} planes from {110} planes and vice versa. The explanation for the high yield stress is then very similar to that of Takeuchi and Kuramoto [20,21] in L1$_2$ alloys. The details of the cross slip process are different however: in B2 alloys in general cross slip would be expected to be easier than in L1$_2$ alloys, there is no Escaig effect closing or opening partial dislocations and there is very little anisotropy of the APB energy to drive the process. Despite these differences cross slip is found to occur in the computer simulation studies of Takeuchi [31]. In these calculations in some instances the two superpartials move on different planes; as indicated in Fig.2, this is the first stage of tube formation.

In the computer simulations of Figs. 4 and 5 slip is restricted to {110} planes and here the only driving forces for cross slip are the resolved components of the applied stress on the three planes and the internal stresses from neighbouring screw dislocations. Tubes are generated in these conditions and they are capable of work hardening the B2 alloys.

The order of magnitude of the hardening to be expected from cross slip generated tubes can be estimated: If the tubes have heights h' and widths w' they will exert a force of $2\gamma(h'+w')$ on the superdislocation. If the tubes are spaced by ℓ along the dislocation, the hardening $\Delta\tau$ is

$$\Delta\tau = (h'+w')\gamma/b\ell \qquad (7)$$

h' and w' are related to w the screw superpartial spacing; writing them both as fw, the hardening is

$$\Delta\tau = fGb/\pi\ell \qquad (8)$$

The critical feature of a future work hardening model will be to calculate ℓ which is related to the frequency of cross slip. From the micrographs of lightly deformed alloys, a value of ℓ of 10^{-7}m is plausible. Putting $f = \frac{1}{2}$, $G = 6\times10^{10}$Pa, $b = 2.5\times10^{-10}$m, equation (8) gives a substantial hardening of 24MPa in Fe 30% at Al.

CONCLUSIONS

1. APB tubes have been observed in large numbers in lightly deformed specimens of both B2 and $L1_2$ ordered alloys.

2. Tubes formed by intersections between primary and forest dislocations may contribute to the extra work hardening of ordered alloys.

3. The temperature and orientation dependences of the proof stress and of the work hardening rate in most $L1_2$ alloys indicate that the same process, thermally activated cross slip onto {001} planes, controls both the proof stress and the work hardening in the region of the proof stress peak.

4. A model of tube formation in B2 alloys involving cross slip of screws satisfactorily accounts for the observed sizes of tubes.

5. The work hardening caused by tubes generated by cross slip is sufficient to contribute substantially to the extra work hardening of ordered alloys. Its temperature and orientation dependences are the same as those of cross slip.

Acknowledgements

We would like to thank Dr.C.T.Chou for many discussions and Dr.S.J.Shaibani for his help with the computer programmes. The work was supported in part by the National Physical Laboratory.

85

References

1. N.S. Stoloff and R.G. Davies, Trans. ASM 57, 247 (1964).
2. N.S. Stoloff, in Strengthening methods in crystals, edited by A. Kelly and R.B. Nicholson (Applied Science, London, 1971), p. 193.
3. D.P. Pope and S.S. Ezz, Int. Met. Revs. 29, 136 (1984).
4. A.E. Vidoz and L.M. Brown, Philos. Mag. 7, 1167 (1962).
5. L.I. Vasilyev and A.N. Orlov, Fiz Metal Metalloved 15, 481 (1963).
6. A.E. Vidoz, Phys. Stat. Sol. 28, 145 (1968).
7. G. Schoeck, Scripta Metall. 2, 283 (1968).
8. G. Schoeck, Acta Metall. 17, 147 (1969).
9. P.B. Hirsch and R.C. Crawford, in Strength of Metals and Alloys, edited by P. Haasen (Pergamon, Oxford, 1979), p. 89.
10. A.E. Staton-Bevan, Philos. Mag. 47A, 939 (1983).
11. E. Kuramoto and D.P. Pope, Philos. Mag. 34, 593 (1976).
12. K. Aoki and O. Izumi, Acta Metall. 26, 1257 (1978).
13. C.T. Chou and P.B. Hirsch, Philos. Mag. A44, 1415 (1981).
14. C.T. Chou and P.B. Hirsch, Proc. Roy. Soc. A387, 91 (1983).
15. C.T. Chou, P.B. Hirsch, M. McLean and E. Hondros, Nature 300, 621 (1982).
16. B.H. Kear, Acta Metall. 14, 659 (1966).
17. C.T. Chou, P.B. Hirsch, P.M. Hazzledine and G.R. Anstis, Proc. of 3rd Asia Pacific Conference on Electron Microscopy, edited by M.F. Chung (Applied Research Corporation, Singapore, 1984), p. 64.
18. B.H. Kear, Acta Metall. 12, 555 (1964).
19. R.C. Crawford, Philos. Mag. 33, 529 (1976).
20. S. Takeuchi and E. Kuramoto, Acta Metall. 21, 415 (1973).
21. V. Paidar, D.P. Pope and V. Vitek, Acta Metall. 32, 435 (1984).
22. M.H. Yoo, Scripta Metall. 20, 915 (1986).
23. A.E.Staton-Bevan and R.D. Rawlings, Philos. Mag. 32, 787 (1975).
24. S.M.L. Sastry and B. Ramaswamy, Philos. Mag. 32, 801 (1975).
25. H.P. Karnthaler (private communication).
26. K. Suzuki, M. Ichihara and S. Takeuchi, Acta Metall. 27, 193 (1979).
27. A. Korner, H.P. Karnthaler and C. Hitzenberger, J. Elect. Microsc. 35, 1569 (1986).
28. Y. Umakoshi, M. Yamaguchi, Y. Namba and K. Murakami, Acta Metall. 24, 89 (1976).
29. M. Yamaguchi and Y. Umakoshi, Phys. Stat. Sol.(a) 43, 667 (1977).
30. Y. Umakoshi and M. Yamaguchi, Scripta Metall 11, 211 (1977).
31. S. Takeuchi, Philos. Mag. A41, 541 (1980).

FIM / ATOM PROBE STUDIES OF
B-DOPED AND ALLOYED Ni$_3$Al

D.D. SIELOFF, S.S. BRENNER AND M.G. BURKE
University of Pittsburgh, Materials Science and Engineering Dept.
Pittsburgh, PA 15261

ABSTRACT

   Field-ion microscopy and atom probe microanalysis have been
used to determine the structure and chemistry of grain boundaries
in ductile, recrystallized Ni$_3$Al containing 0.23 at % B (500 wt.
ppm). The results indicate that the boron concentration
fluctuates along the boundary plane and that the boron enriched
zone is wider than expected from equilibrium-type adsorption.
It was also found that boron lowers the aluminum concentration of
some of the boundary regions.

INTRODUCTION

   The discovery by Aoki and Izumi in 1979 [1] that small
additions of boron markedly improve the ductility of Ni$_3$Al has
generated renewed interest in this and other ordered
intermetallic compounds. C.T. Liu and co-workers [2] who have
been in the forefront of developing ductile, high strength
aluminides found that boron affects the ductility only when
Ni$_3$Al is deficient in aluminum. In the nickel-rich alloys, the
boron changes the fracture mode from intergranular to
transgranular suggesting that the boron increases the cohesion of
the boundaries. Another view expressed by Schulson and co-
workers [3] is that boron helps grain boundaries accommodate
slip during plastic deformation, thus reducing the incidence of
dislocation pile-ups, and consequently crack initiation. While
Auger spectroscopy has clearly shown that boron segregates to the
grain boundaries in Ni$_3$Al [4] and that the amount of segregation
depends on the thermal history of the material [5], there still
exists great uncertainty concerning the mechanism of
ductilization and the effect of stoichiometry on ductility.

   To fully characterize and understand the behavior of the
nickel aluminides requires the use of various techniques and
disciplines. In this study, field-ion microscopy and atom probe
microanalysis were used to investigate the structure and
chemistry of the grain boundary region in Ni$_3$Al of various
compositions. The high spatial resolution of these techniques
and mass independent sensitivity of the atom probe [6] make these
instruments ideally suited for making such measurements. The
major drawback of using this technique for grain boundary studies
is the difficulty of intercepting a grain boundary within the
small volume of material at the tip of the needle specimen. The
use of fine-grained material and the pre-selection of specimens
by means of transmission electron microscopy help to reduce this
problem.

Mat. Res. Soc. Symp. Proc. Vol. 81. ‹ 1987 Materials Research Society

EXPERIMENTAL

Material and Preparation

Most of the material studied was provided by Oak Ridge National Laboratory in the form of either melt-spun ribbons or conventionally-cast bulk material. Work on the melt-spun material was discontinued because of its inhomogeneity and relatively coarse grain size.

The grain size of the as-cast conventionally prepared material was too coarse to prepare needles from it with boundaries near their tips. The ductile material containing boron and 24.5% Al was therefore given a series of thermomechanical treatments and a final recrystallization anneal of either 20 minutes at 800° or 6 minutes at 900°C. This treatment not only resulted in a grain size of less than 5 μm but also completely eliminated the last vestiges of the dendritic structure of the material. Figure 1 shows the microstructure of the material before and after the grain refinement. Unfortunately, this treatment could not be applied to the more brittle material that contained greater amounts of aluminum.

 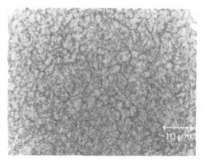

Figure 1.   Optical micrographs of Ni - 24.5 Al - 0.23 B alloy (ORNL designation IC-15) before (left) and after (right) thermomechanical treatment.

Specimen Selection and Preparation

Specimen needles were prepared by the conventional two-stage electropolishing technique described elsewhere [6]. For grain boundary analyses the specimens were first examined by transmission electron microscopy and those specimens that showed grain boundaries within 0.1 μm of the apex were selected for atom probe examination. Figure 2 shows two needles containing boundaries that were oriented either parallel or perpendicular to the specimen axis. Typical micrographs of needles with such boundaries are also presented.

Atom Probe Analysis

Two laboratory-built atom probes of similar design were used in this study. They are of the conventional straight-flight-tube type and have been described in part elsewhere [6]. Optimum operating parameters for producing smooth field evaporation of the aluminide and obtaining reasonably accurate compositions were determined previously [7], and are as follows: specimen temperature 50K, $V_p/V_{dc} = 0.15$ (V_p = evaporation pulse voltage; V_{dc} = standing DC voltage) and neon pressure 2×10^{-8} Torr during field evaporation.

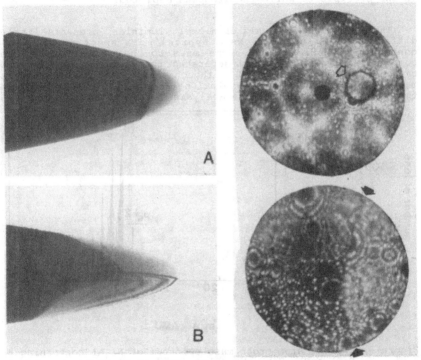

Figure 2. TEM and FIM micrographs of grain boundaries in recrystallized Ni₃Al inclined nearly perpendicular (A) and parallel to the specimen axis (B). Arrows point to grain boundaries.

RESULTS

Field-Ion Microscopy and Atom Probe Analysis - Ductile Material

High resolution micrographs of Ni₃Al can be easily produced by routine field-ion microscopy, as was first shown by Taunt, Sinclair and Ralph [8]. Ni₃Al has an Ll₂ structure with the aluminum atoms located at the corners of the unit cell and the nickel atoms at the face centers. The atom layers perpendicular to the superlattice directions i.e., <100>, <110>, <210> etc.

therefore alternate in composition between 100% Ni and 50% Ni-50% Al. The composition fluctuations give rise to alternating bright and dim images of the atom plane edges [7,8,9].

With one exception, the materials studied were single phase. In one specimen, prepared from as-cast arc-melted material containing 1 at% Hf, 0.5 at% B and 0.5 at% Fe a boride precipitate 20-30 nm in diameter was observed indicating that the boron solubility limit was locally exceeded. Atom probe analysis of this precipitate indicated that it was Ni_3B containing less than 3 at% aluminum [7]. The hafnium was found to be enriched in the precipitate while the iron was depleted.

The atom probe analysis of nickel aluminide presents little difficulty. Figure 3 shows a typical mass spectrum of Ni_3Al containing B, Hf and Fe. The mass-to-charge ratios of the components are sufficiently separated that their concentrations can be accurately determined. Although energy compensating lenses markedly improve the mass resolution of standard probes [10], in this case the accuracy of the analysis would not be affected.

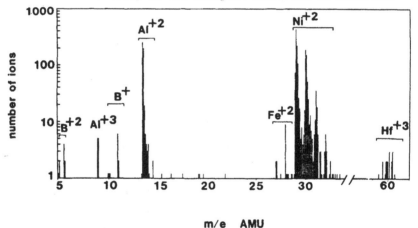

Figure 3. Typical atom probe mass spectrum of Ni_3Al containing B, Fe and Hf. Hf also found in +4 state at m/e = 43.5 - 45.0 which is not shown.

Grain Boundary Analysis

The grain boundaries in the recrystallized material were usually narrow in width, sometimes making their observation difficult. Occasionally, during field evaporation a specimen with a grain boundary would experience a partial fracture after which the grain boundary was heavily decorated. Such a sequence is shown in figure 4. Figure 4a shows the initial boundary with no visible decoration. Just prior to fracture (Fig 4b) small bright spots or clumps of spots appeared at the boundary. After partial fracture the boundary was observed to be heavily decorated (Fig 4c). Similarly decorated boundaries were reported to occur commonly in rapidly solidified material [9].

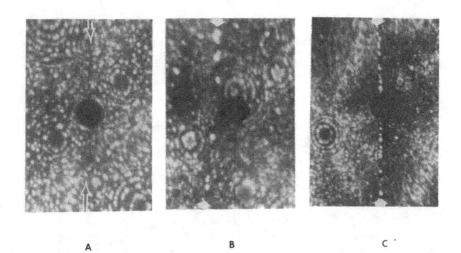

Figure 4. Change in decoration of grain boundary with depth
          during field evaporation. Grain boundary
          appears very narrow with no decoration at start (A).
          Decoration observed (B) just prior to partial
          specimen fracture (C).

    Atom probe analyses of grain boundaries inclined at some
angle to the specimen axis showed maximum boron concentrations of
about 1 to 3 atomic percent, corresponding to an enrichment of
about 5 to 15 times that of the matrix level.   The analyses were
made by centering the image of the boundary over the aperture
(probe hole) of the time-of-flight spectrometer. No systematic
variation in the maximum boron concentration with respect to
boundary orientation has been observed so far.   The measured
concentrations must be considered lower limits since the width of
the enriched zone may be less than the diameter of the probed
area, in which case both the matrix and grain boundary region
contribute to the analysis [11]. However, we have not observed
greatly different results when the diameter of the probe area was
changed from about 1 to 3 nm and conclude that the width of the
enriched zone is considerably broader than a few atom diameters.
Experiments in which the distance of the probed area from the
boundary was systematically varied confirmed that boron
enrichment is not confined to a narrow zone on each side of the
boundary.

    The concentration of boron along the boundary plane
fluctuates widely. Figure 5 shows the sequence of arrival of the
atoms from an area of material centered on the boundary.   Each
line represents the atoms collected from approximately one atom
layer.   The plot thus portrays accurately the change in
composition of the boundary plane as a function of distance from
the starting point.   However, the position of the lattice sites
from which the atoms originated within each layer can not be
determined without knowing more precisely the sequence of field
evaporation. From the diagram it was determined that the boron

92

concentration decreased from 2.6 at% to 0.4 at% after the removal of about 16 atomic layers. The region of low boron concentration continued for more than 10 nm when the concentration again rose prior to the occurrence of a partial fracture, revealing the decoration evident in Figure 4c.

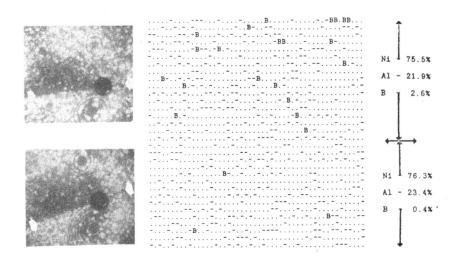

Figure 5. Sequence of arrival of atoms from a grain boundary in recrystallized Ni₃Al + 0.23 B (IC-15). The area of analysis remained centered on the boundary as shown by micrographs. Each row of atoms is approximately equal to one atom layer (~ 0.2 nm) "B" = boron; " - " = Ni; "." = Al.

When the grain boundaries are normal to the direction of field evaporation the analysis can be more precise (see Figure 2b). Only the atoms from the boundary layer are sampled as the last remnant of the grain field evaporates. The results of the analysis of such a "disappearing" grain are shown in figure 6. The maximum concentration of 12 at% was measured over a thickness of only one atomic layer and may not be statistically significant. The overall concentration was approximately 4 at% which persisted over a distance of 10 atomic layers (~20 nm). Although this type of analysis samples the enriched layer on both sides of the boundary, the measured thickness is still greater than the one obtained from Auger spectroscopy measurements [4]. Furthermore, significant boron enhancement was observed as far as 25 atomic layers from the boundary.

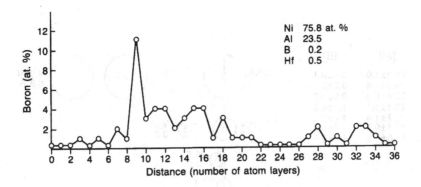

Figure 6. Analysis of boundary perpendicular to specimen axis
           Ni - 23.5 Al - 0.5 Hf - 0.2 B   (IC-50).

Aluminum Concentration of Boundary Region

The ductility of B-doped Ni_3Al greatly depends on the Al/Ni
ratio of the material. The effect of boron on the grain boundary
aluminum concentration may thus have an important bearing on the
ductility effect. Figure 7 shows the results of a series of
analyses in which the position of the area of analysis was varied
as well as the size of the area. In spite of the fluctuations
in the composition of the boundary region that may occur as
discussed above, the results indicate that boron lowers the
aluminum concentration of the boundary region [12]. Far from the
boundary the measured aluminum concentration agreed well with the
nominal one. At some grain boundaries, aluminum concentrations
as low as 18.6 +/- 1.2 at% were observed. The sequence of
measurements indicated a reasonable degree of reproducibility.
The results (summarized in the insert of Figure 7) suggest an
inverse relationship between the boron and aluminum
concentrations. Other experiments gave similar results.

Degree of Order Near Boundary

It might be expected that the degree of order of the boundary
region might be significantly lower or be completely destroyed
when such large deviations from stoichiometry as discussed above
occur, yet this was not found to be the case. The degree of
order of Ni_3Al can be determined conveniently in the atom probe
by measuring the aluminum concentration of the Ni-rich
superlattice planes [7,13]. When a grain boundary traverses a
superlattice plane the ordering near the boundary can be
determined by analyzing the part of the superlattice plane that
abuts the boundary. An analysis made in such a manner indicated
that there was no great change in the degree of order although
small changes could not be ruled out because of the limited data.

94

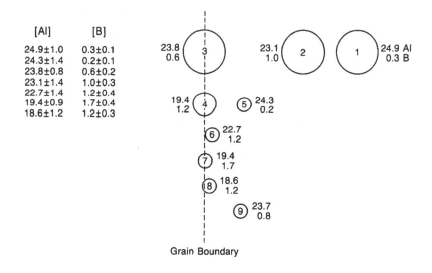

Figure 7. Variation in boron and aluminum concentration near a boundary in Ni - 23.5 Al - 0.5 Hf -0.2 B (IC-50) as a function of relative distance from boundary. Circles show relative size of area of analyses. Diameter of smallest area was ~ 1 nm. The positions of the circles are offset vertically for the sake of clarity. The results (summarized in inset) suggest an inverse correlation between Al and B concentrations. Compositions are listed in at. %.

Boron Clustering in Brittle Ni₃Al

Results of positron annihilation studies of boron-doped stoichiometric Ni₃Al have been interpreted to show the formation of boron clusters as large as 1.7nm in diameter [14]. It was proposed that these clusters are responsible for the loss of ductility in the stoichiometric and Al-rich alloys. To test this hypothesis, alloys with 25.0 at% aluminum and 0.23 at% boron were analyzed extensively. In one experiment 5.9 x 10⁴ atoms were collected from one sample. The measured frequency distribution of the composition of blocks of 500 atoms shown in Figure 8 agreed closely with the expected binomial distribution, indicating a random distribution of boron. The boron concentration was measured to be 0.24 +/- 0.02 at%, in excellent agreement with the nominal value, and indicating that the fraction of the boron atoms which are in the form of clusters can at most be a few percent.

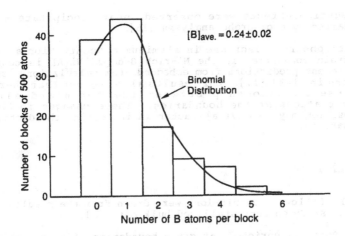

Figure 8. Frequency distribution of the number of B atoms in blocks of 500 atoms collected from Ni-25 Al- 0.23 B. (IC-54) Curve shows expected binomial distribution for a random alloy.

DISCUSSION

The results of this atom probe study, in agreement with those obtained at the Oak Ridge National Laboratory [9,13,15] have clearly confirmed previous Auger spectroscopy results showing that boron is segregated to the grain boundaries of boron- doped Ni_3Al. However, our results suggest a greater degree of complexity and non-uniformity than was indicated by AES. Although the maximum measured boron concentration usually varied by no more than a factor of 2 to 3, greater fluctuations were found to exist along the boundary plane. This suggests there may exist distinct regions of different boron concentration. The distances between these regions can be as great as 15nm. The boron-enriched region also appears to be considerably wider than is expected from an equilibrium-type of solute adsorption. The boron concentration fluctuations appear to be irregular and -unpredictable but the pattern may become clearer as more boundaries are analyzed. The heavily decorated boundaries which we observed in our studies only after partial fractures of the specimen add to the complexities. Miller and Horton [15] have reported that in rapidly solidified material these regions are 1.0 nm wide and contain 5 at% boron and 26 - 31 at% aluminum. They concluded that these regions constituted a distinct second phase. If this is the case, the phase may be discontinuous since the boundary decorations observed in this study changed with depth of field evaporation.

The width of the boron-enriched region measured in the atom probe is greater than that determined by Auger spectroscopy using ion sputtering (about 3 atomic layers). The differences in the results are not clear but may be due to the different dislocation structure near the boundaries of the fine-grained material used in this study and the coarse-grained material normally used in Auger spectroscopy. Similar long-range

96

segregation effects were observed near precipitate - matrix boundaries by atom probe analyses [16].

The observed decrease in aluminum concentration at some of the grain boundaries in the Ni-rich B-doped Ni_3Al is consistent with recent predictions from embedded atom calculations of point defects in Ni_3Al [17]. Our results also suggest that there is an inverse relationship between the boron and aluminum concentrations at the boundaries. These changes in boundary composition may be a vital factor in increasing the ductility of the material.

SUMMARY

The following conclusions were drawn from the results of our study on recrystallized bulk material.

1.   Boron is enriched at grain boundaries of substoichiometric Ni_3Al by a factor of approximately 5 to 20 times the matrix level.

2.   The boron-enriched regions are broader than previously deduced from Auger spectroscopy.

3.   The boron concentration varies along the boundary plane.

4.   There may be two types of segregated boron regions one of which results in heavily decorated boundaries..

5.   The Al/Ni ratio was found to be lower at some of the boundaries than in the bulk. An inverse correlation between the boron and aluminum concentrations holds at some of the boundaries.

6.   Thus far no evidence of large clusters of boron in the matrix of the stoichiometric alloy has been found.

ACKNOWLEDGMENTS

This work was supported by the Office of Basic Energy Sciences, U.S. Department of Energy under Grant No. DE-FG0-85ER45213. We thank Dr. D. Kroeger of Oak Ridge National Laboratory for supplying us with some of the material used in this study.

REFERENCES

1. K. Aoki and O. Izumi, Nippon Kinzoku Gakkaishi,43, 1190 (1979)

2. C.T. Liu and C.C. Koch, Technical Aspects of Critical Materials Used by the Steel Industry, Vol. IIB 83-2679-9, National Bureau of Standards (1983)

3. E.M. Schulson,T.P. Weihs, I.Baker, H.J. Frost and J.A. Horton, Acta Met.,34,1395 (1986)

4. C.T. Liu, C.L. White and C.R. Brooks, Acta Met., 33, 213 (1985)

5. A. Choudhury, C.L. White and C.R. Brooks, Scripta Met., 20, 1061 (1986)

6. S.S. Brenner and M.K. Miller, J. Metals, 35, 54 (1983)

7. S.S. Brenner, D.D. Sieloff and M.G. Burke, J. de Physique C2, 215 (1986)

8. R.J. Taunt, R. Sinclair and B. Ralph, Phys. Stat. Sol. (a) 16,469 (1973)

9. J.A. Horton and M.K. Miller, Acta Metall., 35, 133 (1987)

10. E.W. Muller and S.V. Krishnaswamy, Rev. Sci. Instr. 46, 1053 (1974)

11. R. Herschitz and D.N. Seidman, Scripta Met., 16, 849 (1982)

12. D.D. Sieloff, S.S. Brenner and M.G. Burke, J. de Physique, In Press

13. M.K. Miller and J.A. Horton, Scripta Met., 20, 1125 (1986)

14. A. Das Gupta, to be published

15. M.K. Miller and J.A. Horton, Scripta Met., 20, 789 (1986)

16. R. Moller, S.S. Brenner and H.Grabke, Scripta Met., 20, 587 (1986)

17. S.M. Foiles, to be published this Proceedings

DISLOCATION REACTIONS AT GRAIN BOUNDARIES IN Ll_2 ORDERED ALLOYS

A.H. KING* AND M.H. YOO**
* Department of Materials Science and Engineering, State University of New York, Stony Brook NY 11794-2275.
** Metals and Ceramics Division, Oak Ridge National Laboratory, Oak Ridge, TN 37831-6117.

ABSTRACT

The characteristics of perfect grain boundary dislocations in ordered alloys are discussed, and consideration is given to the possible formation of perfect and imperfect grain boundary dislocations by the impingement of crystal lattice dislocations. It is shown that many dislocation reactions at grain boundaries in ordered alloys are made unfavorable if chemical co-ordination must be maintained in the structure of the grain boundary. This leads to a suggestion that the effect of boron in the grain boundaries of Ni_3Al is to reduce the importance of chemical ordering, and thus to promote deformation by allowing greater freedom for dislocation reactions to occur.

INTRODUCTION

The beneficial effect of boron on the low temperature ductility of aluminum-deficient Ni_3Al is now well established, although its causes remain somewhat mysterious [1,2]. It has been suggested that the boron effect is simply a result of reduced grain boundary energy relative to the surface free energy, consistent with a classical Griffith-type of treatment of the brittle fracture of the undoped material. This type of treatment, however, is somewhat unsatisfactory because it fails to explain the unusual brittleness of the undoped material, seeking instead to explain the strengthening effect of the doping. This is especially problematical in a very strongly ordering material such as Ni_3Al, because at a simple level one might expect that the strong interatomic bonding which gives rise to the ordering would also provide strong cohesive bonds across the grain boundaries. Another problem with this classical approach is that considerable microplastic deformation also occurs during the failure process, as has been demonstrated by in-situ deformation studies of this curious material via electron microscopy [3].

In view of the fact that a large part of the fracture energy is associated with plastic deformation, we have undertaken a study of the possible interactions of grain boundaries in ordered alloys with the deformation process itself.

GRAIN BOUNDARY DISLOCATIONS IN PURE METALS AND SIMPLE ALLOYS

For cubic crystal structures, at least, the properties and behavior of perfect grain boundary dislocations are reasonably well understood. As with crystal lattice dislocations, a perfect dislocation is defined as one which does not disrupt the structure of the material in which it exists, and the possible Burgers vectors of perfect grain boundary dislocations are the translation vectors of the so-called DSC lattice (which are displacements of one crystal with respect to the other that shift the boundary structure, but leave it complete, or unchanged). An example of a perfect grain boundary dislocation in a fcc crystal lattice is shown in Fig.1: notice that the grain boundary plane is stepped at the core of the dislocation, which is a general feature of this type of defect, and is necessary for the preservation of the boundary structure [4]. The DSC lattice is, in fact, the set of vectors which define translations from lattice sites in one crystal to those in

Fig.1. DSC dislocation in a coincidence grain boundary between two fcc crystals. The Burgers vector is $[3\bar{3}2]/22$ and the boundary normal is $[113]$. The value of Σ is 11, meaning that the unit cell of the coincidence site lattice has 11 times the volume of the crystal unit cell. Note that the dislocation preserves the grain boundary structure.

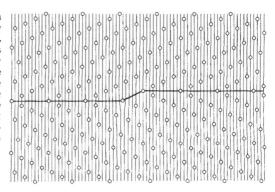

the other, and this has two important consequences for the interaction of slip dislocations with grain boundaries. First, any perfect lattice dislocation can always be dissociated into perfect grain boundary dislocations, with a reduction of strain energy in most cases, because the DSC lattice is relatively fine: such reactions are referred to as "dislocation absorption", and are illustrated schematically in Fig.2a. Second, any perfect dislocation in one crystal can form a perfect dislocation in the other crystal with the creation of a DSC dislocation residue in the boundary plane, as illustrated in Fig.2b: this process is known as "dislocation transmission" and is fundamental to the propagation of slip in polycrystals.

If the impinging dislocations are not perfect dislocations of the crystal lattice, but are partials associated with stacking faults, then the reactions described above are not generally available, as shown by King and Chen [5], because the reaction products may not be perfect dislocations of the grain boundary structure. It is expected that this will have some influence upon the deformation behavior of materials with low stacking fault energy, and indeed there appears to be some correlation between the availability of reactions, as assessed by King and Chen, and the ability of grain boundaries to transmit slip. A third type of reaction was also described by King and Chen, in which the dislocation produced in the second crystal is of a different type than the original dislocation in the first crystal: this is called dislocation transformation, and the conditions on the residue in the boundary plane are the same as for transmission. TEM observations of the interactions of lattice partial dislocations with grain boundaries in silicon [6] are in precise agreement with the analyses by King and Chen.

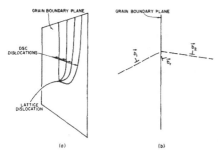

Fig.2. Illustrating the possible reactions between crystal lattice dislocations and grain boundaries. (a) The absorption of a dislocation by dissociation into an integer number of DSC dislocations. (b) The transmission of the dislocation, leaving a DSC residue in the grain boundary plane.

GRAIN BOUNDARY DISLOCATIONS IN ORDERED ALLOYS

If we take the structure shown in Fig.1 and color the atomic sites to correspond to the site occupancy for perfect Ll$_2$ ordering, we find that the dislocation no longer preserves the grain boundary structure, as shown in Fig.3, and that the chemical co-ordination of the grain boundary is disrupted in a manner analogous to the formation of an antiphase boundary (APB) by a crystal lattice dislocation. The "correct" structure (whichever it might be) cannot be recreated on both sides of the dislocation merely by changing the step height, but, just as in the case of the crystal lattice, the structure can be recreated by introducing a second identical dislocation, or by doubling the Burgers vector of the grain boundary dislocation, as shown in Fig.4. The correct Burgers vectors for perfect grain boundary dislocations in Ll$_2$ ordered materials can be determined simply as the DSC vectors applicable to simple cubic crystals, since the Ll$_2$ structure can be correctly described as a simple cubic structure with a basis of four atoms per lattice site (this approach also correctly gives the the structure preserving translations for the lattice itself). The DSC vectors, and other geometrical data, for grain boundaries in cubic crystals are listed in various publications.

Now it is conceivable that for relatively weakly ordered alloys, the structure of the material near the grain boundary may be considered as disordered, or in the case of Ll$_2$ materials, simply fcc, in which case the boundary structure is adequately preserved by the shorter DSC translations associated with the fcc structure, rather than the generally longer ones associated with the simple cubic structure, which preserve both the site occupancy at the boundary and the site positions. The availability of the shorter DSC dislocations has important consequences on the deformation behavior of the material, as we shall see below.

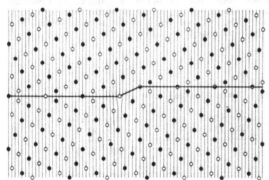

Fig.3. Grain boundary dislocation in a coincidence boundary between two Ll$_2$ crystals. All of the parrameters are identical to those in Fig.1, except that the atomic sites are shaded to indicate the site occupancy. The dislocation no longer has the property of preserving the boundary structure.

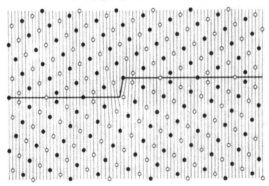

Fig.4. Perfect grain boundary dislocation in the same grain boundary as illustrated in Fig.3. This dislocation has a Burgers vector twice as long as the one illustrated in Fig.3, and enables the interfacial structure to be preserved.

When dislocations impinge upon grain boundaries in ordered alloys, the possible reactions are similar to the ones described above for simple alloys. Dislocation absorption can occur if the impinging dislocation may be dissociated into perfect grain boundary dislocations, and transmission or transformation can occur if the residue in the boundary plane is a perfect dislocation.

We have used the technique described by King & Chen to assess the availability of reactions between dislocations and grain boundaries in ordered alloys, assuming that the impinging dislocations are of the types commonly observed in LI_2 alloys: the APB creating dislocations of type $\langle 110 \rangle /2$, the Shockley partial dislocations, $\langle 112 \rangle /6$, and the super-Shockley partial dislocations of the type $\langle 112 \rangle /3$, yielding a total of thirty different Burgers vectors. This means that for each grain boundary, we have considered 30 possible absorption reactions (the reactions for dislocations in the other grain being identical, on the basis of symmetry) and 2×30^2, or 1800 different transmission or transformation reactions. The factor of 2 in the number of transmission and transformation reactions arises because each incoming dislocation must be tested against every possible product dislocation, including all of their negatives. The numbers of possible reactions are summarized in Table I: these are the only reactions which leave residues in the boundary plane that do not disrupt the boundary structure when they move. The first notable point about the data is that the number of available reactions depends on the value of Σ, just as it was found to do for partial dislocation reactions in the fcc and bcc crystal structures; and the number of reactions is always larger if Σ is an integer multiple of 3.

A more important result, however, is that the number of available reactions is always larger when the DSC lattice against which we test is the one appropriate to fcc crystals, rather than the primitve cubic DSC set. The use of the fcc set in this exercise effectively tests only for preservation of the boundary structure in terms of the atom sites, but not in terms of the site occupancy. When we use the primitive cubic DSC set, we test for the preservation of both site and site occupancy in the grain boundary.

Table I
Numbers of Allowed Dislocation Reactions

	Absorption $\Sigma=3N$	$\Sigma \neq 3N$	Transmission $\Sigma=3N$	$\Sigma \neq 3N$
Disordered (fcc)	12	6	288	216
Ordered (pc)	3	0	72	72
	[Possible 30]		[Possible 1800]	

DISCUSSION

It is not particularly suprising that a greater number of reactions becomes available when we use the fcc DSC set, because this defines a finer lattice than the primitive cubic set, so the probability of any vector being a member of the set is correspondingly higher. The important consequence of this result, as far as the deformation of ordered alloys is concerned, is that for alloys which maintain a strong site preference in the grain boundaries, the number of dislocation reactions at the boundaries becomes severely restricted. This means that the possible deformation modes in the region of a strongly ordered grain boundary may be reduced in number, and that fracture might occur simply because no suitable means of deformation exists.

If this line of reasoning is correct, then the grain boundary brittleness of ordered alloys should be related to the energy difference between the "correct" and "incorrect" site occupancy structures of the grain boundary. Large differences, implying strong ordering, will promote brittle behavior, while small differences will promote ductility, by allowing site-switching dislocations to move in the grain boundaries and thereby allowing many more dislocation reactions to occur. It might be expected that any alloy which is very strongly ordered in the grain interiors will also be strongly ordered at the grain boundaries, so our approach suggests that, at least for Ll_2 alloys, strongly ordering systems should be intrinsically brittle, as is the case for Ni_3Al. Consistent with our line of development, then, would be the conclusion that the manner in which boron affects the mechanical properties is by reducing the energy differences between grain boundary structures with "correct" and "incorrect" site occupancies. There exists some circumstantial evidence that boron is, in fact, associated with less well ordered structures when it segregates to internal surfaces in Ni_3Al [7].

There are further reasons for plasticity to be promoted when the finer DSC lattice determines the Burgers vectors of the grain boundary dislocations produced in any reaction, and these include the following:

1. The smaller Burgers vectors of the product dislocations endow them with smaller strain energies, thereby increasing the driving force (or reducing the energy deficit) for any reaction.

2. Product dislocations in the grain boundary planes will exert smaller back-stresses upon following dislocations in pile-ups, again because of the smaller Burgers vectors.

3. The most important means of back stress relief is the removal of the product dislocations from the path of oncoming dislocations. The relief of back stresses will be more rapid for the weaker dislocations, for two reasons, both associated with the fact that the motion of dislocations in grain boundary planes takes place, in general, by a mixed glide and climb process.

a) For an edge dislocation moving by pure climb, the climb distance, per point defect absorbed, is inversely proportional to the Burgers vector, so for a fixed point defect flux, a weaker dislocation moves further in the boundary plane than a stronger one. Thus for diffusion-limited motion, weaker dislocations move out of the way faster.

b) The finer DSC lattice provides a greater probablitiy of obtaining Burgers vectors that lie in or close to the boundary plane. Dislocations with these Burgers vectors move entirely or predominantly by glide so their velocity is not diffusion limited and the back stresses associated with these reaction products can be relieved almost instantly.

A full understanding of the interactions of lattice dislocations with grain boundaries thus depends not only upon the geometrical availability of the proposed reactions, but also upon the relative energies of the reaction products, and any intermediate defect structures [8]. The interactions between the various dislocations and their individual interactions with the grain boundary planes must also be taken into account. We are currently engaged in such studies, but it is difficult, as yet, to make general statements about the 1,830 possible reactions for each misorientation that we consider. However, if the energy associated with the formation of regions of disordered grain boundary is large compared with the other energies involved in any reaction, then it will influence the process strongly, and it is our suggestion that this is the case for unboronated Ni_3Al, but not if boron is present. The effects described here might also be synergistically linked with the more classical effects associated with changes in the relative values of the surface and grain boundary free energies.

Similar effects might also be expected in other ordered alloys, and we are carrying out geometrical surveys for structures other than Ll_2. Preliminary results suggest that any de-embrittling effects arising from this source should be much weaker in B2 materials, for example, because the difference in the reaction availability when order is enforced in the boundaries is only approximately a factor of two, as opposed to the factor of about four found in the present study, and the Burgers vector lengths are not so significantly reduced.

ACKNOWLEDGMENT

This work was supported by the Division of Materials Sciences, US Department of Energy under Contract #DE-AC05-84OR21400 with Martin Marietta Energy Systems, Inc.

REFERENCES

1. K. Aoki and O. Izumi, Nippon Kinzoku Gakkaishi 43 (1979) 1190.
2. C.T. Liu, C.L. White and J.A. Horton, Acta Met. 33 (1985) 213.
3. I. Baker, E.M. Schulson and J.A. Horton, Proc. 44th. Annual Meeting of EMSA (1986) 864.
4. A.H. King and D.A. Smith, Acta Cryst. A36 (1980) 335.
5. A.H. King and Fu-Rong Chen, Mat. Sci. Eng. 66 (1984) 227.
6. M. El Kajbaji, J. Thiebault-Dessaux and A. Bourret, Proc. XIth International Congress on Electron Microscopy, Kyoto, 1986, p113.
7. M.K. Miller and J.A. Horton, Scripta Met. 20 (1986) 789.
8. A.H. King, Interface Migration and Control of Microstructure, Proc. ASM Intl. Symposium at Detroit, MI, September 1984. p83.

AN ATOM PROBE STUDY OF BORON SEGREGATION TO LINE AND PLANAR DEFECTS IN Ni₃Al

J. A. HORTON AND M. K. MILLER
Metals and Ceramics Division, Oak Ridge National Laboratory,
Oak Ridge, TN 37831

ABSTRACT

Atom probe analyses of rapidly solidified, boron-doped Ni-24 at. % Al subjected to various heat treatments have shown boron segregation to a wide range of linear and planar defects. These include dislocations, superlattice intrinsic stacking faults, antiphase boundaries, twin boundaries, low-angle boundaries (dislocation cell walls), and high-angle grain boundaries. Boron coverage of these features was found to vary along a particular linear or planar defect and from boundary to boundary.

INTRODUCTION

Boron additions to the ordered intermetallic alloy Ni₃Al greatly increase the ductility of the alloy and, thereby, the potential for high temperature structural applications [1]. Pure polycrystalline Ni₃Al is inherently brittle and fractures inter-granularly, whereas boron additions to substoichiometric alloys (<25 at. % aluminum) result in transgranular failure and room temperature ductilities of up to 50%.

Auger electron spectroscopy (AES) of intergranular fracture surfaces has shown boron segregation of up to 10 at. % on grain boundaries and has shown that no embrittling agents are present on the fracture surfaces [2,3]. However, AES can only analyze boundaries that fail or can be made to fail intergranularly, and it averages concentrations over an area defined by the probe size usually > 1 μm. Atom probe field ion microscopy (APFIM) can analyze boundaries that do not fail intergranularly in addition to having excellent spatial resolution and light element sensitivity.

Previous atom probe results have shown boron segregation to antiphase boundaries (APBs), and high-angle grain boundaries in rapidly solidified Ni-24 at. % Al-0.24 at. % B [4,5] and also to twin boundaries in similar cast material [6]. The degree of boron segregation was found to be highly variable. A thin grain boundary phase was also observed on boundaries, although coverage was also highly variable. Boron has not been found to cluster in the matrix [7,8]. In cast material, a slight depletion in aluminum level was found to coincide with the boron segregation [9].

In the previous study, which employed the same starting material as this study, it was shown that the rapidly solidified material fractured transgranularly by ductile rupture [4]. The use of fine-grain-sized material greatly facilitated finding grain boundaries in the atom probe. While this material may not have been completely representative of cast material, boron additions had similar effects on the ductility. It should be noted that boron-free Ni₃Al is generally brittle after rapid solidification while Ni₃Al with boron is generally ductile.

EXPERIMENTAL PROCEDURE

The starting alloy had a nominal composition of Ni-24 Al-0.24 B (designated as IC-15) [1]. Wire specimens were prepared from this alloy by a pendant drop melt extraction technique [10]. This procedure produced wires with diameters of 40 to 80 μm and a grain size of 1 to 4 μm. Various annealing treatments that resulted in various amounts of grain growth were used in order to facilitate atom probe imaging of a range of defects such as dislocation cell walls, see Table I. Previous studies had shown that annealing for short times below 750°C did not remove the APBs [11], while annealing at higher temperatures did. Some of the specimens were extensively cold worked by repeated wrapping the wires around a 1 mm diameter rod. Anneals of these specimens allowed dislocations to form cell walls and low angle boundaries.

Mat. Res. Soc. Symp. Proc. Vol. 81. 1987 Materials Research Society

TABLE I Heat Treatments

Time	Temp °C	Grain Size μm	Comments
unannealed		1 to 4	
1 h	700	1 to 4	APBs still present.
1 h	800	1 to 4	No APBs but SISFs still present.
9 min	1000	6 to 14	Cold worked prior to anneal
30 min	1000	10 to 30	Cold worked prior to anneal, many dislocation cell walls
30 min	1000	10 to 30	Plus 5 h at 800 °C, no prior cold work
1 h	1050	100 to 250	No prior cold work

The atom probe analyses were performed using the ORNL atom probe. Full details of the instrument and the technique can be found elsewhere [12]. The field ion images were obtained using neon as an imaging gas at temperatures from 47 to 80 K. Transmission electron microscopy (TEM) was performed using a Philips EM430T operated at 300 kV. Eighty specimens were surveyed by TEM prior to AP analysis. This resulted in eight specimens containing grain boundaries and three specimens with APBs or dislocations that were successfully imaged in the atom probe.

RESULTS AND DISCUSSION

In the present study, various treatments were carried out to produce specimens with a variety of planar defects. The fracture mode was determined after the longest anneal (1 h at 1050°C) utilized to ensure that no embrittling agents had weakened the boundaries. The transgranular fracture surface, Fig. 1, showed that annealing produced a greater reduction in area and more rupture type failure [4].

One high-angle grain boundary, Fig. 2, showed more extensive boron segregation to the grain boundary after an anneal at 700°C for 1 h than did any of the previously observed ones [4]. In addition, a large boron-containing precipitate was observed on the boundary. This coverage, as in all cases examined, was highly variable. Boron atoms were identified previously as bright spots in a field ion image by a technique of single atom analysis [6].

In bulk material, the APBs were known to anneal out at 800°C [11]. In this material many stacking faults remained. TEM analysis of one specimen performed after an AP analysis determined that superlattice intrinsic stacking faults (SISFs) were present, Fig 3. TEM analyses of similar material by Baker and Schulson [13] has shown the presence of SISFs with widely separated dislocations. In-situ TEM deformation experiments in similar material have shown that dislocation loops containing SISFs are left behind after passage of APB-coupled slip dislocations [14]. The stacking faults in Fig. 3 (b and c) intersect a grain boundary on one side and intersect the free surface elsewhere, therefore there are no terminating dislocations present. The field ion images in Fig. 3 (d and e) show a SISF parallel to a grain boundary. This SISF was probably parallel to the two in the TEM micrographs. At this point in the evaporation sequence, Fig. 3 (d and e), the surface of the specimen has just passed the

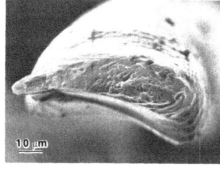

Fig. 1 Scanning electron micrograph of the fracture surface of a rapidly solidified Ni-24 % Al-0.24 % B wire annealed 1 h at 1050°C, then fractured at room temperature. A completely transgranular fracture surface is evident. The reduction in area was > 50%.

intersection of the SISF with the grain boundary and so they appear as parallel lines in the image. There is more boron segregation to the SISF than to the nearby grain boundary, although the boron level on the grain boundary is above the matrix level. The boron was localized to the defect plane of the SISF. It is rather surprising that boron segregates to SISFs since they are a lower energy fault than APBs and contain no nearest neighbor violations as do most APBs [15].

Annealing at 1000°C for 9 min resulted in similar boron coverage of high-angle grain boundaries, with the boron distributed along a band, see Fig. 4(a), as observed previously in unannealed material [4]. Later in the evaporation sequence a dislocation was observed near the grain boundary, Fig. 4(b). At this point there are few boron atoms on the grain boundary, further illustrating how variable the coverage is, whereas the dislocation and associated APB were decorated with boron. Fig. 5 is a field ion image of another dislocation from a specimen with the same history as that shown in Fig. 4. This dislocation also terminated at a 100 pole and had boron segregated to it and to the APB produced on the slip plane. The APB behind the dislocation extends from the pole toward the d marked on the micrograph. At an APB in $L1_2$ ordered alloys, bright rings (mixed planes, Al + Ni) switch to dim rings (pure Ni planes) near 100 poles. This APB then extends nearly to the edge of the micrograph, see arrow, before terminating. The superlattice dislocation spacing measured from the FIM image was 5.5 nm. While this result agrees with the currently accepted value of 5 nm, recent measurements on superlattice dislocations under load in boron-doped Ni_3Al indicate a much wider spacing [16]. Field ion images of superlattice dislocations in Ni_3Al without boron additions have been shown previously by Taunt and Ralph [17]. The plane of the APB fault in Fig. 5 was determined to be {100} by monitoring the trace of the fault during evaporation.

Annealing the deformed material for 30 min at 1000°C yielded specimens with dislocation cell walls and low-angle boundaries, as observed by TEM during the specimen survey. One such example is shown in Fig. 6, which shows two low-angle boundaries (< 1°) well decorated with boron. During evaporation, Fig. 6(a), these boundaries were detectable because of the boron decoration, whereas the image at BIV, Fig. 6(b), reveals the boundary as a row of bright spots (boron atoms). Note that the boron coverage is confined to the plane of the defect as it is with all other planar defects except high-angle boundaries.

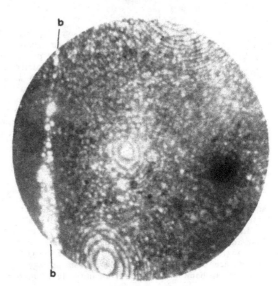

Fig. 2 Field ion image of a specimen annealed for 1 h at 700°C showing a boron-containing precipitate (arrowed) and extensive boron coverage on a high-angle grain boundary, marked b. The boron atoms appear as bright spots.

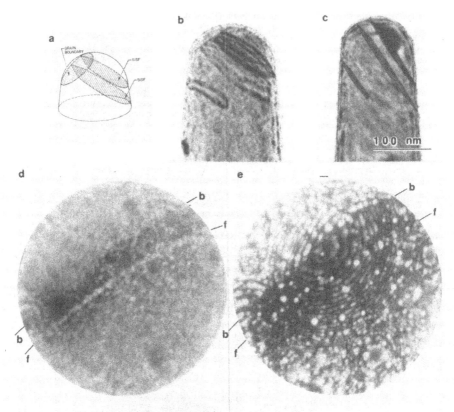

Fig. 3 Sketch (a) and transmission electron micrographs (b and c) at different orientations, made after atom probe analysis, showing two superlattice intrinsic stacking faults (SISFs) intersecting a grain boundary in a FIM specimen of material annealed for 1 h at 800°C. Field ion images (d and e) of a SISF, marked f-f, parallel to a grain boundary, marked b-b, show boron segregation to the SISF localized to the defect plane and show little boron segregation to the nearby grain boundary. The micrograph in (d) is a time exposure recorded during a short evaporation sequence and shows the boron segregation much clearer than the micrograph in (e) which was recorded at best-image-voltage (BIV).

CONCLUSIONS

In this study of Ni_3Al, atom probe field-ion microscopy has shown that boron segregates to many types of planar and line defects including dislocations, superlattice intrinsic stacking faults, and dislocation cell walls (low angle boundaries). Previous studies had shown that boron segregates to antiphase boundaries, twin boundaries, and high-angle grain boundaries. The thin boron-containing phase was observed only on high-angle grain boundaries. On all other planar defects the boron was localized to the defect plane. The level of boron segregation varied in all cases both from boundary to boundary and along an individual boundary.

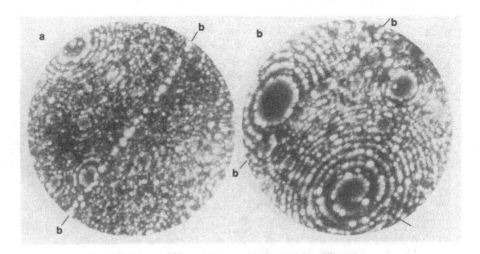

Fig. 4 Field ion images of rapidly solidified Ni₃Al after cold
work and an anneal for 9 min at 1000°C. The micrograph in
(a) shows a high-angle grain boundary (misorientation of
Σ27a), marked b-b, showing extensive boron segregation
present in a band along the boundary. (b) shows the same
boundary later in the evaporation where there was little
boron coverage but with a boron-decorated dislocation
present nearby which terminates at the pole. The slip plane
is marked by d. (c) is a TEM image of same specimen.

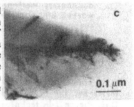

Fig. 5 Field ion image of a
dislocation in a specimen
with the same history as that
in Fig. 4. The dislocation
terminates at a {100} pole
producing a spiral. Its slip
plane extends toward the d
mark. The APB (slip trace)
behind the dislocation also
lies on a {100} plane. The
superlattice dislocation
separation is 5.5 nm.

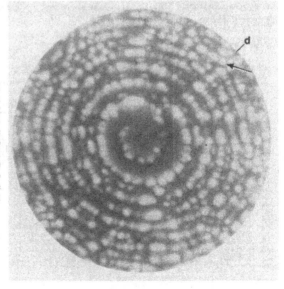

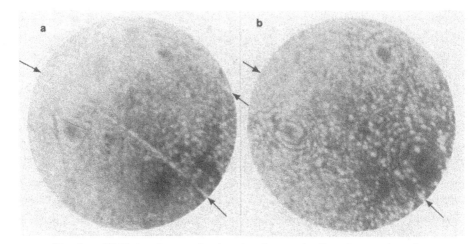

Fig. 6   Field ion images showing two low angle (less than a degree)·
boundaries well decorated with boron, see arrows.   The material was cold
worked and then annealed for 30 min at 1000°C. (a) was recorded during an
evaporation and clearly shows the location of the dislocation cell walls by
the presence of the boron atoms, and (b) was recorded at BIV.

ACKNOWLEDGMENTS

The authors would like to thank Dr. R. Maringer of Battelle Columbus Labora-
tories for the preparation of the rapidly solidified wire, K.F. Russell, J.W. Jones, and
D.H. Pierce for technical assistance, and Drs. C.T. Liu, J. Bentley, and P. Camus for
helpful discussions. This research was sponsored by the Division of Materials Sciences,
U.S. Department of Energy under contract DE-AC05-84OR21400 with Martin Marietta
Energy Systems, Inc.

REFERENCES

1. C.T. Liu, C.L. White, and J.A. Horton, *Acta Metall.* 33, 213 (1985).
2. C.L. White, R.A. Padgett, C.T. Liu, and S.M. Yalisove, *Scripta Metall.* 18, 1417
    (1984).
3. T. Takasugi, E.P. George, D.P. Pope, and O. Izumi, *Scripta Metall.* 19, 551 (1985).
4. J.A. Horton and M.K. Miller, *Acta Metall.* 35, 133 (1987).
5. M.K. Miller and J.A. Horton, *J. de Physique*, C7 (1986), in press.
6. M.K. Miller and J.A. Horton, *Scripta Metall.* 20, 789 (1986).
7. J.A. Horton and M.K. Miller, *J. de Physique*, C2, 209 (1986).
8. S.S. Brenner, D.D. Sieloff, and M.G. Burke, *J. de Physique*, C2, 215 (1986).
9. D.D. Sieloff, S.S. Brenner, and M.G. Burke, *J. de Physique*, C7 (1986), in press.
10. R.E. Maringer and C.E. Mobley, *Wire J.* Jan 1970, 70.
11. J.A. Horton and C.T. Liu, *Acta Metall.* 33, 2191 (1985).
12. M.K. Miller, *J. de Physique*, C2, 499 (1986).
13. I. Baker and E.M. Schulson, *Phys. Stat. Sol.*(a) 85, 481 (1984).
14. I. Baker, E.M. Schulson, and J.A. Horton, to be published *Acta Metall.*
15. D.P. Pope and S.S. Ezz, *Inter. Met. Rev.* 29 (3), 136 (1984).
16. I. Baker and E.M. Schulson, *Phys. Stat. Sol.*(a) 89, 163 (1985).
17. I. Baker, J.A. Horton, and E.M. Schulson, to be published *Phil. Mag. Letters*
18. R.J. Taunt and B. Ralph, *Philos. Mag.* 30, 1379 (1974).

ANNEALING OF COLD WORKED HYPOSTOICHIOMETRIC Ni$_3$Al
ALLOYS USING POSITRON LIFETIME SPECTROSCOPY

S. G. USMAR AND K. G. LYNN
Brookhaven National Laboratory, Department of Metallurgy and Materials
Science, Upton, NY 11973

ABSTRACT

Hypostoichiometric Ni$_3$Al alloys of composition 76.2 Ni:23.8 Al containing impurity levels of boron and hafnium (supplied by Oak Ridge National Laboratory) were either cold rolled or pressed. Rolled and pressed
samples were deformed by 20% and 10% thickness reductions, respectively.
Samples were annealed isochronally at approximately 50°C intervals up to
1050°C. Two major annealing stages were apparent in all three alloys
studied. These could be attributed to vacancy migration to sinks and
annealing of dislocations and(or) recrystallization. The onset of vacancy
migration occurred at approximately 200°C in all three alloys. Annealing
of dislocations started at 650°C to 700°C and was complete at 1000°C for
alloys which contained boron and or hafnium impurities. In the pure alloy
the onset of dislocation annealing occurred at 800°C and was incomplete at
the highest (1050°C) annealing temperatures reached.

INTRODUCTION

Recently the intermetallic alloy on Ni$_3$Al has come into prominence as
a promising high temperature material. The pure alloy has a composition
range extending approximately 2 at.% on each side of stoichiometry and
undergoes intergranular brittle failure over this whole range. Impurity
levels (100 to 500 wt ppm) of boron have been found to improve the ductility [1] of alloys in the hypostoichiometric (25 at.% >Al>23 at.%) composition range. Further hafnium has also been found to improve the high
temperature mechanical behavior of these alloys [2].
    With the above in mind, an investigation of the defect structures of
Ni$_3$Al alloys manufactured by Oak Ridge National Laboratory (ORNL) from
pure commercial grade raw materials was initiated. Positron annihilation
spectroscopy was chosen as the investigative technique because of its
ability to detect and differentiate submicroscopic defects [3]. Here a
brief explanation of the underlying physics of positron annihilation spectroscopy will be useful. A high energy positron (resulting from nuclear
decay) entering condensed matter rapidly ($\sim10^{-11}$ sec) attains thermal
equilibrium. Sometimes later ($\sim2\times10^{-10}$ sec in metals) the positron annihilates with an electron. In the first approximation the positron lifetime is inversely proportional to electron density. Further in defective
solids, notably metals, the coulombic interaction between the positrons
and ion cores results in positrons being trapped at open-volume defects.
Since the electron density at open-volume defects is less than that in the
bulk, positrons trapped at such defects have longer lifetimes than those
which remain in the bulk. Thus, positron lifetime spectra can provide
information pertaining to defect structures in condensed matter.

EXPERIMENTAL

All alloys studied were prepared by drop casting arc melted raw materials into cold copper molds. Each casting then underwent a lengthy heat

treatment (under a vacuum of $\sim 10^{-3}$ Pa) which both homogenized and annealed it. Alloys of composition:

(i) 76.2 at.% Ni:23.8 at.% Al (A)
(ii) 76.1 at.% Ni:23.9 at.% Al+0.24 % B (B)
(iii) 76.3 at.% Ni:23.2 at.% Al:0.5 at.% Hf+0.2% B (C)

where mechanically deformed. Alloy A was pressed with a thickness reduction of 10% while alloys B and C were cold rolled with a 25% thickness reduction. Samples suitable for positron lifetime spectroscopy were prepared from the deformed material.

Positron lifetime spectra were accumulated using a spectrometer employing "Fast-Fast" coincidence counting [4] with a timing resolution of approximately 165 ps FWHM. Isochronal annealing was accomplished by first sealing samples in an evacuated fused silica tube then heating the assembly in a furnace for 1/2 h.

To ensure cleanliness in the annealing the samples were sealed in the fused silica using a two stage process. The tube was first evacuated to 2.6 Pa and back filled with argon gas three times; evacuated to 2.6 Pa and sealed. Several strips of Ta foil were fired at $\sim 1000^\circ C$ for >2.5 min and the samples sealed once more. Further samples were wrapped in Ta foil prior to being sealed in fused silica. At least one positron lifetime spectrum, containing $\sim 1.5 \times 10^6$ counts, was accumulated, at room temperature, after each annealing stage.

RESULTS AND DISCUSSION

Positron lifetime spectra were analyzed, numerically, using an interactive form [5] of the computer program POSITRONFIT [6]. Initial analysis resulted in a "mean lifetime" τ_m the temperature dependence of which is shown in Fig. 1 for alloys B and C. Here it is evident that two annealing stages occur in both alloys. The low temperature stage, whose onset occurs at approximately $200^\circ C$, can be attributed to migration of vacancies to sinks in agreement with the observation of Wang et al. [7]. The high temperature stage is due to migration of dislocations and or recrystallization. An intermediate stage is apparent in C but not in B.

The mean lifetime data indicate that at least two lifetime components were present in spectra up to approximately $950^\circ C$. These components were associated with annihilation of positrons from the bulk and from vacancies or dislocations. In the temperature range $600^\circ C$ to $700^\circ C$ for B and $700^\circ C$ to $800^\circ C$ for C and A the mean lifetime (and therefore the annihilation parameters) remain constant. Here two lifetime components were present thus the pertinent spectra were fitted using two components. The longest lived of these components had a magnitude of 135±5 ps suggesting it be associated with dislocaitons [8,9] in all samples. Now all other lifetime spectra were analyzed using two components, one of which was fixed at 135 ps. The results of these analyses are shown in Fig. 2(a) and (b). In these figures τ_1 (only present above $\sim 500^\circ C$) is the bulk lifetime; τ_2 the lifetime associated with positrons trapped at dislocations and τ_3 the lifetime associated with those trapped at vacancies. The switch, in the 2 component fit, from τ_3 to τ_1 occurred spontaneously (i.e., the initial guesses used when fitting the spectra in which τ_1 first appeared were values close to $\tau_3 \sim 180$ ps). Two component fits, where necessary, were found to be better than single component fits and usually results in variances <1.2. No long-lived component which would suggest void formation was observed.

The detailed numerical analyses reveal that below approximately $450^\circ C$ all positrons annihilated from either vacancies or dislocations, i.e., competitive trapping occurred. Further, the initial ratio of vacancies to

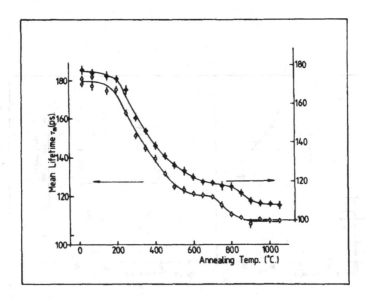

Figure 1. Mean lifetime vs annealing temperature for
for cold rolled alloys; —o—:alloy B, —●—:
alloy C.

dislocations was the same in cold rolled alloys (c.f. Fig. 2(a)). At approximately 200°C vacancies become mobile and migrate to sinks (I_3 decreases) thus the fraction of positrons trapped at dislocations (I_2) increases. At 450°C I_3 has fallen to zero indicating vacancy annealing to be complete. A combination of dislocation density and trap strength conspire to allow some positrons to annihilate in the bulk.

A further increase of temperature resulted in annealing of dislocations. In alloys B and C dislocations become mobile in the temperature range 750°C to 800°C and annealing is complete at 1000°C. In this temperature range I_2 decreases and both I_1 and τ_1 increase in qualitative agreement with the two state trapping model [3].

A close inspection of the results for alloys B and C (c.f. Fig. 2(a)) suggest that the hafnium present in C may interact with both vacancies and dislocations. The solid and dashed lines in Fig. 2(a) emphasize the temperature range in which these interactions are evident. Obviously the effects are small but suggest, tentatively, that Hf stabilizes vacancies and pins dislocation. More work is planned to elucidate this point.

Annealing of A, the pure pressed alloy (c.f. Fig. 2(b)) indicates the defect structures in this sample to differ from those of cold rolled alloys. Pressing results in an initial vacancy: dislocation ratio much smaller than that observed in cold rolled sample. The vacancy annealing stage is, at least qualitatively, the same as that for rolled samples but vacancies persist to higher temperatures. Further dislocations are present up to the highest annealing temperature reached. This incomplete annealing is difficult to explain possibly the defects involved are not dislocations. In this respect a TEM study has been initiated. If the defects are dislocations then the presence of boron impurities in $Ni_{76.2}:Al_{23.8}$ would seem to enhance the mobility of dislocations, a

114

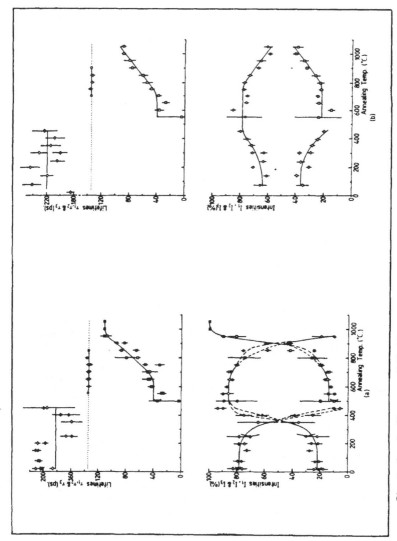

Figure 2. Positron annihilation parameters —□—:I_1,τ_1, —○—:I_2,τ_2 and —◇—:I_3,τ_3 vs annealing temperature for, (a) alloys B(—□—) and C(—■—) and (b) alloy A.

effect not inconsistent with the known mechanical properties of these alloys. Evidently, more work is required to elucidate this behavior.

CONCLUSION

The cold rolled Ni_3Al alloys B and C studied here contained vacancies and dislocations which annealed in two distinct stages at ~200°C and ~750°C, respectively. Further as was found for carbon in iron [10], Hf may stabilize vacancies. Alloy A also contained, after pressing, both vacancies and dislocaions. Here, however, there may be other unknown defects, which persist at high temperatures, present. More work is planned to elucidate these possibilities.

ACKNOWLEDGMENTS

The authors wish to thank W. Tremel for his steadfast technical support and D. Kroger, ORNL, for supplying sample materials. This research was performed under the auspices of the U.S. Department of Energy, Division of Materials Sciences, Office of Basic Energy Sciences under Contract No. DE-AC02-76CH00016.

REFERENCES

1. K. Aoki and O. Izumi, Nippon Kinzku Takkaishi 43, 1190 (1970).
2. C. T. Zui and J. O. Stiegler, Science 226, 636 (1984).
3. R. N. West, in Topics in Current Physics: Positrons in Solids, edited by P. Hautojärvi (Springer-Verlag, New York, 1979), p. 89.
4. W. H. Hardy II and K. G. Lynn, IEEE Trans. Nucl. Sci. 23NS, 229 (1976).
5. C. J. Virtue, R. J. Douglas, and B. T. A. McKee, Comp. Phys. Comm. 15, 97 (1978).
6. P. Kirkegaard and M. Eldrup, Comp. Phys. Comm. 7, 401 (1974).
7. Tian-Min Wang, M. Shimotomai, and M. Doyama, J. Phys. F 14, 37 (1984).
8. M. Doyama and R. M. J. Cotterill in Proc. 5th Intern. Conf. Positron Annihilation, edited by R. R. Hasiguti and K. Fujiwara (The Japan Inst. of Metals, Sendai, 1979) p. 89.
9. L. C. Snedskjaer, M. Manninen, and M. J. Fluss, J. Phys. F 10, 2237 (1980).
10. P. Hautojärvi, J. Johansen, A. Vehanen, and J. Yli-Kauppila, Phys. Rev. Lett. 44, 1326 (1980).

SITE OCCUPATION DETERMINATIONS IN Ni₃Al BY ATOM PROBE

M.K. MILLER AND J.A. HORTON
Metals and Ceramics Division, Oak Ridge National Laboratory,
Oak Ridge, TN 37831-6376

ABSTRACT

The site occupation of three substitutional elements, hafnium, iron and cobalt, in substoichiometric Ni₃Al was determined from atom probe field-ion microscopy. The hafnium was found to have a strong preference for the aluminum sites, the cobalt had a strong preference for the nickel sites, and the iron had a weak preference for the aluminum sites. The atom probe results were in agreement with zone axis electron channeling microanalysis of the same alloys and predictions from the position of the solubility lobes in the ternary phase diagrams.

INTRODUCTION

Small additions of boron to nickel aluminides have produced a fabricable material with good high temperature properties[1]. The properties of these nickel aluminides can be further improved by other additions such as hafnium. One aspect of understanding the influence of these substitutional elements is to ascertain their location in the ordered lattice. This knowledge can then be applied to refining the composition of the alloy and maximizing the degree of order.

The atomic spatial resolution of the atom probe field-ion microscope (APFIM) permits the site occupational probability of any substitutional element to be determined by measuring the composition of successive planes in an ordered lattice[2,3]. This technique has the advantage over more conventional methods of determining the site occupation probability in that it can be used on fine-grained polycrystalline materials. The procedure has no limits as to which elements can be quantified. The technique can be applied to complex alloys containing several substitutional solutes and the measurement is not affected by the presence of antisite or other defects. In addition this approach does not require any special equipment and can be performed on most time-of-flight atom probes designed for metallurgical applications.

EXPERIMENTAL

The nominal compositions of the six alloys used in this investigation are summarized in Table I.

TABLE I

Nominal Composition of the Alloys (Atomic %)

Substitutional element	Al	Ni	B
0.5 % Hf	23.4	75.9	0.24
1.0 % Hf	22.9	75.8	0.24
1.0 % Hf	22.9	75.9	0.10
3.0 % Hf	20.9	75.8	0.24
6.0 % Co	23.9	69.8	0.24
6.0 % Fe	20.9	72.8	0.24

These alloys were based on a nominal composition of Ni - 24.0 at.% Al-0.24 at. % B with additions of hafnium, cobalt, and iron. An additional single crystal sample containing 1.0 % hafnium and 0.1 at. % B was also examined. It should be noted that the aluminum and nickel levels were adjusted assuming that the substitutional additions would prefer a specific site. The presence of the boron additions should not effect the atom probe analyses. The compositions were all within the solubility limits of the $L1_2$ phase fields in their respective ternary phase diagrams and no second phase formation was observed in the transmission electron microscope. The alloys were annealed in the temperature range of 1000 to 1050°C for times between 2 and 5 h.

The APFIM analyses were performed on the ORNL atom probe. A complete description of this instrument and its capabilities can be found elsewhere[4].

The site occupation probabilities were determined by analyzing a cylinder of material normal to a set of planes that alternate in composition, as shown in Figure 1. Unfortunately, the $L1_2$ ordered alloys do not contain any set of planes that alternate between pure A and pure B atoms. The optimum set of planes in the $L1_2$ ordered alloys is the {001} planes since they have the maximum difference in composition, pure A versus equiatomic A+B, and the largest interplanar spacing[3]. The presence of A or nickel sites on both types of layers does not affect the analysis since all the nickel sites are assumed to be equivalent. A schematic diagram of the principle of the technique is shown in Figure 2. The technique relies on the atomic spatial resolution in analysis and the ability of the atom probe to be able to collect atoms from one plane without any contributions from the next layer. The position of the area of analysis must be carefully chosen to fulfill these requirements and not produce artifacts. This position is demonstrated in Figure 1. If the substitutional element substitutes for aluminum then it will be found only on the mixed aluminum plus nickel layer since these are the only layers where aluminum is present. However, if it substitutes for the nickel, then twice as much will be found on the pure nickel layer as on the mixed layer since there are twice as many nickel sites on the nickel layer. A full description of the experimental method and the criteria for accurate analysis are described in detail elsewhere[3].

RESULTS AND DISCUSSION

The difference between the pure nickel layers and the mixed nickel plus aluminum layers of the $L1_2$ ordered lattice can be observed in field-ion micrographs (under certain experimental conditions), as shown in Figure 3a. Atom probe chemical analysis revealed that the dimly imaging rings are the pure nickel layers while the brightly imaging rings are the mixed nickel plus aluminum layers. However, when the field-ion image is taken under Best Imaging Voltage (BIV) conditions, using the same experimental conditions as required for accurate compositional analysis, the contrast between the two types of layers disappears, Figure 3b.

A short section of the raw data over 16 planes is presented for the 3% hafnium material in Figure 4. In this representation of the raw data, known as a ladder diagram, the detection of each nickel atom is plotted as one horizontal step and the detection of each aluminum atom as one vertical step. Therefore the nickel planes are the horizontal sections and the mixed planes are the sections with a 45° slope. The position in the evaporation sequence of the detection of each hafnium atom is indicated by the squares directly below the line. In this material the hafnium is observed to have a strong preference for the mixed nickel plus aluminum planes indicating that hafnium has a preference for the aluminum sites. It should be noted that some of the hafnium is also observed on the horizontal sections indicating that a small portion of the hafnium is on the nickel sites.

The raw data can be quantified, with regards to the probability of finding a solute element on a given site, by counting the number of substitutional atoms detected on each type of plane and comparing it to the relative number of sites of each type

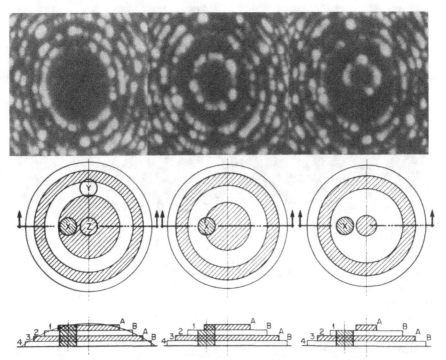

Fig. 1. Schematic diagram of the effective position of the probe aperture with respect to the (100) pole. In position X, atoms from layer 1 are collected before any atoms from layer 2. Position Y is not suitable since two planes are sampled simultaneously. Position Z is avoided because of trajectory abberations. Also shown is a corresponding set of neon field-ion micrographs taken from the 1.0 at. % Hf - 0.1 at. % B alloy.

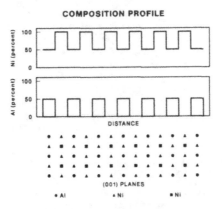

Fig. 2. Schematic diagram of an evaporation sequence perpendicular to a (100) plane. The circles represent aluminum atoms and the squares and triangles the nickel atoms. If the solute element substitutes for the aluminum atoms, then it will only be found on the mixed nickel plus aluminum planes, whereas if it substitutes for nickel atoms, twice as much will be found on the nickel planes than on the mixed planes.

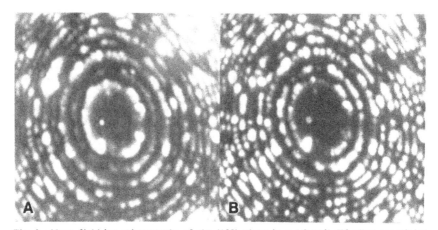

Fig. 3. Neon field-ion micrographs of the (100) plane in a 1.0 at.% Hf alloy containing 0.1 at. % B. Fig. 3a was taken under imaging conditions that revealed the difference between the dimly imaging pure nickel planes and the brightly imaging mixed nickel plus aluminum planes. Fig. 3b was taken under normal Best Image Voltage conditions and does not distinguish between the two types of planes.

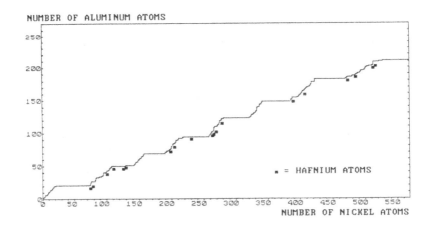

Fig. 4. Ladder diagram of the alloy containing 3.0 at. % hafnium. The horizontal sections are the nickel planes and the sections with the 45° slope are the mixed nickel plus aluminum planes. The squares below the line indicate the position of individual hafnium atoms during the evaporation sequence. This sequence indicates that hafnium has a strong preference for the mixed planes and therefore the aluminum sites.

on those planes. The results from extended analyses of the site occupancy on the aluminum sites are summarized in Table II. The error quoted is for one standard deviation and only reflects contributions from counting statistics.

TABLE II

Aluminum Site Occupation Probabilities of the Alloys

Substitutional element atomic %	Probability %	<100> Zone Axis Electron Channelling from ref. [6]
0.5 % Hf	76.9 ± 8.3	-
1.0 % Hf, 0.24 % B	84.4 ± 3.5	-
1.0 % Hf, 0.1 % B	75.8 ± 2.6	-
3.0 % Hf	62.8 ± 2.7	87
6.0 % Co	0.5 ± 0.5	16
6.0 % Fe	42.6 ± 3.3	51

If the substitutional element has no preference between the two types of sites, then, in perfectly ordered Ll_2 materials, 25% of the solute would be found on the aluminum sites and the remainder on the nickel sites. By comparing the results with this random probability it is evident that, in the composition range studied, the hafnium has a strong preference for the aluminum sites, cobalt has a strong preference for the nickel sites, and iron has a weak preference for the aluminum sites.

While there was some scatter in the results from the four hafnium-containing alloys, they all indicated that the hafnium has a strong preference for the aluminum sites. The site preferences measured in the two 1.0 % hafnium alloys with different boron levels were in good agreement.

The atom probe results are in good agreement with the zone axis electron channeling microanalysis results of Bentley[5,6]. Results for the alloys containing 3.0% hafnium, 6.0% iron, and 6.0% cobalt, which were made with <100> orientated zones and which were corrected for ionization delocalization, are included in Table II for comparison. The site preference determined by the two techniques agreed for all three types of substitutional additions examined.

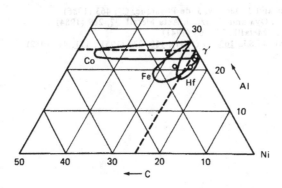

Fig. 5. Section of a semischematic ternary phase diagram for nickel-aluminum-X alloys adapted from Ochiai, Oya and Suzuki[7] that indicate the limits of solubility of the Ll_2 ordered phase field. Cobalt and hafnium data were taken at 1273K, and iron at 1233K. The compositions of the alloys used in this study are indicated by the circles.

The results are also in agreement with the site preference suggested by Ochiai, Oya and Suzuki[7], based on the direction of the solubility lobes in the ternary phase diagrams determined from the experimental data of Schram[8], and Bradley[9]. The applicable section of this data is reproduced in Figure 5. The extent of solubility for the cobalt is observed to follow constant aluminum content indicating that nickel and cobalt are equivalent. In contrast, the extent of the solubility lobe for hafnium follows the constant nickel direction indicating that aluminum and hafnium behave in a similar manner. The solubility lobe for iron is between those of cobalt and hafnium and indicates that it is an intermediate case. The suggested site preferences indicated from the direction of these solubility lobes are in agreement with the atom probe determinations.

CONCLUSIONS

Determination of the site occupation using the atom probe field-ion microscopy is the most direct technique available and is the least subject to error or misinterpretation. The hafnium was found to have a strong preference for the aluminum sites, the cobalt had a strong preference for the nickel sites, and the iron had a weak preference for the aluminum sites.

ACKNOWLEDGMENTS

Research sponsored by the Division of Materials Sciences, U.S. Department of Energy, under contract DE-AC05-84OR21400 with Martin Marietta Energy Systems, Inc. The authors would like to thank K.F. Russell for her technical assistance and Drs. J. Bentley and C.T. Liu for helpful discussions.

REFERENCES

1. C.T. Liu, C.L. White and J.A. Horton, Acta Metall., 33, 213 (1985).
2. M.K. Miller and J.A. Horton, Scripta Metall., 20, 1125 (1986).
3. M.K. Miller, Submitted to J. Microscopy.
4. M.K. Miller, J de Physique, C2, 493 (1985).
5. J. Bentley, in Proceedings of 44th. Meeting of Electron Microscopy Society of America, Albuquerque, ed. G.W. Bailey, San Francisco Press, San Francisco, 1986, p. 704.
6. M.K. Miller and J. Bentley, J. de Physique, C7, 463 (1986).
7. S. Ochiai, Y. Oya and T.Suzuki, Acta Metall. 32, 289 (1984).
8. J. Schram, Z. Metallk, 33, 403 (1941)
9. A.J. Bradley, J.I.S.I., 163, 19 (1941); 168, 233 (1951); 171, 41 (1952).

LATTICE OCCUPATION DETERMINATION OF HAFNIUM IN Ni$_3$Al BY PAC

H.G. BOHN*, R. SCHUMACHER** AND R.J. VIANDEN**
* Institut für Festkörperforschung, KFA Jülich, 5170 Jülich,
  Federal Republic of Germany
** Institut für Strahlen- und Kernphysik der Universität,
  5300 Bonn, Federal Republic of Germany

ABSTRACT

Using the perturbed angular correlations technique we have investigated the intermetallic alloy Ni$_{76}$Al$_{24}$Hf$_1$ in order to determine the lattice location of the Hf in the ordered Ni$_3$Al structure. From the fact that the Hf atoms sense an electric field gradient it is concluded that most of the Hf resides on Ni-sublattice sites.

INTRODUCTION

The cubic ordered intermetallic alloy Ni$_3$Al has a number of attractive high temperature properties. In the past technical application has been limited by the extreme brittleness of the material. This has recently been overcome by the finding that small additions of boron of the order of 0.1 at.% greatly improve the ductility, and that dopants like hafnium have additional benificial effects on the high temperature properties [1]. A fundamental understanding of these features of these so called advanced aluminides requires, among others, information about the lattice site occupation of these dopants.

Using ion channeling combined with analysis for a nuclear reaction it has been shown that the fraction of the boron atoms which is in solid solution in the fcc-Ni$_3$Al matrix occupies the octahedral interstitial sites [2]. Clearly a heavy element dopant like Hf is substitutionally incorporated into the lattice. Here the question is how it is distributed among the nickel and aluminium sites of the ordered L1$_2$ structure.

From the results of both electron channeling and atom probe techniques on material of approximate composition Ni$_{76}$Al$_{21}$Hf$_3$ it was concluded that Hf preferably occupies Al-sites [3,4]. In contrast, Rutherford backscattering (RBS)/ion channeling gave strong evidence for Hf mostly occupying Ni-sublattice sites [2]. The latter experiments were performed on single crystalline material of composition Ni$_{76}$Al$_{23}$Hf$_1$.

In order to shed more light on this open question we have applied another experimental technique, namely the perturbed angular correlations (PAC).

PRINCIPLE OF THE EXPERIMENT

Certain radioactive isotopes decay by emission of a γ - γ cascade in which an intermediate nuclear level with a suitable lifetime is populated. By use of appropriate coincidence techniques, it can be ensured that the radiation detected upon decay to and from the intermediate state originates from the same nucleus. The spatial distribution of the second radiation will then be anisotropic with respect to the direction of the first (angular correlation). If the intermediate nuclear level is split by some hyperfine interaction, for instance between the nuclear quadrupole moment, Q, and an electric field gradient (EFG), the distribution of the second radiation becomes time dependent. This quadrupole interaction can be characterized by the quadrupole frequency ν_Q [5].

From the variation of the intensity versus time in a given direction, and by using the usually known value for Q, one can gain information about the EFG at the site of the probe atom, and thus about the local environment (charge distribution) of the atom under study. An advantage of the technique is that it is not limited to single crystal samples. Although there are only a limited number of nuclei that are suitable for this type of experiment [181]Hf happens to be one of them.

In the ordered $L1_2$ structure, each Al atom is surrounded by 12 nearest neighbour Ni atoms, whereas each Ni atom has 8 Ni and 4 Al nearest neighbours. If the Ni and Al lattice atoms carried exactly the same charge, the lattice as well as the charge symmetry around each atom species would be cubic. Under these circumstances no EFG is expected at either site. If, however, one makes the reasonable assumption that in the alloy a charge shift occurs between the two elements, the cubic charge symmetry around the Ni atoms is destroyed whereas it remains intact around the Al-site. From a simple point charge model calculation it was estimated that for charge transfers of 0.1 electron or more one can expect an observable EFG at the Ni- and none at the Al-site. In this way determination of the sites occupied by the Hf-atoms can be achieved. Detailed analysis of the PAC spectrum also allows for a determination of the fractional occupancy of lattice sites. Details of the method can be found in ref. 5.

EXPERIMENTAL DETAILS

In the present investigation an alloy with a nominal composition $Ni_{76}Al_{23}Hf_1$ + 0.2 at.% B was used. Both a single crystal (actually the same as used in ref. 2) and polycrystalline material was used. The samples were annealed in a vacuum of about 10^{-6}Torr at 1000°C for 1 hour and subsequently at 700°C for 1 day. The suitable [181]Hf isotope was obtained by thermal neutron activation of the natural abundant material at the FRJ-2 nuclear reactor of KFA Jülich. The PAC measurements were performed at room temperature.

RESULTS AND DISCUSSION

Fig. 1 shows the time dependent PAC spectrum, where the exponential decay due to the 10.8 ns lifetime of the intermediate state has been removed. A heavily damped rapid oscillation is clearly seen, indicating the presence of strong but not unique EFGs at the position of the Hf atom. The data can be fitted by assuming two Lorentzian distributions of EFGs leading to $\nu_Q = (625 \pm 130)$ MHz and $\nu_Q = (123 \pm 100)$ MHz with fractions of 19% and 81%, respectively. The solid line represents the best fit of this model to the data. Upon annealing to 550°C or higher for one hour at each temperature in a vacuum of 10^{-6}Torr the first contribution vanishes and the second one is shifted to (75 + 70) MHz with a fraction of about 90%. This result means that essentially all Hf atoms experience an EFG.

The question now is whether this observed EFG can be due to anything other than the symmetry of the host lattice atoms. One possible reason could be the trapping of oxygen by the hafnium, which would give rise to nonuniform EFGs. In view of the extremely low oxygen content of our samples (< 10 ppm) this possibility can be ruled out.

Also during the neutron activation process lattice damage can be introduced. This damage could perturb the cubic symmetry and thus produce an EFG. As mentioned above the high EFG component ν_Q^1 vanishes after the additional anneal treatment and could therefore be attributed to defects. There is little known up to now about the defect properties of this ordered alloy, in particular about the defect recovery. However, the broad distri-

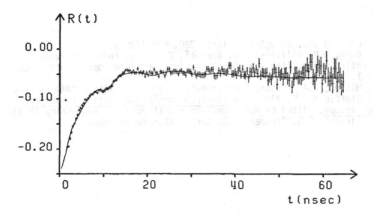

Fig. 1: Time dependent PAC spectrum of ^{181}Hf in Ni$_{76}$Al$_{23}$Hf$_1$. The solid
line represents the least squares fit to the data assuming two
Lorentzian distributions for the EFGs centered at ν_Q = 625 MHz and
ν_Q = 123 MHz, respectively.

bution of the major component ν_Q^2 remains unchanged. The origin of this
distribution is uncertain at the moment. A possible explanation could
be a minor deviation from the perfect order of the lattice. Further investigation along this line are in progress.

From all our experiments on both single crystal and polycrystalline
samples we conclude that the Hf atoms are incorporated into the Ni$_3$Al
lattice on a site which experiences an EFG, and this must be the Ni site.
The fraction of Hf atoms residing on sites without EFG is smaller than
10% within the accuracy of the PAC measurement.

Our result agrees both with the indications obtained from RBS/ion
channeling [2] and a recent atom probe study by Sieloff et al. [6]. It
is in disagreement, however with the results of electron channeling [3]
and atom probe experiment by Miller and Horton [4]. All experiments were
performed on samples with similar nominal composition. There is no obvious
reason for the contradictory results. However, we want to point out that
electron channeling and in particular the atom probe are techniques which
look primarily at the surface. In contrast, RBS/ion channeling looks
considerably deeper into the material and PAC sees all the Hf atoms to
a depth of several hundred microns. Clearly this does not account for
the different results of the two atom probe experiments. Obviously there
is more experimental information needed in order to clarify these conflicting
resuls.

ACKNOWLEDGEMENT

The authors wish to thank Dr. C.T. Liu of the Metals and Ceramics
Division of Oak Ridge National Laboratory for providing the samples.
One of us (H.G.B.) also gratefully acknowledges many stimulating discussions
with him.

REFERENCES

1. C.T. Liu and J.O. Stiegler, Science 226, 636 (1984).
2. H.G. Bohn, J.M. Williams, J.H. Barrett, C.T. Liu, this conference.
3. J. Bentley, Proc. XIth Int. Cong. on Electron Microscopy, Kyoto, 1986, p. 551.
4. M.K. Miller, J.A. Horton, Scripta Metall. 20 1125 (1986); this conference.
5. R.M. Steffen and K. Alder, in Electromagnetic Interaction in Nuclear Spectroscopy, edited by W.D. Hamilton (North Holland, Amsterdam, 1975).
6. D.D. Sieloff, S.S. Brenner, M.G. Burke, this conference.

LATTICE LOCATION OF BORON AND HAFNIUM DOPANTS IN AN ORDERED NICKEL ALUMINIDE BY USE OF ION CHANNELING/NUCLEAR REACTION ANALYSES[†]

H. G. BOHN*, J.M. WILLIAMS**, J.H. BARRETT**, AND C.T. LIU**
* Kernforschungsanlage, Juelich, D5170 Juelich, Federal Republic of Germany
** Oak Ridge National Laboratory, P.O. Box X, Oak Ridge, TN 37831

ABSTRACT

Experiments have made use of Rutherford backscattering (RBS), ion channeling techniques and analyses for nuclear reaction products to study lattice location of B and Hf dopants in an ordered nickel aluminide. Studies were of an alloy single crystal of composition $Ni_{76}Al_{23}Hf_1$ with about 0.1 at. % of B added. Analysis for B was accomplished by detection of alpha particles resulting from the reaction $^{11}B(p,\alpha)^8Be$, in ion channeling experiments in which RBS from the Ni constituent was used to control channeling in the host lattice. Yield of the reaction product from proton interaction with B decreased relative to random for channeling in a <100> direction, but increased relative to random for channeling in a <110> direction. It was concluded that B occupies primarily octahedral interstitial sites. RBS/channeling half-angles for Hf in a <100> direction are somewhat smaller than those for Ni, but nevertheless considerably larger than half-angles expected for Al. It is concluded that the majority of Hf atoms are on the Ni sublattice.

INTRODUCTION

The term "advanced aluminides" refers to ordered hypo-stoichiometric Ni_3Al alloys with interstitial boron dopant and with substitutional dopants such as hafnium. The boron greatly improves the fabricability [1–6], and the substitutional constituent further improves fabricability and strength [3,7,8]. The emergence of these alloys as practical high-temperature materials has prompted renewed interest in research that might lead to a better understanding of the relationship between structure and properties of such ordered alloys. Experiments of the present paper have made use of ion channeling techniques [9,10] in conjunction with RBS techniques [9] and analysis for a nuclear reaction product to study lattice locations of boron and hafnium dopants in an alloy of the nominal Ni_3Al composition. It is known that the mechanism whereby boron improves ductility involves segregation of some of the dopant to grain boundaries [4]. There is nevertheless interest in the lattice location of boron within the interiors of the crystals because of the role that the constituent plays in solid solution strengthening. In addition, lattice location within grains and the closely related issues of diffusion mechanisms and diffusion rates to grain boundaries are part of the total behavior of B. Boron is a small atom, but it expands the lattice when it enters solution [6]; therefore the dopant was presumed to be interstitial at the outset of these experiments. The purpose of the experiment was to determine whether the boron occupies primarily the octahedral or tetrahedral interstitial sites. Because of its low mass, boron is difficult to detect by the RBS technique [9]. Therefore, to identify the boron, the experiment was designed to analyze for the α-particle emitted upon occurrence of the reaction, $^{11}(p,\alpha)^8Be$, in a technique pioneered by Anderson and co-workers [11]. In the channeling

[†]Research sponsored by the Division of Materials Sciences, U.S. Department of Energy under contract DE-AC05-840R21400 with Martin Marietta Energy Systems, Inc.

technique the nuclear reaction is more probable if the target B-atom is in a channel which is otherwise open to incoming channeled protons than it is if the target atom is in line with a row of lattice atoms. As will be seen, this fact enables the octahedral and tetrahedral positions to be distinguished by use of selected channeling directions. The principal question regarding location of the hafnium is whether the hafnium prefers the nickel or the aluminum sublattice. Hafnium in Ni-Al is easily resolved by the RBS technique. Identification of sublattice depends on comparing the angular dependency of the yield of backscattered He-ions from hafnium with the angular dependencies for the yields of host atoms in RBS/ion channeling experiments.

2. EXPERIMENTAL PROCEDURES

All data were obtained for a single-crystal sample of composition $Ni_{76}Al_{23}Hf_1$, which was also doped with B to a concentration of 0.1 at%. The sample was part of a crystal that was provided by the Garret Turbine Engine Company. The crystal was given a homogenizing anneal for 16 h at 1100°C. Then the surface was polished mechanically and electrochemically. It was determined by Laue X-ray analysis that the plane of the disc was 6° off of a <100> axis of the the fcc lattice of this $L1_2$ structure. The Laue spots were somewhat broadened, suggestive of a mosaic spread of about 0.5 to 1°.

The experiments made use of a typical arrangement in backscattering spectrometry [9]. Ions backscattered from the sample, or otherwise produced by reaction in the sample, enter the ion detector at an angle of about 160 degrees with respect to the direction of the beam incident from an accelerator. The sample was mounted on a two-axis goniometer to facilitate alignment of the desired crystal axis with the beam direction. Incident beams of H or He ions were produced by an accelerator of 2.5 MV maximum voltage. Beams had a diameter of 1mm and a divergence of < 0.04 deg as they entered the sample. The silicon surface-barrier detector had an acceptance angle of approximately 2 deg. Data were acquired by use of conventional PHA type of instrumentation.

For the reaction, $^{11}B(p,\alpha)^8Be$, the cross section as a function of energy is peaked at about 620 keV with a FWHM of 300 keV [12]. The incident particle energy was 750 keV. From stopping power information or range-versus-energy data [13], it can be estimated that approximately 75% of the reactions occur within a 2-μm depth of the surface. The Q-value of the reaction is 8.6 MeV, and α particles emitted in the backward direction have energies of approximately 3.7 MeV. This energy is considerably larger than that of the protons which are backscattered from the Ni (0.70 MeV). Nevertheless, the yield of the protons is so much greater than that of the α particles that the detection system would be overwhelmed by background due to coincidence counts in the energy range of interest for the α particles, if most of the backscattered protons could not be excluded. Therefore a filter of Al foil was placed over the particle detector for the experiments involving α-particle detection. The foil thickness of 1.34 mg/cm² (about 6 μm) was designed to let most of the α particles through while rejecting most of the protons.

3. RESULTS

Figure 1 shows RBS spectra for the alloy for the conditions stated. The curve for the rotating random case was obtained by continuously manipulating the crystal during data acquisition so as to present a variety of orientations to the incoming beam. This procedure is intended to produce a spectrum that is more or less representative of amorphous or randomly oriented material. The ledge for the 1 part of Hf (1.83 MeV) is larger than that for the 24 parts of Al (1.12 MeV). This result is because of the Z^2 dependency of the RBS cross section. This small yield for Al, together with the fact that what signal there is is on top of a large background due to Ni, precludes quantitative analyses involving Al by RBS. Because of the small mass and low concentration, the detectability for B is much worse still. Hence the need for nuclear reaction analysis for B. Alignment of the crystal in a <100> channeling direction produced the expected reduction in yield. For Ni the minimum yield, x_{min}, was 0.11. The minimum yield is the yield for the aligned case divided by that for the random case for an energy window that corresponds to scattering events near the surface of the sample (for example 1.4 to 1.5 MeV for Ni for the data of Fig. 1). The fact that the yield for the Hf is approximately proportional to that for the Ni in Fig 1, together with the fact that similar results were obtained for alignment in a <110> direction, means that the Hf is substitutional as expected. This comparison will be presented in more detail later. Aluminum is not detectable for the aligned case.

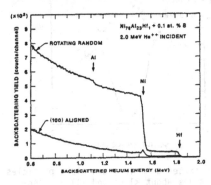

Figure 1. Yield of backscattered He ions incident on a crystal of composition $Ni_{75}Al_{23}Hf_1$. Spectra are shown for the crystal in the "random" and aligned orientations.

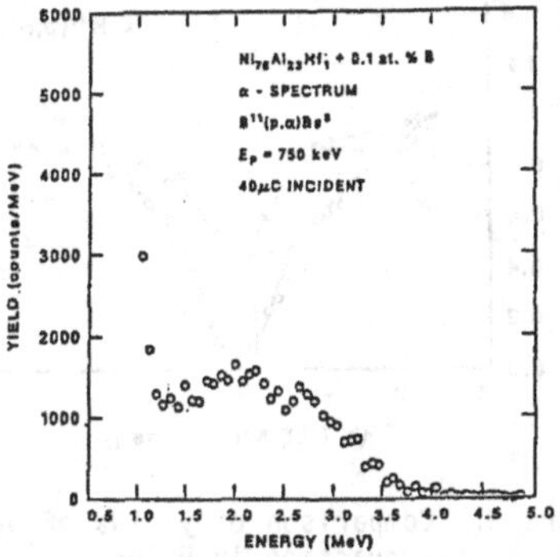

Figure 2. Yield versus energy for the case of 750 keV protons incident and the Al filter in place.

Figure 2 shows yield versus energy for the α-particle detection set-up with protons incident. Above about 1.3 MeV the counts represent α particles that originate at various depths in the sample and then penetrate the filter. The breadth of the peak is due in part to energy straggling incurred over the substantial ranges that the particles have penetrated. The increase below 1.3 MeV represents counts that are somehow attributable to backscattered protons that penetrate the filter (see below). Either the counts are due to protons that straggle through, a possibility that cannot be ruled out from the range calculations used in the filter design, or the filter might have had micropores. Clearly counts at indicated energies of greater than 0.7 MeV are due to pileup events involving protons that came through. These data are quite similar to those of Anderson and coworkers who used a 12-μm mylar film for a filter [11]. To determine the α-particle yield from the B, counts at energies above 1.5 MeV were accepted.

Figure 3 shows angular scans in which normalized yields for 2.0-MeV He ions scattered from the Ni lattice (see Fig. 1) and yields for α particles from the (p,α) reaction are plotted versus tilt angle from a <100> direction and from a <110> direction. As is indicated by the schematic representations above the graphs, octahedral interstitial sites are aligned with rows of lattice atoms for projections in <100> directions. Neither octahedral nor tetrahedral sites are blocked by rows of lattice atoms for projections in <110> directions. The fact that the B yield increases relative to the random yield for alignment in a <110> direction means that the B is interstitial. The reduction in yield coincident with that of the lattice for a <100> scan means that the B occupies mostly octahedral sites.

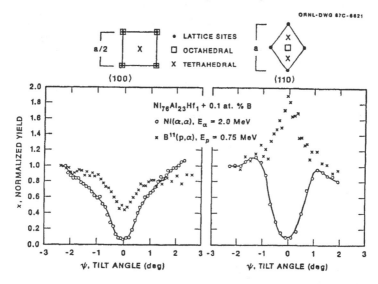

Figure 3. Comparison of yields of backscattered He ions and alpha particles from the p,α reaction in B for angular scans about <100> and <110> directions in the nickel aluminide crystal of the composition indicated. The projections above the graphs illustrate the positions of the interstitial sites relative to the lattice sites. For the <100> projection the tetrahedral sites are exposed in a quadrant of the cube of this Cu_3Au structure, but the octahedral sites line up with rows of lattice atoms. For the <110> projection both types of interstitial sites are visible.

The minimum yield for the B is greater than that for the lattice. We have calculated that at least half of this difference is due to the fact that the proton beam samples B atoms over a very large, but imprecisely known, depth range, whereas lattice yield data were selected from a region of approximately 100 nm deep near the surface of the crystal. The difference is caused by the larger amount of dechanneling associated with the larger penetration depth for the proton beam. Another factor which could contribute to the difference, and whose magnitude is not precisely known, is that the smaller B atoms have larger vibrational amplitudes than the host atoms.

The basis for determination of sublattice occupancy (Ni or Al) for substitutional impurities such as Hf is that dechanneling (and resulting backscattering) of rows of atoms is a result of collective action of the row, and not the property of any given atom [9]. The best possibility for detecting the difference between Ni and Al sublattice occupancy for the present Cu_3Au type of structure is for channeling in a <100> direction. In this direction there are three rows of pure Ni for every one row of pure Al. If Hf occupies mainly Ni sites, the yield versus angle curve for Hf should resemble that for Ni and likewise for the Al sublattice. Unfortunately, as was shown, Al cannot be satisfactorily resolved by Rutherford scattering. Figure 4 shows normalized RBS yields for Hf and for Ni for an angular scan about a <100> direction. The extent of agreement must be evaluated with the aid of theory. The quantity $\psi_{1/2}$ (Fig. 4) is the angle at which the yield, X, is half way between 1 and X_{min}. Table 1 summarizes experimental (Fig.4) and calculated values [14] for $\psi_{1/2}$. Calculations for the Ni and Al rows assume that all of the Hf in the alloy is on that respective type of row, e.g., 1/24 of atoms on Al rows are Hf and 1/75 of atoms on Ni rows are Hf. This assumption does not have a large effect in comparison with calculations for pure rows, however.

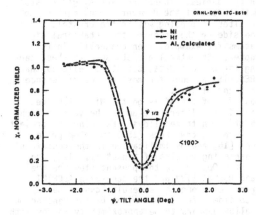

Figure 4. Comparison of RBS yields for Ni and Hf for a scan through a <100> channel for the nickel aluminide crystal.

TABLE I. Certain Calculated and Experimental Values of $\psi_{1/2}$ for the Substitutional Alloy Constituents for a <100> Channeling Analysis.

Alloy Constituent	Experimental $\psi_{1/2}$	Calculated $\psi_{1/2}$
Ni (+Hf)	0.72	0.73
Al (+Hf)	----	0.48
Hf	0.65	----

It is usual for experimental values of $\psi_{1/2}$ to be somewhat less than theoretical ones since most experimental factors such as beam divergence and crystal imperfections result in dechanneling over what would be expected theoretically. If an experimental value were accessible for Al, one would expect it to be less than the calculated value by an amount approximately proportional to the difference for Ni. From Table I it is seen that $\psi_{1/2}$ for Hf differs from that for Ni by about 0.06 deg, whereas the difference for Ni and Al should be about 0.25 deg. Thus, the Hf yield curve is apparently biased toward that of the Ni, so as to suggest that about 3/4 of the Hf is on the Ni sublattice.

No differences in $\psi_{1/2}$ for Ni and Hf were observed for channeling in a <110> direction. These results will not be presented in detail.

DISCUSSION

The most definite conclusion drawn from the experiments is that B occupies the octahedral sites. From space considerations in a hard sphere packing model for the fcc lattice, the octahedral site should be favored; thus the present result is not surprising. In polycrystals the B has a tendency to segregate to grain boundaries. Therefore, for single crystals it is reasonable to speculate that some of the B might be segregating to dislocations or other possible sites of binding within the crystal. If some B atoms were biased to one side or the other of the octahedral side, by the presence of nearby Hf atoms for example, then channeling in the <100> direction would be disrupted. Structural effects such as these cannot be ruled out as influencing the present data, but it is believed that the factors cited above (see RESULTS) account for most of the difference between channeling effects for B and Ni (Fig. 3).

We have inferred that about 75% of the Hf is on the Ni sublattice. There is disagreement at present in the literature among the various techniques on the lattice location of Hf in these alloys. Favoring the present results are the PAC results of Bohn and coworkers [15], the atom probe results of Brenner and coworkers [16] and by implication in the context of the present problem, the ion channeling results of Lin et al. [17], who actually studied Ta instead of Hf. Opposed to the present view are the atom probe results of Miller and Horton [18] and the zone-axis ALCHEMI results of Bentley [19]. Differences among these results cannot be discussed in detail here. As to side effects that might be influencing our own results, we note that the alloy appears to be approximately random with respect to available lattice site occupancy by the Hf. Then the question arises as to whether collision damage produced by the analyzing beam could be producing enough disorder in the structure to "randomize" lattice site occupancy. Again, space does not permit a detailed discussion, but we will state that we have made efforts to control and assess such damage effects. Measures include damage calculations and RBS/channeling measurements during and immediately after in-situ annealing. Furthermore, the results themselves tend to contradict the idea of beam-induced randomization of the alloy. The value of $\psi_{1/2}$ for Ni, in comparison with theoretical, seems to indicate that not much Ni is on Al sites. Also, such difference as there is between Ni and Hf indicates that lattice site occupancy is not totally random. We believe it to be unlikely that damage effects are influencing the present results. Still, until a yield curve for Al has been obtained (by nuclear reaction analysis, for example) some reservation regarding the final conclusion to be drawn from the channeling technique may be appropriate.

ACKNOWLEDGMENTS

The authors are indebted to W. Schilling for first suggesting the study of the boron and to B. R. Appleton for helpful discussions.

REFERENCES

1. K. Aoki and O. Izumi, Nippon Kinzoku Takkaishi 43, 1190 (1979).

2. C.T. Liu and C.C. Koch, in Proceedings of a Public Workshop on Trends in Critical Materials Requirements for Steels of the Future: Conservation and Substitution Technology for Chromium (NBSIR-83-2679-2, National Bureau of Standards, Washington, D.C., June 1983).

3. C.T. Liu, C.L. White, C.C. Koch, E.H. Lee, in High Temperature Materials Chemistry—II, L.A. Munir and D. Cubicciotti, Eds. (Electrochemical Society, Pennington, N.J., 1983), vol. 83, No. 7, p. 32.

4. C.T. Liu, C.L. White, and J.A. Horton, Acta. Met. 33, 213 (1985).

5. A.I. Taub, S.C. Huang, K.M. Chang, Metall. Trans. A 15A, 399 (1984).

6. S.C. Huang, A.I. Taub, and K.M. Chang, Acta. Met. 32, 1703 (1984).

7. C.T. Liu and J.O. Stiegler, Science 226, 636 (1984).

8. C.T. Liu, "Design of Ordered Intermetallic Alloys for High Temperature Structural Use," p. 289 in High Temperature Alloys: Theory and Design, ed. by J. O. Stiegler, AIME publications, 1985.

9. Wei-Kan Chu, James W. Mayer, and Marc-A. Nicolet, Backscattering Spectrometry (Academic Press, New York, 1978).

10. Leonard C. Feldman, James W. Mayer, and S. Thomas Picraux, Materials Analysis by Ion Channeling (Academic Press, New York, 1982).

11. J.U. Anderson, E. Laegsgaard, and L.C. Feldman, Rad. Eff. 12, 219 (1972).

12. L.C. Feldman and S.T. Picraux, p. 109 in Ion Beam Handbook for Material Analysis, ed. by J. W. Mayer and E. Rimini (Academic Press, New York, 1977).

13. H.H. Anderson and J.F. Ziegler, Hydrogen Stopping Powers and Ranges in All Elements, Vol. 3 (Pergamon Press, New York, 1977).

14. J.H. Barrett, Phys. Rev. B3, 1527 (1971), see Eq. 14.

15. H.G. Bohn, R. Schumacher and R.J. Vianden (this volume).

16. S.S. Brenner, D. Sieloff, and M.G. Burke, J. Phys. (Paris) Colloq. C2, Suppl. 3 47, C2-215 (1986).

17. H. Lin, L.E. Seiberling, P.F. Lyman, and D.P. Pope (this volume).

18. M.K. Miller and J.A. Horton (this volume).

19. J. Bentley, in Proceedings of the 44th Annual Meeting of the Electron Microscope Society of America, Albuquerque, New Mexico, August 1986, ed. by G. W. Bailey, San Francisco Press, San Francisco, 1986, p. 704.

KINETICS AND MECHANISMS OF FORMATION OF AL-NI INTERMETALLICS

J. A. Patchett and G. J. Abbaschian
Department of Materials Science and Engineering
University of Florida, Gainesville, Florida 32611

ABSTRACT

The effects of cooling rate and composition on the nucleation and growth kinetics of NiAl, Ni_2Al_3, and $NiAl_3$ intermetallics were studied for alloys containing 25 and 31.5 at.% Ni. For the former composition, the peritectic reaction Ni_2Al_3 + Liquid → $NiAl_3$ was studied over cooling rates from 20 to 10^5 K/s. For the latter composition, the reaction NiAl + Liquid → Ni_2Al_3 was studied at cooling rates ranging form 10 to 600 K/sec. The amounts of constituent phases are shown to depend on the cooling rate, and for the peritectic phases, on the surface area of the primary phase. The nucleation of Ni_2Al_3 and $NiAl_3$, and the ordering of the aluminum-rich NiAl were also examined using transmission electron microscopy. Cooling the NiAl + liquid below the peritectic temperature results in a metastable extension of NiAl with a high dislocation and stacking fault density, followed by the epitaxial nucleation and almost dislocation-free growth of Ni_2Al_3. In contrast, the nucleation of the $NiAl_3$ on Ni_2Al_3 occurs directly without the formation of an intermediate region.

INTRODUCTION

Many intermetallic compounds of commercial interest form by a peritectic reaction during solidification; yet relatively little is known about the kinetics of the reaction. This is in part due to the complexity of the peritectic reaction compared with a eutectic reaction; a peritectic reaction involves dissolution and growth, whereas a eutectic reaction involves only growth. Another difficulty in studying peritectic reactions is that the nucleation behavior of the peritectic phase is affected not only by the thermal conditions but also by the amount, morphology, and nucleation potency of the primary (properitectic) phase. If the peritectic phase is not easily nucleated on the primary phase, the latter may continue to grow and suppress the peritectic reaction to lower temperatures. If, on the other hand, growth of the primary phase is incomplete, which might be the case during rapid solidification or solidification of facet forming intermetallics, the nucleation of the second phase could occur at a composition and a temperature different than the equilibrium.

A study was undertaken to determine the controlling mechanism(s) during peritectic reactions on the aluminum-rich side of the Al-Ni binary. The variables chosen included the cooling rate, both above and below the peritectic transformation temperature (T_p), and the surface area of the primary phase.

EXPERIMENTAL PROCEDURE

Samples with nominal compositions of 25 and 31.5 at.% Ni, each weighing about one gram, were prepared from 99.9999% pure Ni and 99.999% pure Al by first arc melting, then melting in a levitation melting apparatus. Different cooling rates and surface areas were achieved by varying the quenching medium and the time of isothermal holding above T_p. For the latter technique, the samples were held at about 40 K above the peritectic temperature for a period up to 7 minutes, then quenched through the peritectic transformation. A more complete description of the experimental procedure can be found elsewhere [1].

The microstructures of the as-quenched samples were analyzed using optical microscopy, scanning electron microscopy (SEM), and transmission electron microscopy (TEM). For 25 a/o Ni samples, the volume fractions of Ni_2Al_3, $NiAl_3$, and $NiAl_3$ + Al eutectic were determined by an image analyzer and also by using a quantitative X-ray diffraction technique of the powdered samples [2]. The surface area of a primary phase was measured using the standard techniques of stereology [3].

RESULTS

A typical microstructure of a 25 a/o Ni sample solidified at a rate of 20 K/s using helium gas is illustrated in Figure 1a, where three phases can be identified; the light colored Ni_2Al_3 primary phase is surrounded by the faceted $NiAl_3$, followed by the $NiAl_3$ + Al eutectic. The effect of cooling rate on the microstructure can be seen by comparing Figures 1a and 1b. The latter figure is for a sample of the same composition, except solidified at 10^5 K/s. It can be seen that the faceted $NiAl_3$ forms peritectically on Ni_2Al_3 at both cooling rates and there is no evidence of independent nucleation of $NiAl_3$. However, the higher cooling rate reduces the amount of the primary phase with a corresponding increase in the amount of the peritectic phase. The volume fractions of the constituent phases at cooling rates of 20 K/s, 5×10^2, 5×10^3, 2×10^4 and 10^5 K/s as determined by the image analysis are summarized in Figure 2. As the cooling rate increases, the reduction in the volume fraction of the primary phase is balanced by the increase in the secondary phase, and the amount of eutectic remains approximately constant.

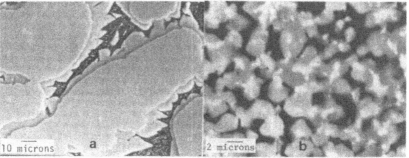

Figure 1   Microstructure of Al-25 a/o Ni; (a) solidified at 20 K/s after isothermal holding, (b) splat cooled at 10^5 K/s.

Detailed examination of the Ni_2Al_3-$NiAl_3$-eutectic junction, as shown in Figures 3a and 3b for samples quenched at 6000 K/s and 20 K/s, respectively, showed the presence of recessions in the primary phase at the junction. The recessions are due to the dissolution of the primary phase during the peritectic reaction. Note also that $NiAl_3$ cells have not grown together to form a continuous layer, indicating that the primary phase liquid contact had persisted until the solidification of the eutectic. Comparing Figures 3a and 3b shows that increasing the cooling rate has increased the number of $NiAl_3$ cells on the Ni_2Al_3 but the width of intercellular regions has remained approximately constant. A TEM image of the Ni_2Al_3/$NiAl_3$ boundary is illustrated in Figure 4, where the arrow shows the dissolution area.

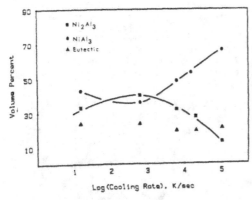

Figure 2
Volume percent of Ni$_2$Al$_3$, NiAl$_3$ and eutectic versus cooling rate for 25 a/o Ni alloys.

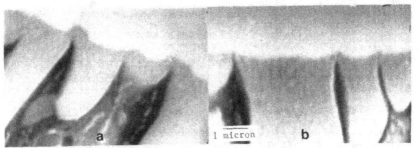

Figure 3 Ni$_2$Al$_3$, NiAl$_3$ and eutectic region for samples isothermally held then quenched at (a) 6000 K/s, and (b) 20 K/s.

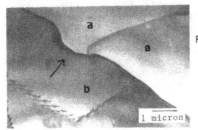

Figure 4 Transmission electron micrograph of the interface region of NiAl$_3$ (a) and Ni$_2$Al$_3$ (b). The arrow indicates the region of dissolution of the primary phase.

The effect of isothermal holding above the peritectic temperature on the microstructure is shown in Figures 5a and 5b. The holding causes the primary Ni$_2$Al$_3$ phase particles to coarsen and become rounded, and eventually coalesce, resulting in a reduction of the primary phase surface area. For example, the surface area per unit volume of the sample decreased from 0.068 to 0.047 $\mu m^2/\mu m^3$ as the holding time was increased from 1/2 to 5 minutes. It should be noted that the isothermal holding not only produces different surface area, but also increases the volume fraction of the primary phase, as the system tends toward equilibrium at the holding temperature. The effect of the surface area on the amount of the peritectic phase is shown in

Figure 6, where the normalized volume fraction of the peritectic phase is plotted versus the surface area of the primary phase for samples quenched after isothermal holding above T_p and for those quenched directly from the liquid at 6000 K/s . Two regions can be described from this figure; at surface areas less than 0.07 $\mu m^2/\mu m^3$, a small change in the surface area produces a very drastic increase in the amount of second phase material while at higher surface areas, the normalized amount of the peritectic phase becomes almost independent of the surface area.

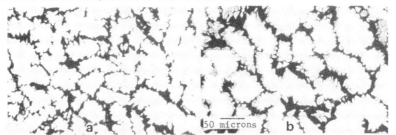

Figure 5   Effect of isothermal holding time on samples of 25 a/o Ni held for (a) one minute, and (b) 4 minutes at 870 °C, then cooled at 20 K/s.

The microstructure of an etched 31.5 a/o alloy isothermally held at 1170 °C then solidified at 6000 K/s is shown in Figure 7. The center of the dendrite is NiAl, which is surrounded by a thin layer of Ni_2Al_3, approximately 2-3 microns thick. This phase is followed by $NiAl_3$ and finally the $NiAl_3$ + Al eutectic. The NiAl phase consists of two regions; the central part which existed at the holding temperature, and the surrounding rim which formed upon quenching of the sample through the peritectic temperature. The combined volume fraction of Ni_2Al_3 + NiAl for 31.5 a/o Ni samples is plotted as a function of the cooling rate in Figure 8, which also shows the volume fraction of Ni_2Al_3 for 25 a/o samples.

TEM examination of the isothermally held alloy also revealed marked differences between the two regions of NiAl. The central portion of the primary dendrites consisted of highly ordered NiAl with a few stacking faults and dislocations. In contrast with the first region, the surrounding layer of NiAl, which formed during quenching of the sample through the peritectic temperature, contained numerous stacking faults and dislocations, as shown in Figure 9. The stacking faults were determined to lie on the cube planes (100) of the NiAl. The Ni_2Al_3 phase, which contained few dislocations or stacking faults, can be seen in the lower portion of the figure.

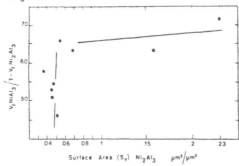

Figure 6
Normalized volume fraction of $NiAl_3$ vs. the surface area for samples quenched from the liquid or after isothermal holding.

rates. Figure 8 shows that for alloys of 25 a/o Ni, the volume fraction of the primary phase passes through a maximum as the cooling rate is increased from 20 to 6000 K/s. At slow cooling rates the amount of Ni_2Al_3 increases because there is ample opportunity for the primary phase to nucleate and grow, but the dissolution of the primary phase also increases which would reduce its amount. The amount of the primary phase dissolved is dependent not only on the time but also on the surface area. The time available for dissolution is governed by the cooling rate and the temperature interval lbetween the peritectic phase nucleation temperature and the eutectic reaction. As shown in Figure 3, dissolution occurs at the junction between the primary phase and the peritectic cells. Thus, for a given initial primary phase surface area, increasing the nucleation rate of the peritectic phase (i.e., increasing the number of cells per unit area) increases the fraction of the interface occupied by the intercellular channels. This is due to the competitive effect of the processes outlined above on the amount of the constituent phases.

As the cooling rate increases, the undercooling prior to nucleation of the peritectic increases. The nucleation rate, which exponentially increases with undercooling, increases the number of nuclei on the surface of the primary phase. The increase in nuclei effectively increases the number of intercellular junctions on the interface, thus increasing the area available for dissolution. Competition between the time and area available can explain the observed variations in volume fraction of the peritectic phase with respect to cooling rate and the surface area of the primary phase.

At cooling rates larger than about 10^3 K/s, the primary phase dissolution seems to play a minor role in the final microstructure. The volume fraction of the primary phase can therefore be estimated from the nucleation and growth rate of the primary, and nucleation temperature of the peritectic phase. Figure 2 shows that the volume fraction of the primary phase decreases by $\varepsilon^{.22}$ at cooling rates above 6×10^2 K/s. Assuming the nucleation temperature of Ni_2Al_3 and $NiAl_3$ to be constant, the volume fraction of the primary phase is then proportional to $t^{1.5}$, where t is time the sample is between the liquidus and the nucleation temperature of $NiAl_3$. This indicates the volume fraction is governed by an Avrami type equation for nucleation and growth of $NiAl_3$.

For alloys of 31.5 a/o, the close similarity in composition and crystal structure between $NiAl$ and Ni_2Al_3 precluded a quantitative analysis of the peritectic reaction involving these two phases. However, TEM results showed that when $NiAl+liquid$ is quenched from the peritectic temperature the solid phase continues to grow by the metastable extension of the phase boundaries below the peritectic temperature until Ni_2Al_3 nucleates. The extended region contained numerous dislocations and stacking faults along the cube faces of $NiAl$. Furthermore, since the formation of Ni_2Al_3 requires an ordering of the nickel vacancies along a cube diagonal of the $NiAl$ lattice, it does not occur directly from the $NiAl$, but requires nucleation.

CONCLUSIONS

In alloys of 25 and 31.5 a/o Ni, as the cooling rate increases, the volume fraction of the primary phase goes through a maximum around 600 K/s. For cooling rates above this value, the decrease in volume fraction of the primary phase was attributed to a reduction in the time available for nucleation and growth of the primary phase prior to the nucleation of the peritectic phase. At the lower cooling rates, the reduction in volume fraction of the primary phase is due to dissolution. The amount of primary phase dissolved is a function of the time and surface available.

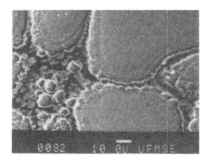

Figure 7
The microstructure of a 31.5
a/o Ni sample held at 1170 °C
then quenched at 6000 K/s

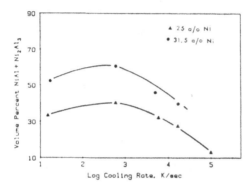

Figure 8
The volume percent
Ni_2Al_3 for 25 a/o Ni and
Ni_2Al_3 + NiAl for 31.5
a/o Ni samples versus
the cooling rate.

DISCUSSION

The volume fraction of the constituent phases during a peritectic reaction is governed by four factors: (1) nucleation and growth kinetics of the primary phase, (2) nucleation and growth kinetics of the

Figure 9   TEM image of the region
between NiAl and Ni_2Al_3 in
a 31.5 a/o Ni sample.

peritectic phase at or below T_p, (3) dissolution of the primary phase by the peritectic reaction, and (4) solid state diffusion between the primary and peritectic phase. The latter, although important in some cases, is not expected to play a major role in the present study because of the cooling

ACKNOWLEDGEMENTS

The support of this research by the Office of Naval Research (Contract No. N0014-81-K-0730) is gratefully acknowledged. The Scientific Officer of the program was Dr. Bruce A. MacDonald.

REFERENCES

1. G. J. Abbaschian, J. A. Patchett, R. Russell, S. P. Abeln, and R. Schmees, Technical Report #4, October 1985, Office of Naval Research Contract N00K1-81-K-0730-Nr031-836.
2. J. A. Sarreal and G. J. Abbaschian, Met. Trans. A <u>17A</u>, 2063 (1986).
3. E. E. Underwood, in <u>Quantitative Microscopy</u>, edited by R. T. DeHoff and F. N. Rhines (McGraw Hill, New York, 1968) Chapter 4.

MICROSTRUCTURES OF RAPIDLY SOLIDIFIED
Ti-15Al-11Mo ALLOY

A. G. JACKSON*, K. TEAL** and F. H. FROES**
*Systems Research Laboratories, Inc., 2800 Indian Ripple Rd., Dayton, OH 45440
**AFWAL Materials Laboratory, Wright-Patterson AFB, OH 45433

ABSTRACT

The microstructural development in a Ti_3Al alloy modified with an addition of 11 w/o Mo has been studied. Under equilibrium conditions (arc-melted) a two phase ordered hexagonal and ordered bcc structure is present. Evidence of omega phase was also observed. By rapid solidification the hexagonal structure is suppressed, which contrasts with the behavior of a Ti_3Al alloy modified with a similar amount of Nb. The mechanism by which the phases present form is discussed in terms of the atomic movements required.

INTRODUCTION

Titanium aluminide alloys are receiving attention because of the potential for high temperature service [1]. Since the alloys are ordered, however, there are difficulties in obtaining the ductilities of a few percent elongation that designers are comfortable with [2]. Additions of beta stabilizing elements to the base alloy improve the ductility somewhat particularly when added in sufficient quantity to move into a two phase hexagonal plus body centered cubic phase field [3]. The presence of the body centered cubic phase allows an improvement in ductility because of the inherently better ductility of this crystal structure in comparison with the hexagonal structure. Even at levels below which the beta phase is stable, the addition of beta stabilizers appears to enhance ductility.

The purpose of the present work was to evaluate the effect of one such beta stabilizer, Mo, on the behavior of a Ti_3Al (α_2) base alloy. This work is part of a much broader program at the authors' laboratory to understand the behavior of the Ti_3Al system, thus allowing development of improved alloys which exhibit a combination of good high temperature creep behavior and ductility levels at ambient temperatures in the 3-5% elongation range. Specific objectives of the present work were to define how the Mo addition affects the microstructural development, particularly under rapid solidification conditions.

EXPERIMENTAL

The alloys used in this work were produced by arc-melting in argon atmosphere with a triple pass, followed by homogenization at 1000°C for 7 days. The rapidly solidified alloy was produced from the arc-melted material by conversion using the vacuum pendant drop melt extraction technique. The transmission microscopy was done using a JEOL 100CX and a 2000FX microscope with Tracor-Northern energy dispersive spectrometer and Gatan energy loss spectrometer.

RESULTS

A TEM image of the microstructure of the arc-melted alloy is shown in Fig. 1. A lenticular phase is observed, the size of which varies in width from 0.5 to 1 micron and varies in length.

In Fig. 2(a) is a TEM image of a grain in the arc-melted alloy, while in Fig. 2(b) a diffraction pattern from the grain is shown, corresponding to a [0001] hcp zone and with planar spacings corresponding to those of α_2. Although the phase is ordered, no antiphase boundaries were observed, suggesting that during the homogenization heat treatment the antiphase domain size has grown to that of the grain.

Heat treatment of the alloy at 660°C for 20 min. followed by a quench, was done in-situ in the microscope to determine whether any decomposition to a two phase structure would occur and as a measure of the effect of the homogenization treatment on the alloy. Figure 3(a) displays a TEM bright field of alloy showing details of the plates and the matrix. A diffraction pattern, [Fig. 3(b)], from a lenticular plate is identified as HCP $[10\text{-}12]\alpha_2$ with a = 0.58 and c = 0.465 nm.

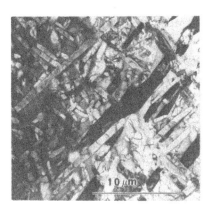

Figure 1.   TEM Bright Field Image of Arc-Melted Ti-15Al-1Mo Alloy.

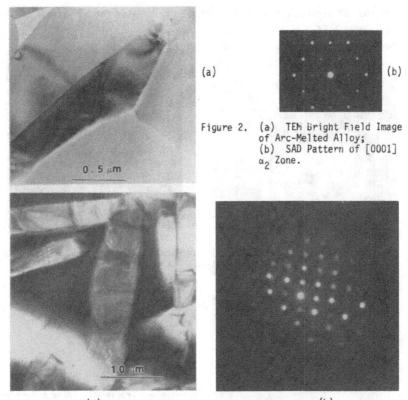

(a)

(b)

Figure 2. (a) TEM Bright Field Image
of Arc-Melted Alloy;
(b) SAD Pattern of [0001]
α_2 Zone.

(a)   (b)
Figure 3. (a) TEM Bright Field Image of α_2 Plate and Matrix; (b)
Convergent Beam Pattern from [10$\bar{1}$2] Zone of α_2.

Figures 4(a) and 4(b) show a TEM image of a plate in the arc-melted
alloy and a diffraction pattern from the plate corresponding to a BCC
structure with a = 0.32 nm. Superlattice spots are present appropriate to a
B2 type lattice oriented along a [012] zone.

The observed microstructures of the melt spun ribbon, as-quenched, displayed microstructures considerably different from the arc-melted alloy.
The ribbon thickness is 55-75 microns, implying a cooling rate in the range
of 7 to $10 \cdot 10^6 °C/sec$ [4].

Examination of the melt spun ribbon, as-quenched, using TEM showed the
presence of a tweed microstructure, antiphase boundaries and featureless
grains, as shown in Fig. 5. In Fig. 6(a), a bright field TEM image using
(200) and (100) beams, and in Fig. 6(b), a dark field image reveal the antiphase boundaries. Figure 6(c) is a diffraction pattern with (200) matrix

and (100) B2 structure beams. Streaking which may be due to strain or the presence of ω [5], is also present.

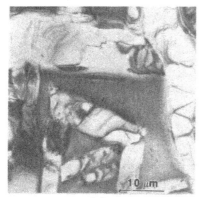

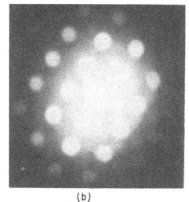

(a)   (b)
Figure 4. (a) TEM Bright-Field Image of α₂ Plates and Matrix; (b) Convergent Beam Pattern from Matrix, Along [012] Zone of B2.

Figure 5. TEM Bright Field from As-Quenched Ribbon Showing Triple Point and Structure at Grain Boundaries.

Figure 7 (a), (b), and (c) contains diffraction patterns from the as-quenched ribbon along the three major bcc zones. Figure 7(a) is [001] BCC and corresponds to B2 with a = 0.31 ± 0.01 nm. Streaking is present along <110>{110}. The source of the streaking may be strain or may be due to imperfectly formed ω, then the orientation of ω/B2 is [1120]/[110], but the lattice parameters are a = 2a_ω and c = 2c_ω. Figure 7(b) is a SAD pattern from <111> B2 which shows streaks along <110> direction and perpendicular to {111}. Thus the streaking appears to be associated with strains in <1120> and <1010> directions in the ω phase HCP structure.

147

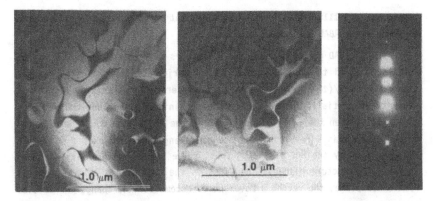

(a)   (b)   (c)

Figure 6. TEM Bright Field (a) and Dark Field (b) Showing Antiphase Domain Boundaries; (c) SAD Pattern with Superlattice Spots Present used for Dark Field in (b).

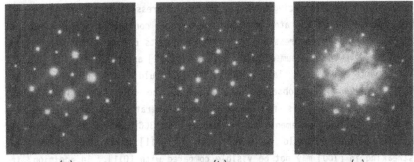

(a)   (b)   (c)

Figure 7. SAD Patterns for (a) [001], (b) [111] and (c) [110] B2 zones Showing Streaking Along (110), (112), (111) Planes.

DISCUSSION

The arc-melted alloy shows a two phase microstructure consisting of lenticular plates of α_2, and a B2 matrix. Energy dispersive spectroscopy data shows the matrix to be richer in Mo than the plates, which is consistent with the B2 structure as determined by selected area diffraction. No α_2 was found. A tweed microstructure was also observed which either arises from strains along {110}[110] or arises from partially formed ω [5].

In disordered Ti alloys ω can form via α to β to ω transformation or from β directly, [6] with orientation relationship [0001]//<111>, {1120}//{110}, and with lattice parameters given by $a_\omega = \sqrt{2}a_\beta$ and $c_\omega = \sqrt{3}a_\beta/2$. The suggested mechanism for ω formation [6-8] is a longitudinal fluctuation of atom positions along <111> with a wave vector of $2/3<111>2\pi/a_\beta$, leading to consolidation of the BC planes sequence to produce the ABAB HCP stacking.

From the SAD patterns presented, ω appears to present. Figure 7 shows streaks parallel to (110) planes and in 4 variants. Assuming the orientation is $(1\bar{1}20)//(110)$, the streaks may be interpreted as arising from the 4 variants of partially formed ω. If the mechanism is the longitudinal wave along <111>, then the (110) planes lying in the <111> zone would be expected to be strained as a result of the change in angle between axes from 90° to 120° required by the HCP lattice. Thus, there are position shifts of atoms in the BCC structure planes. The first is the contraction of positions to form a new plane, the second a movement to accomodate the change in lattice axes.

Such a movement of atoms in {110} is consistent with the longitudinal wave. A compression of the $[\bar{0}11]$ direction occurs as a result of the <111> compression on alternating layers, if this compression is viewed along the [011] zone, and if the atom position movement is considered in terms of the components of the compression, i.e., the movements of two atoms along $[\bar{1}11]$ and $[\bar{1}11]$ viewed as a sum of $[\bar{0}11] + [\bar{1}00] = [\bar{1}\bar{1}1]$ and $[0\bar{1}1] + [100] = [11\bar{1}]$ to produce a new atom layer. Thus strain should also be present on the {100} planes and be observed as streaks in the <100> directions. Such streaking, however, is not observed. Consideration of the components required for atom movements to produce ω from BCC lattice indicates that strain in [100] should be 0.707 that in $[0\bar{1}1]$ direction. Hence the streaking in [100] may not be visible compared with $[0\bar{1}1]$. In addition, if atom movement is along $[1\bar{1}1]$ during initial formation of ω, as seems reasonable, strain in [100] should be small, and streaking in partially formed ω would not be observable.

The rapidly solidified alloy contains ordered beta phase as ordered B2 with partially formed ω phase also present. Hence the rapid cooling rate, in excess of 10^5°C/sec, is insufficient to suppress the ordering reaction or formation of ω. In Ti-Al-Nb rapidly solidified alloys [9] of a similar composition (25 a/o Al-5a/o Nb) α_2 is observed. Comparison of the Ti-Mo and Ti-Nb binary phase diagrams [10] shows that for the equilibrium case, the $\beta/\alpha+\beta$ transus occurs at about 780°C for the compositions used. Therefore both Mo and Nb are expected to be approximately equal β stabilizers, however since no α_2 is detected in the present work it appears that Mo is more potent in suppressing α_2. Thermal treatment of the RS alloy is currently being studied to determine if α_2 will form.

ACKNOWLEDGEMENTS

The production of the alloys was done by J. Paine; foils were prepared by R. Omlor and S. Apt; arc-melts were prepared by R. Sweeney. This research was supported in part by AFWAL Materials Laboratory contract F33615-83-C-5079 and F33615-86-C-5013. Helpful discussions with C. Ward are gratefully acknowledged.

The authors gratefully acknowledge the opportunity to study a prepublication manuscript of the work by Strychor, Williams and Soffa [5].

REFERENCES

1. D. Eylon, P. J. Postans, S. Fujishiro and F. H. Froes, JOM, Nov. 1984, pp. 55-62.

2. H. A. Lipsitt, MRS, 1984.

3. C. Ward, M. Kaufman, R. G. Rowe, T. F. Broderick, A. G. Jackson and F. H. Froes, work in progress.

4. T. F. Broderick, A. G. Jackson, H. Jones and F. H. Froes, Met. Trans, 16A, 1951 (1985).

5. R. Strychor, J. C. Williams and W A. Soffa, accepted for publication in Met. Trans., 1987.

6. S. K. Sikka, Y. K. Vohra and R. Chidambaram, "Omega Phase in Materials" Prog. in Mat. Sci., 27, 245-310 (1982).

7. A. Dubertret and M. Fayard, "Ordering Phenomena in Omega Phases," in Titanium and Titanium Alloys, Volume 2, Eds: J. C. Williams and A. F. Belov, Plenum Press, New York, 1982, p. 1403.

8. T. W. Duerig, G. T. Terlinde, and J. C. Williams, "The Omega-Phase Reaction in Titanium Alloys," in Titanium '80, Volume 2, Eds: H. Kimura and O. Izumi, The Metallurgical Society of AIME, Warrendale, PA, 1980, p. 1299.

9. C. Ward, T. Broderick, A. G. Jackson and F. H. Froes, TMS-AIME Symposium, Fall meeting, Orlando, FL, October, 1986.

10. J. L. Murray, Bulletin of Alloy Phase Diagrams, 2(1), 185 (1981) and 2(2), 53 (1981).

LONG-RANGE ORDERING IN RAPIDLY QUENCHED L1o TiAl COMPOUND ALLOYS

S.H. WHANG, Z.X. LI AND D. VUJIC
Department of Metallurgy & Materials Science
Polytechnic University
333 Jay Street, Brooklyn, New York 11201

ABSTRACT

Rapid solidification has an effect on lattice parameters and long-range order parameter in L1o TiAl compound alloys. In these compounds rapid quenching from the liquid state significantly decreases the long-range order parameter while the tetragonal distortion c/a decreases as a result of rapid quenching. The lattice parameter change due to rapid quenching becomes pronounced with increasing aluminum concentration beyond the stoichiometric alloy composition. The recovery from the disordered state and the relaxation from the tetragonal distortion were found to be two different kinetic processes.

The long-range ordering requires low temperature and long annealing time in comparison with high temperature and short annealing time for the recovery of the tetragonal distortion.

INTRODUCTION

The first article that describes attractive high temperature properties of L1o type TiAl by McAndrew and Kessler appeared two decades ago [1]. Since then, much research effort on TiAl has been focused on plastic deformation, in particular from room to intermediate temperatures, and the ductile-brittle transition behavior [2-4]. The results indicate that the mobility of super-dislocation a[011] and its partial dislocation a/6 [112] is a controlling factor in the plastic deformation of TiAl (Fig. 1). The partial dislocation a/6 [112] which is immobile below 700°C is a source of the brittleness, but becomes unpinned above 700°C, making the compound ductile. However, other L1o type compounds such as CuAu and CoPt having a c/a ratio of less than unity exhibit tensile ductility at low temperatures. The fundamental question that arises is why TiAl which has a higher c/a ratio, larger than unity, behaves differently than those compounds which have a c/a ratio smaller than unity. The problem is not well understood though all L1o type compounds have a contraction along the "c" axis direction where alternative layers of unlike atoms allow a strong interaction between the unlike atoms. Instead, a further query is whether or not the c/a ratio and ductility or plastic deformation are closely related to each other. If this is

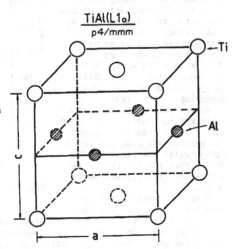

TiAl(L1o)
p4/mmm

Fig. 1 – Schematic Diagram of L1o Ti-Al Structure

so, then any change in the c/a ratio may alter deformation and ductility behavior.

The effect of rapid quenching on properties of titanium aluminides were explored by a number of authors [5,6]. In fact, rapid quenching introduces disordering and decreases long-range ordering in Ni_3Al compounds [8] and improves ductility [9,10]. Nevertheless, such a disordering process may not only change atomic ordering, but also alter tetragonal distortion and plastic deformation behavior in Ll_0 TiAl compounds. This paper deals with some such issues in rapidly quenched TiAl compound alloys.

EXPERIMENTAL

Rapidly quenched alloy foils of binary and ternary Ll_0 titanium aluminides were prepared using the Hammer and Anvil quench technique. The specimen foils of the same thickness with an error of $\pm 1\mu m$ were used for a series of experiments on lattice parameters and long-range order parameters. Identification of metastable phases and general microstructure were carried out by x-ray diffraction and TEM. Microhardness measurements were performed using a Leitz microhardness tester with a load of 100g.

RESULTS AND DISCUSSIONS

Rapidly Quenched Phases

In the binary phase diagram shown in Fig. 2 [11], metastable phases that were formed by rapid quenching are indicated at the bottom of the phase diagram. All these phases are already known phases. No new phase was found in the as-quenched alloys. Metastable Ti_3Al appears in the range of 49-56 at.% Al probably due to the partial interruption of the peritectic reaction near 51 at. % Al. Also, the occurrence of single phase as well as metastable TiAl in the range of 56-62 at.% Al is a result of the suppression of the $TiAl_2$ (oC_{12}) compound during rapid quenching. It is reasonable to expect that the formation of $TiAl_2$ may be readily suppressed during rapid quenching since the peritectoid reaction for $TiAl_2$ (solid reaction) may be slower than the peritectic reaction for TiAl (liquid-solid reaction).

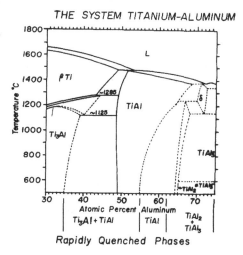

THE SYSTEM TITANIUM-ALUMINUM

Rapidly Quenched Phases

Fig. 2 - Rapidly Quenched Ti-Al Alloys Compared with Ti-Al Equilibrium Phase Diagram [11].

Microstructure of the as-quenched $Ti_{48}Al_{52}$ having dual phases of TiAl + Ti_3Al compounds shows very fine grains of about one micron dia. (Fig. 3a). In contrast, in as-quenched $(Ti_{48}Al_{52})_{92}V_8$ compound, anti-phase domain boundaries, twins and stacking faults were found (Fig. 3b). TEM microstructural study shows that the amount of stacking faults increases with vanadium concentration, indicating that the addition of vanadium to TiAl may decrease the stacking fault energy. The details of alloying effect

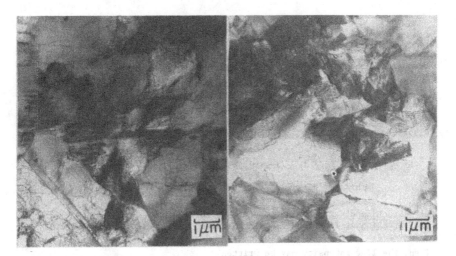

Fig. 3a – TEM Micrograph of As-
Quenched Ti₄₈Al₅₂ Alloy
Foil.

Fig. 3b – TEM Micrograph of As-
Quenched (Ti.48Al.52)92
V8 Alloy Foil.

by vanadium are discussed elsewhere [12].

Long-Range Ordering from the Disordered State

When metals rapidly cool down from the liquid, long-range diffusion is
likely suppressed. Hence, the result is that any highly ordered structure
either is absent in the quenched alloys or remains in the disordered state.
Hence, the titanium aluminides produced by rapid quenching are expected to
be highly disordered.

The long-range order parameter for $L1_0$ TiAl structure may be determined
from the superlattice reflection intensity. The structure factor F for Al
rich TiAl compounds may be written as follows.

The general description of the structure factor is

$$F(hkl) = \sum_{j=1}^{1} G_j \cdot f_j \cdot e^{-2\pi(hu_j+kv_j+lw_j)} \tag{1}$$

where

$u_j, v_j, w_j,$: origins of sublattice j
G_j : geometric structure factor of sublattice j
f_j : average scattering factor of sublattice j

The $L1_0$ TiAl structure may be subdivided into four simple cubic sublattices.
The origins of these sublattices are $0,0,0$, $\frac{1}{2}\frac{1}{2}0$ for Ti; $\frac{1}{2},0,\frac{1}{2}$, $0,\frac{1}{2},\frac{1}{2}$ for
Al.

Therefore, for the stoichiometric TiAl compound, the structure factor
F_0 is

$$F_0 = f_{Ti} \cdot e^{2\pi i \cdot (0)} + f_{Ti} \cdot e^{2\pi i(h/2+k/2)} + f_{Al} \cdot e^{2\pi i(h/2+l/2)}$$

$$+ \ f_{A1} \ e^{2\pi i (k/2+1/2)} \tag{2}$$

Nevertheless, in order to deal with the alloy having an off-stoichiometric composition and a disordered state, two assumptions are made:

1. No vacancy is generated due to rapid quenching in the off-stoichiometric composition.
2. Al atoms fill in vacant Ti lattice sites in Al rich $L1_0$ TiAl compounds.

The resulting structure factor for an off-stoichiometric composition with a disordered state F_D

$$F_D = [1+e^{\pi i (h+k)}] \cdot [\ f_{Ti} \cdot (1-2(x+y)+ f_{A1} \cdot 2(X+y)]$$

$$[e^{\pi i (h+k)} + e^{\pi i (k+1)}] \cdot [\ f_{A1} \cdot (1-2y) + 2 \cdot f_{Ti} \cdot y] \tag{3}$$

x: fraction of excess Al atoms that go to Ti sites.
y: fraction of Ti atoms occupying Al sites due to disordering and vice versa.

Then, the line intensity may be written

$$I = |F_D|^2 \cdot p \cdot \frac{(1 + \cos^2 \Theta)}{\sin^2 \Theta \cdot \cos \Theta} \cdot \exp[-B(\sin \Theta / \alpha)] \tag{4}$$

P: multiplicity factor
B: Debye-Waller factor

The definition of long-range order parameter (LROP) is

$$S_{Ti} = \frac{r_{Ti} - R_{Ti}}{1 - R_{Ti}} \tag{5}$$

where r_{Ti} and R_{Ti} are fraction of Ti atoms at Ti lattice sites and fraction of Ti atoms in the alloy. The long-range order parameter for a given specimen is determined by optimizing the ratio $I_s(ex)/I_f(ex)$ against $I_s(th)/I_f(th)$, where subscripts: s, f, indicate superlattice reflection, fundamental reflection; and ex and th indicate experimental and theoretical, respectively. The determined long-range order parameters for the 50-58 at.% Al range are shown in Fig. 4. The long-range order parameter for the as-quenched TiAl compounds varies from 0.35-0.42 with a maximum near the 55 at.% Al. In other words, the Al rich compound alloys have higher degree of LRO values than those compositions closer to the stoichometric TiAl compound. These LRO values are slightly smaller than 0.55 for rapidly quenched Ni_3Al powder [8]. The discrepancy between the two values may be attributed to either quench rate or different reaction type (Ni_3Al: eutectic reaction).

This behavior of non-uniform LRO values in TiAl compounds may be explained from the equilibrium phase diagram (Fig. 2). The upper limit of the peritectic reaction for $L1_0$ TiAl ($L + \beta \rightarrow \gamma$) is 53 at.% Al. Beyond this point, the melt solidifies directly into $L1_0$ TiAl, i.e., transformation $L \rightarrow \gamma$ takes place. By comparing these two reactions, it is predictable that the latter occurs much faster than the former and therefore, reaches a higher state of ordering. In fact, the dual phase region, $Ti_3Al+TiAl$, extends up to 56 at.% Al instead of 53% at.% Al, indicating that the equilibrium diagram may undergo distortion under rapid quenching.

The long-range ordering reaction is sluggish in these compounds. As-quenched binary TiAl compound alloys which were annealed at 1000°C, 2h (primary annealing) and then isothermally annealed at 800°C for 16h show very little change in LROP.

The assumptions that were employed in these calculations may not be true in a rigorous sense since the density of vacancies is drastically increased in the as-quenched alloys [13-15]. Secondly, the defect density also may not be uniform over the range of off-stoichiometric compositions as reported in Ll_2 Ni_3Al [16]. Nevertheless, the error involved in LROP prediction due to this assumption is nearly negligible. For example, the unfilled sites for Ti lattice sites in $Ti_{46}Al_{54}$ is 4% which is significantly large compared with 0.16 of the vacancy concentration in rapidly quenched aluminum [12].

Another source of the uncertainty in the results may come from a preferred orientation of the grains in the alloy foils. Currently, the problem is being evaluated and will be reported in the future.

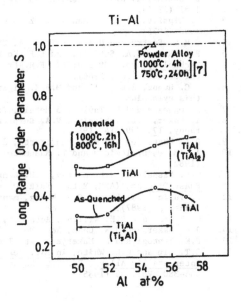

Fig. 4 - Long-Range Order Parameters for As-Quenched Ll_0 TiAl Compound Alloys.

SUMMARY

1. Single phase Ll_0 TiAl occurs in the range of 56-62 at.% Al, away from the peritectic reaction. The peritectic reaction near 50 at.% Al was partially suppressed by rapid quenching resulting in a metastable Ti_3Al compound.
2. Rapid quenching introduces disordering in TiAl alloys. The long-range order parameter in the as-quenched TiAl alloys is in the range of 0.4-0.5. The higher ordering occurs near the 55 at-Al alloy, not at the stoichiometric composition. The long-range ordering at 800°C was shown to be very sluggish.

ACKNOWLEDGEMENT

We thank the continuing support of the Office of Naval Research for the rapidly solidified titanium alloy program at Polytechnic University (Contract No. N00014-85-K-0787). Many thanks are due to Mr. Z.C. Li for his help in calculating long-range order parameters.

REFERENCES

1. J.B. McAndrew and H.D. Kessler, J. Metals, 206, 1348-1353 (1956).

156

2. D. Shechtman, M.J. Blackburn and H.A. Lipsitt, Met. Trans. 5, 1375-1381 (1974).
3. H. Lipsitt, D. Shechtman and R.E. Schafrik, Met. Trans. A, 6A, 1991-1996 (1975).
4. S.M.L. Sastry and H.A. Lipsitt, 8A, 299-308 (1977).
5. R.G. Rowe, J.A. Sutliff, and E.F. Koch, in Titanium Rapid Solidification Technology, pp. 239-248, Ed. F.H. Froes and D. Eylaon, (The Metallurgical Society, Inc. 1986).
6. S.C. Huang, E.L. Hall and M.F.X. Gigliotti, See the proceedings of this symposium.
7. P. Duwez and J.L. Taylor, 5 Metals, Trans. AIME 4, pp. 70-71 (1952).
8. I. Baker, F.S. Ichishita, V.A. Surprenant and E.M. Schulson, Metallography, 17, 299-314 (1984).
9. D.K. Chaterjee and M.G. Mendiratta, J. Met., 33, 5 (1981).
10. A. Inoue, H. Tomioku and T. Masumoto, J. Met. Sci. Lett., 1, 377-380 (1982).
11. T.B. Massalski, J.L. Murray, L.H. Bennett and H. Baker, in Binary Alloy Phase Diagram, (ASM, Metals Park, OH 44073 - 1986) p. 175.
12. D. Vujic, Z.X. Li and S.H. Whang, manuscripts to be submitted to Met. Trans. A (1987).
13. G. Thomas and R.H. Willens, Acta Met. 12, 191-196 (1964).
14. J.C. Baker and J.W. Cahn, Acta Met, 17, 575-578 (1969).
15. P.K. Tastogi and K. Mukerjee, Met. Trans. 1, 2115-2117 (1970).
16. C.T. Liu and C.L. White, in Mat. Res. Soc. Symp. Proc. on Ordered Intermetallic Alloys, Ed. C.C. Koch, C.T. Liu and N.S. Stoloff, pp. 365-380, Vol. 39 (1985).

TWO-PHASE NICKEL ALUMINIDES

P. S. KHADKIKAR[*], K. VEDULA[*] and B. S. SHABEL[**]
* Dept. of Metallurgy and Materials Science, Case Western Reserve University, Cleveland, OH 44106
** Alcoa Laboratories, Alcoa Center, PA 15069

ABSTRACT

The as-extruded microstructures of two alloys in the two phase field consisting of Ni_3Al and NiAl in the Ni-Al phase diagram exhibit fibrous morphology and consist of $L1_2$ Ni_3Al and B2 NiAl. These as-extruded microstructures can be modified dramatically by suitable heat treatments. Martensite plus NiAl or martensite plus Ni_3Al microstructures are obtained upon quenching from 1523 K. Aging of martensite at 873 K results in the recently identified phase Ni_5Al_3 whereas aging at 1123 K reverts the microstructures to Ni_3Al plus NiAl. The microstructures with predominantly martensite or Ni_5Al_3 phases are brittle in tension at room temperature. The latter microstructure does not deform plastically even in compression at room temperature. However, some promise of room temperature tensile ductility is indicated by the Ni_3Al plus NiAl phase mixtures.

INTRODUCTION

The two phase field, consisting of B2 NiAl and $L1_2$ Ni_3Al, in the Ni-Al phase diagram is of potential interest for elevated temperature applications. Since the development of ductile Ni_3Al [1], a considerable amount of research and development effort has been devoted to single phase Ni_3Al alloys (see for example, refs. [2,3]). Single phase NiAl has also been studied to some extent [3], but is plagued by room temperature brittleness. However, the two phase field between these two phases has not received much attention. The thrust of the present investigation has been to obtain a better understanding of the physical metallurgy of this two phase field, so that attempts can be made to control the microstructures and properties of the alloys of interest. The martensitic transformation in Ni-rich NiAl has been studied in some detail [5]. A new phase, Ni_5Al_3, in the composition range of the two phase field consisting of Ni_3Al and NiAl has recently been identified and studied to some extent [6,7,8]. Singleton et al. [9] have incorporated this new phase in the Ni-Al phase diagram, the relevant portion of which is shown in Figure 1. It is the purpose of this paper to illustrate the microstructures that are obtained with various heat treatments in this interesting and

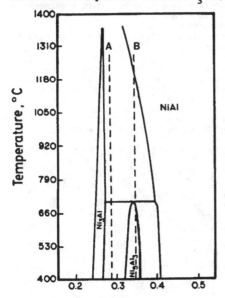

Atom fraction, Al

Figure 1. Portion of the Ni-Al Phase Diagram [9].

complex two phase field. The mechanical properties of selected alloys at room temperature in tension and compression are also presented.

EXPERIMENTAL

Two alloys, indicated in Figure 1, containing 28.8 (alloy A) and 34.7 at% (alloy B) aluminum were studied. These alloys were obtained by hot extrusion of blends of Ni$_3$Al (24 at% Al, 0.2 at% B and 0.4 at% Hf) and NiAl (44 at% Al and 1 at% Hf) powders canned in 50.8 mm diameter mild steel containers at NASA Lewis Research Center, Cleveland, Ohio. The extrusions were made at 1400 K (1127°C) and a 16:1 area reduction ratio. The chemical compositions of the two alloys are shown in Table I.

TABLE I. Chemical compositions of the two alloys studied (at%).

Alloy	Al	Hf	B	Ni	O_2 wtppm	N_2
A	28.8	0.8	0.1	balance	187	95
B	34.7	0.9	<0.1	balance	193	2

The mild steel cans were removed from the extrusions by centerless grinding before any heat treatments. Solutionizing treatment for various times was carried out at 1523 K (1250°C) in flowing argon for both alloys. The specimens were water quenched from this temperature. Aging treatments in flowing argon and at temperatures of 1123 K (850°C) and 873 K (600°C) were employed to obtain the various combinations of Ni$_3$Al, NiAl and Ni$_5$Al$_3$. Following heat treatments to obtain the desired microstructures, the tensile (3.2 mm dia. and 30.0 mm G.L.) and compression (4.7 mm dia. and 9.5 mm G.L.) specimens were prepared by centerless grinding and electrolytic polishing. Mechanical properties at room temperature were studied in tension (where possible) and compression on an Instron testing machine at initial strain rates of 2.8 x 10^{-5} and 8.9 x 10^{-5} sec^{-1} respectively. The as-extruded materials were also tested in tension at 1073 K (800°C) at the initial strain rate of 2.8 x 10^{-5} sec^{-1}. Hardness measurements were made on Rockwell A scale. Normal metallographic means were employed to examine the microstructures by optical microscopy. X-ray diffraction was employed to identify the phases present.

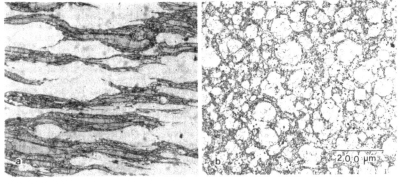

Figure 2. As-extruded microstructures of alloy B (a) Longitudinal section, (b) Transverse section.

RESULTS AND DISCUSSION

Microstructures

As-Extruded

As-extruded microstructures exhibit fibrous morphology and consist of B2 NiAl and Ll_2 Ni_3Al. As an example, transverse and longitudinal sections of the alloy B are shown in Figure 2. Two phases, Ni_3Al (light etched) and NiAl (dark etched), are observed. The NiAl phase forms the continuous matrix since it is softer and more ductile at the extrusion temperature. Elongated prior particles and the diffusion zone near phase boundaries are observed. Both NiAl and Ni_3Al exhibit equiaxed grains, although, the grain boundaries in the Ni_3Al phase are not resolved in Figure 2. The grain size appears to be much finer near phase boundaries. Vickers microhardness measurements at room temperature indicate that the NiAl phase is harder than the Ni_3Al.

Homogenized, and Homogenized and Aged

At the homogenization temperature of 1523 K (1250°C), the alloy B completely transforms to Ni-rich NiAl. The alloy A transforms to a mixture of Ni_3Al plus NiAl phases with the compositions of the respective phase boundaries (Figure 1). Upon rapid cooling from this temperature, the Ni-rich NiAl transforms to 3R martensite. Therefore, the alloy A transforms to a two phase mixture of martensite plus NiAl (Figure 3a) whereas, the alloy B transforms completely to martensite (Figure 3d).

Aging at 1123 K (850°C) transforms martensite or martensite plus Ni_3Al microstructures to Ni_3Al plus NiAl phase mixtures as one would expect from the phase diagram (Figure 1). These microstructures are shown in Figure 3b (alloy A) and Figure 3e (alloy B). Aging at 873 K (600°C) results in the Ni_5Al_3 phase. The alloy B transforms to Ni_5Al_3 plus NiAl (Figure 3f) whereas, the alloy A transforms to Ni_5Al_3 plus Ni_3Al (Figure 3c). The exact volume fractions of each of the phases and the compositions of the Ni_5Al_3 phase are in question since the phase boundaries in Figure 1 [9] are not exactly known and are based on only a few data points [7]. The proportions of the two phases vary and approach equilibrium with aging time. The various heat treatments employed and the resultant microstructural phases detected by X-ray diffraction are summarized in Table II.

A detailed X-ray diffraction study of the Ni_5Al_3 phase was carried out using alloy B, which was rapidly cooled from 1523 K (1250°C) after homogenization for four hours and aged at 823 K (550°C) for 15 days [8]. The structure of this new phase was confirmed to be orthorhombic D_{2h}^{19} (Cmmm) by matching the observed (hkl) peak intensities with those calculated for the structure. The changes in the crystal structure during the transformation from NiAl to martensite to Ni_5Al_3 can be quantified by comparing the c/a ratios for the tetragonal orientation of B2 NiAl, the tetragonal Ll_0 martensite and the tetragonal approximation for the Ni_5Al_3 structure. These c/a ratios were found to change from 0.707 to 0.853 to 0.900. The lattice parameters of the Ll_0 martensite were found to be $a_o = 0.3804$ nm and $c_o = 0.3244$ nm and that of orthorhombic Ni_5Al_3 were $a_o = 0.7475$ nm, $b_o = 0.6727$ nm and $c_o = 0.3732$ nm.

Mechanical properties

As-Extruded Materials

In as-extruded condition both the two phase alloys fail in a brittle manner (Figures 4a and 4b). This is not unexpected since NiAl is brittle at room temperature. However, even Ni_3Al phase that contains boron fails in a brittle intergranular fashion as evidenced by the fracture surface (Figure

160

Alloy A   Alloy B

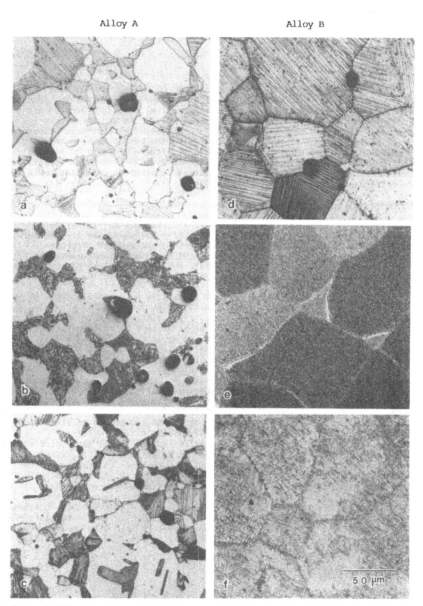

Figure 3. Microstructures of alloys A and B after various heat treatments.
(a) and (d) 1523 K/24 hr, water quenched, (b) and (e) 1523 K/24 hr, water
quenched plus 1123 K/24 hours, (c) and (f) 1523 K/24 hr, water quenched plus
873 K/24 hr.

TABLE II. A summary of the heat treatments and the resultant phases
along with the Rockwell A hardness measurements.

Alloy	Heat Treatment	Phases detected by X-ray diffraction	Hardness Rockwell A
A	As–extruded	$Ni_3Al+NiAl$	67
A	1523 K/24hr*	$Ni_3Al+Martensite$	67
A	1523 K/24hr+873 K/24hr	$Ni_3Al+Ni_5Al_3$	71
A	1523 K/24hr+873 K/7days	$Ni_3Al+Ni_5Al_3$	71
A	1523 K/24hr+1123 K/24hr	$Ni_3Al+NiAl$	67
A	1523 K/24hr+1123 K/7days	$Ni_3Al+NiAl$	67
B	As–extruded	$Ni_3Al+NiAl$	72
B	1523 K/24hr	Martensite+NiAl(trace)	67[**]
B	1523 K/24hr+873 K/24hr	Ni_5Al_3+NiAl	80[**]
B	1523 K/24hr+873 K/7days	Ni_5Al_3+NiAl	81[**]
B	1523 K/24hr+1123 K/24hr	$NiAl+Ni_3Al$	75
B	1523 K/24hr+1123 K/7days	$NiAl+Ni_3Al$	74

* All specimens water quenched from the homogenization temperature of 1523 K.

** Cracks near indentation marks were observed.

5a). Ductile tearing, which is restricted to the centers of larger Ni_3Al
particles, indicates that diffusion is responsible for this loss in
ductility. The interdiffusion of either boron (resulting in the depletion of
boron), or aluminum (making Ni_3Al phase aluminum-rich) may have caused this
embrittlement.

At 1073 K (800°C, the temperature at which serious intermediate
temperature embrittlement has been reported in alloys based on Ni_3Al [3]) the
as-extruded materials show significant plastic deformation. The tensile
stress–strain curves show unusual cyclic flow behavior (Figure 4c). The
inset optical micrograph of the section of a failed specimen reveals a number
of cracks normal to the tensile direction. Most of these cracks were
observed to be in the Ni_3Al phase. These cracks decrease the load bearing
area and explain the decrease in stress with increasing strain. The unusual
cyclic flow behavior, however, is believed to be the result of dynamic
recrystallization of the NiAl phase [10].

Homogenized, and Homogenized and Aged Alloys

Alloy A was tested in tension and compression at room temperature in
homogenized, and homogenized and aged conditions. It was not possible to
test alloy B in tension as specimens could not be machined due to extreme
brittleness of the phases. The tension and compression curves for the alloy
A and the compression curves for the alloy B are shown in Figures 4a and 4b.
Alloy A shows ductility in compression, but fails in a brittle fashion in

162

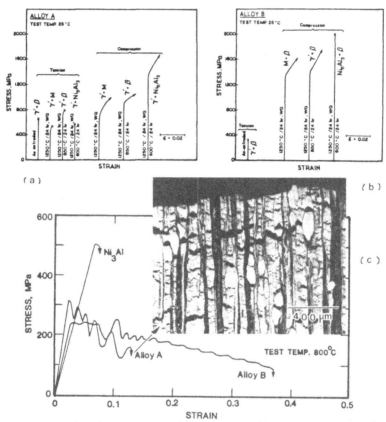

(a)

(b)

(c)

Figure 4. Engineering stress-strain curves for alloys A and B. (a) and (b) room temperature tests, (c) 1023 K tensile tests for as-extruded alloys A and B, and Ni₃Al (24 at% Al) with 0.2 at% B. Inset optical micrograph shows the longitudinal section of the fractured specimen.

tension. When alloy A is heat treated to obtain the phase mixtures of Ni$_3$Al plus NiAl, and tested in tension, it does show a deviation from linearity, although it fails immediately afterwards in a brittle manner.

Alloy B in water quenched condition (martensitic microstructure) and when aged at 1123 K (850°C) exhibits ductility in compression. However, when it is aged at 873 K (600°C) to obtain Ni$_5$Al$_3$ plus NiAl microstructure it fails in brittle fashion even in compression with a very high fracture strength.

The Rockwell A hardness results, listed in Table II, agree in general with the tension and compression properties. The hardness values for predominantly martensitic and Ni$_5$Al$_3$ microstructures do not reflect the correct hardness because these microstructures crack upon the application of indentation loads.

The results indicate that Ni$_5$Al$_3$ is particularly brittle but has a very high fracture strength. It deforms in compression when present with Ni$_3$Al.

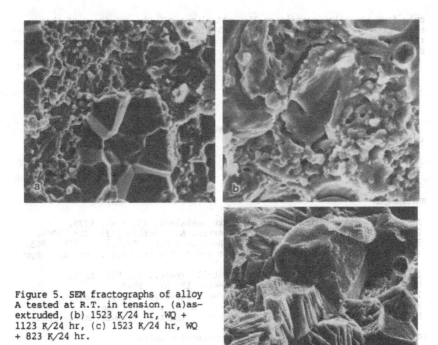

Figure 5. SEM fractographs of alloy
A tested at R.T. in tension. (a)as-
extruded, (b) 1523 K/24 hr, WQ +
1123 K/24 hr, (c) 1523 K/24 hr, WQ
+ 823 K/24 hr.

However, the specimen with Ni_5Al_3 plus NiAl microstructure fractures without
any indication of deformability even in compression. Some promise of
obtaining a two phase microstructure with tensile ductility is indicated only
by the two phase mixtures of Ni_3Al plus NiAl.

The SEM fractographs of alloy A tested in tension are shown in Figure 5.
The SEM fractograph of the alloy A specimen (Ni_3Al plus NiAl) fractured in
tension which indicated deviation from linearity is shown in Figure 5b. It
shows some tearing type fracture regions, indicating potential for some
ductility at room temperature. A distinct lamellar cleavage type fracture is
observed for the Ni_5Al_3 phase in a mixture of Ni_5Al_3 plus Ni_3Al
microstructure (Figure 5c).

CONCLUSIONS

A study of the two alloys containing 28.8 and 34.7 at% aluminum in the
two phase field consisting of Ni_3Al and NiAl indicates that a variety of
microstructures can be obtained by suitable heat treatments. Microstructures
consisting of either martensite and Ni_3Al or martensite and NiAl are obtained
upon quenching from the homogenization temperature of 1523 K. These
as-quenched microstructures transform to Ni_5Al_3 plus either Ni_3Al or NiAl
when aged at 873 K. Only Ni_3Al and NiAl phases are obtained upon aging at
1123 K. All these alloys deform plastically in compression at room
temperature, except for the Ni_5Al_3 plus NiAl microstructure which exhibits a
high fracture strength in compression. Tensile specimens could be machined

only from the alloy containing 28.8 at% aluminum. The specimens of this alloy with the microstructures of Ni_3Al plus NiAl indicate some potential for room temperature tensile ductility. [3] The as-extruded alloys exhibit extensive plastic deformation and unusual cyclic flow curve in tension at 1023 K.

ACKNOWLEDGEMENTS

The support of ALCOA Technical Center and NASA Lewis Research Center are gratefully acknowledged.

REFERENCES

1. K. Aoki and O. Izumi, Nippon Kinzaku Gakkaishi, 43, 1190, 1979.
2. C. T. Liu, C. L. White and J. A. Horton, Acta Metall., 33, 213, 1985.
3. C. T. Liu and C. L. White, Mater. Res. Soc. Symp. Proc., 39, 365, 1985.
4. K. Vedula, V. Pathare, I. Aslanidis and R. Titran, Mater. Res. Soc. Symp. Proc., 39, 365, 1985.
5. S. Chakravorty and C. M. Wayman, Metall. Trans., 7A, 555, 1976.
6. K. Enami and S. Nenno, Jpn. Inst. Met., 19, 571, 1978.
7. I. M. Robertson and C. M. Wayman, Metallography, 17, 43, 1984.
8. P. S. Khadkikar and K. Vedula, submitted to J. Mater. Res.
9. M. F. Singleton, J. Murray and P. Nash, submitted to Bull. Alloy Phase Diagrams.
10. P. S. Khadkikar and K. Vedula, unpublished results.

LATTICE LOCATION OF Ta IN Ni₃Al BY ION CHANNELING AND NUCLEAR REACTION ANALYSIS

H. LIN*, L. E. SEIBERLING**, P. F. LYMAN** AND D. P. POPE*
* University of Pennsylvania, Department of Materials Science and Engineering, Philadelphia, PA 19104
** University of Pennsylvania, Department of Physics, Philadelphia, PA 19104

ABSTRACT

We have investigated the lattice location of Ta in Ni_3Al using Rutherford backscattering with channeling, and nuclear reaction analysis. An 8 MeV ^{4}He ion beam was directed along the <100> crystallographic axis of the $Ni_{75}Al_{24}Ta$ single crystal. A silicon surface barrier detector was used to analyze ^{4}He ions backscattered from Ni and Ta atoms. Neutrons generated from Al by the $^{27}Al(^4He,n)^{30}P$ reaction were detected by a large volume liquid scintillator placed outside of the scattering chamber. Essentially all of the Ta atoms were found to be substitutional, as determined by the Ta channeling minimum yield. A comparison of the width of the channeling angular scan for Al, Ni and Ta indicated that the Ta atoms are predominantly distributed on the Ni sites. This result is in conflict with expectations based on the ternary phase diagram.

INTRODUCTION

A large number of ternary elements show substantial solubility in Ni_3Al [1], and some of these additions can cause dramatic changes in the mechanical properties of the alloy [2]. Furthermore, the magnitude of the change, e.g., the increase in critical resolved shear stress for dislocation motion, appears to depend very strongly on whether the ternary element substitutes for Ni or for Al [3,4]. The traditional technique for determining the lattice site of the ternary additions has been based on the shape of the solubility lobe in the ternary phase diagram [1]. More recently, attempts have been made to directly determine the atom sites by atom probe measurements [5] and ion channeling [6]. The atom probe experiments agree with expectations based on phase diagrams for the three cases studied (Hf, Co and Fe) [5]. However, the ion channeling study of a Ni_3Al + Hf alloy [6] indicates that Hf substitutes for Ni. This result is in conflict with both the atom probe results, and phase diagram information.

We have undertaken a study of Ni_3Al + Ta using both Rutherford backscattering with channeling and nuclear reaction analysis techniques. Based on the ternary phase diagrams, both Ta and Hf are expected to substitute for Al. Because a previous ion channeling study indicated that Hf substitutes for Ni [6], we wished to investigate another, similar ternary element using ion channeling. As a channeled ion passes through a crystal, it is gently steered away from each atomic row that it approaches by the screened Coulomb field of the row of atoms (essentially a uniform line of charge). Thus, if an impurity atom sits in a substitutional site in a particular row, the close encounter probability of the incident ion with the impurity atom will be determined by the average atomic number of the row in which it sits. The width of the channeling angular scan is determined by the close encounter probability as a function of tilt angle of the crystal. Thus, the width of the channeling angular scan of the impurity should be the same as that of the host row in which it lies [7]. In addition to the standard Rutherford backscattering analysis of the Ta and Ni channeling angular widths, it was hoped that a measurement of the Al angular width using the $^{27}Al(^4He,n)^{30}P$ reaction would allow a more precise determination of the Ta lattice site. Rutherford backscattering analysis of the same crystal using both 8 MeV ^{4}He ions and 5.9 MeV ^{9}Be ions was performed to assure that beam damage effects were not important.

166

EXPERIMENTAL

A single crystal was prepared with composition $Ni_{75}Al_{24}Ta$. The structure of Ni_3Al is FCC-based, so that each row along the $<100>$ direction is comprised entirely of one element, with three Ni rows for each Al row. The (100) face of the sample was first metallographically polished using standard techniques, finishing with 0.3 μm alumina powder, then annealed in vacuum at 1000° C for 64 hours. Final polishing was done on a cloth-faced lapping wheel with fine colloidal silica (Syton HT40) as the polishing medium. The quality of the surface was checked by determining the asterism of the back-reflection Laue spots using a 45 keV X-ray beam at grazing incidence (5°) to the sample surface. This results in a penetration depth of several μm into the crystal, comparable to the maximum depth from which backscattered ions are observed in the Rutherford backscattering spectrometry (RBS) experiments.

The ion beams used for analysis were produced by the Penn FN tandem Van de Graaff accelerator and collimated to an angular divergence of less than 0.1° before entering the scattering chamber. Figure 1 depicts the geometry of the experimental apparatus. A precision, two-axis goniometer was used to orient the sample relative to the ion beam. A magnetically driven beam stop with 20% duty cycle was used to measure the ion dose. Ions scattered from a thin gold layer on the beam stop into a surface barrier detector were counted, and the number of counts were taken to be proportional to the ion dose. This procedure routinely yields current integration to better than 5% accuracy. All measurements were made with the sample at room temperature.

Experiments were conducted with two different incident ions. A 5.9 MeV ^{9}Be beam was used to obtain backscattering results for Ta and Ni. An 8.0 MeV ^{4}He beam was chosen to detect Al via the nuclear reaction $^{27}Al(^4He,n)^{30}P$, in addition to Ni and Ta by elastic backscattering. The threshold energy of the nuclear reaction corresponds to a 3 MeV ^{4}He in the laboratory frame, but the cross section rises rapidly for energies up to approximately 7 MeV [8,9]. Backscattered ions were analyzed using a surface barrier detector at 150° from the incident beam direction, subtending 2.15 msr. Neutrons from the $^{27}Al(^4He,n)^{30}P$ nuclear reaction were detected using a large volume liquid scintillator [10], placed outside the scattering chamber at a forward angle of 55°. Both neutrons and gamma rays generated in the target arrive at the scintillator with little attenuation. The neutron signal was separated from the gamma signal by exploiting the difference in the response of the scintillator over time to neutrons versus gamma rays using a standard pulse-shape discrimination method [11].

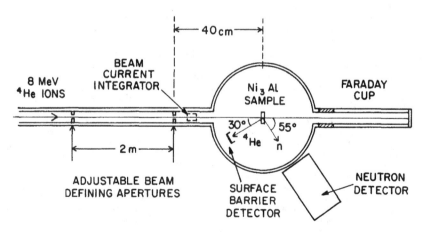

FIGURE 1   Geometry of the experimental apparatus.

RESULTS

Figure 2 shows normalized RBS yields of Ta and the host Ni for an angular scan about the <100> axis with the 5.9 MeV ⁹Be beam. The data were selected from a region having a depth of about 200 nm near the surface of the crystal. The Ni curve gives a minimum yield (χ_{min}) of approximately 12%, and has a half-angle (the angle at which the decrease of the yield from the random level is half of the maximum decrease) of 0.52°. The Ta data show a minimum yield of about 15%, indicating that nearly all Ta atoms are substitional for Ni and/or Al. The Ta half-angle is estimated to be 0.47° ± 0.03°, slightly narrower than the Ni curve. After completing an entire angular scan, the minimum yields were remeasured on the same spot, and the same value (within experimental uncertainty) was found. This indicates that little or no ion-induced damage occurred in the region of study.

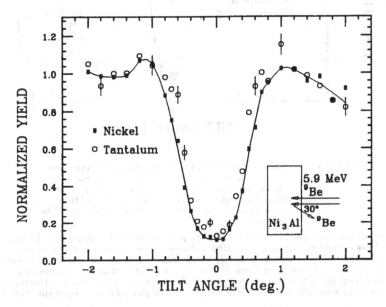

FIGURE 2   Normalized RBS yields of Ta and host Ni for an angular
scan about the <100> axis with the 5.9 MeV ⁹Be beam.
The data were selected from a region with depth of about
200 nm near the surface of the crystal.

An angular scan on a fresh beam spot about the same axis with the 8.0 MeV ⁴He beam produced the results in figure 3. In addition to the normalized RBS yields of Ta and Ni, normalized neutron counts from the ²⁷Al(⁴He,n)³⁰P reaction are displayed. The RBS data were selected from a region from approximately 150 to 700 nm below the surface. The narrowing of the angular dips is due to the higher beam energy, but the Ta and Ni results are quite similar to those from the ⁹Be scan. The χ_{min} of the Ni curve is about 10%, and the half-angle is 0.31°. The Ta minimum yield is about 13%, and the estimated half-angle is 0.28° ± 0.03°. The ratio of the Ta to the Ni half-angle (0.90) is in excellent agreement with the ratio from the ⁹Be scan (0.90). The curve from the Al events has a half-angle of 0.20° with a minimum yield of about 46%.

The increase in the minimum yield for the Al curve over the others is due in part to the larger depth over which the Al atoms were probed, because the dechanneled component of the ion flux grows with depth. Since the detection system does not measure the

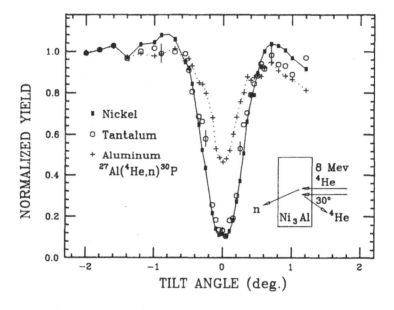

FIGURE 3   Normalized RBS yields of Ta and Ni and normalized
neutron counts from the $^{27}Al(^4He,n)^{30}P$ reaction from
an angular scan about the <100> axis with the 8.0
MeV ^{4}He beam. The RBS data were selected from
approximately 150 to 700 nm below the surface of the
crystal.

energy of the neutrons, the depth scale of the probe is set by the range over which the ^{4}He
ions have enough kinetic energy to initiate the nuclear reaction. Another consideration is
the efficiency of the neutron detector, which falls to zero rapidly for neutron energies
below 2.5 MeV. Based upon these factors, it is estimated that the neutron counts originated from as deep as 8 μm into the crystal. Due to the falling nuclear reaction cross
section, proportionally fewer of the neutron counts originated from deep in the crystal.

DISCUSSION

    If the sole effect of probing more deeply into the crystal was to increase the number
of events originating from dechanneled ions, the half-angles of the angular scans would
not depend on the probe depth. However, as the ion flux penetrates the crystal more and
more deeply, small deflections of ions that arise from electronic collisions begin to become
appreciable. After many such collisions, the flux becomes less sharply peaked in the
center of the lattice channel. It is the flux distribution that determines the half angle, and
a broader distribution produces a smaller half angle. Therefore, a more deeply-probing
angular scan would have a smaller half-angle. Analysis of the Ni events originating from
as deep as 3 μm bears out these assertions. A slight decrease in half-angle was observed
at greater depths. Therefore, the Al and the Ni data cannot be reliably compared on an
equal footing. The experimental width of the Al angular scan quoted herein can be taken
as a lower limit to its true value. We are confident that the ^{4}He beam energy and detector
parameters can be adjusted so that an Al angular scan originating from the top 0.5 μm of
the sample can be measured.

If an alloy has atomic rows along the channeling direction that are composed entirely of one of the constituent elements, then the ratio of the half-angles of the angular scans of the two different elements is predicted to be the square root of the ratios of their atomic numbers [7]. For a pure Ni_3Al crystal the calculated ratio of the Al half-angle to that of the Ni is 0.681. If a small amount of a substitutional impurity is added to the crystal, it may alter the average charge per unit length of the row in which it is located. Because of the disparate atomic numbers of Ta and Al, and the paucity of Al atoms compared to Ni atoms, just one percent Ta would raise the average Z of atoms on the Al row from 13 to 15.4 if the Ta occupied only Al lattice sites. Similarly, if the Ta were to substitute only for Ni atoms, the average atomic charge on a Ni row would be raised from 28 to 28.6. If the Ta atoms were to be distributed to lattice sites randomly (*i.e.* 75% on Ni sites and 25% on Al sites), the average atomic numbers of Al and Ni rows would be raised to 13.6 and 28.45, respectively. A crystal having Ta atoms located only on Ni sites would then be predicted to have a ratio of Al to Ni half-angles of 0.674, that with a random distribution, 0.691, and that with Ta only on Al sites, 0.742. A more accurate estimate of this ratio can be obtained by Monte Carlo methods. A Monte Carlo simulation on a Ni_3Al crystal with 1% Hf revealed that the ratio of the half-angles of Al to Ni would be 0.72 if the Hf occupied Ni lattice sites, and 0.79 if they occupied Al sites [12]. As Ta differs in atomic number from Hf by merely one unit, we expect these estimations to be applicable to the present crystal. The experimental value is 0.65. This represents a lower limit on this ratio, owing to the narrowing of the Al angular scan relative to the Ni scan due to the different depths from which the scans were obtained.

A reliable comparison can be made between the half-angles of the Ni and Ta angular scans. An atom occupying a substitutional lattice site and sharing the vibrational amplitude of its host will have the same half-angle as the element for which it substitutes. If its vibrational amplitude is greater than that of the host, its half-angle will be smaller than that of the host element. At temperatures below the Debye temperature, it is possible for a more massive substitutional atom to have an average vibrational amplitude slightly greater than that of the lighter host element, but it is unlikely that the heavier element will have a smaller amplitude. If the Ta atoms were to occupy Ni sites, the ratio of the Ta half-angle to the Ni half-angle should be equal to (or slightly less than) one. If the Ta atoms were to be on the Al sites, the ratio of the Ta to Ni half-angles should be close to the calculated Al to Ni half-angles, or 0.742. If the Ta were randomly distributed, this ratio should be close to 0.922. The experimental value for both curves is 0.90, with experimental uncertainties of approximately $\pm$ 0.06 for the ^9Be data, and $\pm$ 0.10 for the ^4He data. Consequently, we conclude that the Ta atoms are predominantly distributed on the Ni sites. We cannot say whether the Ta is entirely on the Ni sites or 75% on the Ni sites.

CONCLUSIONS

Using the ion channeling technique with RBS and nuclear reaction analysis, we have found, as expected, that Ta atoms occupy substitutional lattice sites in a Ni_3Al + Ta crystal. We have considered three possible substitutional configurations: the Ta is all on Al sites, the Ta is all on Ni sites, and the Ta is distributed randomly between Ni and Al sites. Although expectations based on the phase diagram are that the Ta occupies exclusively Al sites, we find this configuration highly unlikely. We conclude that the Ta predominantly occupies Ni sites. We cannot resolve the case in which the Ta atoms substitute only for Ni atoms from the case in which they substitute mostly for Ni and partially for Al.

The nuclear reaction $^{27}Al(^4He,n)^{30}P$ has been used to measure the channeling half angle of Al in Ni_3Al + Ta. This promises to be a useful way to study substitutional elements in Ni-Al alloys with channeling techniques. Adjustment of the ^4He beam energy and neutron detector parameters should allow an Al angular scan to be measured from the top 0.5 μm of the sample. In this way, accurate comparisons of the Al, Ni and substitutional element angular scans can be made. It is clear that a high atomic number substitutional element in concentrations of as little as 1% can have a major effect on the angular scans of the host elements. A precise knowledge of the width of the angular scans is crucial in determining the lattice site of the impurity with high accuracy. For this reason, obtaining an accurate Al angular scan, in addition to a Ni and Ta scan, is very important. A study of the variation of the Al and Ni angular scans with Ta concentration (from zero to 4%) is planned, and should provide independent confirmation of the lattice location of the Ta.

ACKNOWLEDGEMENTS

The authors would like to thank J.M. Williams for discussions motivating the present investigation, D.P. Balamuth for his expertise regarding nuclear reaction analysis, and R.L. Headrick for helpful discussions during the planning stages of this inquiry. The expert aid of S.F. Pate regarding the acquisition of the data is acknowledged, as is the diligent efforts of D. Dimlich during the experimental phase. Ongoing discussions of the results with J.H. Barrett and E.J. Mele were greatly appreciated.

This work was supported by NSF MRL program under the grant DMR-85-19059 through the Laboratory for Research on the Structure of Matter, and by the NSF under the grant PHY-82-13598 through the Tandem Accelerator Laboratory, both of the University of Pennsylvania.

REFERENCES

1. S. Ochiai, Y. Oya and T. Suzuki, Acta Metall. 32 , 289 (1984).

2. D.P. Pope and Salah S. Ezz, Int. Metals Rev. 29 , 136 (1984).

3. R.D. Rawlings and A. Staton-Bevan, J. Mater. Sci. 10 , 505 (1975).

4. K. Aoki and O. Izumi, Phys. Stat. Sol.(A) 32 , 657 (1976).

5. M.K. Miller and J.A. Horton, Scr. Metall. 20 , 1125 (1986).

6. J.M. Williams, H.G. Bohn, J.H. Barrett, C.T. Liu and T.P. Sjoreen, Mat. Res. Soc. Symp. Proc., this session.

7. L.C. Feldman, J.W. Mayer and S.T. Picraux, "Materials Analysis by Ion Channeling", Academic Press, New York (1982).

8. P.H. Stelson, F.K. McGowan, Phys. Rev. B 133 , 911 (1964).

9. D.S. Flynn, K.K. Sekharan, B.A. Hiller, H. Laumer, J.L. Weil and F. Gabbard, Phys. Rev. C 18 , 1566 (1978).

10. NE13, a commercial Xylene-based scintillator, is appropriate for neutron detection.

11. T.K. Alexander and F.S.Goulding, Nucl. Instr. Meth. 13 , 244 (1961).

12. J.H. Barrett, to be published.

PART III

Mechanical Behavior

DEFORMABILITY IMPROVEMENTS OF L1$_2$-TYPE INTERMETALLIC COMPOUNDS

OSAMU IZUMI[*] AND TAKAYUKI TAKASUGI[*]
[*]The Research Institute for Iron, Steel and Other Metals, Tohoku
University, Sendai, Japan

ABSTRACT

Intermetallic compounds are usually so brittle that they simply cannot
be fabricated into useful shape components. Recent developments of physical and metallurgical principles for the structural materials have led to
the conclusion that the deformability of several intermetallic systems can
be substantially improved. Among them, in this lecture, the nature of
grain boundary structure in L1$_2$-type intermetallics is first discussed.
Then, the possibilities and instances to overcome the brittleness are
shown. Polycrystalline L1$_2$-type Ni$_3$Al, even prepared from high purity
metals, exhibits brittle intergranular fracture. On the other hand, single
crystals of Ni$_3$Al are ductile even at room temperature. Thus, it is said
that the grain boundaries of this alloy are intrinsically weak. Our recent
analyses showed that there exist several configurations of atomic bonds,
resulting in the heterogeneous electronic environments which can be
regarded as "cavities". The drastic ductility improvement of Ni$_3$Al alloy
by a small boron addition, which was found in 1979, can be interpreted as
segregated boron atoms at grain boundaries to modify the electronic environments of boundaries and to reinforce boundary cohesion, thus suppressing intergranular fracture. From a quite similar point of view, modification of boundary structures can be achieved by substituting proper elements
for the constituents of compounds or by selecting atom combinations of
compounds. Thus, the ductility of Ni$_3$Al could be improved by substituting
a few percent of Mn or Fe for Al. Also, by selecting atom combination,
several L1$_2$-type alloys such as Co$_3$Ti, Cu$_3$Pd, Ni$_3$Mn and Ni$_3$Fe have been
found as ductile intermetallics.

I Introduction

An interest in intermetallic compounds has recently been arisen from
their peculiar behaviors such as heat-resistant, superconducting and/or
magnetic properties, which are expected to encourage their development as
new materials for promising application. For instance, certain intermetallic compounds have unique potential as new high temperature structural alloys since they occasionally show an increase in flow strength with increasing temperature and since the restricted atomic mobility generally
results in excellent corrosion and creep resistance. Also, attempts to
develop for functional applications such as energy-conversion and -stored
materials, electric devices and magnetic materials have been carried out
extensively. However, they are usually so brittle that they simply cannot
be fabricated into useful shape components. This is a main reason why the
compounds have been alienated from practical usage for a long time.

The main causes of the brittleness can usually be attributed to an insufficient number of slip systems and grain boundary weakness. Commonly,
the crystal structures of intermetallics have large unit cells and complex
constructions compared with usual metals and alloys, resulting in a higher
Peierls stress and lack of operative slip systems. Therefore, it has been
considered basically impossible to expect the deformability for intermetallic compounds. Recent developments of physical and metallurgical principles for the structural materials have shown that the deformability of
several intermetallic systems can be substantially improved.

New knowledge on the character of ordered structures, the behavior of superlattice dislocations, the configuration of grain boundary structures, and so on, have made it gradually possible to overcome the brittleness of intermetallics. Thus, at present, ductility improvements have already succeeded in several compounds.

The polycrystalline $L1_2$-type intermetallics often show an intergranular embrittlement which seems to be enhanced with increasing ordering energy. Although no systematic investigation has been reported for this phenomenon, for instance, Ni_3Al, Ni_3Ge, Ni_3Si and Fe_3Ga with high ordering energy unexceptionally show the severe grain boundary fragility. Since single crystals of this structure have been shown to exhibit high ductilities, this embrittlement has been attributed to the inherent weakness of grain boundaries themselves.

With regard to the mechanism of the brittleness of the polycrystalline $L1_2$-type structure, several major factors have been demonstrated; for instance, the restriction of the number of available slip systems or cross-slip [1] and/or the segregation of impurities to grain boundaries [2]. The former is related to the superlattice structure and therefore provide the reasonable cause for more brittleness in the ordered state than the disordered state. However, it is not applicable to explain the grain boundary brittleness observed in a large number of $L1_2$-type alloys, since the enough number of slip systems is operative in those alloys. Concerning the latter which is not necessarily related to the superlattice structure, no direct evidences indicating the segregation of impurity or constituent atoms have been reported. The microscopic (atomistic) mechanism of brittle intergranular fracture in the disordered alloys [3] revealed that the crack usually nucleates at structural inhomogeneity such as inclusion or triple junction of grain boundaries, then the energy is absorbed by bond stretching and breaking off along grain boundary. Therefore, the interpretation for the fracture energy required to cut off the atomic bonds becomes very important. It is thus supposed that the grain boundary fracture directly relates to the structure and bond environment of boundary itself.

A first theoretical analysis was made by Marcinkowski [4] with respect to the low angle tilt boundary configurations in the superlattice alloys and it was recognized that the boundary was effectively disordered. Recently, the present authors [5] studied the grain boundary structure of $L1_2$- and $L2_0$-type intermetallics from the viewpoint of the coincidence site lattice (CSL) theory, and recognized the defect structure of bonds on the grain boundary which might be related to the embrittlement. This geometrical investigation has been extended further to the analyses of the atomistic structure and related electronic environment of the grain boundary region which can offer the explanation of the intergranular strength and failure behavior in the $L1_2$-type superlattice alloys [6,7]. Thus, it becomes possible to seek the way to suppress the grain boundary failure and to improve the deformability of $L1_2$-type intermetallic compounds.

This paper is a summary of our systematic research work on the deformability of $L1_2$-type intermetallics.

II. Grain Boundary Structure of $L1_2$-type Intermetallics---Geometrical and Electronical Aspects

The present authors [5] recently studied the grain boundary structure of the $L1_2$-type ordered alloys in terms of the CSL model. Many kinds of atom configuration were constructed geometrically on a given boundary orientation. The $L1_2$-type A_3B structure has twelve unlike atoms A in the nearest neighbor sites around one B atom, and this ordered atomic arrangement has to be destroyed at a boundary region. It was found that the A-B (fair) bonds drastically decreased while A-A (wrong) bonds increased at boundaries. Thus, as a result, the defect structures of bond were introduced into grain boundaries. An example is shown in Fig.1.

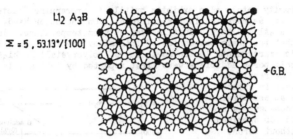

$L1_2 A_3B$

$\Sigma = 5 , 53.13°/[100]$

←G.B.

Fig. 1 Schematic representation of the effect of covalent bonds
between the A and B atoms at a grain boundary of the $L1_2 A_3B$-alloys.

Now, the intergranular fragility of the $L1_2$-type A_3B alloys has to be
considered by both concepts of the crystal structure and electronic environment at grain boundaries, because of a large difference in bond nature
between A-B bonds and others. It is generally suggested that in the $L1_2$
-type alloys the A-B pair more or less has the character of a covalent
bond. It is apparently illustrated in Fig.1 that in grain interiors the
covalent bonds between A and B atoms build up the three-dimensional cell
structure, suggestive of superior strength, while in the grain boundary
region those are scarce and distorted, therefore revealing inferior
strength. In other words, the A-B covalent bonds perpendicular to the
boundary plane which are supposed to sustain the strength of grain boundary
are not found so easily in the grain boundary. Next, most reliable fracture mechanism, which is also related to the environment of the covalent
bonds at the grain boundary, is similar to the mechanism by which the
embrittlement was proposed in the boundary segregated by impurity atoms.
From these models based on the quantitative calculations, it is suggested
in our grain boundary that when the B atom is strongly covalent with the
transition A atom, the B atom draws charge from the A atom onto itself.
The result is that less charge is available to participate in the A-A, i.e.
metal-metal bonds and they are weakened. This modification for the
electronic environment is enhanced as the character of covalent bonds becomes stronger, in other words, as the valency difference between the A and
B atoms becomes larger. Thus, the heteropolarity of charge due to the A-B
bond at the boundary plane introduces the environment somewhat like a penny
like cavity as shown in Fig.1. In addition, inward drawing of charge from
the A-B pairs of the matrix might enhance less charge available to participate in the A-A metal bonds which are dominant at the boundary. On the
other hand, if the covalent character of the A-B bond becomes weaker, the
drawing of charge from the A-A bonds may be reduced, then the suppression
of grain boundary embrittlement will be expected. Furthermore, it is also
promising to improve the ductility of this type of alloy if the charge
deficiency of A-A pairs is filled up by adding the elements acting as
electron donors or if the heterogeneity of charge distribution along grain
boundaries is homogenized by alloying with the proper elements.

III. Deformability Improvements

1. Addition of Interstitial Elements

In the past, though the $L1_2$-type intermetallic compound Ni_3Al (so
called γ' phase) which is the major strengthening phase of Ni-base superalloys has excellent high temperature characteristics, it is extremely
brittle not only at room temperature but also at high temperature and its
use by itself was considered impossible. Polycrystalline Ni_3Al prepared

from high-purity metals also exhibits brittle intergranular fracture, even though impurity segregation is insignificant. On the other hand, single crystals of Ni_3Al are highly ductile even at room temperature. Thus, it is concluded that the grain boundaries of this alloy are intrinsically weak.

Drastic improvements in the ductility of polycrystalline Ni_3Al resulting from small boron additions were first reported by Aoki and Izumi in 1979 [8]. The result is shown in Fig.2 in which the tensile stress-strain curves at room-temperature are compared for specimens with and without boron. It was observed that a transition of fracture mode from intergranular to transgranular type occurred. This beneficial effect of boron was attributed to the structural change of the grain boundary itself. Afterwards, this behavior was also confirmed by other researchers [9,10]. Auger spectra showed that boron segregates predominantly to grain boundaries, but not to free surfaces, suggesting that boron re-inforces the atomic bond at grain boundary regions [10,11,12]. As mentioned in the previous section, our analyses on grain boundary structure of $L1_2$-type intermetallics showed that there exist several kinds of configuration of atomic bonds, resulting in the heterogeneous electronic environments which can be regarded as "cavities". It has been suggested that, in nickel, boron atoms act as electron donors and thereby strengthen the atomic bond [13,14]. Thus it is said that segregated boron atoms at the grain boundaries of Ni_3Al may modify the electronic environments of grain boundaries and reinforce boundary cohesion, then suppressing intergranular fracture. It is interesting that the beneficial effect is most pronounced in alloys containing 24at%Al, i.e. Ni-rich side, and becomes less effective at higher aluminum concentrations. Through this change of concentration, the fracture mode changes from transgranular to intergranular. Stoichiometry effects on the ductility will be discussed in the following section.

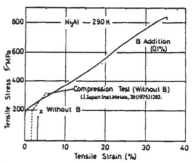

Fig. 2 Tensile stress-strain curves at room temperature, showing the effect of boron addition on the elongation of Ni_3Al. Dashed lines shows the compressive stress-strain curve of Ni_3Al.

2.Alloying of Substitutional Elements

$L1_2$-type Ni_3Al is a compound composed of elements of group VIII and IIIb. At the stoichiometric composition, the Ni and Al atoms occupy the face center and cube corner sites, respectively. The preferential site occupation of substitutional ternary elements in Ni_3Al was systematically studied by Guard and Westbrook [15] and Suzuki et al [16], which are summarized in Table 1. The elements from group Vb to IIIb, i.e., Sb, Si, Ge and Ga which are similar in electronic nature to Al atoms substitute exclusively on the Al sites. The elements of groups Ib and VIII, i.e. Cu and Co, which are similar in nature to Ni atoms substitute exclusively on the Ni sites. The elements from group VIII to VIa, i.e. Fe, Mn and Cr which are also similar to Ni atoms tend to substitute evenly on both or predominantly on the Al sites. The elements from group VIa to IVa, i.e. Mo, W, V, Nb, Ta and Ti again substitute on the Al sites. Those ternary elements were added to Ni_3Al with stoichiometric composition at the expense of nickel or aluminium content, respectively, depending on the preferential site occupancy of the elements described above. Followings are the results of our experiments summarized briefly.

Table 1 Values of the valency of the third element itself and the valency difference between the third element and constitutive solvent atom substituted by the third element. These values are referred in Delehouzee et al. [17] and Elliott et al. [18] (described in the parentheses), respectively.

Third element	Electronic nature	Substitution site	Valency of third element	Valency difference† between solvent and third element
Sb	Vb	Al	5	−2
Si	IVb	Al	4	−1
Ge	IVb	Al	4	−1
Al	IIIb	—	3	—
Ga	IIIb	Al	3	0
Cu	Ib	Ni	? (0)	? (+0.25)
Ni	VIII	—	0 (0.25)	—
Co	VIII	Ni	0.8 (0.72)	−0.8 (−0.47)
Fe	VIII	Al	0.2 (0.92)	+2.8 (+2.08)
Mn	VIIa	Al	0.9 (1.35)	+2.1 (+1.65)
Cr	VIa	Al	1.2 (1.69)	+1.8 (+1.31)
Mo	VIa	Al	3	0
W	VIa	Al	3	0
V	Va	Al	1.5 (2.19)	+1.5 (+0.81)
Nb	Va	Al	?	?
Ta	Va	Al	?	?
Ti	IVa	Al	4.2 (3.92)	−1.2 (−0.02)
Sc	IIIa	Ni	?	?

+ Defined as (valency of constitutive solvent-valency of ternary element).

The grain morphology in the arc-melted condition of each alloy exhibited similar coarse columnar grain structures. Ductility was compared by bend and tensile tests at room temperature. Addition of the b-subgroup elements, i.e. Sb(Vb), Si(IVb), Ge(IVb) and Ga(IIIb) resulted in no visible plasticities, and failure modes were all the brittle intergranular fracture similar to the binary Ni$_3$Al. The ternary alloys with true metal elements, i.e. Cu(Ib), Co(VIII), Fe(VIII) and Mn(VIIa) showed quite interesting results; the samples with additions of Cu or Co did not show any ductility while addition of Fe and Mn did cause quite extensive deformability. Fractography revealed smooth grain boundary facets (intergranular fracture) in the former alloys and predominantly transgranular fracture surfaces in the latter alloys. Since apparent deformabilities were found in this group of alloys, the systematic investigations were carried out particularly as a function of the ternary elements. In the former two ternary alloys, even further additions of the third elements did not alter the feature of the intergranular failure. On the other hand, in the latter two ternary alloys, the deformability, i.e. the transgranular fracture mode, began to appear at about 5at.% addition and then became more obvious with increasing solute contents, examples of which are shown in Figs.3–4. It should be noted that the favorable elements, Fe and Mn substitute for Al sites, while the unfavorable ones, Cu and Co do for Ni sites. The ternary alloys with the transition elements of VIa(Cr, Mo and W), Va(V, Nb and Ta), IVa(Ti) and IIIa(Sc) were fractured intergranularly.

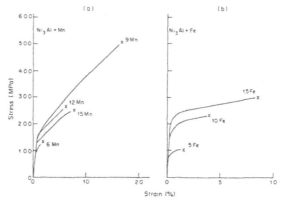

Fig. 3 Stress-strain curves of ternary Ni₃Al added (a) with Mn
and (b) with Fe, respectively.

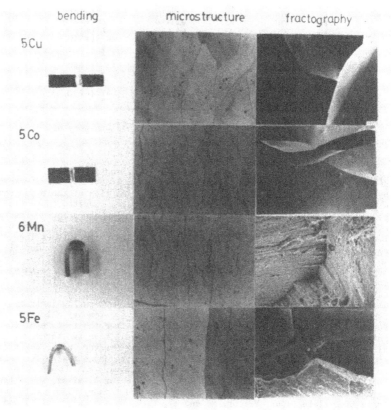

Fig. 4 Summaries of the bending test, metallography and fractography observed for ternary Ni₃Al added with the elements of groups Ib, VIII and
VIIa.

By the addition of third elements, part of the bonds between Ni and Al are replaced with the bonds between third element and one of the constitutive elements depending on the site occupation and concentration, respectively. Therefore, the effect of third elements is discussed from the point of view of the alternation of electronic nature of bonds between Ni and Al atoms. The valence differences between the third element and the constitutive solvent element substituted by the former element were evaluated and shown in Table 1. This criterion correctly predicts the degree of the intergranular strength and related failure behaviors. The valency differences in the alloys added with Fe and Mn were found to be the greatest, i,e. +2.8 and +2.1, respectively. On the other hand, those with Cu and Co were found to be +0.25 and -0.8, respectively, which are close to zero and not effectable on the nature of electronic environment. It is thus predicted that the grain boundary strength is enhanced when the valency difference is adequately positive. Thus, most ternary elements in b-subgroup are inherently "brittler atoms" in the Ni_3Al compound because of their valency differences. Other transition elements may be less effective or rather make the boundary weaker.

The ternary atoms added to Ni_3Al occupy preferred sites not only in the matrix but also in the grain boundary region. Consequently, part of bonds between nearest neighbor atoms adjoining at grain boundary are alternated with the ternary atoms. Depending on the electronic character and substitution nature of the ternary atoms, three types of grain boundary structure are schematically illustrated in Fig.5, where one of the configurations constructed in Σ =5, 53.13° boundary with <100> axis is shown. Fig.5(a) corresponds to the ternary alloys with b-subgroup elements and with a part of transition elements (Va and IVa). It is seen that the newly introduced stronger covalent bonds produce a more heterogeneous electronic charge distributions through the boundary plane, accounting for the reduced grain boundary strength. Fig.5(b) corresponds to the alloys with Cu or Co atoms, and the similar covalent bond to the original one does not appear to modify the electronic charge distribution, resulting in no improvement of boundary strength. However, in Fig.5(c) which corresponds to the alloys with Fe or Mn atoms, it is recognized that the newly alternating metal-metal bonds create the more homogeneous charge distribution through the boundary plane, responsible for the improved grain boundary strength in these alloys.

The fact that the present phenomenon depended on the substitution nature in addition to the electronic nature of the third elements implies that the crystallographic (atomistic)

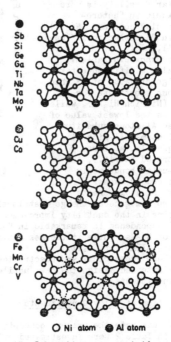

Sb
Si
Ge
Ga
Ti
Nb
Ta
Mo
W

Cu
Co

Fe
Mn
Cr
V

O Ni atom   ● Al atom

Fig. 5 Schematic representations showing the effect of both the substitution and electronic nature of the third element on the grain boundary of Ni_3Al.

180

accommodations in the grain boundary region are much less in comparison
with those of disordered alloys. If the accommodation of the grain boundary structure is fully established, the third element with the similar
electronic nature should have the similar effect on the fracture behavior.
In other words, the present result suggests that each sub-lattice site of A
and B atoms is still distinguished in the grain boundary region.

3.Effect of Non-stoichiometry

It is well known that non-stoichiometry in the ordered intermetallics
is a very important crystallographic parameter affecting their physical and
mechanical properties. It was reported that the drastic improvement of
ductility by boron addition was found only in Ni-excess Ni_3Al (8). The
present authors have recently observed that, in the $L1_2$-type A_3B alloys, as
the A atoms (metal atom of group VIII) became excess from stoichiometry,
yield strength, fracture strength and elongation always increased and the
transgranular fracture modes
became dominant. Fig.6 represents the change of elongation of the $L1_2$-type alloys
as a function of composition
of B atoms. It is clearly
seen that the compositional
dependence on the inter-
granular fragility is more
remarkable in "hetero-
electronic" compound (Ni_3Al)
than in "homoelectronic" one
($Ni_3(Al,Mn)$).

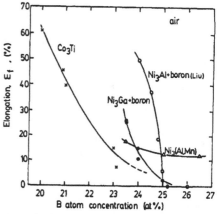

Fig. 6 Effect of deviation from stoichiometry on the elongation of A_3B
type alloys.

As shown in Table 1,
the valence difference takes
the highest value of +3 when
Ni atoms substitute for Al
atoms (Ni-rich Ni_3Al), while
it takes the lowest value of
-3 when Al atoms substitute
for Ni atoms (Al-rich Ni_3Al).
Consequently, as already
illustrated in Fig.5, the
metal-metal bonds newly
introduced by excess
A atom (Ni atom) create
the more homogeneous charge distribution through the boundary plane,
resulting in the ductility improvements in Ni-rich side. This compositional dependence is remarkable in "heteroelectronic" ordered alloys like
Ni_3Al or Ni_3Ga, while it becomes ambiguous in more "homoelectronic" alloys
like $Ni_3(Al,Mn)$ or Co_3Ti.

Those geometrical and electronical concepts would open the way further
to estimate and design the new ductile intermetallics by selecting the alloy constituents.

4.Selection of Alloy Constituents

To evaluate the present concepts, several kinds of $L1_2$-type A_3B alloy
shown in Table 2 were investigated by mechanical tests and metallographical
observations. The specimens consist of different kinds of electronic bond
nature; that is, the character of covalent bond becomes stronger if B atoms
change from group VIII to IVb or VIII to IVa in the periodic table. The
metallography observed in each alloy exhibited quite different features, in
particular with respect to the grain size and morphology. Here, an important result is, however, that irrespective of the grain morphology

Table 2 Values of the valence of the B atom itself and the absolute
valence difference between the A atom and B atom. These values are
referred from Delehouzee's [17] and Elliott's [18] reports (the latter is
described in the parentheses).

Alloy	Valency of B atom	Valency difference between A and B atoms
Ni_3Ge	4 (4)	4 (3.75)
Ni_3Si	4 (4)	4 (3.75)
Ni_3Ga	3 (3)	3 (2.75)
Fe_3Ga	3 (3)	2.8 (2.08)
Ni_3Al	3 (3)	3 (2.75)
Co_3Ti	4.2 (3.92)	3.4 (3.20)
Cu_3Pd	–	–
Ni_3Mn	0.9 (1.35)	0.9 (1.10)
Ni_3Fe	0.2 (0.92)	0.2 (0.67)

and size, the specimens showed no visible plasticities in the alloys containing b-subgroup element as the B atom whereas the obvious ductilities
were seen in the alloys with a-subgroup elements (transition elements) as
the B atom. The fractographies indicated the intergranular fracture surfaces in the former group of alloys and the dimple or ripple transgranular
fracture surfaces in the latter group. The stress-strain curves obtained
in the tensile
tests on the latter
alloys are shown
in Fig.7. It is
interesting that as
the difference in
electronic chemical
bond nature between
two constituent
atoms becomes
smaller, the strain
hardening rate and
resultant ultimate
tensile stress were
lowered and then
elongation increased.
    Those phenome-
nological aspects in
the A_3B alloys clearly
indicate that the
electronic parameter
of the B atom, that
is, the valency (or
electron to atom ratio)
of the B atom has
to be taken into con-

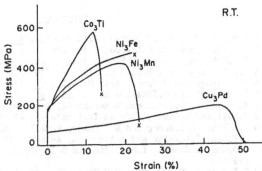

Fig. 7 Stress-strain curves of the binary A_3B
alloys with elements of group VIII and a-
subgroup as the B atom. Note that the strain-
hardening rates and tensile strength decreased
and then the elongations increased as the
difference of electronic chemical bond nature
between two constituent atoms becomes smaller.

sideration. These valency values are shown in Table 2. The valency difference in the binary $A_3B(A=Ni)$ alloys is larger when B atom is b-subgroup
element than when B is a-subgroup element. Thus, as already discussed, it
seems that the grain boundary of the former type of alloys is inherently
weak, that is, brittle, while that of the latter type is inherently strong,
that is, ductile. In the nickel-based $A_3B(Ni_3B)$ alloys, the degree of
grain boundary fragility may be in the following sequence;
$Ni_3Ge=Ni_3Si>Ni_3Ga=Ni_3Al>Ni_3Mn>Ni_3Fe$. However, when the valency differences
are compared among the alloys with different A elements, the quantitative
evaluation will be less successful. For example, four alloys of Ni_3Ga,
Fe_3Ga, Ni_3Al and Co_3Ti, whose valency differences were derived to be about

3.0 with fluctuation of 0.4, showed a discrepancy for the failure behavior; Ni_3Ga, Fe_3Ga and Ni_3Al failed intergranularly while Co_3Ti failed transgranularly. This result appears to arise from the fact that the electronic structure constructed between two constituent atoms is slightly different among these four alloys even if the A atom of nickel, cobalt or iron belongs to the same group VIII. In addition, the size of B atom would be expected as the secondary effect. It is thus considered that the size effect of B atom and the slightly different electronic structure of A atom could not predict the definite valency difference at which ductile to brittle transition of A_3B alloys occurs.

IV Summary and Conclusion

In this paper, the grain boundary structures of $L1_2$-type A_3B inter-metallic compounds were constructed by a geometrical procedure. Even this geometrical concept gave us insight into several peculiarities in grain boundary of ordered alloy; there exist a multiplicity of structures and a heterogeneous electronic or chemical environment at the grain boundary region. In particular, the latter characteristics might be the fundamental origin of the lack of ductility in the ordered intermetallic compounds. From the point of view described above, the weak grain boundary in $L1_2$-type alloys due to heterogeneous charge distribution could be improved by modifying the electronic environments; that is, by the addition of inter-stitial elements (boron), alloying with substitutional elements (Fe, Mn) and the selection of alloy constituents (Co_3Ti, Cu_3Pd, Ni_3Mn, Ni_3Fe).

References

1 N.S.Stoloff and R.G.Davies, Acta Metall., 12, 473 (1964).
2 J.H.Westbrook and D.L.Wood, J.Inst.Metals., 91, 174 (1962-63).
3 C.J.McMahon, Jr. and V.Vitek, Acta Metall., 27, 507 (1979).
4 M.J.Marcinkowski, Phil.Mag., 17, 159 (1968).
5 T.Takasugi and O.Izumi, Acta Metall., 31, 1187 (1983).
6 T.Takasugi and O.Izumi, Acta Metall., 33, 1247 (1985).
7 T.Takasugi, O.Izumi and N.Masahashi, Acta Metall., 33, 1259 (1985).
8 K.Aoki and O.Izumi, J.Japan Inst.Metals., 43, 1190 (1979).
9 A.I.Taub, S.C.Huang and K.M.Chang, Metall.Trans.A., 15A, 399 (1984).
10 C.T.Liu, C.L.White and J.A.Horton, Acta Metall., 33, 213 (1985).
11 T.Ogura, S.Hanada, T.Masumoto and O.Izumi, Metall.Trans.A., 16A, 441 (1985).
12 T.Takasugi, E.P.George, D.P.Pope and O.Izumi, Scripta Met., 19, 551 (1985).
13 C.L.Briant and R.P.Messmer, Phil.Mag., B42, 569 (1980).
14 R.P.Messmer and C.L.Briant, Acta Metall., 30, 457 (1982).
15 R.W.Guard and J.H.Westbrook, Trans.Metall.Soc.AIME, 215, 807 (1959).
16 S.Ochiai, Y.Oya and T.Suzuki, Acta Metall., 32, 289 (1984).
17 L.Delehouzee and A.Dernyttere, Acta Metall., 15, 727 (1967).
18 R.P.Elliott and W.Rostoker, Trans.Am.Soc.Metals, 50, 617 (1957).

AN OVERVIEW OF THE TEMPERATURE DEPENDENCE
OF THE STRENGTH OF THE Ni₃Al SYSTEM

JOHN K. TIEN, SANDRA ENG, and JUAN M. SANCHEZ
Center for Strategic Materials, Henry Krumb School of Mines
Columbia University, New York, NY 10027

Abstract

Many $L1_2$ ordered alloys including the Ni_3Al intermetallic system are
noted for their anomalous temperature dependence of strength. It is
also generally accepted that this dependence is due to a thermally assisted cross-slip, work hardening based model [1,2]. An alternative
antiphase boundary (APB) based model has long been dismissed by the
research community because prior calculations of APB energy, and some
measurements, have shown that the appropriate APB energy of Ni_3Al should
remain constant with temperature [3]. These aspects will be reviewed
briefly and will serve as a basis for a presentation of some more recent
results. These will include the strain rate insensitivity of strength
versus temperature in the increasing strength temperature region, a
result that is, in our view, rather contradictory to the thermally assisted cross-slip model. Some very recent calculations of equilibrium
APB energies will also be reviewed in the context of the strength dependence issue. These results show that the equilibrium APB energy
increases with temperature. The question of what role, if any, equilibrium APB plays in the deformation and strengthening process will be
discussed.

Introduction

The coherent gamma prime precipitation strengthened superalloys have
been and still are the preferred structural materials in the hot sections of gas turbines and in jet engines. Long ago it was recognized
that the high temperature limit for these alloys is determined by the
sudden strength drop-off observed over and over again at about the 700°
to 800°C range [4,5]. More recently, the gamma prime phase has become
popular as a potential weight effective matrix or monolith for elevated
and high temperature structural applications [6-9]. Its high temperature limit also pivots about a not so surprisingly similar strength
drop-off at the same temperature range as the superalloys. Other similarities abound, albeit in different degrees. The engineering yield
strength of the higher strength superalloys increases slightly with temperature prior to the drop-off. The engineering yield strength of the
gamma prime monolith increases much more sharply. At first glance,
these trends appear to be caused by the same dislocation mechanism. Superdislocation pairs, separated by an antiphase boundary (APB), which
force the nickel and aluminum to move from ordered $L1_2$ coordination to
homospecies clustering, are known to be mobile in both systems [10,11].
Despite the similarities in the two systems, there are also differences, some subtle and others not. The gamma prime phase by itself work
hardens quite rapidly with strain [8], while the superalloys do not
[12]. The gamma prime exhibits a microyield strength that is weak and
apparently not very temperature sensitive [8,13], while the superalloys
work harden very little between their proportional limit and the engineering or 0.2% yield strengths [12]. (See Fig. 1.)
In this paper, we have critically assessed the highlights of what
the international community knows about the fundamental strengthening
mechanisms of the gamma prime intermetallic itself and the gamma prime

184

strengthened superalloys. We pivot the assessment on such questions as:
· Do the intermetallic and the superalloy share the same basic
  strengthening mechanism?
· Is there a dilemma in the contrasting behavior of gamma prime
  microyield and engineering yield? Ditto superalloy's?
· How reliable is the generally accepted cross-slip model for the
  temperature dependence of the macroyield of γ'?
· What role, if any, is there for the equilibrium APB's?
· Why the sudden and cataclysmic fall-off in strength of γ' and γ/γ'
  superalloys at high temperatures?
There is a need to know the answers to these questions to increase
the strength of superalloys further, to exploit the intermetallics, and
if possible to forestall the strength fall-off at what appears to be a
700°-800°C temperature barrier.

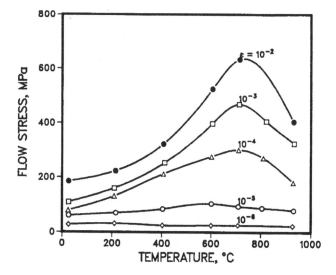

Fig. 1: Strain dependence of the yield strength of Ni₃Al as a function
of temperature [8].

Microyield of Gamma Prime

By definition, microyield occurs when, on the average, mobile dislocations move a minute but finite distance without returning. Perhaps the
best empirical definition of microyield is the point at which internal
friction becomes strain dependent as a consequence of damping caused by
the permanent movement of dislocations [14,15], or the point when x-ray
lines begin to broaden. Microyield strength and macroyield (engineering
yield) strength are numerically equivalent if little or no work hardening occurs during, say, the first 0.2% of permanent offset. This is certainly not the case for the gamma prime system, in which work hardening
appears to increase with temperature until the fall-off [8].
It is not difficult to assess and postulate a microyield mechanism.
It requires only an understanding of the interaction between the just
starting to move dislocations and the nearest, or most uniformly distributed, or most densely distributed pinning points. In the gamma

prime monolith, these pinning points are basically the orderliness of
the L1$_2$ crystal structure itself.

As Koehler and Seitz [16] proposed and as Marcinkowski [17] first
observed, the stable dislocation configuration in a superlattice is a
superdislocation, which consists of a pair of equal dislocations separated by an antiphase boundary. The dislocations themselves may further
separate into intrinsic and/or extrinsic pairs [18]. According to Brown
[19], the critical stress required to move the pair of superpartial dislocations A and B moves to positions A' and B' on the same slip plane
is:

$$\tau_c = (E_{BB'} - E_{AA'})/2b \qquad (1)$$

where b is the Burger's vector and $E_{BB'}$ and $E_{AA'}$ are the surface
energies produced by the slip of the leading and trailing dislocation,
respectively. In the case of a superdislocation moving from its initial
position, it is assumed that the superdislocation prior to the onset of
deformation is at thermal equilibrium with the lattice. (See Fig. 2a.)

What distinguishes a nonequilibrium APB from an equilibrium APB is a
difference in long range order in the vicinity of the APB's. Nonequilibrium APB's are generated by the rapid shearing of two planes in the
wake of a dislocation. Because there is no time for diffusion to occur
in the extreme case, the creation of the APB will disturb the symmetry
of the superlattice only at the two planes forming the APB. On the
other hand, during the formation of an equilibrium APB, diffusional
effects lead to a decrease in the LRO parameter not only at the APB surface, but up to n parallel planes away from the APB [20]. As a result,
the surface energy of an equilibrium APB will be affected by changes in
LRO over a volume of n layers parallel to the APB [21]. Although Brown
recognized the difference in LRO encountered by the leading and trailing
dislocations at the APB surface, he did not consider the difference
between the APB energies in the equilibrium and nonequilibrium cases.

When the leading superpartial, B, slips in γ', it requires energy
$E_{BB'}$, to generate a nonequilibrium or instantaneous antiphase boundary
(BB') which is restricted to two atomic planes. The trailing superpartial, A, annihilates the original planar defect, restoring energy $E_{AA'}$
without restoring order to the disordered volume of the equilibrium APB.
(see Fig. 2b.) Because of the difference in the long range order parameter near the leading and trailing superpartials, the energies
required to move both dislocations cannot be equal. How to model and
then calculate this nonequilibrium APB energy present an interesting
challenge.

(a)   (b)

Fig. 2: (a) Superpartials A and B are at rest and separated by an APB
at thermal equilibrium with the L1$_2$ lattice. (b) Upon loading, leading
superpartial B slips to position B', creating a nonequilibrium APB in
its wake, while trailing superpartial A moves to position A', annihilating the original planar defect. Immediately following annihilation, the
LRO near the original APB is still lower than the LRO of the bulk.

Macroyield of Gamma Prime

Kear and Wilsdorf [9] have most elegantly postulated that unusual work hardening takes place during the initial 0.1% or 0.2% of plastic deformation to give rise to the macroyield or engineering yield strength of gamma prime. They have further proposed that this work hardening can take the form of the leading dislocation cross-slipping from the (111) slip plane onto the low and intermediate temperature sessile (010) plane, driven by the fact that the (010) or cube plane APB is much less than that of the (111) or octahedral plane APB. Being thermally activated, this maneuver will be facilitated as the temperature increases; hence, the positive and sharp temperature dependence of the macroyield strength. The model received valuable qualification by Takeuchi and Kuramoto [1]. This contribution derived the thermally assisted kinetics of the cross-slip by introducing a simple activation analysis of dislocation motion and activated volume. The resulting expression for the yield strength has a reasonably sharp temperature dependence. In the Takeuchi- Kuramoto model, the increase in resolved shear stress (RSS) at temperatures below the peak flow stress temperature is expressed by the relationship:

$$\Delta\tau_{pb} = A \cdot \exp\left[(-H + \tau_{cb} \cdot V)/3kT \right] \qquad (2)$$

where A is a constant, p is the primary plane, the (111) plane, b is the Burgers vector, H is the activation enthalpy, τ_{cb} is the stress component on the (100) plane in the direction of the Burgers vector, V is the activation volume, k is Boltzmann's constant, and T is temperature. This model also accounts for the experimentally observed violation of the Schmid law, i.e., the yield strength depends on both τ_{111} and τ_{100}.

This analysis was further refined and quantified by Pope and coworkers [22], and more recently by Paidar, Pope and Vitek [2]. Paidar et al. took into consideration the effects of dislocation core geometry and transformations on the ability to cross-slip and quantified the driving force for cross-slip to be a geometric anisotropy ratio of $\sqrt{3}$ between the {111} and {001} planes. The Paidar, Pope, Vitek (PPV) model attributes this geometric anisotropy ratio to the anisotropy of APB energies between the {111} and {001} planes. For the core transformation to become operative, this ratio must exceed $\sqrt{3}$. These refinements qualitatively explain the observed tension-compression asymmetry in yield stress.

In assessing this kinetic model for the temperature dependence of cross-slip, we recognize that the sharp and positive temperature dependence of the macroyield would naturally be consistent with a negative strain rate dependence. As temperature increases, the cross-slip model stipulates that cross-slip and sessile locking become easier. Similarly, at any temperature, as strain rate decreases, there is more time for cross-slip to occur and, therefore, it should result in higher strength at lower strain rates and vice-versa. Consistent with the work of Leverant et al. [23], our findings show a positive strain rate effect on the yield strength of directionally solidified gamma prime compound, IC-72, both before and after the peak yield temperature, although the magnitude of the strain rate effect was much greater above the peak yield temperature, Fig. 3. In view of this apparent discrepancy, it can be argued that dislocations can also penetrate the sessile cross-slipped dislocation segments by another thermally assisted process. Hence, the rationale can be that this process will cancel the other thermally assisted process of cross-slip, giving rise to no strain rate sensitivity. The perhaps fatal weakness of this rationalization is that by the same reasoning the temperature dependence will also be negated, but which is just not the case.

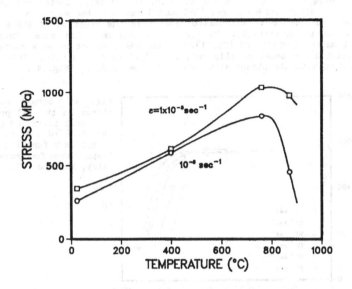

Fig. 3: Strain rate effect on the yield stress of gamma prime as a
function of temperature.

The Takeuchi and Kuramoto derivation, because of its simplifying
assumptions in terms of dislocation velocity, cannot help resolve the
dilemma since it does not explicitly or implicitly include a strain rate
parameter. Indeed, this derivation and the refinements are not even sensitive to strain so that they cannot explain the strain dependent work
hardenings that are observed within the 0.2% plastic range [1].

Although the proposed driving force for cross-slip, the anisotropy
of APB energies, was originally modeled over twenty-five years ago, limited experimental verification has been reported for the Ni₃Al system.
Data reported by Veyssiere et al. [24] show that the ratio of APB
energies between the {111} and {010} planes is closer to unity than to
the √3 factor needed to drive cross-slip in the Paidar et al. mechanism.
Hence, the APB energy reduction driving force for core transformations
may not be the controlling issue. In view of the disparity between
theoretically and experimentally determined APB energies, Yoo recently
proposed an alternative driving force for the cross-slip pinning mechanism. In his model, the cross-slip driving force is a combination of an
elastic anisotropy factor and an APB energy anisotropy term [25]. This
development is providing new impetus to the understanding of the problem.

Microyield and Macroyield of Superalloys

The onset and early stage of slip in gamma prime precipitate strengthened superalloys is different from that in the gamma prime monolith.
In high γ' volume fraction superalloys, the fine cube oriented gamma
prime particles are within submicron distance of each other. Hence,

cutting of the γ' precipitates by the leading superpartial dislocation, is expected right at the onset of slip. Accordingly, Copley and Kear [10] and Gleiter and Hornbogen [26] postulated that the major and first encountered barrier to dislocation motion is the creation of APB's in the gamma prime particle. Furthermore, consistent with some limited experimental observations [23,27], is the small amount of work hardening between the onset of slip and, say, 0.2% plastic strain. This is consistent with the planar slip nature of superalloys. (Fig. 4.)

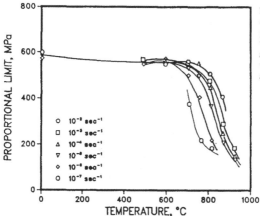

Fig. 4: Strain rate effect on the (a) proportional limit and (b) the 0.2% offset yield stress as a function of temperature for superalloy, Nimonic 115.

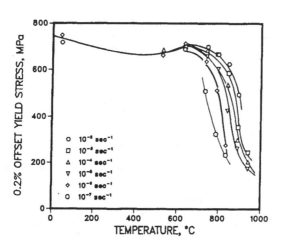

Accordingly, at this stage it is reasoned that both microscopic and macroscopic yielding in high volume fraction gamma prime strengthened superalloys are caused by a common mechanism controlled by the nucleation energy from the formation of APBs. This is in contrast to the gamma prime monolith, which displays unusually high work hardening. In particular, the yield strength, τ_y, in high volume γ' precipitation strengthened superalloys below the temperature of peak yield strength is basically [10]

$$\tau_y = \gamma_{APB}/2b \tag{3}$$

The γ_{APB} term, however, is a nonequilibrium value like E_{BB}, of Eq. (1) because APBs are not present prior to deformation as in the case of γ' alone. Whereas Eq. (1) predicts a low value for the microyield of gamma prime, Eq. (3) accurately predicts a high value for the strength of superalloys, since nothing is subtracted from it.

Shah and Duhl recognized that superalloys derive most of their strength below the peak yield strength temperature from the shearing of γ' particles, and developed a model for the yield strength of single crystal superalloys [28]. The deformation behavior can be quantified by the equation:

$$\sigma_y \cdot m = A \cdot [\gamma(T)]/b + B \cdot (Gb)/R + C \cdot \exp\ (-H_0 + V_1 N \sigma_y m - V_2 Q \sigma_y m)/kT \tag{4}$$

In the above equation, the critical resolved shear stress, $\sigma_y \cdot m$, is the yield strength times the Schmid factor. The first term represents the primary strengthening term, where $\gamma(T)$ is the APB energy as a function of temperature, b is the Burgers vector, and A is a constant. The second term, where G is the γ' shear modulus, B a constant, and R is a geometrical hardening term inversely proportional to γ' size. In the third term, Shah and Duhl attribute an increasing resistance of dislocation shearing of γ' particles to thermally activated cross slip, where H_0 is the thermal activation energy, V_1 and V_2 are activation volumes for cube cross slip and constriction stress interactions respectively, N and Q are Schmid factor ratios, K is Boltzmann's constant, and C is a constant [28].

If both the microscopic and macroscopic yield strength of superalloys are, indeed, due to APB energy, the temperature dependence of APB energy must be characterized. Nonequilibrium APB models like that of Flinn and Brown [3,19] do not address the effect of high temperature in their models. Again, a fuller understanding of this problem is of value not only for its own sake, but also as preparation to tackle the still unexplained quantitative aspects of the fall-off in strength at the higher temperatures.

In view of the above, we have conducted a study to characterize the nature of APB's with temperature in the L1$_2$ superlattice. We have begun the study by thermodynamically modeling equilibrium {111} APB's slip planes and both conservative and nonconservative APB's along the {100} planes [21,29] The detailed results are reported elsewhere in this symposium [21]. Although equilibrium APB's are not truly reflective of an APB created immediately following rapid dislocation slip, they can perhaps indicate the behavior of APB's formed at high temperatures or as the result of very slow deformation rates. The modeling of equilibrium APB's will serve as a foundation to develop a kinetic APB model in the future. To model the equilibrium antiphase boundary energies, the degree of LRO near the APB was calculated by two methods, the Bragg-Williams approximation and the Cluster Variation Method (CVM) in the tetrahedron approximation using interatomic pair interactions.

Calculations obtained by both methods for equilibrium APB's on the (111) as well as the (100) plane reveal that antiphase boundary energy increases with temperature. Unlike the nonequilibrium APB's generated by slip too quickly for entropy to be a factor, our calculations show both the (111) and (001) equilibrium antiphase boundaries become increasingly diffuse with temperature. Contributing to the thickening of the APB with temperature is the effect of configurational entropy on the APB. Experimental APB observations by Veyssiere, Liu, et al. allude to the possibility of the diffusivity of APBs [30,31]. As temperature increases, the LRO parameter decreases faster near the APB than in the bulk. Since the unpinning of an equilibrium superdislocation pair is expected to be thermally activated, one would expect that microyield would also be strain rate dependent.

In addition to being a function of temperature, γ_{APB} was found to be theoretically and experimentally sensitive to ternary additions. As can be seen in Figure 5, our theoretical calculations show the ternary addition Mo increases γ_{APB} more so than W, and thus, Mo should be a better APB energy strengthener than W. The results of our theoretical calculations were verified experimentally by modifying the refractory content of commercially used high volume fraction γ' superalloys. Preliminary tensile and STEM-EDS results suggest that Mo is a better gamma prime phase strengthener than W [32]. In modifying the refractory content of these superalloys, the heat treatments of the alloys were modified according to solidus changes, in order to maintain constant γ' volume fraction, size and distribution. Nevertheless, definite conclusions cannot be drawn since the refractory content affects both γ_{APB} and lattice mismatch. In-situ x-ray studies are currently being completed on the modified superalloys to determine the effect of refractory chemistry on γ/γ' lattice parameter mismatch.

In addition to understanding the effects of substituting various refractory elements in high volume γ/γ' superalloys on coherency strain strengthening, we are also interested in determining the lattice positions of these substituted elements in the $L1_2$ superlattice. The lattice occupancies are currently being determined by anomalous dispersion experiments performed with a synchrotron radiation source [33]. These will provide valuable input data to our calculations in order to fine tune pair interaction energies.

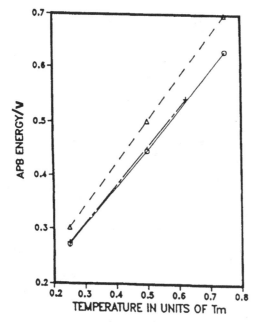

Fig. 5: Effect of ternary additions on the theoretically calculated APB energy of Ni_3Al as a function of temperature.

Legend
○ BINARY Ni_3Al
+ TERNARY(1) Ni_3Al,W
△ TERNARY(2) Ni_3Al, Mo

Creep-Fatigue Aspects

Differences in the strength and work hardening behaviors of γ' and superalloys should be related to their creep-fatigue behavior. The creep-fatigue resistance of directionally solidified (D.S.) and single crystal (S.C.) $Ni_3Al(B,Hf)$ has recently been shown to be superior at 760°C to Nimonic 115, a conventional, wrought, high-strength (45% gamma prime) nickel-base superalloy [34]. In this manner, the creep-fatigue resistance of D.S. and S.C. (within 7° of <001>) $Ni_3Al(B,Hf)$ was shown to be greater than Nimonic 115. This behavior was attributed to the inherent LRO of the material and perhaps enhanced by the lower number of grain boundaries in the intermetallic compound. However, there appears to be an orientation dependence for S.C. specimens of $Ni_3Al(B,Hf)$ with orientations greater than 7° from the <001> axis [35]. This orientation dependence of cyclic strength is currently being related to tensile properties. However, the lack of strain dependence in current tensile theories for $L1_2$ ordered structures makes it difficult to apply these theories to the case of cyclic loading.

The Strength Fall-Off

Long recognized as the temperature limit for the application of superalloys and γ' intermetallics is the strength fall-off between 700°C and 800°C (Figs. 1, 3, 4). This fall-off, occurring at the same range for both, is indeed the strict limit for applications where ambient temperature level brute strengths, and related properties, are required. Where loading conditions are less demanding, the application metal temperature may be as high as 1050°C. However, above the fall-off the design luxury of having a reasonably temperature independent strength is no longer there. The question has been and still is, What causes this fall-off?

That the fall-off is kinetically controlled is indicated by the available but limited strain rate data on superalloys, Figure 3, and on γ' itself [12,23]. It is reported that the strength peak temperature is pushed to higher temperatures when the strain rate is increased. Conversely, at any given temperature in this range, the higher the strain rate, the higher the strength.

There has been no detailed empirical or theoretical treatment of this limiting and scientifically exciting phenomenon. One favored point of view is based on the extension of the cross-slip model [2]. It is argued that at high enough temperatures, like those at the fall-off range, rampant cubic slip will occur, causing the strength to fall off. Further, such cubic slip being thermally assisted, will be kinetically suppressed at, say, higher strain rates. The potential problem with this rationale is that even if it applies to the monolithic case, it may not apply to the superalloys, since $γ_{APB}$ and not cross-slip work hardening is the major strengthening mode in superalloys.

A theory more consistent with the superalloys case is the proposal of vacancy roughening of the APB at higher temperatures [4]. It is argued that APB energy is decreased by the segregation of excess vacancies to the APB at the higher temperatures. Strain rate consistency enters because as strain rate is increased, less time is available for said segregation, and ergo, less segregation and roughening, and higher strength. Both calculations and experimental work indicate that APB's can indeed serve as sites for disordering [30,31]. The kinetic feasibility of such a model remains to be demonstrated. Nature usually being simple may dictate that whichever the fall-off mechanism, it would pivot about the same key element for both the superalloys and the monolithic γ' cases.

Concluding Remarks

At present, the strengthening mechanisms in monolithic γ' is generally understood in terms of anisotropy driven cross slip whereas in high
volume fraction superalloys it is APB pinning. However, several questions concerning the microscopic aspects of such models remain unan-
swered.

The experimental evidence points towards a clear distinction between
micro and macroyield in the γ' intermetallics, while such distinction
appears unnecessary in the two-phase superalloys. The interaction of
superdislocation pairs with APB's is thought to be responsible for the
yield behavior in both systems. The similarities end there, however,
since the structure of the APB's involved in the deformation process of
the intermetallics and of the superalloys are different. We have seen
that by recognizing this difference, one can qualitatively explain the
weak temperature dependence of the yield strength in superalloys as well
as the comparatively low strength of the intermetallics at low temperature.

The cross slip work hardening model postulated to operate in the γ'
has been very successful in explaining the observed tension-compression
asymmetry and the apparent violation of Schmid law. Elastic anisotropy
effects introduced recently have also helped in reconciling the cross
slip model with the relatively low APB anisotropy observed experimentally in Ni_3Al. At present, however, the model does not address the
effect of strain, expected of any work hardening model, or the effect of
strain energy, that should be present in any thermally activated process. These are complex and subtle phenomena, the explanation of which
may require the treatment, on the same footing, of the thermodynamic of
APB, elastic anisotropy and the characterization of the dislocation core
structure.

Acknowledgements

This research was funded by the Office of Energy Systems Research, a
division of Energy Conversion and Utilization Technologies, (ECUT),
under subcontract MRTTA 19x-89664c, through Martin Marietta Energy Systems, Inc. Portions of this work were also partially funded by the
Office of Naval Research under contract N00014-83K-0223. Directionally
solidified gamma prime IC-72 crystals were kindly supplied by Garrett
Turbine Engine Co.

References

1. S. Takeuchi, E. Kuramoto, Acta. Met. 21, 415 (1973).
2. V. Paidar, D.P. Pope, V. Vitek, Acta. Met. 32, 435 (1984).
3. P.A. Flinn, Trans. AIME 218, 145 (1960).
4. John K. Tien,"On The Celestial Limits Of Nickel-Base Superalloys"
   in Superalloys Processing, Metals and Ceramics Information Center,
   September 1972,W1.
5. J.H. Westbrook, Trans. AIME 209, 898, (1957).
6. D.P. Pope, S.S. Ezz, Int. Metals Rev., 29, 136, (1984).
7. N.S. Stoloff. "Ordered Alloys for High Temperature Applications",in High-Temperature Ordered Intermetallic Alloys, Materials
   Research Society, Boston, 1984, pp. 3-27.
8. P.H. Thornton, R.G. Davies, T.L. Johnston, Met. Trans. 1, 207
   (1970).
9. B.H. Kear, H.G.F.Wilsdorf, Trans. AIME 224, 382 (1962).
10. S.M. Copley, B.H. Kear, Trans AIME 239, 984 (1967).

11. B.H. Kear, D.P. Pope, "Role Of Refractory Elements In Strengthening Of Y' And Y' Precipitation Hardened Nickel-Based Superalloys", in Refractory Alloying Elements in Superalloys, ASM Press, Metals Park, OH, 1984, pp. 135-152.

12. R.R. Jensen, J.K. Tien, Met. Trans. A, 16A, 1049 (1985).

13. R.A. Mulford, D.P. Pope, Acta. Met., 21, 1375 (1973).

14. L.E. Willertz, JTEVA, 2, 478 (1974). 15. R.D. Carnahn, J. Of Metals, Dec. 1964, 990. 16. J.S. Koehler, F. Seitz, J. App. Mech. 14, A217 (1947).

17. M.J. Marcinkowski, N. Brown, R.M. Fisher, Acta. Met. 9, 129 (1961).

18. B.H. Kear, A.F. Giamei, J.M. Silcock, R.K. Ham, Scripta. Met. 2, 287 (1968).

19. N. Brown, Phil. Mag. 4, 693 (1959).

20. L.E. Popov, E.V. Kozlov, N.S. Gosolov, Phys. Stat. Sol. 13, 569 (1966).

21. J.M. Sanchez, S. Eng, Y.P. Wu. J.K. Tien, "Modeling of Antiphase Boundaries in L1$_2$ Structures," same symposium.

22. C. Lall, S. Chin, D.P. Pope, Met. Trans. A 10A, 1323 (1979).

23. G.R. Leverant, M. Gell, S.W. Hopkins, Mater. Sci. Engn., 125 (1971).

24. P. Veyssiere, J. Douin, P. Beauchamp, Phil. Mag. 51, 469 (1985).

25. M.H. Yoo, Scripta Met. 20, 915 (1986).

26. H. Gleiter, E. Hornbogen, Phys. Stat. Sol. 12, 235 (1965).

27. R.R. Jensen, Ph.D. Thesis, Columbia University, New York, NY, 1984.

28. D.M. Shah, D.N. Duhl, in Superalloys 1984, edited by M. Gell, et al. (The Metallurgical Society of AIME, Warrendale, PA, 1984), p. 105.

29. Y.P. Wu, J.K. Tien, J.M. Sanchez, Columbia University, unpublished research.

30. P. Veyssiere, Phil. Mag. A 50, 189 (1984).

31. J.A. Horton, C.T. Liu, Acta Met. 12, 2191 (1985).

32. G.E. Vignoul, S. Eng, J.M. Sanchez, J.K. Tien, presented at the 1986 Annual AIME Meeting, New Orleans, LA.

33. D.F. Langiulli, J.K. Tien, Columbia University unpublished research.

34. R.S. Bellows, E.A. Schwarzkopf, J.K. Tien submitted for publication in Met. Trans. A.

35. R.S. Bellows, J.K. Tien, in press, Scripta Met.

THE STRENGTH AND DUCTILITY OF INTERMETALLIC COMPOUNDS:
GRAIN SIZE EFFECTS

E.M. Schulson, I. Baker and H.J. Frost
Thayer School of Engineering, Dartmouth College, Hanover, New Hampshire 03755

INTRODUCTION

Since writing on this subject two years ago [1], a number of developments have occurred, particularly in relation to the mechanical properties of the L1$_2$ nickel aluminide Ni$_3$Al. Some elucidate the nature of the yield strength and the extraordinarily beneficial effect of boron on low-temperature ductility. Some others expose, at least in part, the nature of the marked reduction in ductility at elevated temperatures. Another considers the mechanisms dominating creep deformation. Also during this period, contradictions have appeared: the relationship between the yield strength and the grain size, d, at room temperature has been contested, and opposing views of grain refinement on ductility have been reported.

This paper reviews these developments. Although broadly directed at intermetallic compounds, the discussion is specific to Ni$_3$Al. The hope is that the knowledge and understanding gained about this compound will benefit the class as a whole.

YIELD STRENGTH

Figure 1 shows the yield strength [2], σ_y, versus d$^{-0.5}$ for stoichiometric Ni$_3$Al with (0.35 at.%) and without boron strained at temperatures from 77K to 1023K at a rate of 10^{-4}s^{-1}. At the lower temperatures (< 673K) yielding occurs discontinuously and σ_y is taken as the average stress to propagate Lüders bands along the gauge section of the test specimens. At the higher temperatures, σ_y is taken as the flow stress at an offset corresponding to 0.2% elongation. The data show that at the lower temperatures ($\leq$ 873K), and particularly at 673K and below, grain refinement strengthens the alloys. At the highest temperature it weakens them owing to the onset of grain boundary sliding (discussed later). At 295K, where the data are extensive, the curves for both alloys show positive curvature, implying that neither displays Hall-Petch behavior at this temperature. Similar curvature is present in the plot of the 77K data. At the higher temperatures (673K and 873K) the curves appear to be linear, but owing to the absence of data on large grained material (d> 100μm) cannot be concluded to be so. Regression analysis of the room-temperature data reveals the relationship (regression coefficient > 0.98):

$$\sigma_y = \sigma_0 + k_y d^{-0.80 \pm 0.05} \tag{1}$$

where σ_0 = 93 and 241 MPa, respectively, for Ni$_3$Al and Ni$_3$Al + B and where k_y = 2080 and 1200 MPa $\cdot$ μm$^{0.8}$, respectively.

This unusual dependence has not been seen by others. Liu et al. [3] and Hanada et al. [4] plotted the yield strength of Ni-24 at.% Al + B and Ni-25 at.% Al, respectively, versus d$^{-0.5}$ and drew straight lines through the data. Oya et al. [5] did the same thing for Ni-rich material containing boron. While it is possible that this dependence changes with deviations from stoichiometry, it is unlikely that minor deviations (i.e., Hanada's alloy vs. ours) can account for the difference. One reason for the apparently different dependence may be the difference in the range of grain sizes involved; another may be the method of analysis. The grain size in our work varied by a factor of 550 (2μm to 1100μm) and the data were analyzed statistically. The grain sizes in the other works varied by factors of 27 (9μm to 240μm, Hanada et al.), 11 (19μm to 215μm, Liu et al.) and 69 (4μm to 275μm, Oya et al.) and the data were not analyzed statistically. Another reason may be a difference in the stress taken as the yield strength. We took the average stress to propagate the Lüders band, because it was reproducible and varied by only ±5 MPa during the propagation. The others did not specify their procedure.

Our view, therefore, is that the d$^{-0.8}$ - dependence at room-temperature is real. To explain it, we invoke a work-hardening model of yielding [6]. Accordingly, the yield stress, or the Lüders band propagation stress, is equivalent to the applied stress needed to continue plastic flow through a matrix which has been strained by the composition independent but grain-size dependent Lüders strain, $\varepsilon_L = \lambda d^{-m}$, and is thus hardened by an amount $\Delta \sigma_L$; i.e.,

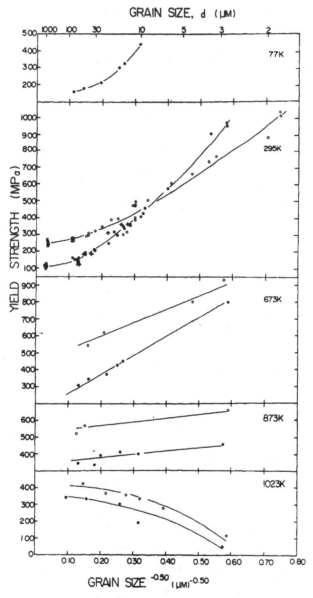

Fig. 1. Yield strength versus $d^{-0.5}$ for stoichiometric Ni_3Al and for Ni_3Al +0.35 at. % B at temperatures from 77K to 1023K. $\varepsilon = 10^{-4}s^{-1}$. (From ref. 2). All data are from consolidated rapidly solidified powder except those for $d > 100\mu m$ which were obtained from homogenized and compressed ingots.

● Ni_3Al tension	O Ni_3Al + B tension
-●- Ni_3Al compression	-O- Ni_3Al + B compression
✖ Ni_3Al ingots	✗ Ni_3Al + B ingots

$$\Delta \sigma_L = \alpha\, Gb\, \rho_L^{\,1/2} = \alpha\, Gb \left(\frac{M\varepsilon_L}{b\beta d}\right)^{1/2} = \alpha\, Gb \left(\frac{M\lambda\, d^{-m}}{b\beta d}\right)^{1/2} = \alpha\, G \left(\frac{bM\lambda}{\beta}\right)^{1/2} d^{-(m+1)/2} \quad (2)$$

where α and β are geometrical constants of order unity, G is the shear modulus, b is the Burgers' vector and M is the Taylor orientation parameter; λ is an experimentally determined Lüders constant. Experimentally, $m = 0.55$ with and without boron [7], Figure 2, leading to a predicted d-dependence of $d^{-0.78}$. The product of coefficients of the d-term in equation (2), incidentally, agrees within a factor of two of measurement, using reasonable values of the parameters [6,7].

The other interesting points about the room-temperature data are the values of σ_0 and k_y (equation 1). The addition of boron raises σ_0 from 93 MPa to 241 MPa, implying that boron increases the lattice resistance to slip by ≈ 400 MPa/at.%. Such potency was noted earlier [3,8] but was underestimated in magnitude owing to the facts that the measurements were made on relatively fine-grained material and that no consideration was given to a possible effect (i.e., reduction) of boron on k_y. As seen from the data, boron lowers k_y from 2080 MPa • $\mu m^{0.8}$ to 1200 MPa • $\mu m^{0.8}$, implying that grain boundaries in the boron-bearing alloy are less effective barriers to slip. This effect is so marked that it leads to boron-induced weakening of the more finely grained ($d < 10\mu m$) aggregates. Hardness measurements [2] versus grain size confirm both the lattice hardening and the grain boundary softening imparted by boron, Figure 3. (Instead of exhibiting $d^{-0.8}$ dependence, however, hardness shows $d^{-0.5}$ dependence, because the plastic strain is independent of grain size; i.e., ≈ 0.08.) When interpreted in terms of the above model of yielding (equation 2), the reduction [6] is attributed to a reduction in the work hardening caused by dislocation pileups at grain boundaries (see below) and, hence, to a decrease in the parameter α.

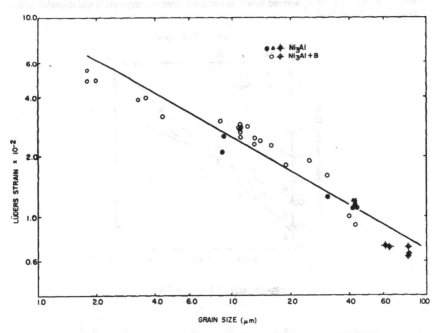

Fig. 2.   Lüders strain versus grain size for stoichiometric Ni$_3$Al with (0.35 at.%) and without boron. The open symbols are for Ni$_3$Al + B. (From ref. 6).

198

The reduction in k_y, we believe, is important not only in relation to the yield strength, but also in relation to ductility. For this reason the effect is discussed below in terms of the grain boundary accommodation of slip.

When the data in Figure 1 are replotted versus temperature [2] three other points stand out, Figure 4. First, the yield strength of the more coarsely-grained polycrystals increases with temperature up to 873K, beyond which it decreases. The yield strength of the most finely-grained material, on the other hand, decreases. In other words, the yield strength-temperature relationship over intermediate temperatures changes from one having a positive slope to one having a negative slope as the grain size decreases, reminiscent of the behavior of another Ll_2 aluminide Zr_3Al [9]. This effect is a manifestation of the increase in lattice resistance to slip (note the extrapolated yield strength at $d^{-0.5} = 0$, Figure 1) and of the decrease in the effectiveness with which grain boundaries impede slip (note the slopes of the curves in Fig. 1) as temperature rises. The effect is less evident in the alloy containing boron, because the grain boundary impediment is less thermally dependent in this alloy.

Another point is that the peak temperature, T_p, decreases with decreasing grain size, from 873K for d = 85μm to ≈ 700K for d = 10μm (Figure 4). This effect is probably related to grain boundary sliding (see below) which is expected to begin at lower temperatures in the more finely-grained material. It might also be related to the transition from octahedral to cubic slip which begins at temperatures around T_p [10], because the lowest temperature at which cubic slip is activated probably decreases with increasing applied stress (i.e., with decreasing grain size).

The third point is that the temperature, T_v, at which the yield strength shows a minimum, increases with decreasing grain size, from < 295K for d = 85μm to ≈ 295K for d = 10μm. This effect is a direct result of the opposing effects of temperature on the lattice resistance and on the grain boundary impediment to slip.

Both of the latter effects, when added to the reduction in the rate of thermal strengthening with decreasing grain size, lessen the difference between the peak strength and the valley strength. Collectively, they imply that an attempt [11] to ascribe fundamental significance to plots of σ_y vs T and to T_v derived fom data obtained from polycrystals is questionable unless grain size is considered.

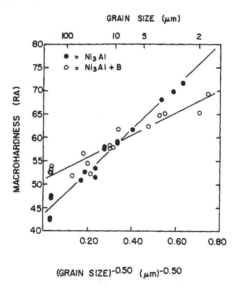

Fig. 3.   Hardness versus $d^{-0.5}$ for stoichiometric Ni_3Al and for Ni_3Al + 0.35 at.% + B. Note that boron hardens the more coarsely grained (d >10μm) aggregates, but softens the more finely grained aggregates. (From ref. 2).

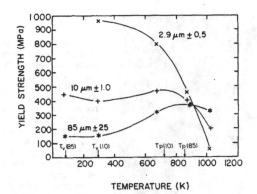

Fig. 4.   Yield strength versus temperature for stoichiometric Ni₃Al of three different grain
sizes. Note that as the grain size decreases, thermal hardening changes to thermal
softening and the temperatures of the peak (T_p) and valley (T_v) decrease and
increase, respectively.

GRAIN BOUNDARY ACCOMMODATION OF SLIP

Returning to the reduction in k_y at room temperature through the addition of boron, it is
thought that the effect reflects the fact [3,12] that boron segregates to grain boundaries and, in
so doing, eases the accommodation of slip at the head of dislocation pileups. Such features
occur early in the deformation history of the material, as revealed through in-situ straining
within a TEM [13] Figure 5, and intensify strain at localized regions on the grain boundary.
Generally, the misorientation across a boundary is such that, when a slip propagation event
occurs, the Burgers' vector of dislocations leaving a boundary is different from that of
dislocations entering it. The vector difference creates a grain boundary dislocation. To allow
other lattice dislocations to enter the boundary on the same slip plane, the boundary dislocations
must move away from the head of the pileup. Otherwise, a crack may nucleate. Boron, we
think, eases this motion and thereby lowers the effectiveness with which the boundaries impede
the transmission of slip.

In ascribing the boron-induced reduction in k_y to changes in the properties of the grain
boundaries, we ruled out a more traditional interpretation in terms of lattice properties, discussed
by Johnston et al. [14]. They proposed from work on both the ordered (L1₂) and the
disordered forms of Ni₃Mn that k_y decreases as the propensity for cross slip increases. We are
unable to invoke this possibility here because the TEM studies [13] show that slip in Ni₃Al with
boron is as planar as it is in the unalloyed material.

The question now is: Why should boron ease the accommodation of boundary slip? We
believe that the answer resides in the structure of the grain boundaries in L1₂ alloys [15, 16] and
in a relaxation of the constraints which long-range atomic ordering imposes upon the atomic
configurations. In more detail, modeling the atomic structure of grain boundaries reveals that
boundary dislocations experience drag from three sources: (i) the difference in energy between
the structure of the original and the slipped boundary; (ii) frictional resistance, analogous to the
Peierls force for a lattice dislocation; and (iii) atomic reshuffling to carry a grain boundary ledge
along with the dislocation. If we assume no ledge at all and assume that the frictional resistance
is negligble, then the change in energy, $\Delta \varepsilon^s_{GB}$, associated with a change in boundary structure
dominates. In L1₂ alloys, $\Delta \varepsilon^s_{GB}$ is expected to be greater than in f.c.c. metals and alloys not
possessing a superlattice, owing to the greater number of possible configurations introduced by
the different possible arrangements of Ni and Al at the boundary. Generally, configurations rich
in Al-Al nearest neighbors will have a higher energy than those that avoid Al-Al neighbors,
because the tendency towards L1₂ ordering is equivalent to reducing the number of Al-Al
nearest neighbors. A particular shear that would recreate the original boundary structure in a
system not possessing a superlattice may produce a change in the degree of order in the Ni₃Al
boundary [16]. Thus, grain boundary accommodation of slip is expected to be more difficult in
Ni₃Al than, for example, in nickel.

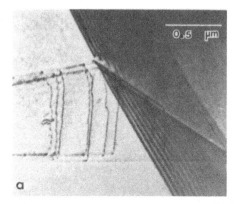

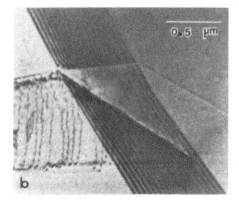

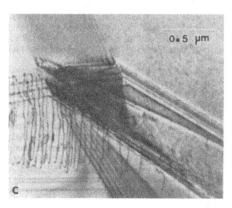

Within this framework, boron is viewed as lowering the difference in energy between different boundary configurations. One speculative possibility is that boron could bond with aluminum in such a way as to reduce the aluminum-aluminum replusion, resulting in a reduction in the energy of those configurations which are rich in Al-Al pairs. However, if boron bonds preferentially to aluminum, then we would expect an increase in boron segregation as the Al concentration increases. In fact, the opposite trend is observed [3]. This then leads to another possibility. Boron may bond preferentially with Ni and lead to an enhancement of Ni and to a depletion of Al at the boundary. A boundary depleted in Al would form fewer high energy Al-Al pairs when it is sheared. The reduced drag could thus account for the reduction in k_y.

Fig. 5. Sequence of transmission electron micrographs showing the development of a dislocation pile-up in stoichiometric Ni_3Al + 0.35 at.% B strained in-situ. In (a) some dislocations are forming the pile-up and others have entered the boundary.

In (b) a slip band has formed in the adjacent grain at the head of the pile-up. In (c) both bands develop upon additional strain.
(From ref. 13).

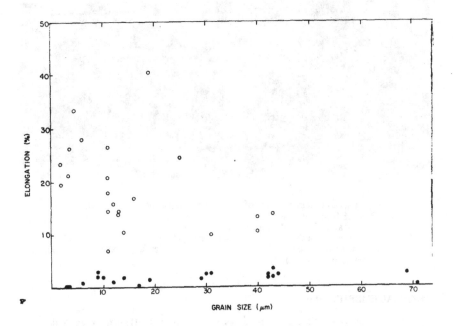

Fig. 6.   Elongation versus grain size for stoichiometric Ni₃Al (closed symbol) and for Ni₃Al + 0.35 at.% B (open symbol) at 295K. (From ref. 2).

DUCTILITY

Speculative though the boundary accommodation model is, it helps to explain another effect of boron; viz., the extensive low-temperature ductility [3,17] imparted to an otherwise brittle solid. This effect has previously been explained [3] in terms of grain boundary cohesion and, given that boron segregates to grain boundaries in Ni₃Al, to a boron-induced improvement in the cohesion. However, it seems that more than boundary cohesion may influence the ductility of polycrystals, because the cohesive energy, γ_c, of unalloyed (i.e., brittle) Ni₃Al ($\gamma_c = 2\gamma_s$ - $\gamma_{g.b}$, where γ_s is surface energy and $\gamma_{g.b}$ is grain boundary energy), judging from calculation [18], is about the same as that for ductile elemental nickel.

If slip accommodation at grain boundaries is relevant to global plasticity, then a brittle to ductile transition is expected upon lowering the grain size to below a critical value. This expectation is based upon the assumption that cracks in Ni₃Al, as in other materials, are proportional in length to the grain size. In fact, such a transition is seen in the room-temperature ductility of stoichiometric Ni₃Al with boron, Figure 6, [2] . Correspondingly, the fracture mode changes from predominantly intergranular to predominantly transgranular cleavage. A transition, although expected, is not seen in unalloyed material, possibly because the size of the finest grains ($\approx$ 3μm) exceeds the critical grain size in this rather fragile material.

Curiously, Hanada et al. [4] report that the room-temperature ductility of stoichiometric Ni₃Al decreases with decreasing grain size, from $\approx$8 % for d = 240μm to $\approx$1 % for d = 9μm. This trend contradicts our results and the generally observed relationship between the grain size and the ductility of low-toughness solids. We cannot explain the difference.

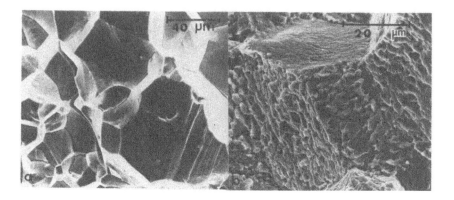

Fig. 7.   Scanning electron micrograph of the fracture surfaces of stoichiometric Ni_3Al +
0.35 at.% B broken in tension:
a) at 295K; d = 38 μm; 10.6% elongation.
b) at 948K; d = 40 μm; 4.8% elongation.
Note the appearance of microvoids on the intergranular facets exposed at the higher
temperature. (From ref. 2).

THERMAL EMBRITTLEMENT

Turning to behavior at elevated temperatures, it is now clear [19-21] that the tensile ductility
of Ni_3Al microalloyed with boron drops to almost zero at temperatures around 873K to 1073K.
Liu and White [20] termed the effect "dynamic embrittlement" and owing to its suppression at
873K under a reduced pressure (10^{-3}Pa) and to the subsequent bending at room temperature of
test specimens broken in air at 873K, attributed it to an oxygen-induced reduction in the strength
of atomic bonds at the tip of propagating cracks. Taub et al. [21] have confirmed the deleterious
role of oxygen. While it is clear that oxygen exacerbates the phenomenom, it is not clear that
oxygen is the only factor: even at the reduced pressure of 10^{-3}Pa significant reduction in
ductility still occurs when the temperature is raised further [21].

The fracture mode accompanying the drop in ductility is intergranular. However, its
character [2] is different from the intergranular fracture generally reported, at least for
stoichiometric Ni_3Al + 0.35 at.% B. Rather than the featureless facets seen after fracture at
low temperatures, the fracture surfaces of the "embrittled" material contains microvoids, Figure
7. This suggests that grain boundary cavitation is a significant factor.

While unambiguous statements are premature, it appears that grain boundary sliding (GBS) may
be important also, possibly as the mechanism of crack nucleation. There is now clear
metallographic evidence [2], at least for stoichiometric Ni_3Al with (0.35 at.%) and without
boron, that GBS occurs at temperatures as low as 673K, Figure 8. The relevant point is that the
sliding leads to wedge cracks which must concentrate stress. If long enough, the attendant
mode-I stress intensity factor, K_I, may be as large as K_{IC} in which case fast fracture may ensue
with little global plasticity. Possibly, the change [10] from {111} to {100} slip, which results
in a reduction from five to three independent slip systems, exacerbates the growth rate.

Two remedies, both related to grain size, suggest themselves. One is to refine the grains to
the point that the cracks are less than supercritical. The other is to increase the length to diameter
ratio of the grains in order to reduce the total grain boundary area normal to the direction of
maximum principal tensile stress. We have explored the first approach using equiaxed
aggregates, and have found [2] it to be successful in stoichiometric Ni_3Al with (0.35 at.%) and
without boron, at least at temperatures from 673K to 1023K, Figure 9. We are now exploring
the second approach.

The improvement in ductility is not without costs, however. The price, at least of the first
approach, is a loss in strength. As noted above, at high temperatures fine-grained material with
and without boron is weaker than coarse-grained material (Figure 1).

Before leaving this topic, one other point is noted. The embrittlement considered here is not unique to Ni_3Al. It is a well known phenomenon in Ni-based superalloys which bear many of the characteristics noted above [22]. An interesting point, yet to be studied, however, is the degree to which the grain boundary cavitation in the $L1_2$ superlattice differs (exceeds?) that within random solid solutions.

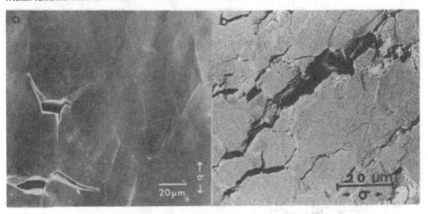

Fig. 8.   SEM micrographs of the gauge sections of stoichiometric Ni_3Al alloys strained to fracture. The arrows denote the tensile axis:
        a) $Ni_3Al + 0.35$ at.% B;   d = 39 µm; T = 673K;   21% elongation.
        b) Ni_3Al;   d = 15 µm; T = 1023K;   8% elongation.

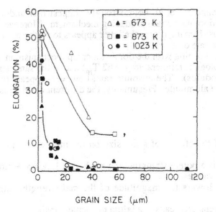

Fig. 9.   Elongation versus grain size of stoichiometric Ni_3Al (closed symbols) and of Ni_3Al + 0.35 at.% B (open symbols) at 673K, 873K and 1023K. Note the brittle to ductile transition upon reducing grain size. (From ref. 2).

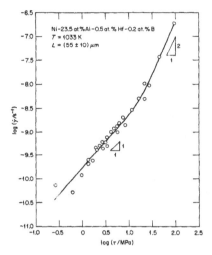

Fig. 10.   The creep rate versus shear stress of a Ni₃Al-based alloy of 55 μm grain size crept at
1033K. (From ref. 23).

CREEP

To our knowledge the only other mechanical property of Ni₃Al - based alloys that has been
examined as a function of grain size is the creep resistance. Schneibel et al. [23] studied the
creep of coils of Ni-23.5 at.% Al-0.5 at.% Hf-0.2 at.% B at temperatures from 873K to 1033K
and at shear stresses from 0.3 to 100MPa. The grain size was varied from ≈12μm to ≈150 μm.
At 1033K and at stresses below ≈ 10 MPa, the creep rate is almost directly proportional to
the applied stress, Figure 10. Also, it is inversely proportional to d $^{-1.9}$, Figure 11 *.

These results, as Schneibel et al. note, strongly suggest that under the conditions noted
diffusional creep is the dominant strain producing mechanism. Moreover, of the possibilities of
Coble creep and Nabarro-Herring creep, the latter appears to be the more likely, given the near
linear dependence of the rate on d $^{-2.0}$.

These findings are in keeping with the dominant creep mechanism in elemental nickel when
crept at the same homologous temperature (0.62 T$_m$) and at the same stress (normalized with
respect to the shear modulus). The absolute rate, however, appears to be about an order of
magnitude lower for the aluminide. Presumably, the difference is the result of a lower volume
diffusion coefficient.

CONCLUSION

From a review of the effects of grain size on the mechanical properties of Ni₃Al it is
concluded that:
(i)   Grain refinement is a potent strengthening method at low temperatures but weakens the
       alloy at high temperatures.
(ii)   Grain refinement lowers the magnitude of thermal strengthening, leading to thermal
       weakening.
(iii) Grain refinement generally leads to a brittle to ductile transition.
(iv)   The addition of boron results in lower grain boundary strengthening, suggesting that it aids
       the accommodation of slip at grain boundaries.
(v)   Creep at intermediate temperatures and low stresses is controlled by the Nabarro-Herring
       mechanism.

---

* Figures 10 and 11 were taken from ref [23] where L is used as grain size.

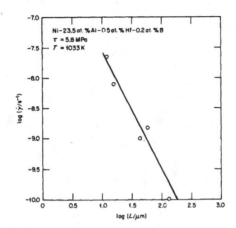

Fig. 11. The creep rate versus grain size for the same Ni$_3$Al-based alloy as in Fig. 10, crept at 1033K under a shear stress of 5.8 MPa. (From ref. 23).

ACKNOWLEDGEMENTS

The authors acknowledge the support of the Office of Basic Energy Sciences of the U.S. Department of Energy, Grant No. DE-FG02-86ER45260 and the support of the SHaRE program of the Oak Ridge Associated Universities, contract No. EY-76-C-05-0033.

REFERENCES

1. E.M. Schulson, Mat. Res. Soc. Symp. Proc., 39, 194 (1985).
2. T.P. Weihs, V. Zinoviev, D.V. Viens and E.M. Schulson, Acta Met. (in press).
3. C.T. Liu , C.L. White and J.A. Horton, Acta Met., 33, 213 (1985).
4. S. Hanada, S. Watanabe and O. Izumi, J. Mat. Sci., 21, 203 (1986).
5. Y. Oya, Y. Mishima, K. Yamada and T. Suzuki, Report Tokyo Institute of Technology, March 1984, pp. 37-49.
6. E.M. Schulson, T.P. Weihs, I. Baker, H.J. Frost and J.A. Horton, Acta Met., 34, 1395 (1986).
7. E.M. Schulson, T.P. Weihs, D.V. Viens and I. Baker, Acta Met., 33 1587 (1985).
8. S.C. Huang, A.I. Taub and K.-M. Chang, Acta Met., 32, 1703 (1984).
9. E.M. Schulson, Int. Met. Rev., 29, 195 (1984).
10. A.E. Staton-Bevan and R.D. Rawlings, Phys. Stat.Sol., 29, 613 (1975).
11. T. Suzuki, Y. Oya and S. Ochiai, Met. Trans. 15A, 173 (1984).
12. M.K. Miller and J.A. Horton, Scripta Met. 20, 789 (1986).
13. I. Baker, E.M. Schulson and J.A. Horton, Acta Met. (in press).
14. T.L. Johnston, R.G. Davies and N.S. Stoloff, Phil. Mag., 12 , 305 (1965).
15. T.Takasugi and O. Izumi, Acta Met., 31, 1187 (1983).
16. H.J. Frost, Acta Met. 35 (1987) 519.
17. K. Aoki and O. Izumi, Nippon Kinzoku Gakkaishi, 43, 1190 (1979).
18. S.P. Chen, A.F. Voter and D.J. Strolovitz, Scripta Met., 20, 1389 (1986).
19. A.I. Taub, S.C. Huang and K.-M. Chang, MRS Symp. Proc., 39, 221 (1985).
20. C.T. Liu and C.L. White, Acta Met. (in press); also Scripta Met., 19, 1247 (1985).
21. A.I. Taub, K.-M. Chang and C.T. Liu, Scripta Met., 20, 1613 (1986).
22. D.A. Woodford and R.H. Bricknell, Treatise on Materials Sci. and Tech., "Embrittlement of Structural Materials ," ed. C.L. Briant, Acad. Press, N.Y., 25, 157 (1983).
23. J.H. Schneibel, G.F. Petersen and C.T. Liu, J. Mat. Res., 1, 68 (1986).

EFFECTS OF ELASTIC ANISOTROPY ON THE ANOMALOUS YIELD BEHAVIOR OF CUBIC ORDERED ALLOYS[*]

M. H. YOO
Metals and Ceramics Division, Oak Ridge National Laboratory, Oak Ridge, TN 37831-6117

ABSTRACT

The positive temperature dependence of yield stress in certain $L1_2$ and $B2$ alloys, e.g. Ni_3Al and β'-CuZn, is analyzed on the basis of the nature of dislocation dissociations predicted by anisotropic elasticity theory. In the case of Ni_3Al, the torque due to the tangential component of the elastic interaction between two superpartials is a major driving force for either the cross-slip pinning model or the force couplet model. The corresponding torque term is relatively unimportant in the B2 structure. The core transformation of individual superpartials may be important for the anomalous increase of yield stress in β'-CuZn.

INTRODUCTION

The anomalous increase in yield and flow stresses with increasing temperature is a unique characteristic of the ordered intermetallic compounds (e.g. Ni_3Al and β'-CuZn). Current understanding of the mechanisms responsible for the anomalous yield and flow behavior can be summarized from the recent papers by Vitek [1], Pope and Ezz [2], Yamaguchi and Umakoshi [3], and Tien [4]. The existing theories belong to two basic categories: intrinsic effects and the cross slip model.

In $L1_2$ alloys, the anisotropy of antiphase boundary (APB) energy, $\gamma_1/\gamma_0 > \sqrt{3}$ where γ_1 and γ_0 are the APB energies on $\{111\}$ and $\{100\}$ planes, respectively, was considered as the necessary condition [5] for the cross-slip pinning model [6]. Recently, Yoo [7] modified the condition by the combined anisotropy of the elastic interaction force and the APB energy, $(1 + f_1\sqrt{2})\gamma_1/\gamma_0 > \sqrt{3}$ where f_1 is an anisotropy factor. In the case of Ni_3Al, the elastic anisotropy, 1.9 with $f_1 = 0.6$, is larger than the APB energy anisotropy, $\gamma_1/\gamma_0 \approx 1.2$ [8,9]. The purpose of this paper is to examine the role of elastic anisotropy in the physical mechanisms for the temperature dependence of yield stress in cubic ordered alloys by considering both intrinsic effects and the cross slip model.

INTERACTION TORQUE

(a) $\underline{L1_2 \text{ structure}}$. The radial and tangential components, $F_r = f_r/2\pi r$ and $F_\theta = f_\theta/2\pi r$, of the interaction force between a pair of superpartial screw dislocations of Burgers vector $\underset{\sim}{b} = [\bar{1}01]/2$ and interspacing r are given by [7,10]

$$f_r = K_s b^2 \tag{1}$$

$$f_\theta = K_s b^2 f(\theta) \tag{2}$$

---

[*]Research sponsored by the Division of Materials Sciences, U.S. Department of Energy under contract DE-AC05-84OR21400 with Martin Marietta Energy Systems, Inc.

208

$$f(\theta) = \frac{(A - 1)\sin 2\theta}{2(A\cos^2\theta + \sin^2\theta)} \quad , \tag{3}*$$

where K_s is the energy factor, $K_s = \sqrt{c_{44}(c_{11} - c_{12})/2}$, A is Zener's ratio of shear anisotropy, $A = 2c_{44}/(c_{11} - c_{12})$, and θ is the angle of rotation about the screw axis between the habit plane of dislocation pair and the reference (010) plane, i.e. θ = 0 for the (010) plane. The elastic constants, c_{11}, c_{12}, and c_{44}, for cubic ordered alloys at room temperature [11] are listed in Table 1. Figure 1 shows the variation of $f(\theta) = F_\theta/F_r$ for Ni$_3$Al and Cu$_3$Au.

Table 1. Elastic constants of ordered cubic intermetallic compounds at room temperature [11]

| Type | Alloy | 10^{11} N/m^2 | | | A | R |
		c_{11}	c_{12}	c_{44}		
	Ni$_3$Al	2.23	1.48	1.25	3.34	1.12
L1$_2$	Ni$_3$Fe	2.46	1.48	1.24	2.53	1.07
	Cu$_3$Au	1.87	1.35	0.68	2.60	1.08
	NiAl	2.12	1.43	1.12	3.28	1.12
B2	AgMg	0.84	0.56	0.48	3.47	1.13
	CuZn	1.29	1.10	0.82	8.49	1.43

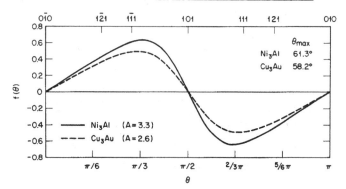

Fig. 1. Variation of $f = F_\theta/F_r$ with respect to the orientation of a pair of screw dislocations along [$\bar{1}$01].

(b) B2 structure. The Burgers vector, b = [111]/2, and the energy factor for a screw dislocation [12], $\bar{K}_s = R/S_{44}$ and R = $\sqrt{S_{11}S_{44}/(S_{11}S_{44} - S_{15}^2)}$, where S_{ij}'s are the modified elastic compliance constants [13] with reference to the dislocation coordinate axes. The ratio of the tangential and radial components is given by [14]

$$h(\theta) = \frac{1 - R^2}{\cot 3\theta + R^2\tan 3\theta} \quad , \tag{4}$$

---

*A typographical error on $\cos^2\theta$ and $\sin^2\theta$ was found in Eq. (1) of Ref. [7].

where θ is measured from the (1̄10) plane. Figure 2 shows the variation of h(θ) = F_θ/F_r for β´-CuZn and NiAl. The anisotropy factors, A and R, relevant to Eqs. (3) and (4) are listed in Table 1.

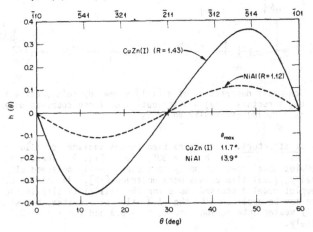

Fig. 2. Variation of h = F_θ/F_r with respect to the orientation of a pair of screw dislocations along [111].

FORCE COUPLET MODEL

(a) L1$_2$ structure. The value of f(θ = 54.74°) for the (111) primary slip plane is $f_1 = \sqrt{2}$ (A − 1)/(A + 2), which gives f_1 = 0.62 for Ni$_3$Al. In order for a pair of superpartial screw dislocations, with the interaction torque on them, to be in equilibrium there must exist a distribution of "force couplets" across the APB interface as is depicted in Fig. 3(a). This would give rise to bending of the APB about the axis parallel to the dislocation line. As such, the nonplanar nature of core splitting of a superdislocation may be classified as an intrinsic effect [2], and it is related to the glide and climb dissociation of fractional edge dislocation pairs of opposite sign [10]. This is analogous to the glide-climb dissociation of a pair of mixed superpartials on {010} planes envisaged by Veyssiére [15]. In either case, the climb force can be induced not only by the elastic interaction, but also by the chemical potentials of the constituent atoms near the APB interface.

The anisotropy factors, A and f_1, of Ni$_3$Al increase slightly with increasing temperature [7]. Therefore, in the case of Ni$_3$Al, super-dislocations lying in the (111) plane are pinned intrinsically by the nonplanar nature of the (111) APB over a wide range of temperature. The temperature dependence of the chemical forces on the fractional edge dislocations are closely related to the local state of order at the APB. It was predicted for thermal APB [16,17] that the long-range order parameter, S, is much smaller on the APB than in the matrix, and the thickness of the APB increases with increasing temperature. Since the elastic aniso-tropy increases with decreasing S [10], the extent of climb dissociation should increase and hence the intrinsic resistance to superdislocation motion increases with increasing temperature. On the other hand, at elevated temperatures where self-diffusion occurs, the mobility of super-dislocations is enhanced by synchronous motion of the fractional edge dislocation in conjunction with local diffusive atomic rearrangement.

210

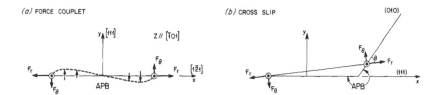

Fig. 3. Schematic description of a $[\bar{1}01]$ screw superdislocation in the $L1_2$ structure. (a) A distribution of force couplets under $\tau_a = 0$, (b) cross slip under $\tau_a = \tau_{yz}$.

(b) B2 structure. The interaction torque vanishes on $\{1\bar{1}0\}$ and $\{11\bar{2}\}$ planes, i.e. h_1 ($\theta = 0°$) = h_2 ($\theta = 30°$) = 0 in Fig. 2. Therefore, the force couplet model does not apply to these commonly observed slip planes. Occasionally, $\{12\bar{3}\}$ slip planes were observed [18]. In this case, the force couplet model discussed above for the $\{111\}<\bar{1}01>$ slip in the $L1_2$ structure may be applicable to the $\{12\bar{3}\}<111>$ slip system, but with relatively weaker interaction torque, $h_3 = 0.3$ and 0.1 for β'-CuZn and NiAl, respectively.

CROSS SLIP MODEL

(a) $L1_2$ structure. Figure 3(b) shows schematically the cross-slip process originally proposed by Kear and Wilsdorf [19]. Takeuchi and Kuramoto [6] proposed a "cross-slip-pinning" model, which describes dynamic breakaway of superdislocations from cross-slipped segments. Paidar et al. [5] prescribed the form of activation enthalpy for the thermally activated process of a double kink formation. The driving force for this process is [7]

$$F = \frac{\gamma_1}{\sqrt{3}} (1 + f_1\sqrt{2}) - \gamma_0 + \tau_c b ,$$ (5)

where τ_c is the resolved shear stress on the (010) cross-slip plane. The following necessary condition results from the combined anisotropy of elastic shear and APB energy:

$$\left(\frac{3A}{A + 2}\right) \frac{\gamma_1}{\gamma_0} > \sqrt{3} .$$ (6)

As was demonstrated earlier [7], the positive temperature dependence of the critical resolved shear stress of (111)$[\bar{1}01]$ slip of Ni$_3$Al [20] is explained very well with the cross-slip pinning model (see Fig. 4). The comparison with the isotropic approximation (A=1, f_1=0) shown in Fig. 4 indicates the importance of the interaction torque as driving force for the cross-slip pinning process.

(b) B2 structure. To interpret their experimental data of anomalous yield behavior in β'-CuZn single crystals, Umakoshi et al. [21] suggested a cross slip model analogous to the one proposed for the $L1_2$ structure [6]. The leading superpartial <111>/2 screw dislocation is to cross slip from $\{1\bar{1}0\}$ plane to $\{11\bar{2}\}$ plane. Though small as the difference between the two as may be, only 15% in the B2 structure [22], the $\{11\bar{2}\}$ APB energy is higher than the $\{1\bar{1}0\}$ APB energy. Therefore, additional sources of driving force for the thermally activated process of cross slip are required, such as a high resolved shear stress on the $\{11\overline{2}\}$ plane in the twinning sense [18], and the interaction torque due to elastic anisotropy. The latter is zero in both $\{1\overline{1}0\}$ and $\{11\overline{2}\}$ planes, as shown in Fig. 2.

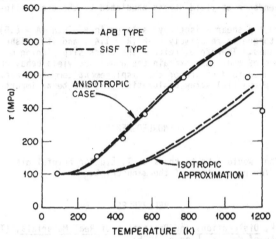

Fig. 4. Anomalous increase of yield stress in Ni_3Al. Data points are from Ref. [20].

DISCUSSION

In the case of Ni_3Al, the relationship between the combined elastic and APB anisotropy, Eq. (6), and the positive temperature dependence of yield stress is relatively well known (See Table 2). The recent atomistic calculations of γ_1/γ_0 based on the embedded atom method [23,24] indicate a sensitive dependence of the calculated results on the input parameters related to the so-called embedding energy. Referring to the experimentally determined value of $\gamma_1/\gamma_0 \approx 1.2$ [8,9], one can conclude that the interaction torque due to the elastic anisotropy plays a major role in the cross-slip pinning process which gives rise to the positive temperature dependence of yield stress.

Table 2. Anisotropy factors and the temperature dependence of yield stress in Ll_2 ordered alloys

Alloy	Elasticity $\left(\dfrac{3A}{A+2}\right)$	APB energy $\left(\dfrac{\gamma_1}{\gamma_0}\right)$	[Ref.]	Yield stress $\dfrac{\Delta\tau}{\Delta T}$
Ni_3Al	1.9	1.2	[8,9]	+
		$1.7 - 3.4$	[23,24]	
Ni_3Fe	1.7	4.5	[25]	? (+)
Ni_3Ga	? (A > 0.2)	6.5	[26]	+

212

A new model, "force couplet" model, which predicts intrinsic effects of both positive and negative temperature dependences of yield stress, is introduced. The nonplanar dissociation of a superdislocation in Ni_3Al is directly related to the interaction torque between two superpartials. These two intrinsic effects should be superposed on the positive temperature dependence of yield stress predicted by the cross-slip-pinning model. Although the shear anisotropy is large in β'-CuZn (A = 8.5), the interaction torque is relatively weak (R = 1.4), and it vanishes on $\{\bar{1}10\}$ and $\{11\bar{2}\}$ planes. Therefore, neither the cross-slip-pinning model, nor the force couplet model can explain the anomalous yield behavior in β'-CuZn. Rather, the three-fold nature of the displacement component normal to the individual superpartial screw dislocation [27] may be an important factor in this case.

ACKNOWLEDGMENTS

The author would like to thank C. T. Liu for helpful discussions and C. L. Dowker for preparation of the manuscript.

REFERENCES

1. V. Vitek, Dislocations and Properties of Real Materials, (The Institute of Metals, London, 1985), p. 30.
2. D. P. Pope and S. S. Ezz, Int. Metal. Rev. 29, 136 (1984).
3. M. Yamaguchi and Y. Umakoshi in The Structure and Properties of Crystal Defects, edited by V. Paidar and L. Lejcek, (Elsevier Science Publ., New York, 1984), p. 131.
4. J. K. Tien, in High Temperature Ordered Intermetallic Alloys (in this issue).
5. V. Paidar, D. P. Pope, and V. Vitek, Acta Metall. 32, 435 (1984).
6. S. Takeuchi and E. Kuramoto, Acta Metall. 21, 415 (1973).
7. M. H. Yoo, Scr. Metall. 20, 915 (1986).
8. J. Douin, P. Veyssiére, and P. Beauchamp, Philos. Mag. 54, 375 (1986).
9. J. A. Horton and C. T. Liu, Acta Metall. 12, 2191 (1985).
10. M. H. Yoo, Acta Metall. (in press).
11. M. H. Yoo, to be published.
12. A. K. Head, Phys. Status Solidi 6, 461 (1964).
13. A. N. Stroh, Philos. Mag. 3, 625 (1958).
14. J. P. Hirth and J. Lothe, Phys. Status Solidi 15, 487 (1966).
15. P. Veyssiére, Philos. Mag. A50, 189 (1984).
16. R. Kikuchi, J. Phys. Chem. Solids 27, 1305 (1966).
17. L. E. Popov, E. V. Kozlov, and N. S. Golosov, Phys. Status Solidi 13, 569 (1966).
18. M. Yamaguchi in Mechanical Properties of BCC Metals, edited by M. Meshii (TMS-AIME, Warrendale, PA, 1981), p. 31.
19. B. H. Kear and H. G. F. Wilsdorf, Trans. TMS-AIME 224, 382 (1962).
20. S. S. Ezz, D. P. Pope, and V. Paidar, Acta Metall. 30, 921 (1982).
21. Y. Umakoshi, M. Yamaguchi, Y. Namba, and K. Murakami, Acta Metall. 24, 89 (1976).
22. P. A. Flinn, Trans. TMS-AIME 218, 145 (1960).
23. S. P. Chen, A. F. Voter, and D. J. Srolovitz, Scr. Metall. 20, 1389 (1986).
24. S. M. Foiles and M. S. Daw, J. Mater. Res. (in press).
25. G. Inden, S. Bruns, and H. Ackermann, Philos. Mag. 53, 87 (1986).
26. K. Suzuki, M. Ichihara, and S. Takeuchi, Acta Metall. 27, 193 (1979).
27. M. H. Yoo and B. T. M. Loh, J. Appl. Phys. 43, 1373 (1972).

STRENGTHENING OF Ni_3Al BY TERNARY ADDITIONS

F. HEREDIA AND D. P. POPE.
University of Pennsylvania, Department of Materials Science and
Engineering, 3231 Walnut Street, Philadelphia, PA 19104

ABSTRACT

It is important for a number of reasons to have a basic
understanding of the mechanisms by which ternary additions strengthen
Ni_3Al. First of all, since the basic strength-controlling mechanisms in
pure Ni_3Al are different from those in pure metals and substitutional
solid solutions, it is expected that the mechanisms of solid solution
strengthening will also be different in Ni_3Al. Secondly, such an
understanding will provide valuable insights into the properties of
nickel-base superalloys in which Ni_3Al is a key constituent. In addition,
since new alloys based on an ordered Ni_3Al matrix are being developed, it
is important to understand the strengthening mechanisms in such alloys. In
the present study, flow stress measurements have been performed on single
crystals of Ni_3Al containing additions of Hf and Ta, and on binary Ni
rich, Ni_3Al used as a reference alloy. The data have been collected over a
wide range of temperatures, for different orientations within the unit
triangle, and as a function of the sense of the applied uniaxial stress.
The effect of such additions on the critical resolved shear stress (CRSS)
for octahedral slip has been determined and combined with previous data.
An attempt is then made to clarify whether a lattice parameter/modulus
mismatch effect or a dislocation core effect is the dominant mechanism for
the strength increase with compositional changes. It appears that a
lattice parameter/modulus mismatch is the dominant mechanism for
orientations in which the tension/compression flow stress asymmetry
disappears.

INTRODUCTION

At most temperatures the flow of Ni_3Al is controlled by the motion
on (111) planes of dislocations having a <110> Burgers vector. In order to
reduce the total dislocation energy, the $[\bar{1}01](111)$ dislocation
dissociates into two superpartials separated by an anti-phase boundary
(APB), a localized planar region where the order has been disrupted by the
leading dislocation. Since in the Ll_2 structure the Burgers vector of the
dislocation is twice as long as in the fcc materials, more possibilities
of dislocation dissociation exist in Ll_2 materials than in fcc materials.
This means that there are more possibilities for the strength of Ni_3Al to
be affected by the nature of dislocation core dissociations. It is
important to remark at this point that the APB fault energy is
anisotropic, and it has been found by means of nearest-neighbour
interaction calculations, to be a minimum on (010) planes (no nearest-
neighbour violations) and much higher on (111) planes [1-7]. Such a
difference in the APB energy promotes cross slip of screw dislocations
from (111) onto (010) planes, after which the dislocations become sessile.
Since the process is thermally activated, a positive temperature
dependence of the flow stress is expected, a fact which has been
experimentally confirmed many times [8-14]. In addition, the process has

214

been shown to also depend on the resolved shear stress (RSS) on the cube slip system [10] and on the effects of dislocation core dissociation, which explain, among other things, the existence of a tension/compression asymmetry [6].

The flow stress of Ni_3Al is known to be affected by additions of ternary elements that are soluble in the γ' phase (Ni_3Al) but whose substitutional behavior is still controversial since contradictory results have been obtained depending on the experimental technique used to determine the sites [15,16,17,18]. Most experiments have been performed on polycrystalline material, although some work on single crystalline material has been reported, but in either case the positive effect of ternary additions on the yield strength is a common fact. Additions of elements like Ti, Zr, Nb, Hf, and Ta to polycrystalline Ni_3Al result in a remarkable increase of the flow stress at low temperatures as well as a more rapid increase with increasing temperature [12,15,18,19,20,21,22]. Curwick [24], working with single crystals, measured the CRSS of Ni_3Al with additions of Mo, Nb, Ta, Ti, and W as a function of temperature and orientation of the compression axis. He found that all the ternary additions increase the CRSS for octahedral slip and that it depended both on temperature and concentration. Pope and coworkers [1,4,7] found similar effects of Nb and Ta on single crystals for a wider range of orientations of the tension/compression axis. There are two points of view regarding the mechanisms of strength increase: One group assumes that the effect is due to the usual lattice parameter and modulus mismatch mechanisms which occur in disordered systems. The other group holds that the flow strength increase is primarily due to changes in the frequency of cross slip, i.e., it is related to the anisotropy of the APB energy, specifically the difference between the APB energies on (111) and (010) planes, the main driving force for (111) to (010) cross slip. This latter viewpoint is contained in the model proposed by Wee et al. [25,26].

In the current investigation, which still in progress, we are attempting to separate the effects of lattice and modulus misfit from the cross slip effect. The tensile and compressive CRSS of single crystalline samples of ternary $Ni_3(Al,X)$ alloys are measured as a function of temperature, composition, and orientation. Since the difference between the tensile and compressive CRSS can only be due to cross slip, it can be taken as a measure of the cross slip rate and the effect of the ternary additions on the cross slip rate can then be weighted.

EXPERIMENTAL

Single crystal bars of binary nickel rich Ni_3Al and Ni_3Al containing additions of Hf and Ta were obtained from the TRW Aircraft Components Group. The chemical composition of the binary Ni_3Al was 76.6 at.% Ni and 23.4 at.% Al. Two compositions were obtained for Ta: (a) 24.0 at% Al, 1.0 at% Ta, and 75.0 at% Ni; (b) 22.5 at% Al, 2.5 at% Ta, and 75.0 at% Ni. Two compositions have been used with additions of Hf: (a) 23.2 at% Al, 1.0 at% Hf, and 75.8 at% Ni; (b) 21.9 at% Al, 3.3 at% Hf, and 74.8 at% Ni. The bars were heat treated for 24 hours in vacuum at 1477K. Oriented samples with 10/32 threaded grips, a 6.5mm gauge length and a 2.5mm gauge diameter were manufactured from the homogenized single crystals by grinding. The orientation of these samples are shown in Fig. 1 in the standard [001]-[011]-[$\bar{1}$11] unit triangle. Some Schmid factors are listed in table I. Due to limitations in size of the single crystal bars only two orientations were obtained from the bars with 2.5 at% Ta and 1.0 at% Hf.

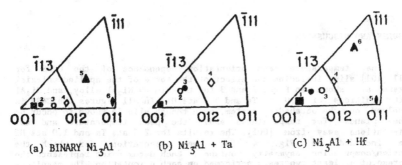

(a) BINARY Ni₃Al   (b) Ni₃Al + Ta   (c) Ni₃Al + Hf

Fig. 1. Orientations of the samples tested in the current study. The numbers refer to the orientation column in table I.

TABLE I. Schmid factors of the tested samples.

Composition	Orientation	(111)[$\bar{1}$01]	(001)[$\bar{1}$10]
Binary	1	0.449	0.087
	2	0.463	0.137
	3	0.491	0.200
	4	0.499	0.270
	5	0.479	0.417
	6	0.429	0.363
1.0 at% Ta	1	0.440	0.078
	2	0.478	0.264
	3	0.489	0.319
	4	0.492	0.402
2.5 at% Ta	1	0.435	0.061
	2	0.478	0.234
1.0 at% Hf	1	0.431	0.070
	2	0.474	0.186
3.3 at% Hf	1	0.453	0.104
	2	0.475	0.227
	3	0.489	0.278
	4	0.478	0.369
	5	0.435	0.376
	6	0.412	0.488

The tension/compression tests were performed in an Instron Testing Machine at a nominal strain rate of 1.3×10^{-4} s^{-1} at temperatures from 77 to 1250K. Tests from room temperature to 1250K were carried out in air with the tension/compression system enclosed in a furnace. The samples were loaded up to a 0.2% nominal plastic strain by the same technique described in a previous paper [27].

RESULTS AND DISCUSSION

The temperature and orientation dependence of the CRSS for (111)[101] slip, including the effect of the sense of the applied uniaxial stress are shown in Figs. 2 and 3 for the binary Ni₃Al alloy, and Ni₃Al with additions of 1.0 at% Ta and 3.3 at% Hf. In all figures the CRSS is resolved on the octahedral system, even though slip is known to occur on cube planes for temperatures above the peak temperature and for orientations away from [001]. The results for 2.5 at% Ta and 1.0 at% Hf are included in Fig. 4 for the orientation for which the tension/compression asymmetry vanishes. Each datum point represents the average of at least two tests performed on each orientation. The magnitude of the CRSS at temperatures below the peak temperature depends on the orientation of the tension/compression axis, but the effect is accentuated with additions of Ta and Hf. As can be seen from figures 2 and 3, the value of the CRSS is always increased by the addition of the ternary elements and the samples oriented near [001] show the largest improvement in strength, and the magnitude of the tension/compression asymmetry also increases.

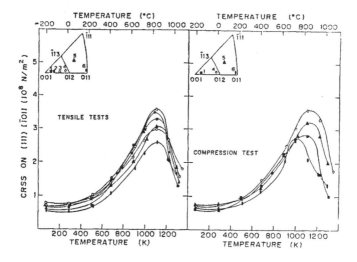

Fig. 2. The CRSS for (111)[101] slip of binary Ni₃Al as a function of temperature and orientation of the stress axis.

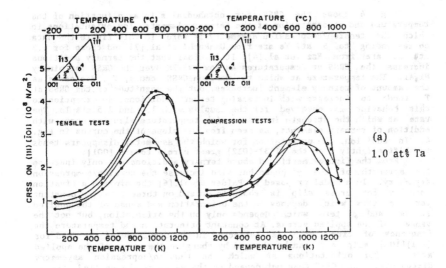

(a)

1.0 at% Ta

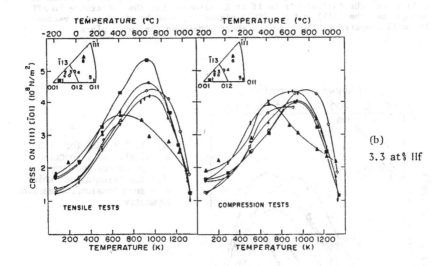

(b)

3.3 at% Hf

Fig. 3. The CRSS for (111) [Ī01] slip as a function of temperature and orientation of the stress axis for Ni$_3$Al with additions of (a) 1.0 at% Ta, and (b) 3.3 at% Hf.

Fig. 4 shows the CRSS for octahedral slip as a function of the temperature and concentration of the ternary elements, for orientations in which the tension/compression asymmetry is zero. In this figure the data corresponding to 5 at% Ta are from Umakoshi et al.[7] and those for 4.3 at% Nb are from Ezz et al.[4]. It is clear that the ternary additions increase the CRSS at temperatures as low as 77K over the CRSS for binary Ni₃Al. The temperature at which the maximun CRSS occurs, T_p, decreases as the amount of ternary element increases, but the magnitude of the CRSS at T_p tends to increase with increasing ternary additions. An exception to this behavior is observed for the samples with 1.0 and 2.5 at% Ta. The rate at which the strength increases with temperature also increases with addition of ternary elements, as seen from the slope of the curves in Fig. 4. In addition, the orientation for which the asymmetry disappears tends to shift slightly from the [Ī13]-[012] great circle toward [001].

From the figures mentioned above ternary additions not only increase the strength of the γ' phase but also increase the tension/compression asymmetry. In a model proposed by Paidar et al.[6] the minimun activation enthalpy for cross slip is shown to depend on three terms, a constant term, a term which depends on the orientation and sense of the applied stress, and a term which depends only on the orientation, but not the sense, of the applied stress. At constant orientation and temperature the frequency of cross slip is reflected in the magnitude of the CRSS for (111)[Ī01] slip and can be changed by changing the sense of the applied stress. For orientations at which the tension/copression asymmetry disappears, the CRSS does not depend on the sense of the applied stress. For these orientations, equation (23) in Paidar et al.[6] can be shown to reduce to the following form if it is assumed that the difference in APB energy on the (111) and the (001) planes (divided by the Burgers vector) is small compared to the resolved shear stress on the (001) plane:

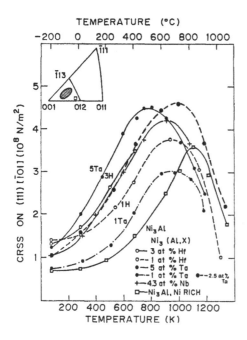

Fig. 4. The CRSS for octahedral slip as a function of temperature and composition for orientations which show zero tension/compression asymmetry.

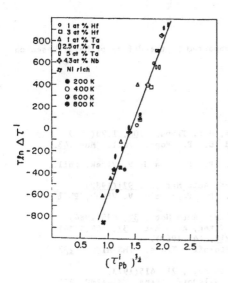

Fig. 5 . $T\ln\Delta\tau^i$ as a function of composition and RSS for a given temperature.

$$T\ln\Delta\tau^i = A - B(\tau_{pb}^i)^{1/2} \qquad (1)$$

where $\Delta\tau^i$ is the difference between the RSS at a determined temperature with respect to the RSS at 77K for the same ith ternary addition; τ_{pb}^i is the resolved shear stress on the primary (111) plane in the direction of the [$\bar{1}$01] Burgers vector, also for the same ith ternary addition, and A and B are constants. Fig. 5 shows the result of plotting $T\ln\Delta\tau^i$ vs. $(\tau_{pb}^i)^{1/2}$. A straight line is obtained in which seven different compositions and four different temperatures are represented. Further analysis of these results is necesary, but for now the results seem to indicate that the APB energy differences do not depend on temperature or composition, i.e., only ordinary solid solution strengthening effects are important.

CONCLUSIONS

According with the results obtained so far, the core dissociation effect is an important mechanism for the strenthening of $Ni_3(Al,X)$ for these orientations in which there is a tension/compression asymmetry but where it disappears, solid solution strengthening effects are the dominant ones. The increase of the strength due to ternary additions is related to the orientation of the stress axis with the [001]-oriented samples the most affected. In general there is agreement regarding the global effects of ternary additions to Ni_3Al with those found in polycrystalline and single crystalline material. Further studies with additions of elements like Zr to single crystalline material will improve the present understanding of the strengthening mechanisms taking place in $Ni_3(Al,X)$.

ACKNOWLEDGEMENT

The present work is being supported by the Office of Naval Research under the grant no. 5-21233.

REFERENCES

1. C. Lall, S. Chin and D. P. Pope, Met. Trans., 10A, 1323(1979
2. M. Yamaguchi, V. Paidar and D. P. Pope, Phil. Mag.,43, 1027(1981)
3. M. Yamaguchi, V. Paidar, D. P. Pope and V. Vitek, Phil. Mag.,45, 867(1982)
4. S. Ezz, D. P. Pope and V. Paidar, Acta Met.,30. 921(1982)
5. V. Paidar, M. Yamaguchi, D. P. Pope and V. Vitek, Phil. Mag.,45, 883(1982).
6. V. Paidar, D. P. Pope and V. Vitek, Acta Met., 32, 435(1984).
7. Y. Umakoshi, D. P. Pope and V. Vitek, Acta Met., 32, 449(1984).
8. P. A. Flinn, Trans. TMS-AIME, 218, 145(1960).
9. R. G. Davies and N. S. Stoloff, Trans. TMS-AIME, 233, 714(1965).
10. S. Takeuchi and E. Kuramoto, Acta Met., 21, 415(1973)
11. B. H. Kear and H. F. G. Wilsdorf, Trans. TMS-AIME, 224, 382(1962).
12. P. H. Thornton, R. G. Davies and T. L. Johnston, Met. Trans., 1A, 207(1970).
13. R. A. Mulford and D. P. Pope, Acta Met., 21, 1375(1973).
14 S. M. Copley and B. H. Kear, Trans.AIME, 239, 977(1967).
15. S. Ochiai, Y. Oya and T. Suzuki, Acta Met., 32, 289(1984).
16. M. K. Miller and J. A. Horton, present conference.
17. J. M. Williams, H. G. Bohn, J. H. Barrett, C. T. Liu and J. P. Sjoreen, present conference.
18. H. Lin, E. Seiberling, P. Lyman and D. P. Pope, present conference.
19. R. D. Rawlings and A. Staton-Bevan, J. Metals Sci., 10, 505(1975).
20. K. Aoki and O. Izumi, Phys. Stat. Sol.(a), 38, 587(1976).
21. K. Aoki and O. Izumi, Phys. Stat. Sol.(a), 32, 657(1975).
22. O. Noguchi, Y. Oya and T. Suzuki, Met. Trans., 12A, 1647(1981).
23. Y. Mishima, S. Ochiai, M. Yodogama and T. Suzuki, Trans. Jap. Inst. Met., 27, 41(1986).
24. L. R. R. Curwick: Ph.D. Thesis, University of Minnesota (1972).
25. D. M. Wee and T. Suzuki, Trans. JIM, 20, 634(1979).
26. D. M. Wee, O. Noguchi, Y. Oya and T. Suzuki, Trans. JIM, 21, 237(1980).
27. F. E. Heredia and D. P. Pope, Acta Met., 34, 279(1986).

EFFECTS OF Fe AND Mn ADDITIONS ON THE DUCTILITY AND FRACTURE OF Ni₃Al

DENNIS M. DIMIDUK*, VICKY L. WEDDINGTON**, AND HARRY A. LIPSITT**
*AFWAL Materials Laboratory, AFWAL/MLLM, Wright-Patterson AFB, OH 45433-6533
** Wright State University, Dayton, OH 45435-0001

ABSTRACT

Recent studies suggested that additions of either Fe or Mn to polycrystalline Ni3Al without boron, result in a ductile alloy. In the present study the effects of Fe and Mn additions on the strength and ductility of Ni₃Al were investigated at 25°C, 600°C in air and at 600°C in vacuum. Metallographic examination showed that with 4 at.% solute a single phase structure existed but at higher solute levels the alloys were two phase. Single phase materials tested in bending showed virtually no plastic strain and low bend strengths at all temperatures. The two phase materials tested at 25°C and 600°C in vacuum showed marked ductility. However, the same alloys tested at 600°C in air exhibited brittle failure. These results suggest that Fe and Mn have no effect on the high temperature dynamic environmental embrittlement known to occur in Ni₃Al.

INTRODUCTION

Takasugi, et al. [1], demonstrated that room temperature ductility could be achieved in alloys of Ni₃Al by additions of the substitutional solutes Fe and Mn without boron present. These results were rationalized on the basis of their coincident site lattice, quasi-chemical modeling of the grain boundary structures in L1₂ alloys [2], along with their correlations of valence differences between the constituents of a compound and the observed ductile or brittle behavior of that compound [3]. In the present study, the elevated temperature behavior of the Fe and Mn modified alloys identified by Takasugi, et al. [1], was investigated along with further evaluation of the 25°C behavior. The effects of Fe and Mn additions are interpreted in light of the microstructural condition of the alloys.

EXPERIMENTAL

Ni-Al-Fe and Ni-Al-Mn compositions were prepared by arc melting. The nominal compositions of the Fe modified alloys are shown in Fig. 1 on the 1200°C isotherm of the Ni-Al-Fe system. Alloy compositions are referred to in atomic percent in the text. The isotherm shown has been constructed from the data of Lipsitt [4] and the microstructural evaluations of the present study. Both the Fe and the Mn compositions studied are listed in Table I. The 200 gm arc melted buttons were given initial homogenization and ordering heat treatments (in air) which were otherwise identical to those reported by Takasugi, et al., [1]. For convenience, these samples are referred to as the "T" heat treated samples throughout the paper. Additional heat treatments, as described in the text were used to achieve other microstructural variations in two of the Fe containing alloys. The heat treated buttons were sectioned into bending specimens with the approximate dimensions of 3 mm x 6 mm x 25 mm. All of the bending specimens had the elongated cast grain structure aligned roughly parallel to the 6 mm dimension of the specimen in a comparable way to the specimens of Takasugi, et al.[1]. Grain dimensions of several hundred micrometers in diameter by several millimeters

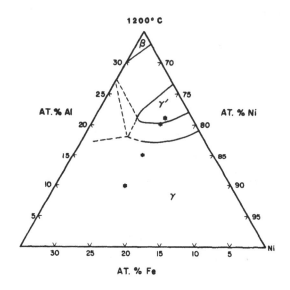

Figure 1. 1200°C Isothermal Section of the Ni-Al-Fe Phase Diagram.
Asterisks indicate compositions evaluated in present study.

TABLE I - NOMINAL COMPOSITION, a/o

Ni	Al	Fe	Mn
75	21	4	-
75	20	5	-
75	15	10	-
75	10	15	-
75	21	-	4
75	16	-	9
75	10	-	15

in length existed in all of the samples. Four point bending tests were
conducted in air at 25°C and 600°C and in a vacuum of 5 x 10⁻⁵ Torr at
600°C. All tests were performed with a cross head speed of 0.25 mm/min.
Optical and scanning electron microscope (SEM) examinations were
carried out on as-polished metallographic specimens and on specimens etched
with either Kallings, Glyceregia, or Marbles etchants. Transmission electron microscope (TEM) samples were prepared by twin-jet electropolishing in
a solution of 2% perchloric acid in butoxy-ethanol at 0°C and 80 volts.

RESULTS AND DISCUSSION

Alloy Microstructures

Metallographic examination, including compositional images from backscattered primary electrons in the SEM, revealed that a segregated,
dendritic microstructure persisted in all of the alloys after the T heat

treatments. Further, all of the T treated alloys, were multiphase. The 4% Mn alloy and the 4% and 5% Fe alloys were expected to be single phase below 1125°C based on available phase diagrams, however, some two phase regions were observed. These observations indicate that the maximum temperatures of 1000°C and 1050°C used in the T treatment are inadequate for complete homogenization in reasonable times. Detailed analysis of the two phase structure was performed in the Fe containing alloys. Figure 2(a) shows the dendritic structure of the 10% Fe alloy after the T heat treatment. Figure 2(b) reveals a two phase precipitate structure in the same alloy. The bright field TEM image in Fig. 2(c) and the dark field image in Fig. 2(d) show a bimodal distribution of the coherent precipitate phase. Electron diffraction studies revealed that the precipitate phase has the $L1_2$ ordered structure, however, the structure of the continuous matrix phase could not be determined unequivocally. The presence of an extremely fine (approximately 10 nm in diameter), uniform distribution of the ordered phase [Fig. 2(d)] prohibited sampling of the matrix region alone, even with a micro-diffraction beam. There was no evidence from the dark field microscopy studies to suggest ordering of the matrix phase. Figure 3 shows a similar dendritic, two phase microstructure in the 9% Mn alloy.

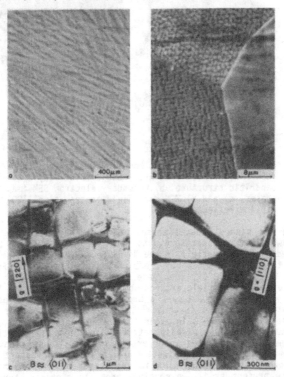

Figure 2.   Microstructure of the 10% Fe Alloy. "T" heat treatment: 1000°C/24 hr. + 600°C/24 hr. + 500°C/72 hr. a) Optical image of dendritic structure, b) secondary electron SEM image of precipitates, c) TEM bright field two-beam image near <011> zone axis, and d) TEM dark field image using {110} superlattice reflection.

224

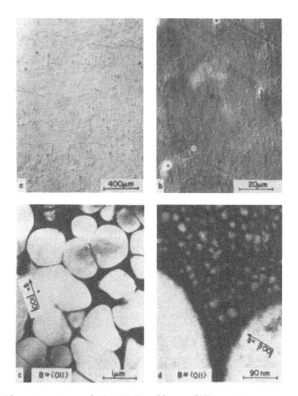

Figure 3.   Microstructure of the 9% Mn alloy. "T" heat treatment:
1050°C/24 hr. 800°C/24 hr. + 500°C/72 hr. a) Optical image of
dendritic structure, b) secondary electron SEM image of two
phases, c) and d) TEM dark field images using {100} superlattice reflection.

While the binary ordered phase, Ni_3Fe, is known to exist at temperatures below 503°C, no literature data could be found to show the extent of
the ordered Ni_3Fe phase field in the Ni-Al-Fe system. The final temperature
for the T treatments was 500°C. The binary phase field of Ni_3Fe has a width
of less than 2% at 500°C [5]. Phase diagrams by Bradley [6] indicate that
the disordered γ-phase field extends nominally parallel to the Ni-Fe axis of
the ternary Ni-Al-Fe system over a broad range of temperatures. Such data
implies that Al must have an unusually profound effect on the width of the
Ni_3Fe phase field at 500°C in order for the Fe containing alloys considered
in this study to have an ordered matrix. Similar phase relationships must
hold in the Ni-Al-Mn system in order for the Mn alloys to have an ordered
matrix, however, no literature data could be found to support this. Based
on the known phase relationships in these systems and the microstuctural
data of the present study, we conclude that the microstructures of the 10
and 15% Fe alloys and the 9 and 15% Mn alloys after the T heat treatment
consist of continuous γ-phase with a varying volume fraction of the Ll_2
γ'-phase; a microstructural condition which is common to the nickel base
superalloys. Both of the 4% solute alloys are single phase γ' with a few
isolated regions of γ-γ' structure, after the T heat treatments. These are

considered single phase. The T treated 5% Fe alloy has a slightly greater amount of the γ-γ' regions.

Bend Tests Results, T Heat Treatments

The T treated alloys were tested in four point bending at 25°C and 600°C in air, and at 600°C in vacuum. The bending strengths of the Fe and Mn containing alloys are reported in Fig. 4 and indicate the mean of the maximum elastic load obtained from at least two bending tests for each condition. The associated bending ductilities are summarized in Fig. 5. The bending strengths are generally poor in comparison to other two phase nickel base alloys, however, as reported below, these strengths are dominated by grain boundary failure. The alloys which exhibit notable ductilities at 25°C are two phase after the T heat treatment. Examination of the fracture surfaces of the Fe containing alloys after 25°C testing indicated that intergranular failure occurred in all but the 15% Fe alloy which did not fracture during bending. SEM fractographs revealed only a few small regions of transgranular failure in the 10% Fe alloy.

At 600°C in air both the single phase and two phase alloys, failed with plastic strains of less than 0.3%. This behavior is similar to the behavior of single phase boron containing Ni₃Al alloys as reported by others [7,8]. At 600°C in vacuum the two phase alloys, with the exception of the 5% Fe alloy, exhibit ductilities similar to those observed at 25°C in air. The single phase alloys and the 5% Fe alloy exhibited brittle failure in vacuum at 600°C. These results suggest that the presence of γ-phase in the Ni₃Al alloys is not sufficient to impart ductility at all temperatures.

In order to further establish the effects of γ-phase on 25°C strength and ductility, an additional set of heat treatments and bending tests were conducted and the results of these tests are reported below.

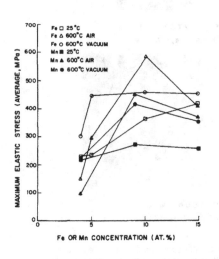

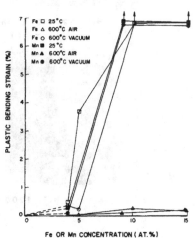

Figure 4. Maximum Elastic Bending Stress in "T" Treated Alloys vs. Fe or Mn Content.

Figure 5. Average Plastic Bending Strain in "T" Treated Alloys vs. Fe or Mn Content.

Second Phase Effects

Samples of the 4% and 5% Fe alloys were homogenized at 1200°C for 100 hr. in air, followed by air cooling. Following homogenization, one set of samples from the 4% Fe alloy received a heat treatment of 1275°C for 2 hr. followed by a water quench, and another set was water quenched from the 1200°C homogenization treatment. The microstructures resulting from these two heat treatments are shown in Figs. 6(a) and 6(c), respectively. Figure 6(a) reveals a continuous grain boundary film of γ-phase. TEM specimens were prepared from this alloy and examined. Grain boundary regions were not found in the thin foils and therefore it is not known whether or not the grain boundary films contain only γ-phase or a two phase γ-γ' mixture. A two phase mixture is expected in these regions given that nucleation of the γ'-phase cannot be suppressed even by rapid solidification.

Four point bending tests were conducted at 25°C on the samples described above. In the 4% Fe alloy, the two phase samples exhibited a 16 fold increase in the maximum elastic stress over the single phase samples.

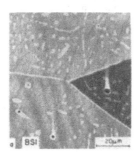 

H.T. 1200°C/100hr + 1275°C/2hr/W.Q.

σ =117 MPa; ε,~0.3%

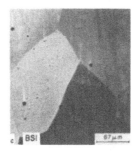 

H.T. 1200°C/100hr/W.Q.

σ =7 MPa; ε,~0

Figure 6.   Effects of Microstructure on Fracture in 4% Fe Alloy; a) back-scattered electron SEM image after homogenization and quenching from γ + γ' two phase field, b) fracture surface from two phase sample, c) back-scattered electron SEM image after homogenization and quenching from single phase γ' field and d) fracture surface from single phase sample.

227

The single phase samples were quite brittle and could be readily broken by
hand. Both the single phase and the two phase samples failed intergranularly after little or no plastic strain, as shown in Figs. 6(b) and
6(d). The grain boundary film was resolved in the fracture surfaces and the
intergranular facets in the two phase samples were distinctly rougher than
those of the single phase samples. These results suggest that a second
phase film on the grain boundaries provides an effective increase in the
grain boundary cohesion, but does not enhance the ability to accommodate or
transmit slip sufficiently to exhibit high ductility. Such behavior might
be expected if the grain boundary film is precipitation hardened.

In a similar fashion, samples of the homogenized 5% Fe alloy were heat
treated at 1150°C for 24 hr. and water quenched, while another set was heat
treated at 1100°C for 48 hr. and water quenched. The microstructures of
these 5% Fe alloys are shown in Figs. 7(a) and 7(c). A two phase microstructure was obtained after the 1150°C treatment, however, continuous grain
boundary films were not observed. The 1100°C heat treatment eliminated all
but a very small amount of the γ-phase, as shown in Fig. 7(c). The residual
γ-phase was not present in all of the grains, was located only intrag-
ranularly, and could not be detected in the grain boundary regions in any of
the samples.

H.T. 1200°C/100hr + 1150°C/24hr/W.Q.
σ =166 MPa; ε,~2%

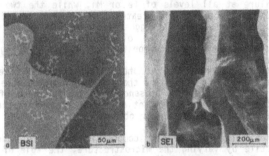

H.T. 1200°C/100hr + 1100°C/48hr/W.Q.
σ =61 MPa; ε,~0.1%

Figure 7.   Effects of Microstructure on Fracture in 5% Fe Alloy; a)
back-scattered electron SEM image after homogenization and
quenching from γ + γ' two phase field, b) fracture surface
from two phase sample, c) back-scattered electron SEM image
after homogenization and quenching from single phase field,
and d) fracture surface from single phase sample.

The 5% Fe alloys demonstrated an increase in bending ductility when the second phase was present. The elastic bending strengths were a factor of three greater in the two phase samples and were about the same magnitude as the values found for the two phase 4% Fe samples. Figures 7(b) and 7(d) show the fracture surfaces of the two phase and single phase samples, respectively. A distinct fracture surface roughness which exhibited the same general morphology as the γ-γ' regions in the bulk was observed in the two phase samples.

SUMMARY AND CONCLUSIONS

1. The same Ni_3Al alloy compositions and microstructural conditions in which Takasugi, et al. [1], observed 25°C ductility without boron additions, were examined at 25°C and at 600°C in air and in vacuum. The alloys were segregated and contained a second phase. Bending tests at 25°C confirmed the existence of ductility in the two phase samples, however, less than 0.1% ductility was measured in the single phase alloys.

2. Alloys tested at 600°C in air exhibited brittle intergranular failure at all levels of Fe or Mn, while the two phase alloys showed plastic strains greater than 7% when tested at 600°C in vacuum. These results suggest that neither Fe nor Mn additions alone are sufficient to eliminate the dynamic environmental embrittlement problem common to Ni_3Al alloys in a high temperature air environment.

3. Both single phase and two phase microstructures were produced in the 4 and 5% Fe alloys and these microstructures were evaluated in bending at 25°C. The presence of a small amount of γ-phase was shown to have a significant effect on bending strength, however, intergranular failure was observed even with the second phase present.

4. Since alloys of the same composition could be made brittle or ductile by varying the microstructure, the role of valence differences between the alloy species in affecting ductile behavior, as presented by Takasugi, et al., [3,1], is not clear. Hanada, et al. [9], have shown that the presence of γ-phase is associated with ductile behavior in binary Ni_3Al without boron. Such behavior appears to have masked any effects due to valence differences suggested by Takasugi, et al. [1].

REFERENCES

1. T. Takasugi, O. Izumi, and N. Masahashi, Acta Met. 33, 1259 (1985).
2. T. Takasugi and O. Izumi, Acta Met. 31, 1187 (1983).
3. T. Takasugi and O. Izumi, Acta Met. 33, 1247 (1985).
4. H. A. Lipsitt (private communication).
5. F. A. Shunk, Constitution of Binary Alloys, Second Supplement, (McGraw-Hill, Inc., New York, 1969) p. 335.
6. A. J. Bradley, J. Iron Steel Inst. 163, 19 (1949).
7. C. T. Liu, C. L. White, and E. H. Lee, Scripta Met. 19, 1247 (1985).
8. A. I. Taub, K. M. Chang, and C. T. Liu, Scripta Met. 20, 1613 (1986).
9. S. Hanada, S. Watanabe, and O. Izumi, J. Mater. Sci. 21, 203 (1986).

DEFORMATION OF SINGLE CRYSTALLINE Mn$_3$Sn

J. Y. LEE, D. P. POPE and V. VITEK
University of Pennsylvania, Department of Materials Science and Engineering,
3231 Walnut Street, Philadelphia, PA 19104 USA

ABSTRACT

Single crystalline Mn$_3$Sn was chosen as a model material for a study of
the deformation machanism of DO$_{19}$ ordered alloys. Single crystals were
grown by the Bridgman method in an argon atmosphere. Compression test
specimens of six different orientations were cut from the bulk single
crystals by electric discharge machining and polished for slip trace
analysis. Compression tests were carried out in an Instron testing machine
at a strain rate of 1.2 X 10^{-3} s^{-1} at temperatures ranging from 77 $^{\circ}$K to
1180 $^{\circ}$K. Tests at temperatures above 400 $^{\circ}$K were performed in an argon
atmosphere while those of lower temperatures were performed in air.

Preliminary results were obtained and analyzed in order to determine
the temperature and orientation dependence of the operating slip systems
and the yield stress. Basal slip was observed to be the dominant slip
mechanism. Slip lines were observed on specimen surfaces compressed at low
temperatures but such observations were difficult on specimens deformed at
temperatures above 1000 $^{\circ}$K due to surface contamination.

An anomalous positive temperature dependence of the critical resolved
shear stress for basal slip was found, similar to that reported for octahedral slip in many intermetallic compounds with the Ll$_2$ structure.

INTRODUCTION

There have been a large number of investigations dealing with the
deformation behavior of intermetallic ordered alloys. However, most of
those studies have been carried out on alloys with the cubic crystal
structures (Ll$_2$, B2, etc.) while relatively little work has been done on
materials with the hexagonal DO$_{19}$ crystal structure. The DO$_{19}$ structure
has the same atomic arrangement on the close packed (0001) basal plane as
on the {111} type octahedral slip planes of Ll$_2$ alloys, only the stacking
sequence is different in the two structures. A super dislocation in the
Ll$_2$ structure is unique in that the energy of the mobile configuration on
the {111} plane can be considerably higher than that of the much less mobile
configuration on the {100} plane, since the antiphase boundary energy on the
{111} plane produced by $\frac{1}{2}$<110> shear is higher than that on the {100} plane
[1]. Takeuchi and Kuramoto proposed a model in which the strength anomaly

in Ll_2 compounds is caused by the thermally activated cross-slip of screw dislocations from a glissile configuration on the {111} plane to a sessile configuration on the {100} plane [2]. The driving force for this cross-slip is the anisotropy of the APB energy and the applied stress.

It is expected that similar anisotropies of the APB energy will occur in the DO_{19} structure, but in this case the high energy mobile dislocation configuration is expected to occur on the basal plane while the low energy sessile dislocation configuration occurs on the prism plane [3,4].

Compression tests were carried out on Mn_3Sn single crystals in order to investigate the temperature and orientation dependence of the yield stress and operating slip systems.

EXPERIMENTAL PROCEDURES

The Mn_3Sn alloy was produced by mixing the proper amounts of pure Mn (purity 99.9 %) and Sn (purity 99.999 %) to obtain alloys of composition 23.75 atomic % Sn. An alloy with the stoichiometric composition (25 atomic % Sn) is not single phase [5]. Polycrystals of Mn_3Sn were produced by melting and slow cooling of the Mn:Sn mixture in an argon atmosphere. Losses due to evaporation were compensated for by the addition of extra manganese. From such polycrystals, single crystals were grown by the Bridgman method in an alumina crucible under an argon atmosphere. A slightly positive argon pressure was maintained to minimize the material loss caused by evaporation at high temperature. After the crystal mold passed through the temperature gradient, the furnace was cooled from 1100 to 900 °C over a period of two days to allow time for odering.

The single crystals (about 3 cm long and 2 cm in diameter) produced by this method grew in a variety of orientations, and compression test specimens (3 mm X 3 mm X 7 mm) of different orientations were cut from the homogenized bulk single crystals by electric discharge machining.

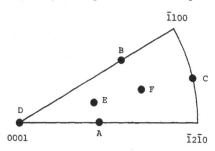

Figure 1. Crystallographic orientations of the compressive axes.

The sample faces were polished for slip trace analysis. Crystallographic orientations of the compressive axes are shown in figure 1. The specimens were annealed before testing at 800 °C in an evacuated quartz capsule.

Compression tests were carried out in an Instron testing machine at a strain rate of 1.2 X 10^{-3} s^{-1} at temperatures ranging from 77 °K to 1180 °K. Tests at temperatures above 400 °K were performed in an argon atmosphere while those of lower temperatures were performed in air. Slip traces were produced in the samples by an axial plastic deformation of 0.2 %, and the slip planes were identified by two surface slip trace analysis. The temperature dependence of the 0.2 % offset yield stress was also investigated for each orientation.

RESULTS AND DISCUSSION

The ductility of Mn_3Sn single crystal is very low as can be demonstrated by the cleavage that occurs when a single crystalline sample is deformed with a 1/16 inch diameter spherical hardness indenter. The cleavage plane was identified as the (0001) plane by the Laue back reflection method.

In orientation A, slip lines were fine and parallel to the trace of the (0001) plane, since the (0001) <1$\bar{2}$10> slip system is most favorable in this orientation. The temperature dependence of the 0.2 % offset axial yield stress is shown in figure 2. Note that the yield stress increases with increasing temperature. A similar behavior was observed by Takeuchi and Kuramoto [6,7], but there are some important differences. The yield stress values are larger than those reported by Takeuchi and Kuramoto at high temperatures, i.e., we observed a higher and sharper peak at 1100 °K in the yield stress vs. temperature dependence.

In orientation B, fine slip lines were also observed over the entire specimen length as in the case of orientation A. Most of the slip lines were identified as basal slip although some prismatic slip was also observed.

Samples of orientation C deformed by inhomogeneous {10$\bar{1}$0} <1$\bar{2}$10> slip. Massive deformation was concentrated near one end of the sample leading to local crushing of the sample. Note that the RSS on the (0001) plane is zero for these samples because the basal plane is parallel to the compression axis. This difficulty in activating prismatic slip is in agreement with the observed dominance of basal slip in samples of orientation B.

Samples of orientation D fractured catastrophically into small pieces at test temperatures from room temperature to 700 °K. For this orientation, the compression axis is perpendicular to the basal plane and therefore basal slip is not possible. Apparently pyramidal slip does not occur at temperatures below 700 °K.

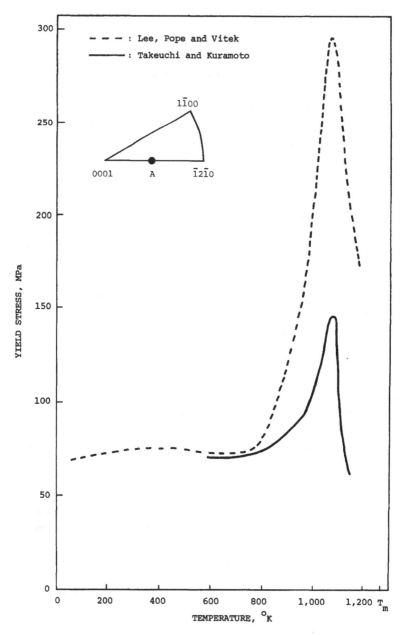

Figure 2. Temperature dependence of yield stress

SUMMARY

1. The ductility of Mn_3Sn is very low as evidenced by the (0001) cleavage at room temperature.

2. Basal slip is the dominant slip mode in Mn_3Sn crystal.

3. The only available slip systems at temperatures below $700^\circ K$ are the (0001) $<1\bar{2}10>$ and $\{10\bar{1}0\}$ $<1\bar{2}10>$ systems. Since samples of orientation D underwent brittle failure, no other slip systems are available.

4. An anomalous temperature dependence of the yield stress is observed in Mn_3Sn similar to that seen in materials with the Ll_2 structure.

ACKNOWLEDGEMENT

This research was supported by the National Science Foundation no. DMR 85-01974.

REFERENCES

1. P. A. Flinn, Trans. Met. Soc. AIME v.218, 145 (1960).

2. S. Takeuchi and E. Kuramoto, Acta Met. v.21, 415 (1973).

3. M. J. Marcinkowski : Electron Microscopy and Strength of Crystals, G, Thomas and J. Washburn, eds, p. 333, Interscience, New York, 1963.

4. R. G. Davies and N. S. Stoloff : Trans. TMS-AIME, 1964, v.230, pp. 390-395.

5. M. Hansen : Constitution of Binary Alloys, 2nd ed., p.954, McGraw Hill Book Co., New York, 1958.

6. S. Takeuchi and E. Kuramoto, Acta Met. v.22, 429 (1974).

7. S. Takeuchi and E. Kuramoto, Met. Trans. v.3, 3037 (1972).

THE EFFECT OF STRAIN RATE ON THE YIELD BEHAVIOR OF IRON-BASED
LRO ALLOYS AT ELEVATED TEMPERATURES

H. T. Lin, R. C. Wilcox and B. A. Chin
Materials Engineering, Auburn University, Alabama 36849

ABSTRACT

The sensitivity of iron-based LRO alloys to strain rate at elevated
temperatures was investigated. The tensile results indicate that the
iron-based LRO alloys exhibit a positive temperature dependence of yield
behavior, but are insensitive to variations in strain rate between 10^{-5} and
10^0 sec^{-1}. These observations, along with SEM and TEM results, suggest
that previous proposed diffusion-controlled mechanisms are not responsible
for the increase in yield strength with increasing temperature.

INTRODUCTION

The yield strength of many Ll$_2$ type long range ordered alloys exhibits
an unusual temperature dependence: the yield strength increases with
increasing temperature as temperature below critical ordering temperature.
The anomalous flow behavior of Ll$_2$ alloys was first observed by Westbrook
[1] who measured the hardness of Ni$_3$Al as a function of temperature. This
observation was subsequently confirmed by Flinn [2] in a study of flow
stress of Ni$_3$Al as a function of temperature. Flinn also proposed that the
anomalous yield behavior is due to diffusion-controlled changes of the
dislocation configuration. If diffusion-controlled mechanisms are
responsible for the anomalous flow behavior, then the flow behavior would
be expected to be strain rate dependent. Davies and Stoloff [3] have shown
that the flow stress of Ni$_3$Al is not strain-rate dependent and hence a
number of alternate models have been hypothesized to explain the anomalous
yield behavior of Ll$_2$ alloys. Some of these models are based on intrinsic
lattice defects [3,4,5], and others based on changes in long range order
[6]. However, models based on variations of the Flinn model are thought by
many investigators to be the most plausible [7,8,9].
       The mechanism to explain the positive temperature dependence of the
yield strength of iron-based LRO alloys is hypothesized to be caused by a
thermally-activated process rather than a change in the degree of order
with temperature [10]. In an effort to provide a better understanding of
this anomalous flow behavior, this study investigated the effect of strain
rate on the yield behavior of (Fe,Ni)$_3$V LRO alloys at elevated
temperatures. Previous work has been presented by the authors on the room
temperature strain rate behavior and the temperature dependence of the
UTS and YS of the same alloys (LRO-37) [11,12].

236

EXPERIMENTAL PROCEDURE

The tensile specimens of iron-based LRO alloy (designation LRO-37) with gage section dimensions of 12.7 x 2.8 x 0.76 mm were fabricated by Oak Ridge National Laboratory. The nominal chemical composition of LRO-37 is 37Fe-39.5Ni-22.4V-0.4Ti (in wt. %). The specimens were solution-treated for 20 minutes at 1100°C, followed by an ordering treatment involving step cooling from 600 to 500°C. All specimens were electro-polished in a solution of 40% distilled water, 40% nitric acid and 20% hydrofluoric acid. Tensile tests conducted at strain rates raging from 10^0 to 10^{-5} sec^{-1} were performed with an MTS hydraulic testing machine on both LRO-37 and 20% CW 316 SS at temperatures from 400 to 750°C. The 316 SS material which is known to be strain rate insensitive was included as a check on testing procedures. Both extension and load as a function of time were measured using a digital oscilloscope at the high strain rates to insure accurate knowledge of both strain rate and stress. To perform tensile tests at elevated temperatures, a furnace with an argon inert gas atmosphere was used. Selected specimens were examined with a JEOL 840 scanning electron microscope and a JEOL 1200EX transmission electron microscope (TEM) to determine microstructural changes.

RESULTS AND DISCUSSION

The mechanical properties of the iron-based LRO alloys as a function of strain rate and temperature are given in Figures 1, 2 and 3. Generally, the ultimate tensile strength decreases with increasing temperature and with decreasing strain rate. The decrease in tensile strength is apparently due to the mechanisms of annealing and recovery of dislocation structures at elevated temperatures resulting in a decreased strain-hardening capability of the material. In addition, the longer the testing time, the longer the time for dislocation structure changes to occur by diffusion, resulting in a softening of the material. The ultimate tensile strength of 316 SS shows a similar tendency, but is insensitive to strain rate except for 750°C where creep in 316 SS has assumed a significant role in the deformation process.

The tensile results indicate that the yield strength of the LRO-37 alloys increases with increasing temperature, rather than decreasing like that of a conventional alloy such as 316 SS. An important observation is that the yield behavior of LRO-37 alloys, at all temperature tested, is insensitive between strain rates of 10^{-4} and 10^0 sec^{-1} which is similar to that of 316 SS. Only at the very lowest strain rate and highest temperatures tested ($2x10^{-5}$ sec^{-1} and 600-750°C) is there a hint of strain rate sensitivity. This result is not in accord with the mechanism of thermally-activated processes, since diffusion-controlled climb effects would be enhanced as the strain rate is lowered at all temperatures.

Additional information from previous work [11,12], which investigated the effect of aging at 600°C on the structural stability of LRO-37, also suggest that thermally-activated processes are not responsible for the

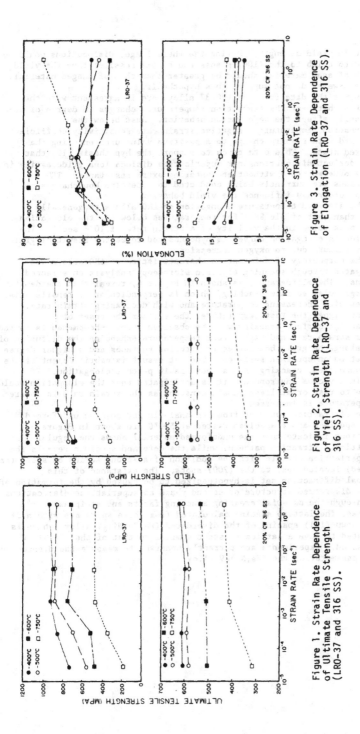

Figure 1. Strain Rate Dependence
of Ultimate Tensile Strength
(LRO-37 and 316 SS).

Figure 2. Strain Rate Dependence
of Yield Strength (LRO-37 and
316 SS).

Figure 3. Strain Rate Dependence
of Elongation (LRO-37 and 316 SS).

increase in yield strength. During the thermal age, dislocations would be expected to climb to the (100) planes and become sessile; thus the yield strength of aged material should be greater than that of unaged material. The data however do not support this hypothesis.

However the data from the LRO-37 alloys are in agreement with the hypothesis that near the transition temperature, changes in degree of order may strongly affect the deformation behavior. Just below the transformation temperature a negative strain rate sensitivity coefficient is measured with a sharp change to a positive value upon entering the disordered state. This data tends to support the hypothesis [10] that neither perfect dislocations nor superlattice dislocations glide easily in a weakly ordered state without encountering extra resistance. TEM examinations are currently being conducted to determine how changes in degree of order may influence the yield behavior.

The elevated temperature elongation of LRO alloys is generally greater than that of 316 SS. For temperatures below T_c, the elongation shows a peak value in the vicinity of a strain rate of 10^{-3} sec^{-1}. The peak value in elongation curves is hypothesized to be the result of grain boundary failure due to oxygen ingression.

The morphology of deformation characteristics of LRO alloys was investigated through scanning electron microscopy analysis on selected specimens. The micrographs, as shown in Figure 4, reveal that the density of intergranular surface cracking, which is perpendicular to tensile axis, increases with increasing temperature and with decreasing strain rate. Also, the higher the test temperature, the earlier the onset of intergranular surface cracking will be observed. This phenomenon is caused by lower strain rate and higher test temperature enhancing the diffusion of oxygen along the grain boundaries to interior surfaces and a higher degree of oxidation on specimen surfaces. The net result is embrittlement of the grain boundaries, cracking, and a decrease in percent elongation. If tested in a vacuum environment, it is anticipated that the elongation would continue to rise or at least remain constant as the strain rate is lowered and temperature is increased.

The diffraction patterns from TEM analysis for both an undeformed ordered-specimen and a specimen tested at 500°C are shown in Figure 5. Observations indicate that the undeformed material shows the regular superlattice diffraction pattern, while the deformed specimen reveals the same superlattice pattern with the addition of a second series of spots (as indicated) located next to the (200) spots. The existance of this additional diffraction set is hypothesized to be caused by the formation of either a disordered structure or second phase as superlattice dislocations sweep through the material creating stacking faults and antiphase boundaries. The lattice spacing calculated from these spots are within 3% of the known (111) spacing of the disordered (Fe,Co,Ni)$_3$V alloy, which is anticipated to have a lattice constant similar to that of the (Fe,Ni)$_3$V alloy tested. Experiments are currently underway to measure the disordered lattice parameter of the (Fe,Ni)$_3$V alloy.

Figure 4. Photomicrographs Showing Strain Rate Dependence of
Deformation Mechanism: LRO-37.
a) Specimen Tested at 400°C and 10^{-4} sec^{-1}.
b) Specimen Tested at 500°C and 10^{-3} sec^{-1}.

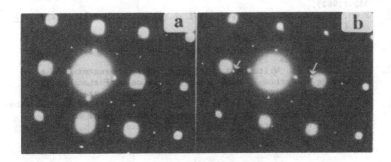

Figure 5. Photomicrographs Showing TEM Diffraction Patterns: LRO-37.
a) Undeformed Ordered LRO Alloy.
b) After Tensile Testing at 500°C and a Strain Rate of
10^{-5} sec^{-1}.

CONCLUSIONS

The results of this study lead to the following conclusions:

● The yield strength of iron-based LRO alloy, designated LRO-37, increases with increasing test temperature as temperature below T_c, but is insensitive to strain rate.

● Insensitivity to strain rate and lack of an aging effect of the yield strength suggest that the diffusion-controlled mechanism is not responsible for the anomalous yield behavior of LRO-37.

● The density of intergranular surface cracking increases with increasing temperature and with decreasing strain rate indicating the ingress of oxygen along grain boundaries may be a potential mechanism of embrittlement.

REFERENCES

[1].   J. W. Westbrook, Trans. Metall. Soc. AIME 209, 898 (1957).
[2].   P. A. Flinn, Trans. Metall. Soc. AIME, 218, 145 (1960).
[3].   R. G. Davies and N. S. Stoloff, Trans. Metall. Soc. AIME, 233, 714 (1965).
[4].   T. L. Johnston, A. J. McEvily and A. S. Tetelman, in High Strength Materials, ed. V. F. Zackay, p. 363-381, Wiley, New York, 1965.
[5].   S. M. Copley and B. H. Kear, Trans. AIME, 239, 977 (1967).
[6].   D. P. Pope, Philos. Mag. 25, 917 (1972).
[7].   B. H. Kear and H. G. F. Wilsdorf, Trans. Metall. Soc. AIME, 224, 382 (1962).
[8].   P. H. Thornton, R. G. Davies and T. L. Johnston, Metall. Trans., 1A, 207 (1970).
[9].   S. Takeuchi and E. Kuramoto, Acta Metall., 21, 415 (1973).
[10].   C. T. Liu, Paper Presented at the AIME Annual Meeting, Las Vegas, Nevada, 1980.
[11].   H. T. Lin, R. C. Wilcox and B. A. Chin, Paper Presented at the Proceeding of Second International Conference on Fusion Reactor Materials and Published in J. Nucl. Mater. (in Press).
[12].   H. T. Lin, R. C. Wilcox and B. A. Chin, Paper Presented and Published at the proceeding of ASTM 13th International Syposium on the Effect of Radiation on Materials (in Press).

THE SEGREGATION OF BORON AND ITS EFFECT ON THE FRACTURE OF AN Ni₃Si BASED ALLOY

W. C. Oliver* and C. L. White**
*Oak Ridge National Laboratory, Oak Ridge, TN 37831
**Department of Metallurgical Engineering, Michigan Technological University, Houghton, MI 49931

ABSTRACT

It is now well established that microalloying additions of B to Ni_3Al drastically reduce low temperature grain boundary fracture and consequently increase the ductility of this intermetallic compound. One possible explanation for such effects involves the relationship between boron segregation to grain boundaries and free surfaces, and the resulting effect of such segregation on the cohesive energy of the grain boundaries. This study involves the extension of these concepts to an alloy based on Ni_3Si. Auger spectroscopy has been carried out on fractured grain boundaries, grain interiors, and free surfaces to determine how B segregates in $Ni_3(Si,Ti)$. The consequences of the segregation of B on the cohesive energy of grain boundaries in Ni_3Si based alloys are discussed.

INTRODUCTION

The effect of microalloying additions of boron to Ni_3Al is well documented [1]. With addition of as little as 0.02 wt % of B the ductility of Ni_3Al is increased from a few percent to as high as 50% elongation. In addition, the fracture path changes from intergranular to transgranular. The reasons for these effects are not completely understood; however, it is clear that the effect of B on the properties of grain boundaries is a key factor. One reasonable explanation involves the effect of B on the cohesive energy of the grain boundaries. A model, put forward by J. R. Rice [2], considers the effect of segregation on the free surface energy and grain boundary energy and predicts the resulting changes in the cohesive energy of the boundary. One of the predictions of the model is that if an element segregates to grain boundaries but not to free surfaces, then the minimum cohesive energy of the boundary will be increased by the segregation. The segregation behavior of B in Ni_3Al has been shown to be consistent with this model [3]. This study has been undertaken to determine the applicability of these effects and explanations in the Ni-Si system.

Ni_3Si is an Ll_2 intermetallic compound with many similarities to Ni_3Al. These similarities included increasing yield strength with increasing temperature [4], low room temperature ductility [5] and excellent oxidation resistance. Taub et al. showed that rapidly solidified ribbons of Ni-23 Si doped with 0.5% Boron and properly heat treated demonstrate some ductility; however, they do not report results for the alloy without boron [12]. Although alloys based on Ni_3Si could be useful intermediate temperature structural alloys, their lower melting temperatures result in lower high temperature strength when compared with Ni-Al based alloys.

A unique property of Ni_3Si is its ability to resist corrosion in certain acidic, aqueous solutions [6]. It is the basis for several commercial alloys designed for this purpose (i.e., Hastelloy D®); however, due to the low room temperature ductility of these alloys they are only used in the cast condition.

The room temperature ductilities of alloys with high volume fractions of Ni_3Si were greatly improved through an alloy development program undertaken by K. J. Williams and co-workers [7,8]. Through the addition of Ti, the room temperature ductilities of such alloys, in the cast and homogenized form, were greatly improved. Titanium was found to replace Si in the Ni_3Si crystal structure; however, the reasons for the ductility improvements associated with Ti additions are not completely understood.

In this study we use one of the single phase Ll_2 alloys developed by Williams et al. and determine the effect of microalloying additions of B on the ductility of the alloy. In addition, Auger electron spectroscopy is used to determine the segregation behavior of B in the alloy, and the consequences of the findings concerning the cohesive energy of the grain boundaries are discussed.

EXPERIMENTAL

The composition of the alloys used in this study is Ni-18.9 Si-3.2 Ti (atomic %) and the same alloy doped with 0.02 wt % boron. The alloys were made up from carbonyl Ni, electronic grade Si, commercial purity Ti rod, and a Ni-4% B master alloy. The materials were arc melted in an argon atmosphere and cast into a water cooled copper mold. The castings were homogenized for 4 h @1000°C. The sheet used to make the tensile specimens for this study was fabricated by successively cold working 10% and annealing for 4 h @1000°C a sufficient number of times to reduce the 12.5 mm thick castings to 750 μm. The specimens were then punched from the sheet materials and given a final anneal of 1000°C for 4 h. The tensile tests were performed on a screw driven Instron mechanical test system. The tests were performed at constant crosshead speed with an initial strain rate of $3.33 \times 10^{-3} sec^{-1}$.

Specimens for AES fracture analysis were notched and fractured in tension under a vacuum of approximately 2×10^{-8} Pa. The boron containing specimen was fractured in a liquid nitrogen cooled fixture. Following analysis of the as fractured surface, the specimens were sputter etched and reanalyzed. Specimens for free surface segregation studies were metallographically polished, and sputter etched prior to heating. They were then heated from room temperature to 400°C within a few minutes, and held at that temperature for about 90 min. The temperature was then increased to 500°C for 90 min, then 600°C for 90 min. Three Auger spectra, taken at 30 min intervals, were generally obtained while the specimen was held at each temperature. Detailed discussion of the AES analysis is given in reference 9, and specific aspects of the in-situ heating studies are described in reference 10.

We should note that the temperatures reported for the surface segregation studies are probably not highly accurate. Attempts to heat specimens above 600°C resulted in discrepancies between the thermocouple and the optical pyrometer readings. One factor that may be significant in this respect is that the thermocouple itself appeared to constitute a significant heat sink, with the region near the thermocouple being noticeably less incandescent than the rest of the specimen.

RESULTS AND DISCUSSION

An inspection of the microstructure of the Ni-18.9 Si-3.2 Ti alloy used in the study in the annealed condition showed that the alloy is single phase Ll_2. The grain size of the alloy is approximately 70 μm. The addition of B to the alloy did not change the microstructure.

Table 1 shows the results of the tensile tests performed on each alloy. The scatter in the experimental results is small. The yield

strength of the two alloys are not significantly different. In each
tensile test there was a very short region (elongation of 1%) of very low
work hardening. This could be evidence of a yield point. The work
hardening rates in all of the tests are identical. The addition of 0.02%
wt % B did increase the ductility of Ni-18.9 Si-3.2 Ti from 17.6% to 29.1%.
In addition, the UTS of the alloy was increased.

Table 1

Ni-18.9 Si-3.2 Ti

Strength, MPa (ksi)		Elongation To fracture, %
Yield	Ultimate	
553(80.3)	1173(170.2)	14.4
576(83.6)	1403(203.6)	19.7
570(82.8)	1385(201.0)	18.7
Average:		
566(82.2)	1320(191.6)	17.6

Ni-18.9 Si-3.2 Ti + 0.02 wt % B

Strength, MPa (ksi)		Elongation To fracture, %
Yield	Ultimate	
541(78.5)	1544(224.1)	27.8
569(82.5)	1584(229.9)	29.5
567(82.3)	1578(229.0)	30
Average:		
559(81.1)	1567(227.3)	29.1

Figures 1A and 1B show the fracture surfaces from room temperature
tensile tests of the two alloys. The addition of B to the alloy changed
the fracture path from predominantly brittle intergranular fracture to 100%
ductile transgranular. There is some transgranular fracture in the alloy
without B. The mixed mode of fracture could be responsible for the
increased scatter in the measurements of ductility and UTS in this alloy.
Figure 1C shows the fracture surface of the B containing alloy after fracturing near liquid nitrogen temperature in the AES system. The reduced
temperature has resulted in some intergranular fracture. This allowed us
to examine grain boundaries in the B containing alloy.

Clearly, as in the case of Ni_3Al, the addition of B to the alloy
Ni-18.9 Si-3.2 Ti [$Ni_3(Si,Ti)$] has a dramatic effect on ductility, fracture
path and UTS. The yield strength and work hardening rate are unaffected,
indicating the B is affecting the properties of the grain boundaries in the
alloy.

A typical Auger spectrum from the undoped specimen is given in
spectrum A of Fig. 2. This spectrum shows peaks only from Ni, Si and Ti.
There is no evidence of O, C or B. Similar spectra from intergranular and

244

transgranular regions on the as-fractured surface of the boron doped sample
are shown in spectra B and C respectively. Both of these latter spectra
contain Auger peaks arising from carbon and oxygen, as well as the Ni, Si
and Ti in spectrum A. In addition, spectrum B (intergranular region) has a
peak due to boron at 180 eV. While carbon and oxygen could be present in
the alloy, and segregate to the grain boundaries, they are also components
of the residual atmosphere in the fracture and analysis chambers (H_2O,
CO, CO_2, CH_4) and can be present on the fracture surface as a result of
contamination by these gasses. Two differences between spectrum B (boron
doped, intergranular) and the rest of the spectra that do seem significant
are the presence of boron, and a lower lever of silicon (as evidence by
the intensity of the 88 eV Si peak). We also note that after sputter
etching a few atom layers (less than 10) from the as-fractured surface, the
boron peak is gone and the silicon peak is approximately same as for the
other spectra.
    In-situ heating generally resulted in extensive segregation of sulfur
to the free surface. Figure 3, spectra A shows the third Auger spectrum
taken at 600°C from the undoped surface segregation specimen. In addition
to the very large sulfur peak, this spectrum also has a smaller 88 eV Si
peak and larger titanium peaks, than do spectra A or C in Fig. 2. A similar analysis for the boron doped sample is shown in spectrum B of this
figure. In this latter spectrum, the S peak is smaller and the titanium
peak is similar in size to that for spectrum B. Particularly notable in
the spectrum B is the absence of any boron peak at 180 eV.
    This brief investigation of grain boundary and surface segregation
behavior in $Ni_3(Si,Ti)$ with and without 0.02 wt % B reveals several similarities to B in Ni_3Al. The first similarity to Ni_3Al, is the absence of
any significant level of embrittling impurity on the grain boundaries of
the undoped alloy, suggesting that this Ll_2 alloy is also intrinsically
brittle. This result is in agreement with the results previously reported
for Ni-Si binary alloy [11]. Equally important is the evidence that, like
Ni_3Al, the ductilizing effect of boron in $Ni_3(Si,Ti)$ is associated with its
segregation to grain boundaries. Finally, the fact that boron appears not
to segregate strongly to free surfaces is consistent with Rice's thermodynamic theory of grain boundary cohesion that has previously been used to
explain the beneficial effects of boron segregation to grain boundaries in
boron doped Ni_3Al [2].

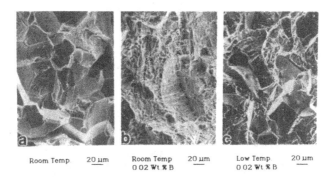

Room Temp   20 μm   Room Temp   20 μm   Low Temp   20 μm
                         0 02 Wt.% B   0 02 Wt.% B

Fig. 1. Scanning electron micrograph of fracture surfaces of Ni-18.9
Si-3.2 Ti (a) room temperature fracture; (b) room temperature fracture of
alloy doped with 0.02 wt % B; (c) B doped alloy fractured at low temperature.

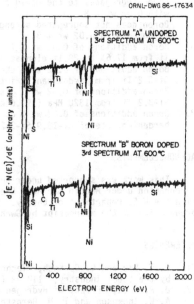

ORNL-DWG 86-17634

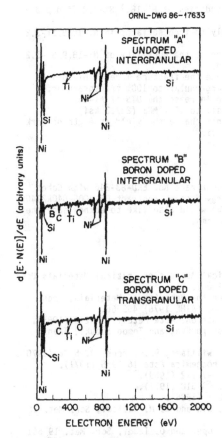

ORNL-DWG 86-17633

Fig. 3. Auger spectra from the in-situ heating experiments on $Ni_3(Si,Ti)$. Spectrum A is from the undoped alloy and Spectrum B is from the boron doped alloy. Details of the heating schedule are given in the text.

Fig. 2. Auger spectra from "as-fractured" Ni-18.9 Si-3.2 Ti specimens. Spectrum A is from an intergranular region on the undoped alloy. Spectrum B is from an intergranular region on the boron doped alloy. Spectrum C is from a region of transgranular failure on the boron doped alloy.

CONCLUSIONS

1. Ni-18.9 Si-3.2 Ti is a single phase Ll_2 alloy.
2. The grain boundaries in this alloy are inherently weak. Grain boundary fracture occurs without any detectable level of grain boundary impurities as determined with AES.

3. Boron segregates to the grain boundaries of Ni-18.9 Si-3.2 Ti doped with 0.02 wt % B.
4. Boron does not segregated strongly to free surfaces in Ni-18.9 Si-3.2 Ti doped with 0.02 wt % B.
5. Boron additions of 0.02 wt % increase the ductility of Ni-18.9 Si-3.2 Ti from 17.6% to 29.1% elongation.
6. Boron additions of 0.02 wt % change the fracture mode of Ni-18.9 Si-3.2 Ti from predominantly intergranular to 100% transgranular.
7. Boron additions of 0.02 wt % also increase the UTS of Ni-18.9 Si-3.2 Ti from 1320 MPa (191.6 ksi) to 1567 MPa (227.3 ksi).
8. Boron additions of 0.02 wt % do not change the yield strength or work hardening rate of Ni-18.9 Si-3.2 Ti.

ACKNOWLEDGMENTS

This work was supported under Subcontract No. ERD-83-346 with Cabot Corporation under Martin Marietta Energy Systems contract DE-AC05-84OR21400 with the U.S. Department of Energy. We would also like to acknowledge the preparation of the manuscript by Gwen Sims.

REFERENCES

1. C. T. Liu and C. C. Koch, "Technical Aspects of Critical Materials Used by the Steel Industry," 11B 42 (1983).
2. J. R. Rice, "Effect of Hydrogen on the Behavior of Materials," eds. A. W. Thompson and I. M. Bernstein, (AIME 1976) 455–66.
3. C. T. Liu, C. L. White and J. A. Horton, Acta Metall. 33 213 (1985).
4. Dang-Moon Wee, Osami Naguchi, Yoshiro Oya, and Tomoo Suzuki, Trans. JIM 21 237 (1980).
5. W. Barker, T. E. Evans, and K. J. Williams, Br. Corros. J. 5 76 (1970).
6. T. E. Evans and A. C. Hart, Electrochemica Acta 16 1955 (1971).
7. K. J. Williams, J. of Inst. Mets. 97 112 (1969).
8. K. J. Williams, J. of Inst. Mets. 99 310 (1971).
9. C. L. White, Cer. Bull. 69(12) 1571 (1985).
10. C. L. White, R. A. Padgett, C. T. Liu and S. M. Yalisov, Scr. Met. 18 1417 (1985).
11. T. Takasugi, E. P. George, D. P. Pope, and O. Izumi, Scr. Met. 19 551 (1985).
12. A. I. Taub, C. L. Briant, S. C. Huang, K. M. Chang and M. R. Jackson, Scr. Met. 20 129 (1986).

FATIGUE OF INTERMETALLIC COMPOUNDS

N.S. STOLOFF, G.E. FUCHS, A.K. KURUVILLA and S.J. CHOE
Materials Engineering Department, Rensselaer Polytechnic Institute,
Troy, New York 12180-3590, USA

ABSTRACT

The fatigue behavior of intermetallic compounds is reviewed. The
effects of long range order, stoichiometry, test temperature and test
environment on crack initiation, high cycle fatigue lives and crack growth
rates are emphasized. In the case of Ni_3Al+B stoichiometry affects
high cycle lives largely through the influence of aluminum on ductility
and notch sensitivity. High cycle fatigue behavior of Fe_3Al is dependent
upon stoichiometry and temperature in a complex way which is connected
with the formation of superlattice dislocations and with phase changes
during high temperature exposure. Oxygen and hydrogen are shown to
be detrimental to high cycle fatigue and crack growth in several compounds.

INTRODUCTION

It was first reported some twenty years ago that long range order
can enhance high cycle fatigue resistance [1]. However, the lack of
tensile ductility manifested by most ordered alloys with potential for
structural applications precluded serious attention of researchers to
cyclic behavior. Rather, the emphasis has been upon increasing ductility
through alloying or improved processing techniques. Therefore, only
a few studies, especially on high cycle fatigue, were reported in the
literature in the past two decades. However, recent progress in producing
ductile ordered alloys has led to renewed interest in fatigue behavior
and, in particular, to the effects of long range order on crack initiation
and propagation, respectively. The purpose of this paper is to review
the literature on cyclic deformation of ordered alloys, with emphasis
upon recent work on $(Fe,Ni)_3V$, Ni_3Al and Fe_3Al-type alloys.

GENERAL FEATURES OF ORDERED SYSTEMS

Direct studies of the influence of long range order on cyclic behavior
can only be carried out in systems that can be disordered by heat treatment.
These include $L1_2$ superlattices such as Cu_3Au, Ni_3Fe and Ni_3Mn, B2 superlattices such as FeCo-V and DO_{19} superlattices such as Mg_3Cd. Other,
indirect, information on the role of long range order in fatigue has
been provided by studies of Ni_3Al and Ni_3Ge ($L1_2$), Fe_3Al (DO_3 or partial
B_2) TiAl ($L1_0$) and Ti_3Al (DO_{19}), none of which can be completely disordered
by heat treatment. For convenience, cyclic test results on all systems
will be described under the headings of high cycle fatigue and crack
propagation, which are usually carried out under stress control, or
low cycle fatigue, which is usually performed under strain control.
Under the latter heading will be included also results of cyclic hardening
tests. Crack initiation, which may be studied under either stress or
strain control, is discussed in a separate section.

HIGH CYCLE FATIGUE

The influence of long range order on stress-controlled high cycle

248

fatigue was first established for Ni₃Mn and FeCo-V [1]. In both cases
fatigue lives were substantially increased, in spite of a decrease in
yield stress with order for FeCo-V. Later, Cu₃Au also was shown to
exhibit increased life with order, although the effect was small [2].
The beneficial effects of order in Cu₃Au are maintained at temperatures
to 373°C [3]. In each case it was hypothesized that planar slip, brought
about by the presence of superlattice dislocations, was responsible
for delayed crack initiation. In most alloys studied to date, order
favors planar slip by inhibiting cross slip and/or multiple slip, either
of which are generally considered to be necessary for cracks to initiate
readily. The one ordered system in which planar slip is favored by
disorder is Mg₃Cd. Whitehead and Noble [4] showed that in this system
poor fatigue properties were exhibited by both ordered and disordered
material.

Later, high cycle fatigue studies were extended to other systems:
(Fe,Ni)₃V [5], and TiAl [7] polycrystals, and Ni₃Al single crystals
[8] as a function of test temperature, and to β brass as a function
of ternary alloy content (to influence slip planarity) [6]. The results
of these studies are summarized in Table I, together with data for some
commercial alloys that are not ordered. It is clear that the ratio
of the fatigue limit to yield strength exceeds 0.5 for most of the ordered
alloys, while in the disordered condition that ratio decreases significantly.
For TiAl, the ratio varied with temperature, ranging from 0.8 at room
temperature to 0.5 at 900°C [7].

Recently, the influence of Al content on high cycle fatigue of
Ni₃Al [9] and Fe₃Al-type [10] alloys has been examined. For Fe₃Al,
an additional variable is DO₃ vs partial B2-type order (the latter produced
by quenching from above Tc, the critical ordering temperature). Typical
results for Ni₃Al+B alloys are shown in Fig. 1a). The fatigue resistance
of Ni-26%Al is much less than that of Ni-24%Al, even when normalized
for differences in yield stress, Fig. 1b). Differences in fatigue lives
of the two alloys are reflected also in the fracture mode: faceted
stage I growth in Ni-24%Al (Fig. 2a) and intergranular or interdendritic
paths in Ni-26%Al, (Fig. 2b). At 500°C, fatigue lives are nearly unchanged
relative to 25°C for Ni-24%Al, but are considerably higher for Ni-26%Al.
This appears to be a consequence of increased yield stresses with temperature
and increased ductility (together with reduced notch sensitivty) for
Ni-26%Al. Ni₃Al single crystals tested at the same stress levels show
relatively constant fatigue lives with temperature, see Fig. 3 [8].
However, P/M polycrystals reveal a sharp drop in fatigue lives with
increasing temperature, Fig. 4, paralleling a drastic loss of tensile
ductility in the same temperature regime. Fracture paths in specimens
tested at temperatures of 600°C and above revert to intergranular, as
in tensile tests. An environmental effect undoubtedly is responsible
for the loss of tensile ductility [11], and is probably responsible
for loss of fatigue resistance as well. Note that the applied stress
range of Δσ = 500MPa is above the tensile strength of Ni₃Al+B at temperatures
above 500°C.

High cycle fatigue results for Fe₃Al-type alloys reveal an unexpected
reversal in properties between Fe-23.7%Al and Fe-28.7%Al tested at 25°C
and at 500°C [10]. Fig. 5 shows that the hypostoichiometric alloy is
less fatigue resistant at 25°C, but is more resistant at 500°C. This
appears to arise from an aging effect, since a two-phase, coherent α
+ DO₃ structure develops in Fe-23.7%Al at 500°C, see Fig. 6a)-c), but
not in Fe-28.7%Al. Hardness data for Fe-23.7%Al, Fig. 7, confirm the
conclusion of Inouye [12] that an aging reaction occurs after exposure
at 500°C or 560°C, but not after exposure to 600°C. It has been concluded
that enhanced fatigue resistance of the higher-Al containing alloy at
25°C is due to the presence of superlattice dislocations, which cause
crack initiation to be more difficult than in the hypostoichiometric
alloy; the latter deforms by the motion of unpaired dislocations [13].

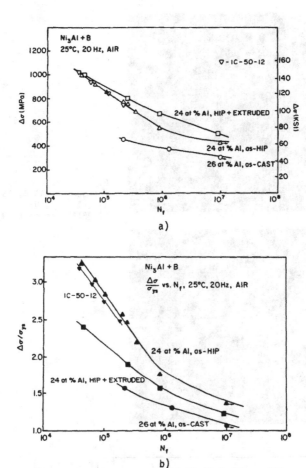

Fig. 1 High cycle fatigue of several Ni₃Al+B alloys [9]
a) σ-N data   b) σ-N data normalized for differences
in yield stress.

CRACK INITIATION

For β brass, crack initiation at room temperature occurs at grain boundary triple points, while at 78°K transgranular initiation from extrusions and intrusions usually was noted [14]. Triple point initiation in β brass probably arises from the marked elastic anisotropy of this alloy. In single crystals cycled at room temperature, cracks were usually associated with regions of coarse slip together with some cross slip. Cracks were not associated with extrusions. Unfortunately, β brass cannot be disordered at low temperatures; therefore, it is not possible to directly link the presence of order with the observed phenomena.

In an effort to determine the influence of long range order on crack initiation, studies of the FeCo-2%V alloy (CsCl structure) are now underway. Transgranular crack initiation, along slip bands, has been observed in both ordered and disordered material, see Fig. 8a)

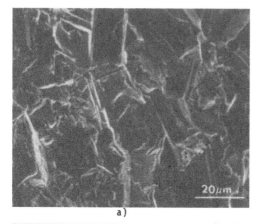

a)

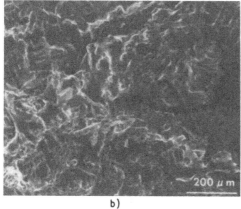

b)

Fig. 2 Crack paths in
fatigued Ni₃Al+B, 25°C
a) 24%Al, as HIP
b) 26%Al, as cast [9].

Fig. 3 Effect of temper-
ature on high cycle
fatigue of Ni₃Al single
crystals [8].

TABLE I

High Cycle Fatigue Data, $R \stackrel{\sim}{=} 0.1$

Alloy (a%)	T°C	Environment	$\frac{\Delta\sigma 10^6}{\sigma_{ys}}$	$\frac{\Delta\sigma 10^7}{\sigma_{ys}}$	Ref.
Fe-24Al-DO$_3$	25	Air	0.84	0.71	10
Fe-24Al-B2	25	Air	0.76	0.62	10
Fe-29Al-DO$_3$	25	Air	1.02	0.89	10
Fe-29Al-B2	25	Air	0.67	0.65	9
Ni-24Al HIP	25	Air	1.79	1.38	9
Ni-24Al HIP+Ext	25	Air	1.57	1.20	9
Ni-26Al Cast	25	Air	1.32	1.05	9
Ni$_3$Al (crystal)	25	Air	0.72	0.56	8
LRO-37 (Fe,Ni)$_3$V	25	Air	2.08	---	5
Nitac 14B	25	Air	1.25	1.12	32
MarM-200DS (crystal)	25	Vac	0.55	0.47	8
Waspaloy	25	Air	0.57	0.48	35
IN617	25	Air	1.53	1.44	35
Hastelloy C	25	Air	---	1.05	35
Fe-24Al-DO$_3$	500	Air	1.08	0.83	10
Fe-29Al-DO$_3$	500	Air	1.01	0.83	10
Fe-24Al-B2	560	Air	0.68	---	10
Fe-29Al-B2	600	Air	0.95	---	10
Ni-24Al HIP+Ext	500	Air	0.98	0.92	9
Ni$_3$Al crystal	425	Air	0.42	0.33	8
Ni$_3$Al crystal	760	Air	0.36	0.28	8
LRO-1-3 (Fe,Co,Ni)$_3$V	650	argon	1.28	0.98	5
LRO-37 (Fe,Ni)$_3$V	400	argon	1.73	---	5
Nitac 14B	825	argon	1.30	1.11	32
U500	650	air	0.61	0.47	35
Waspaloy	800	air	0.50	0.46	35
IN718	650	air	0.56	0.44	35

and b), respectively, whether cycling is carried out in air or in 5×10^{-7} torr vacuum. (The development of cracks seems to be delayed somewhat in vacuum.) However, as pointed out earlier, high cycle fatigue lives are increased substantially when this alloy is ordered. Since crack propagation is much more rapid in the ordered condition, see below, it is concluded that ordering substantially delays crack initiation, while not changing the location of the earliest cracks.

In all Fe$_3$Al alloys, independent of the type of order, crack initiation occurs at grain boundaries near or at the specimen surface, Fig. 9a), and the crack then propagates transgranularly [10]. Crack initiation in wrought Ni$_3$Al occurs along slip bands, but in P/M Ni$_3$Al cracks generally initiated at pores or inclusions, see Fig. 9b), and then propagated transgranularly in the stage I mode at both 25°C and 500°C [9].

CRACK PROPAGATION

Although a significant number of studies of crack growth in ordered alloys have been carried out, see Table II, relatively little is known

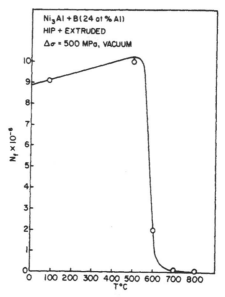

Fig. 4   High cycle fatigue life vs temperature
for Ni_3Al+B, $\Delta\sigma$=500MPa [9].

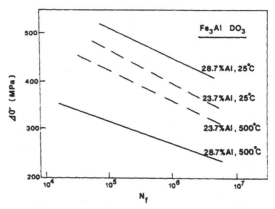

Fig. 5   Effects of composition and temperature on
high cycle fatigue of Fe_3Al polycrystals
at 25°C and 500°C [10].

about mechanisms. For example, order reduces growth rates in $(Fe,Ni)_3V$
[15], a ductile alloy, but increases it in FeCo-V [16], which is brittle
when ordered. Cracks in Ni_3Al+B grow at about the same rate as in $(Fe,Ni)_3V$
at 25°C (Fig. 10a) and somewhat more rapidly than in $(Fe,Ni)_3V$ at 600°C,
Fig. 10b). For both alloys crack growth rates increase with temperature,
e.g., Fig. 11a) for $(Fe,Ni)_3V$ and Fig 11b) for Ni_3Al+B, in spite of

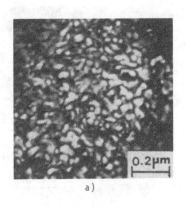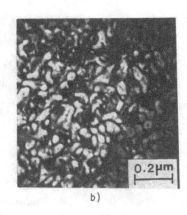

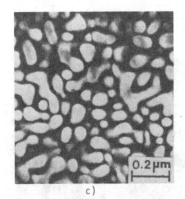

Fig. 6 Development of two phase α+DO₃ microstructures in Fe-23.7a%Al, fatigued at 500°C. Dark field TEM [10].

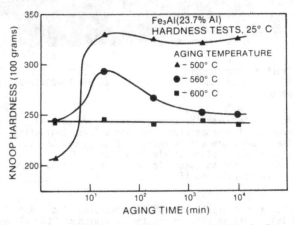

Fig. 7 Hardness vs aging time, Fe-23.7%Al [33].

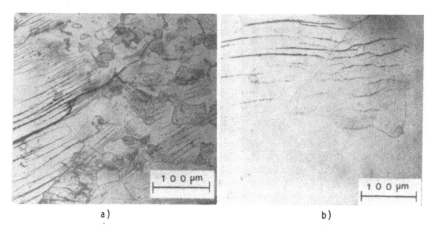

a)   b)

Fig. 8   Crack initiation along slip bands in FeCo-2%V, total strain amplitude = 1% [13]   a) disordered, N=8000   b) ordered, N=600

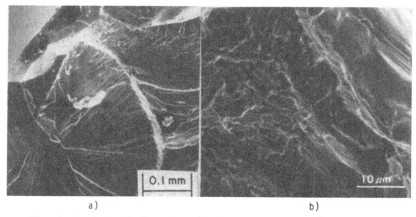

a)   b)

Fig. 9   Crack initiation sites in aluminides, 25°C   a) Fe-23.7a%Al B2, at grain boundaries [10]   b) as HIP Ni-24%Al+B, at pore (arrow) [33].

increased yield stresses over the same temperature range. Plastic zone sizes are reduced at high temperatures in both alloys, indicating that a portion of the temperature effect noted in Fig. 11 may be due to a shift from plane stress to plane strain conditions.

In many models of stress controlled crack propagation, the constant C in the Paris-Erdogan [17] equation:

$$da/_{dN} = C_\Delta K^m \tag{1}$$

is postulated to vary inversely with yield stress, modulus and/or tensile ductility [18,19]. This is clearly not the case with the LRO alloys and Ni_3Al, suggesting that one or more atomic species in the gaseous environment of moderate vacuum or argon causes an increase in growth rates at elevated temperature. Chang, et al. [20] suggested, in fact,

TABLE II

Crack Growth in Intermetallics

Alloy	Temp. °C	Structure	Environment	Crack Path	da/dN ΔK=40MPa (m/cycle)
IC-50 ($Ni_3Al+B+Hf$)	25	Ll_2	air	TG	10^{-8}
Ni-24Al+B	400	Ll_2	air	IG	6×10^{-5}
Ni-24Al+B	500	Ll_2	vac	TG	5×10^{-7}
Ni-24Al+B	600	Ll_2	vac	TG	8×10^{-7}
FeCo-V (d)	25	bcc	air	TG	2×10^{-9}
FeCo-V (S=0.7)	25	B2	air	TG	---
Fe-25.05Al	25	B2	air	TG	1×10^{-6}
Fe-25.05Al	25	DO_3	air	TG	7×10^{-6}
Ti_3Al	649	DO_{19}	air	IG	---
Ti_3Al	649	DO_{19}	vac	TG	---
$Ni_3Fe(o)$	25	Ll_2	air	IG	$>10^{-4}$
$Ni_3Fe(d)$	25	fcc	air	IG	$>10^{-4}$
$(Fe,Ni)_3V*$	25	Ll_2	argon	TG	5×10^{-10}
$(FeNi)_3V*$	600	Ll_2	argon	TG	7×10^{-7}
$(FeNi)_3V*$	25	Ll_2	hydrogen-gas	IG	3×10^{-7}
$(FeNi)_3V*$ (d)	25	fcc	hydrogen-gas	TG	3×10^{-7}

* LRO 42
d=disordered
o=ordered
s=degree of order

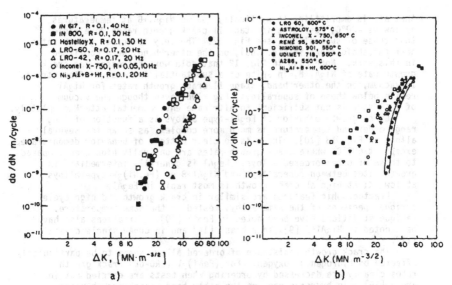

Fig. 10 Crack propagation rates in $(FeNi)_3V$ and $Ni_3Al+B+Hf$ (IC-50) compared to commercial alloys a) 25°C b) 600°C.

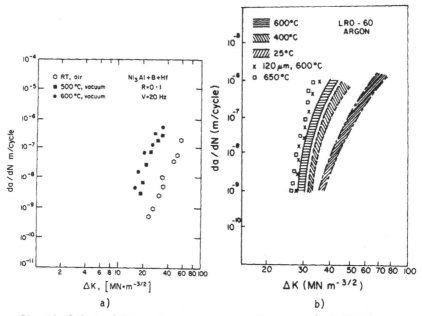

Fig. 11 Effect of temperature on crack growth rate   a) Ni₃Al+B+Hf
(IC-50) [9]   b) (Fe,Ni)₃V [15].

that air is embrittling to cyclically loaded Ni₃Al+B at temperatures
as low as 400°C, resulting in stage II crack growth rates more rapid
than those of superalloys, see Fig. 12. The very low crack growth rates
near threshold shown in Fig. 10a), on the other hand, were confirmed
in this work. Note also in Fig. 12 that cold work increases the crack
growth rate of Ni₃Al+B, in spite of a substantial increase in flow stress.
In vacuum, on the other hand, stage II crack growth rates for Ni₃Al
remain below those of superalloys, Fig. 10b, even though the vacuum
of 10^{-5} torr is not sufficient to eliminate environmental attack completely.

Crack growth behavior of Fe₃Al-type alloys as a function of long
range order and temperature is much more complex, as shown for several
alloys in Fig. 13 [10]. There is no clear pattern of behavior demonstrated,
except at 600°C, where crack growth rates are unusually high when compared
to those of other ordered alloys. Fe₃Al is generally intermediate in
growth rates between superalloys and Ni₃Al+B or (Fe,Ni)₃V-type alloys
at low ΔK; at high ΔK crack growth is most rapid in Fe₃Al.

Fractographic features are similar in crack growth and high cycle
fatigue specimens of the same alloy, tested at the same temperature.
Fatigue striations have been noted in Fe₃Al [10]. Striations also have
been noted in Ni₃Al+B [9], in β brass [14] and in Cu₃Au single crystals
[21].

The crack growth resistance of ordered alloys seems to be particularly
affected by hydrogen or oxygen. For (FeNi)₃V (LRO-60) crack growth
rates clearly are decreased by ordering when tests are carried out in
argon [16]. In hydrogen gas, on the other hand, crack growth rates
in ordered and disordered material are identical, and much greater sensitivity
to hydrogen is displayed in the ordered condition.

The FeCo-V alloy has been studied in the ordered and partially

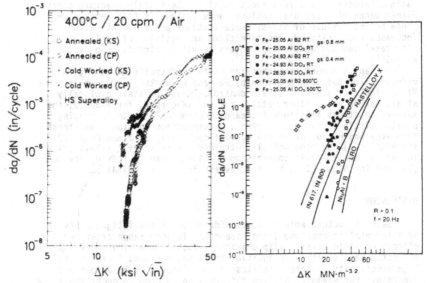

Fig. 12 Comparison of crack growth
rates of Ni₃Al+B and a nickel-
base superalloy tested in air
at 400°C [20].

Fig. 13 Effects of aluminum content
and temperature on crack
growth of Fe₃Al [10].

ordered (S=0.4) conditions in vacuum and in hydrogen gas [16]. The
partially ordered condition showed somewhat more susceptibility to cracking
than did disordered material, again suggesting a link between ordering
and hydrogen embrittlement. Fully ordered FeCo-V could not be tested
in either vacuum or hydrogen because of its severe brittleness.

Insofar as oxygen is concerned, the previously cited observations
of Chang et al [20] on Ni₃Al+B vs superalloys at 400°C, Fig. 12, and
the rise in crack growth rates with increasing temperature in Ni₃Al+B
cycled in moderate vacuum, Fig. 11b), point to a likely embrittling
effect of oxygen. Further, crack growth rates in Ti₃Al at 649°C are
increased by about ten times in air relative to vacuum except near threshold,
where oxide induced crack closure causes crack arrest in air [36].

LOW CYCLE FATIGUE

Few detailed studies of strain-controlled cycling in ordered alloys
have been reported. For Cu₃Au single crystals, Chien and Starke [21]
found that order had little effect on room temperature life and fracture
mode. The Manson-Coffin equation was obeyed in both the ordered and
disordered conditions, with slightly higher slope for the latter. Ordered
material, however, hardened then softened during cycling, while disordered
material continually hardened. These results were attributed to strain-
induced disordering of initially ordered materials.

Studies on Ti₃Al in the range 500-800°C showed pronounced hardening
at temperatures to 700°C, the extent of hardening increasing with increased
plastic strain amplitude [34]. Fatigue hardening decreased between
700°C and 800°C, while fatigue ductility increased monotonically between
500°C and 800°C, see Fig. 16 [34]. At temperatures up to 700°C Ti₃Al

fatigue hardens to a stress about equal to the tensile fracture stress. Planar bands of basal dislocations cause fatigue in this temperature range by the buildup of stress concentrations prior to saturation. Increased cross slip, dislocation climb and activity of dislocations with basal and non-basal slip vectors result in high fatigue ductility at 800°C.

Perhaps the most comprehensive picture of cyclic hardening in an ordered alloy has been provided by Pak and co-workers [22], who studied Ni_3Ge single crystals in strain control. These crystals cyclically strain harden as do other metals, but unlike copper, for example, the saturation stress amplitude increases monotonically with increasing plastic strain amplitude. Further, a compression-tension flow stress asymmetry was noted, as in the case of cyclically strained Cu_3Au [21] and Ni_3Al [23] and monotonically strained two-phase $\gamma+\gamma'$ alloys [24,25]. The stress amplitude in compression is higher than in tension for Cu_3Au [21] and for Ni_3Ge [22], but the reverse is true for Ni_3Al [23].

DISCUSSION

Several factors enter into consideration of the fatigue behavior of intermetallic compounds; most important is long range order and its effects on yield stress, slip character, ductility at the crack tip and notch sensitivity. In addition to these, one must factor in the high sensitivity of intermetallics to impurities such as oxygen and hydrogen. For example, Ni_3Al+B suffers a sharp drop in ductility above 500°C [11,26], probably due to the dynamic adsorption of oxygen [11]. At low temperature hydrogen is known to embrittle Ni_3Al+B [27], $(Fe,Ni)_3V$ [28], FeCo-V [28] and Co_3Ti [29], all in tension. Therefore, oxygen, hydrogen or water vapor in the environment could be expected to adversely influence fatigue life or crack growth rate. However, this is an extrinsic effect, and therefore is less important to this discussion than is consideration of intrinsic ordering effects.

When long range order is present, notch sensitivity tends to be high even for alloys such as binary Ni_3Al+B (see Table III), which exhibits very high tensile ductility in unnotched tests. Crack advance in notch-sensitive material can, therefore, consist of two components: crack blunting and static fracture (e.g., by cleavage). In notch-insensitive alloys such as $(Fe,Ni)_3V$, on the other hand, crack blunting alone need be considered. The latter process must depend upon slip character at the crack tip. Dislocations on slip bands intersecting the crack tip may move reversibly, as in planar slip material, or irreversibly, as in alloys which deform on many slip systems or with easy cross slip. Long range order increases planarity of slip in virtually all intermetallic compounds (Mg_3Cd is the exception) [30]. The result is expected to be delayed crack initiation and slower crack propagation. Direct evidence for order extending high cycle fatigue lives has been obtained in several systems (FeCo-V [1], Ni_3Mn [1] and Cu_3Au [2]); similarly, lower crack growth rates are noted in ductile, notch-insensitive $(Fe,Ni)_3V$ alloys when ordered, see Fig. 11a) [15]. However, in FeCo-V, crack propagation rates in the fully ordered condition are too rapid to measure. Even partially ordered FeCo-V displays considerably higher growth rates than does disordered material [16]. The influence of varying degrees of order on tensile ductility of unnotched FeCo-V parallels that pattern of crack growth behavior [31]. Cyclic crack growth in partially ordered material is by transgranular cleavage, indicating that static fracture modes are dominant in this material. In ordered $(Fe,Ni)_3V$, on the other hand, stage I (slip-band) cracking is observed, indicating that crack blunting processes control.

Normalized high cycle fatigue data for the aluminides, see Table I, are comparable to results for conventional single phase alloys.

TABLE III

Notch Sensitivities of Intermetallics

	T°C	NSR*
Ni-24Al (HIP+Ext)	25	0.56
Ni-24Al (HIP+Ext)	500	0.92
Fe-22%Al (DO$_3$)	25	0.57
Fe-22%Al (B2)	25	0.79

* NSR = Notch Sensitivity Ratio = $\dfrac{(\sigma_{UTS})\ \text{notched}}{(\sigma_{UTS})\ \text{unnotched}}$

Precipitation hardened nickel-base superalloys, titanium-base alloys and wrought austenitic stainless steels, on the other hand, display markedly inferior normalized fatigue properties to the aluminides; coarse planar slip is characteristic of the former materials. The one class of materials that demonstrates high cycle fatigue properties superior to the intermetallics are the DS eutectics, of which Nitac 14B is the most fatigue resistant [32].

Perhaps the least understood aspects of cyclic behavior of intermetallic compounds are the processes leading to crack initiation. Studies on polycrystals show that grain boundaries, pores, inclusions and/or slip bands can be preferred sites for initiation, depending upon the alloy system and processing technique. Slip bands are preferred sites in (Fe,Ni)$_3$V, FeCo-V and wrought Ni$_3$Al+B, while grain boundaries are the dominant sites in Fe$_3$Al.

Single crystal studies have been sparse. Cracks in both ordered and disordered Cu$_3$Au initiated at the intersections of slip bands and deformation bands near the grips [21]. Stage I propagation was noted through out the fatigue test in both conditions. There is need for work on single crystals of other ordered alloys to establish the relations among ordering energy, crystal orientation and crack initiation sites. Also, the role of cyclic hardening (or softening) in crack initiation remains unclear. In non-ordered metals, initiation is connected with the onset of saturation. However, in the few studies carried out to date on ordered alloys, wide differences in cyclic hardening behavior have been noted. For example, while a flow stress asymmetry has been noted in single crystals of Cu$_3$Au [21], Ni$_3$Al [23] and Ni$_3$Ge [22], the effect is very small in Cu$_3$Au, and is not always in the same direction in the three systems. For Cu$_3$Au and Ni$_3$Ge the flow stress is higher in compression than in tension, while in Ni$_3$Al the reverse is true. Further, in Ni$_3$Ge the strength difference between compression and tension is almost constant with number of cycles and no cyclic softening is observed [36], while in Cu$_3$Au softening occurs at a number of cycles that decreases with increasing strain amplitude. Differences in ordering energy among the various intermetallics, as well as different orientations of crystals in each investigation, could be responsible for a portion of the discrepancies, but clearly much more work is needed in this field.

SUMMARY AND CONCLUSIONS

Long range order has been shown to increase the high cycle fatigue lives of several alloys, presumably by delaying crack initiation. However,

260

detailed studies of the relationships among superlattice dislocations,
cyclic hardening (or softening), formation of persistent slip bands
and crack initiation are still lacking. To date, there is no evidence
that ordering is beneficial under strain controlled, low cycle fatigue
conditions. The factors controlling crack propagation are complex:
notch sensitivity, tensile ductility, yield stress and environment all
seem to play a role in crack growth behavior. Insufficient data are
available to determine which of these are most important. However,
for the $L1_2$ alloys $(Fe,Ni)_3V$ and Ni_3Al+B, hydrogen is undoubtedly detrimental to fatigue resistance at low temperatures, and oxygen appears
to be damaging at high temperatures.

ACKNOWLEDGMENT

The author is grateful for financial support by the National Science
Foundation under Grant. No. DMR84-09593 and to the Office of Naval Research
under Contract No. N00014-84-K-0276.

REFERENCES

1. R.C. Boettner, N.S. Stoloff and R.G. Davies, Trans. AIME, 236,
    131 (1966).
2. G. Rudolph, P. Haasen, B.L. Mordike and P. Neumann, Proc. First Int. Conf.
    on Fracture, Sendai, Japan, 2, 501 (1965).
3. A. Gittins, Met. Sci. J, 2, 114 (1968).
4. R.S. Whitehead and F.W. Noble, J. Mat. Sci. 5, 851 (1970).
5. S. Ashok, K. Kain, J. Tartaglia and N.S. Stoloff, Met. Trans. A, 14A,
    1997 (1983).
6. H. McI. Clark, Nature, 209, 193 (1966).
7. S.M.L. Sastry and H.A. Lipsitt, Acta. Met. 25, 1279 (1977).
8. J.E. Doherty, A.F. Giamei and B.H. Kear, Met. Trans. A, 6A, 2195 (1975).
9. G.E. Fuchs, A.K. Kuruvilla and N.S. Stoloff, submitted to Met. Trans. A
    (1986).
10. S.J. Choe, G.E. Fuchs and N.S. Stoloff, RPI, unpublished.
11. C.T. Liu and C.L. White, Abstract, p. 14, J. of Metals, 37, Nov. (1985).
12. H. Inouye, in High Temperature Ordered Intermetallic Alloys, MRS Symposia
    39, Materials Res. Soc., Pittsburgh, PA, 22 (1985).
13. S.J. Choe and N.S. Stoloff, Rensselaer Polytechnic Inst., Troy, NY
    unpublished (1986).
14. H.D. Williams and G.C. Smith, Phil. Mag. 13, 835 (1966).
15. A.K. Kuruvilla and N.S. Stoloff, Met. Trans. A, 16A, 815 (1985).
16. A.K. Kuruvilla and N.S. Stoloff, Proc. 7th Int. Conf. on Strength of
    Metals and Alloys, Montreal, Canada 2, 1335 (1985).
17. P. Paris and F. Erdogan, J. Basic Eng., Trans. ASME, Series D, 85, 528
    (1963).
18. A.J. McEvily, Jr. and T.L. Johnston, Intl. J. Fract. Mech. 3, 45 (1967).
19. L.H. Burck and J. Weertman, Met. Trans. A, 7, 257 (1976).
20. K.M. Chang, S.C. Huang and A.I. Taub, General Electric Co., Rept.
    86CRD202, Oct. (1986).
21. K.H. Chien and E.A. Starke, Jr., Acta. Met. 23, 1173 (1975).
22. H-R. Pak, L-M, Hsiung and M. Kato, in High Temperature Ordered
    Intermetallic Alloys, MRS Symposia 39, Materials Res. Soc.,
    Pittsburgh, PA, 239 (1985).
23. S.S. Ezz and D.P. Pope, Scripta Met. 19, 741 (1985).
24. S.S. Ezz, D.P. Pope and V. Paidar, Acta. Met. 30, 921 (1982).
25. D. Jablonski and S. Sargent, Scripta Met., 15, 1003 (1981).
26. A.I. Taub, S.C. Huang and K.M. Chang in High Temperature Ordered Inter-

metallic Alloys, MRS Symposia $\underline{39}$, MRS, Pittsburgh, PA 221 (1985).

27. A.K. Kuruvilla and N.S. Stoloff, Scripta Met. $\underline{19}$, 229 (1985).

28. A.K. Kuruvilla and N.S. Stoloff, in High Temperature Ordered Intermetallic Alloys, MRS Symposia $\underline{39}$, Materials Res. Soc., Pittsburgh, PA, 229 (1985).

29. T. Takasugi and O. Izumi, Acta. Met. $\underline{34}$, 168 (1986).

30. N.S. Stoloff and R.G. Davies, Trans. ASM $\underline{57}$, 247 (1964).

31. N.S. Stoloff and R.G. Davies, Prog. Mat. Sci., $\underline{13}$, no. 1, 1 (1966).

32. K.A. Dannemann, N.S. Stoloff and D.J. Duquette in Deformation of Multiphase and Particle Containing Materials, Proc. 4th RISO Symp., 205 (1983).

33. G.E. Fuchs, PhD Thesis, Rensselaer Polytechnic Institute (1986).

34. S.M.L. Sastry and H.A. Lipsitt, Met. Trans. A, $\underline{8A}$, 299 (1977).

35. Aerospace Structural Materials Handbook.

36. S. Venkataraman, T. Nicholas and L.P. Zawada, Abstracts, ASM Materials Week '86, Orlando, FL, 25 (1986).

CREEP IN TERNARY B2 ALUMINIDES AND OTHER INTERMETALLIC PHASES

I. JUNG, M. RUDY AND G. SAUTHOFF, Max-Planck-Institut für Eisenforschung GmbH., D-4000 Düsseldorf, Federal Republic of Germany

ABSTRACT

This paper examines the possibilities for obtaining a high creep resistance in intermetallic phases. In a first section, the creep mechanisms are discussed with respect to the effects of stress, temperature and composition in the light of recent experimental results on the creep behaviour of the B2 phase (Fe,Ni) Al between 650 °C and 1100 °C. The familiar models for describing the creep of conventional disordered alloys apply to intermetallic phases, too, and describe the observed behaviour in a quantitative way. In the second section the creep of various ordered bcc phases - (Fe,Ni)Al and (Co,Ni)Al with B2 structure and the Heusler-type Ni_2AlTi with $L2_1$ structure - and ordered fcc phases - Ni_3Al with $L1_2$ structure, Fe_3AlC and Ni_3AlC with $E2_1$ structure and $NbAl_3$ with DO_{22} structure - are examined. Large increases in creep strength are possible by the transition to phases with more complex structures as well as by appropriate alloying. Much further work is necessary for making more reliable predictions on the obtainable creep strengths.

INTRODUCTION

The intermetallic phases are of interest for high-temperature applications because of their outstanding high-temperature stability which results from the tight binding between the unlike metal atoms. This is examplified by the classical B2 phase NiAl which has a higher melting point than the components. From this it is expected that intermetallic phases have a comparatively high creep resistance in parallel with their high hardness. Furthermore, creep is coupled with diffusion, and diffusion is more difficult in ordered alloys than in disordered alloys.

However, it was found that, e.g., the Ni aluminides Ni_3Al and NiAl show a poor creep resistance - at least in their unalloyed state - relative to that of commercial Ni-base high-temperature alloys [1, 2]. The reasons for this unexpected behaviour have not been quite clear because only few studies on the creep of intermetallic phases were undertaken in the past, and the physical understanding of the creep mechanisms in intermetallic phases is still insufficient in spite of the appreciable efforts to develop new high-temperature materials on the basis of intermetallic phases [1-3].

Therefore a larger programme was started for studying the creep behaviour of intermetallic phases in more detail. The aim is a better understanding of the creep mechanisms on one hand and the study of the possibilities for developing materials with high creep resistances on the other hand. Both aspects are presented in the following, i.e., in the first part, the creep mechanisms are discussed in the light of the recent findings, and in the second part the data of various phases and structures are compared with each other.

CREEP MECHANISMS IN THE B2 ALUMINIDES

The B2 structure forms in the binary or multinary phases MAl where M is
Fe, Ni, Co or a combination of these metals. As an example, Fig. 1
shows the ternary Fe-Ni-Al system. The bcc B2 structure (or $L2_0$ in the
ternary case) is stable in the (Fe,Ni)Al phase up to the melting point
for all Fe/Ni ratios and deviations from stoichiometry (50 at. % Al)
down to 40 at. % Al and less depending on the Fe/Ni ratio and
temperature. In spite of this compositional variability, (Fe,Ni)Al is
much more stable than the neighbouring phases Fe_3Al with DO_3 structure
which is derived from B2, and Ni_3Al with the fcc $L1_2$ structure. Because
of this stability and variability the (Fe,Ni)Al phase is rather suited
for studying the creep behaviour as a function of temperature and
composition.

Dislocation creep

Fig. 2 shows representative stress - creep rate curves for
stoichiometric (Fe,Ni)Al with various Fe/Ni ratios [4]. The data are
described by straight lines in the double logarithmic plot - at least
for not too low stresses - which indicates dislocation creep according
to a power law. Dislocation creep is a well-known creep mechanism in
pure metals and conventional disordered alloys and is described by the
familiar Dorn equation

$$\dot{\varepsilon} = A \ (DGb/kT) \ (\sigma/G)^n \qquad\qquad (1)$$

where $\dot{\varepsilon}$ = secondary strain rate, A = dimensionless factor, D =
effective diffusion coefficient, G = shear modulus, b = Burgers vector,
k = Boltzmann's constant, T = temperature, σ = applied stress, and the
exponent n is usually 3, 4 or 5 [5].
   In dislocation creep of disordered alloys the dislocations glide
and climb consecutively. If climb is the slower step as in pure metals,
the creep rate is controlled by dislocation climb which gives rise to a
well-defined subgrain structure, and the stress exponent is 4 or 5.
This is characteristic for the class-II alloys [6]. Otherwise viscous
dislocation glide is rate controlling which leads to dislocation
tangles without subgrain formation and a stress exponent 3 as
characteristics of the class-I alloys.
   The creep behaviour of the intermetallic (Fe,Ni)Al shows analogous
characteristics. In the Ni-rich phase and in the binary NiAl a
well-defined substructure is found after creep as is illustrated by
Fig. 3 a and b. The subgrain size is of the order of 10 μm, and the
dislocation density within the subgrains is about 10^7 cm^{-2} in this
particular case. In agreement with this, stress exponents between 4 and
4.5 have been found for the Ni-rich phase, i.e., this phase behaves
like class-II alloys with dislocation climb controlling the creep [4,
7]. In the Fe-rich phase and in FeAl, however, no subgrain formation
has been observed even after long creep times (Fig. 3 c, d). The
dislocation density remains high - about 10^{10} cm^{-2}, and the stress
exponent varies between 3 and 3.6. This indicates class-I behaviour,
i.e., here the creep is controlled by the viscous glide of the
dislocations [4, 7].

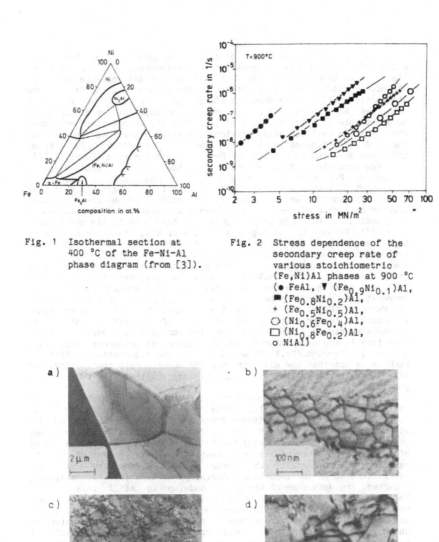

Fig. 1 Isothermal section at
400 °C of the Fe-Ni-Al
phase diagram (from [3]).

Fig. 2 Stress dependence of the
secondary creep rate of
various stoichiometric
(Fe,Ni)Al phases at 900 °C
($\bullet$ FeAl, $\blacktriangledown$ (Fe$_{0.9}$Ni$_{0.1}$)Al,
$\blacksquare$ (Fe$_{0.8}$Ni$_{0.2}$)Al,
+ (Fe$_{0.5}$Ni$_{0.5}$)Al,
$\bigcirc$ (Ni$_{0.6}$Fe$_{0.4}$)Al,
$\square$ (Ni$_{0.8}$Fe$_{0.2}$)Al,
o NiAl)

Fig. 3 TEM micrographs of the
dislocation structures
in (Ni$_{0.83}$Fe$_{0.17}$)$_{1.2}$
Al$_{0.8}$ (a,b) and
(Fe$_{0.8}$Ni$_{0.2}$)Al (c,d)
after creep (10^{-7}s^{-1})
at 900 °C.

The reasons for such an effect of composition on the creep mechanism and on the microstructure development are not yet quite clear. It is true that two types of dislocations are possible in the B2 structure, (a/2)<111> superlattice dislocations and perfect a<100> dislocations [1]. Furthermore the APB energy is much lower for FeAl than for NiAl which makes the formation of superlattice dislocations more probable in FeAl than in NiAl [7]. However, theoretical estimates show that in FeAl the a<100> dislocation still has a lower energy than the (a/2)<111> dislocation [4], and indeed only a<100> dislocations have been observed in Ni-rich and Fe-rich stoichiometric (Fe,Ni)Al [7].

The formation of a subgrain structure results from recovery processes. For this a driving force is necessary as well as a sufficient dislocation mobility. The driving force originates from the total dislocation energy which is proportional to the shear modulus. Elastic data even of the binary B2 phases are scarce in the literature, in particular for high temperatures, and they show little consistency (see [8]). Because of the higher stability of NiAl its shear modulus at high temperatures is expected to be higher than that of FeAl which would mean a higher driving force for subgrain formation in NiAl. As to the dislocation mobility, it has been observed by TEM that a second phase forms in Fe-rich (Fe,Ni)Al at lower temperatures (<750 °C) [4]. This is confirmed by the remarkably high 0.2 %-proof stress at lower temperatures [9] and by the special stress dependence of the creep rate of Fe-rich (Fe,Ni)Al at lower temperatures (Fig. 4) which contrasts with that at higher temperatures and that of Ni-rich (Fe,Ni)Al and which corresponds to that of alloys with strengthening particles [10]. From this it may be concluded that at higher temperatures clustering occurs in Fe-rich (Fe,Ni)Al which reduces the dislocation mobility to such an extent that the dislocation glide becomes rate controlling, and the subgrain formation with the already small driving force is not possible.

It follows from the above discussion that the mechanisms of dislocation creep in intermetallic phases – at least in B2 phases – do not differ from those in conventional disordered alloys, i.e., the special structure of the intermetallic phase does not lead to special creep effects. Then it should be possible to describe the dislocation creep of such intermetallic phases in a quantitative way by means of the Dorn equation which contains the shear modulus G and the diffusion coefficient D as the important material parameters (eq. 1). As already remarked, the data base of the shear modulus is still insufficient. Therefore only the effect of the diffusion coefficient is discussed now. It is known that the diffusion coefficient depends in a sensitive way on the degree of atomic order. Thus deviations from stoichiometry which give rise to constitutional disorder increase the diffusion coefficient from which a reduced creep resistance results according to eq. 1. This has indeed been observed and discussed repeatedly and had been found for (Fe,Ni)Al [3, 7, 9].

For stoichiometric (Fe,Ni)Al the creep resistance increases with decreasing Fe/Ni ratio according to Fig. 2. However, the maximum creep resistance is shown by $(Fe_{0.2}Ni_{0.8})Al$ as is visible more clearly in Fig. 5 [9]. The reason for this behaviour is again the effect of composition on the diffusion coefficient as is demonstrated by the corresponding plot of the diffusion coefficient (Fig. 6) which has been calculated from data of Moyer and Dayananda [11] and which shows a distinct minimum for $(Fe_{0.2}Ni_{0.8})Al$ [7].

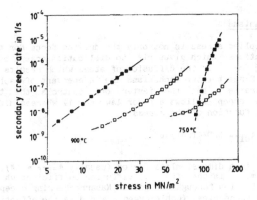

Fig. 4 Stress dependence of the
secondary creep rate of
$(Fe_{0.8}Ni_{0.2})Al$ (■)
and $(Ni_{0.8}Fe_{0.2})Al$
(□) at 750 °C and
900 °C.

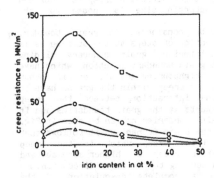

Fig. 5 Creep resistance
(at $10^{-7}s^{-1}$) as a
function of the iron
content for stoichiometric
(Fe,Ni)Al at 750 °C (□),
900 °C (○), 982 °C (◇),
and 1027 °C (△) (from
[7]).

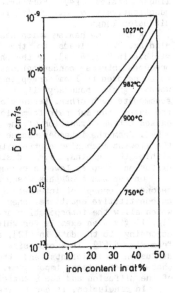

Fig. 6 Estimated interdiffusion
coefficient $\bar{D}$ of stoichiometric (Fe,Ni)Al as a
function of the iron
content for various tem-
peratures (from [7]).

Other Mechanisms

The applied stress is not only the driving force for the migration of dislocations which gives rise to dislocation creep, but it is also the driving force for the diffusion of atoms which results in diffusion creep [5]. Since the two mechanisms are independent of each other, they work in parallel, and the faster one controls the total creep. Dislocation creep follows a power law (eq. 1) whereas diffusion creep is a linear function of the stress:

$$\varepsilon_{diff} = A_{diff} \, (\Omega D/kT \, d^2) \, \sigma \qquad (2)$$

where A_{diff} is a dimensionless factor (usually $A_{diff} = 14$), Ω is the atomic volume, D is the effective diffusion coefficient which considers both the diffusion through the grain (Nabarro-Herring creep) and along the grain boundaries (Coble creep), and d is the effective diffusion length [5]. Thus at higher stresses dislocation creep is observed, and only at low stresses diffusion creep becomes dominant.

At lower stresses Fig. 2 indeed shows a deviation of the observed creep from the power law of dislocation creep (straight line in the double logarithmic plot). The contribution of diffusion creep is larger the slower the dislocation creep is, i.e., the larger the shear modulus is, as follows from the comparison of eqs. 1 and 2. Correspondingly the Ni-rich (Fe,Ni)Al at the lower temperature 750 °C shows a clear transition from dislocation creep to diffusion creep (Fig. 7 with linear scales) [4]. Furthermore, this diffusion creep is described quantitatively by eq. 2 if the subgrain size is chosen for the diffusion length.

A third mechanism which has to be considered is grain boundary sliding. This leads to the formation of steps at the surface as is visible in Fig. 8 [4]. In the phase interior, grain boundary sliding produces stress concentrations at grain boundary junctions which have to be relaxed by local creep in the neighbouring grains, or in other words: grain boundary sliding and the creep within the grains have to accomodate each other. Because of the interaction between the creep processes the subgrain formation starts at the grain boundaries and it causes a transfiguration of the grain boundaries as can be seen in Fig. 9 [4]. From the interdependence of the creep mechanisms it follows that the slowest mechanism controls the total creep which is the creep within the grains, i.e. dislocation creep at higher stresses and diffusion creep at lower stresses according to the above discussion.

On the basis of this discussion a complete description of the secondary creep of intermetallic phases is possible with eqs. 1 and 2 as constitutive equations. Then deformation maps can be calculated which allow the interpolation and extrapolation of measured creep data. Fig. 10 gives an example for which the shear modulus was estimated according to the data in [12], and the diffusion coefficient was taken from Fig. 6 [7]. This emphasizes the fact that the knowledge of the diffusion coefficients and the elastic constants of intermetallic phases is of primary importance, i.e., there is great need for studies of the diffusion and the elasticity of intermetallic phases.

In conclusion, it can be stated that the models for describing the creep of conventional disordered alloys are applicable to intermetallic phases, too, and can be used for a complete description of the observed creep. The special structure of the intermetallic phases does not lead

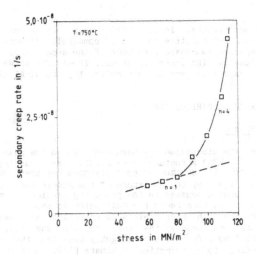

Fig. 7. Stress dependence of the secondary creep rate of $(Ni_{0,8}Fe_{0,2})Al$ at $750\ °C$.

Fig. 8 Micrograph of $(Ni_{0,8}Fe_{0,2})Al$ after 1.5 % creep strain at 900 °C with grain boundary sliding.

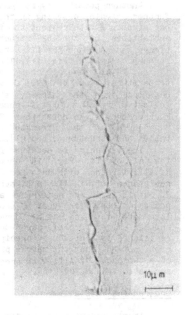

Fig. 9 Micrograph of $(Ni_{0,5}Fe_{0,5})Al$ after 5.9 % creep strain at 900 °C with subgrain formation at grain boundaries.

to special creep effects. Indeed, the creep resistance of the Ni aluminides is poor relative to that of commercial Ni-base superalloys because the special two-phase structure of the superalloys gives rise to an exceptionally high creep resistance. These effects were discussed in [10] with respect to ferritic superalloy-type Fe-Ni-Al alloys.

CREEP RESISTANCE OF VARIOUS STRUCTURES

Ordered bcc Phases

Up to now the discussion has been centered on the phase $(Fe,Ni)Al$ with the B2 (or $L2_o$) structure which results from ordering in the bcc lattice (Fig. 11). The maximum creep resistance has been obtained with $(Fe_{0.2}Ni_{0.8})Al$, i.e., by the admixture of the softer FeAl to the harder NiAl which has been discussed in the preceding section. A much higher creep resistance has been found for CoAl which is shown in Fig. 12 [4]. The reason for the high creep resistance of CoAl is not clear since the melting points are nearly the same for CoAl and NiAl, and the APB energy of CoAl (360 mJ/m^2) is even slightly less than that of NiAl (400 mJ/m^2) according to a theoretical estimate [13]. Apart from this, the creep behaviour of the CoAl-base ternary B2 phase is quite similiar to that of $(Fe,Ni)Al$. In particular, the creep resistance again increases with the admixture of the softer NiAl as is shown for $(Co,Ni)Al$ in Fig. 12 [4].
Another possibility for establishing a higher creep resistance is further atomic ordering in the bcc lattice. If ordering occurs in the bcc structure with respect to four sublattices instead of the two sublattices of the B2 structure, the DO_3 structure results for the binary case A_3B and the $L2_1$ structure (known as Heusler type) for the ternary case A_2BC (Fig. 11). The DO_3 phase in the Fe-Ni-Al system (Fig. 1) is Fe_3Al, and it has indeed a higher strength at elevated temperatures (up to 540 °C which is the critical temperature of ordering) than the already strong B2 phase $(Ni_{0.8}Fe_{0.2})Al$ as is indicated by the 0.2 % proof stresses of these phases in Fig. 13. Furthermore, it is noted that Fe_3Al shows an anomalous temperature dependence of the strength which has been observed for some other phases and structures, too [3].
The $L2_1$ phase is obtained from NiAl by the substitution of half of the Al by Ti. Fig. 14 shows the creep resistance of the resulting Ni_2AlTi together with data for B2 and other phases. Indeed, the creep resistance of Ni_2AlTi is higher than that of NiAl by a factor of about 2, but it is in the same range as that of CoAl or the ternary B2 phases. This means that by proper alloying the creep resistance of an intermetallic phase can be increased as much as by the transition to another more complex structure. A still larger increase in creep resistance is achieved by precipitating a second phase which may be even softer than the matrix phase. Fig. 14 shows the creep resistance of a two-phase Ni_2AlTi with precipitated NiAl which equals that of the superalloy MAR-M 200 [18].

Ordered fcc Phases

Atomic ordering in the fcc lattice produces the $L1_2$ structure (Fig. 11), and its most eminent representative is the familiar γ'phase Ni_3Al. It is similar to its counterpart Fe_3Al in the Fe-Ni-Al system

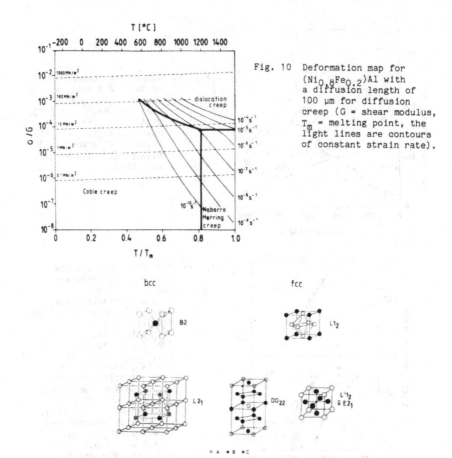

Fig. 10 Deformation map for $(Ni_{0.8}Fe_{0.2})Al$ with a diffusion length of 100 μm for diffusion creep (G = shear modulus, T_m = melting point, the light lines are contours of constant strain rate).

Fig. 11 Ordered bcc structures B2 and $L2_1$, and ordered fcc structures $L1_2$, $L'1_2$ and DO_{22}.

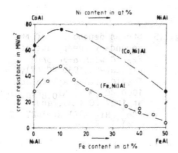

Fig. 12 Composition dependence of the creep resistance (at $10^{-7}s^{-1}$) at 900 °C of (Co,Ni)Al and (Ni,Fe)Al.

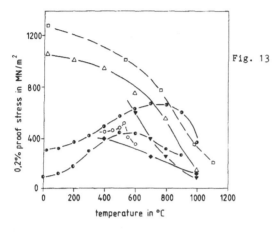

Fig. 13 Temperature dependence of the 0.2 % proof stress for $(Ni_{0.8}Fe_{0.2})Al$ (♦) [9], Fe_3Al (0) [17], Ni_3Al (•) [15], advanced Ni_3Al (◐) [16], Fe_3AlC with graphite (▼), Ni_3AlC (△), and MA 6000 (◻) [14].

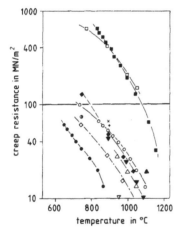

Fig. 14 Temperature dependence of the creep resistance (at $10^{-7}s^{-1}$) for single-crystalline NiAl (●) [18], polycrystalline NiAl (◇) [4, 9], $(Ni_{0.8},Fe_{0.2})Al$ (♦) [4, 7, 9], CoAl (+) [4], $(Co_{0.8}Ni_{0.2})Al$ (X) [4], Ni_2AlTi (○) [18], Ni_2AlTi with precipitated NiAl (■) [18], MAR-M200 (◻) [18], advanced Ni_3Al (◐) [19], Fe_3AlC with precipitated α-Fe (▽), Ni_3AlC (△), Fe_3AlC with precipitated graphite (▼), and $NbAl_3$ (▲).

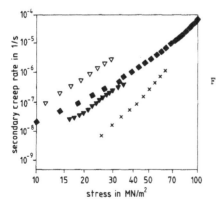

Fig. 15 Stress dependence of the secondary creep rate of Fe_3AlC with α-Fe precipitate at 950 °C (▽), Fe_3AlC with graphite at 1000 °C (▼), $Ni_{0.8}Fe_{0.2})Al$ at 1027 °C (◇), and $(Co_{0.8}Ni_{0.2})Al$ at 1027 °C (X).

with respect to the strength at elevated temperatures and its anomalous temperature dependence in spite of the difference in structure. The alloying of Ni_3Al with B and further elements has led to the development of ductile "advanced aluminides" with an increased high-temperature strength at higher temperatures, and an example is given in Fig. 13 [16, 20]. However, the creep resistance of an advanced Ni_3Al compares less favourably with those of ordered bcc phases according to the data in Fig. 14. In particular, the creep resistance of the advanced Ni_3Al at 760 °C is lower than that of the B2 phase $(Ni_{0.8}Fe_{0.2})Al$ and also that of the L2$_1$ phase Ni_2AlTi.

Ni_3Al is the only stable L1$_2$ phase in the Fe-Ni-Al system (Fig. 1) since the counterpart Fe_3Al has the DO$_3$ structure. However, the dissolution of C in Fe_3Al stabilizes the L1$_2$ structure in Fe_3Al, too. The structure of the resulting phase Fe_3AlC is quite analogous to that of Ni_3Al with dissolved C, i.e., the substitutional atoms form a L1$_2$ lattice with the C atoms in the interstices, and it is called L'1$_2$ or E2$_1$ (perovskite-type) [21]. (Usually such phases are regarded as complex carbides, i.e. they are on the border between the intermetallic phases and the interstitial compounds.) The creep behaviour of such phases is quite normal (Fig. 15) and the creep resistance of Fe_3AlC with precipitated graphite particles is similar to that of $(Ni_{0.8}Fe_{0.2})Al$ whereas Fe_3AlC with disordered α-Fe on the grain boundaries is much softer (Figs. 14 and 15). The high-temperature strength of Ni_3AlC (without a second phase) is still larger than that of Fe_3AlC with graphite particles, i.e., it is between that of Ni_3Al and "advanced" Ni_3Al, and furthermore it increases with decreasing temperature in contrast to Ni_3Al (Fig. 13). It is noted that Fe_3AlC and Ni_3AlC show a sufficient oxidation resistance during the creep experiments.

A further atomic ordering in the L1$_2$ lattice produces the DO$_{22}$ structure which is present in $NbAl_3$ [22]. The creep resistance of $NbAl_3$ is 20 MPa at 1100 °C according to Fig. 14 which compares rather favourably with advanced Ni_3Al and also with the L2$_1$ phase Ni_2AlTi. It is expected for $NbAl_3$ as well as for the other comparatively unknown phases that the creep resistance of the unalloyed phase can be increased significantly by alloying with further elements in a similar way as in the case of the B2 aluminides NiAl and CoAl (Fig. 12). The next step would be the formation of two-phase structures for a further increase of the creep resistance. The exploitation of these possibilities necessitates much further fundamental work which should be directed to the determination of the respective deformation maps.

The problem of brittleness has not been discussed here since it is beyond the scope of this paper. Apart from the Ni_3Al-base advanced aluminides which have been ductilized by means of the B effect, all the discussed phases are brittle. The ductilization of NiAl appears to be possible by grain refinement to below a critical grain size [23]. As to the other phases, the possibilities for ductilization still have to be explored.

ACKNOWLEDGEMENTS

The financial support of the Bundesministerium für Forschung und Technologie and of the Deutsche Forschungsgemeinschaft is gratefully acknowledged.

REFERENCES

1. N.S. Stoloff, in High-Temperature Ordered Intermetallic Alloys, edited by C.C. Koch, C.T. Liu and N.S. Stoloff (Materials Research Society, Pittsburgh, 1985), p. 3.
2. P.R. Strutt and B.H. Kear, in ref. 1, p. 279.
3. G. Sauthoff, Z. Metallkde. 77, 654 (1986).
4. M. Rudy, Dr.-Ing. thesis, RWTH Aachen, 1986.
5. H.J. Frost and M.F. Ashby, Deformation - Mechanism Maps (The Plasticity and Creep of Metals and Ceramics) (Pergamon Press, Oxford, 1982).
6. O.D. Sherby and P.M. Burke, Progr. Mat. Sci. 13, 325 (1968).
7. M. Rudy and G. Sauthoff, Mat. Sci. Eng. 81, 525 (1986).
8. M.R. Harmouche and A. Wolfenden, J. Testing and Evaluation 13, 424 (1986)
9. M. Rudy and G. Sauthoff, in ref. 1, p. 327.
10. I. Jung and G. Sauthoff, in Strength of Metals and Alloys (ICSMA7), edited by H.J. McQueen, J.-P. Bailon, J.I. Dickson, J.J. Jonas and M.G. Akben (Pergamon Press, Oxford, 1985), p. 731.
11. T.D. Moyer and M.A. Dayananda, Metall. Trans. 7A, 1035 (1976).
12. R.J. Wasilewski, Trans. AIME 236, 455 (1966).
13. G. Inden (private communication).
14. Inco Alloys Data Sheet MA 6000.
15. N.S. Stoloff and R.G. Davies, Progr. Mat. Sci 13, 3 (1966).
16. C.T. Liu, W. Jemain, H. Inouye, J.V. Cathcart, S.A. David, J.A. Horton, and M.L. Santella, ORNL report 6067, 1984.
17. N.S. Stoloff and R.G. Davies, Acta Met. 12, 473 (1964).
18. R.S. Polvani, W.S. Tzeng, and P.R. Strutt, Met. Trans. 7A, 33 (1976).
19. J.H. Schneibel, G.F. Petersen, and C.T. Liu, J. Mater. Res. 1, 68 (1986).
20. C.T. Liu and C.L. White, in ref. 1, p. 365.
21. L.J. Huetter and H.H. Stadelmaier, Acta Met. 6, 367 (1958).
22. D. Shechtman and L.A. Jacobson, Met. Trans. 6A, 1325 (1975).
23. E.M. Schulson, in ref. 1, p. 193.

DEFORMATION OF THE INTERMETALLIC COMPOUND Al₃Ti
AND SOME ALLOYS WITH AN Al₃Ti BASE

M. YAMAGUCHI, Y. UMAKOSHI AND T. YAMANE
Osaka University, Department of Materials Science and Engineering,
Faculty of Engineering, 2-1 Yamada-Oka, Suita, Osaka 565, Japan

ABSTRACT

In this paper, firstly, the results of deformation experiments which
have been made on Al_3Ti and Al_3Ti-base alloys are reviewed. Secondly, the
possibility of the improvement of ductility of Al_3Ti by alloying processes
is discussed, and then possible alloying elements for this purpose are
suggested. Thirdly, the oxidation resistance of Al_3Ti is described and
then possible future technological applications of Al_3Ti are briefly
discussed.

INTRODUCTION

There are three intermetallic compounds or ordered structures in the
aluminum-titanium system based on the compositions Al_3Ti, $AlTi$ and Ti_3Al.
The deformation behavior of the last two compounds has been studied
extensively and proved to be very attractive materials with extensive
potential applications for high-temperature structural materials of high
strength/density ratio [1-16]. The compound TiAl and some dilute alloys
with a TiAl base, for example, have been shown to exhibit a higher creep
rupture strength/density ratio than conventional nickel-base superalloys
[17]. The development of the studies on the mechanical properties of Ti_3Al
and TiAl has been reviewed by Lipsitt at the Materials Research Society
Symposium on High Temperature Ordered Intermetallic Alloys in 1984 [18].
In comparison with Ti_3Al and TiAl, little attention has been paid to

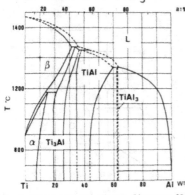

Fig.1 Binary Ti-Al phase diagram [19].

Table.1

Al_3X compounds with the DO_{22} and
DO_{23} structures and their lattice
parameters (Å)

Compounds and Structures		a	c	c/a
Al₃Ti	DO₂₂	3.85	8.596	2.234
Al₃Zr	DO₂₃	4.013	17.320	4.316
Al₃Hf	DO₂₃	3.989	17.155	4.301
	DO₂₂	3.893	8.925	2.293
Al₃V	DO₂₂	3.780	8.321	2.202
Al₃Nb	DO₂₂	3.845	8.601	2.237
Al₃Ta	DO₂₂	3.842	8.553	2.226

Mat. Res. Soc. Symp. Proc. Vol. 81. 1987 Materials Research Society

Al_3Ti. Al_3Ti crystallizes in the DO_{22}-type structure. Although Al_3Ti possesses a lower melting point, 1340°C and a much narrower composition range than TiAl and Ti_3Al(see Fig.1), it has a lower density and much better oxidation resistance than TiAl and Ti_3Al because of its aluminum content. Such attractive characteristics make Al_3Ti also a potential candidate for a high-temperature structural material with a low density. In addition to Al_3Ti, the Al_3X-type compounds of aluminum with elements of groups IV_A and V_A in the Periodic Table are known to have a structure of the DO_{22}-type or the type closely related to DO_{22} (Table 1)[20]: only on Al_3Nb, a deformation study has been made by Shechtman and Jacobson [37].

Recently, we have performed some deformation experiments on Al_3Ti [21], $Al_3(Ti,X)$ where X are mainly elements of groups IV_A and V_A [22] and Al_3Ti-base two-phase alloys containing either small amounts of Al or TiAl [23], and have studied their stress-strain behavior, deformation modes and deformation structures. In this paper, we first present a detailed description of the results of the above-mentioned deformation experiments. Secondly, the possibility of the improvement of ductility of Al_3Ti by alloying processes is discussed and then possible alloying elements are suggested. Finally, the oxidation resistance of Al_3Ti is described and possible future technological applications of Al_3Ti is briefly discussed.

DEFORMATION OF Al_3Ti

Crystal Structure

The DO_{22}-type structure is derived from the $L1_2$ structure (Fig.2a), which is one of the simplest f.c.c.-based ordered structures, by introducing an antiphase boundary (APB) with a displacement vector of the type $1/2\langle 110\rangle$ such as that indicated by an arrow on every (001) plane. The unit cell of the DO_{22}-type structure is tetragonal (Fig.2b). The lattice parameters of Al_3Ti has been reported to be a=0.384 nm and c=0.8596 nm [20]. The density calculated from these lattice parameters is 3.37 g/cm^3. For the sake of simplicity, the Miller indices for crystallographic planes

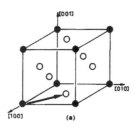
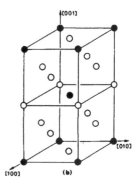

Fig.2 Crystal structures of types $L1_2$ (a) and DO_{22} (b).

and directions in this paper are given using f.c.c. notations.

Among intermetallic compounds and ordered alloys with the DO_{22}-type structure, Ni_3V has been known to exhibit good compression ductility at room temperature [24,25]. Its deformation proceeds by propagation of partial dislocations resulting in a profuse micro-twinning which does not disturb the DO_{22}-type long range order [24,25]. This suggests that the related DO_{22} structure formed in Al_3Ti may exhibit an observable ductility at room temperature.

Stress-Strain behavior and Yield Stress

Figure 3 shows examples of compressive stress-strain curves for polycrystalline specimens with an average composition of $Al_{3.04}Ti_{0.96}$ and with dimensions approximately 3 mm x 3 mm x 7 mm in the temperature range 25-860°C. Specimens contain a small volume fraction of dispersed secondary phase. The obtained values of yield stress are also plotted in the figure as a function of temperature. The yield stress remains almost constant up to about 330°C and then decreases rapidly as the temperature increases to about 630°C. Above 630°C it decreases more slowly with increasing temperature. At temperatures below 620°C, after yielding, stress begins to decrease and soon fracture occurs. At room temperature, fracture often precedes yielding as seen in the case of curve A in Fig.3. Such a small fracture strain observed at temperatures below 620°C, however, does not necessarily mean that the DO_{22} structure formed in Al_3Ti does not exhibit any plasticity in this temperature range. In fact, even on specimens deformed at room temperature, intensely deformed regions are observed (Fig.4). A small fracture strain at temperatures below 620°C would probably be because deformation is concentrated into such regions and cracks are initiated at microstructural irregularities such as inclusions

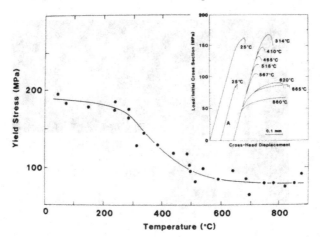

Fig.3 Stress-strain curves and yield stress of Al_3Ti
at 25-860°C [21].

278

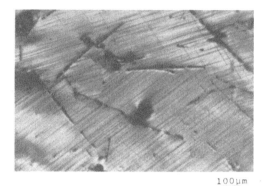

100μm

Fig.4 Deformation markings on Al₃Ti at 25°C [21].

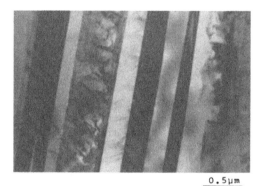

0.5μm

Fig.5 Deformation twins in Al₃Ti deformed at 25°C [21].

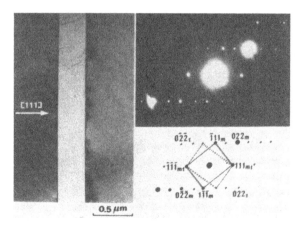

Fig.6 A deformation twin in Al₃Ti deformed at 760°C and the corresponding diffraction pattern [21].

and grain boundaries in the intensely deformed regions almost at yielding.

Recently, we obtained single crystals consisting of the single phase of Al_3Ti and deformation experiments using these single crystals are now in progress.

Deformation Mode

The major mode of deformation of Al_3Ti is twinning of the type $(111)[11\bar{2}]$ which does not disturb the DO_{22} symmetry of the lattice. Figure 5 shows such deformation twins observed in a specimen deformed at $25°C$. Figure 6 shows a twin in a specimen deformed at $760°C$ whose twin plane lies almost parallel to the electron beam. The diffraction spots due to the matrix and the twin are readily indexed as shown in the figure, similarly to the case of a $(0\bar{1}1)$ pattern with a (111) twin plane in the f.c.c. system. This indicates that the twin in Fig.6 is of neither the $(111)[\bar{2}11]$-type nor the $(111)[1\bar{2}1]$-type, but of the $(111)[11\bar{2}]$-type since the diffraction pattern can only be obtained for this specific mode of twinning which does not disturb the DO_{22} symmetry of the lattice. Such specific twins are called ordered twins and are exactly the same as those observed in $Ni_3V[24,25]$ and $Al_3Nb[37]$. Such ordered twins have been found not only in Ni_3V, Al_3Nb and Al_3Ti but also in CuAu [26], Ag_3Mg [27] and TiAl [2] which have tetragonal superstructures based on the f.c.c. lattice. The deformation markings observed in Fig.4 would be mostly traces of twins and micro-twins of this type.

The fact that the major mode of deformation of Al_3Ti is the ordered twinning of the type $(111)[11\bar{2}]$ indicates that the pure stacking fault led by Shockley partials of the type $b=1/6[11\bar{2}]$ has a much lower energy than the other possible faults on $\{111\}$ planes, for example, APB with displacement vector of the type $1/2[1\bar{1}0]$ and complex stacking fault (CSF) with those of the types $1/6[1\bar{2}1]$ and $1/6[\bar{2}11]$. According to recent atomistic studies on the stability of planar faults in the $L1_2$ and f.c.c.-based tetragonal superstructures [28,29], the CSF and APB on $\{111\}$ planes in the DO_{22}-type structure can even be unstable. Therefore, well-defined dislocation dissociation involving the APB and/or CSF may not be expected to occur on $\{111\}$ planes in Al_3Ti. Unless such a dislocation dissociation can occur, the magnitude of Burgers vector of dislocations associated with slip on $\{111\}\langle110\rangle$, which is common to f.c.c. and f.c.c.-based structures, is too large for such dislocations to be mobile. This is the reason why slip of the type $\{111\}\langle110\rangle$ is not observed in Al_3Ti.

At higher temperatures, not only does the ordered twinning occur but also slip vectors of the types [110], [100] and [010] become operative. Therefore, more than five independent slip systems are provided. Figure 7 shows a typical deformation structure at a higher temperature. Dislocations denoted by A, B, C have Burgers vectors of [110], [010] and [100], respectively and they often form nodes via the reaction, [100] + [010] = [110]. The micro-twin at M in the figure lies on $(\bar{1}\bar{1}1)$ and the fault fringes inside the twin are due to the faults with a displacement vector of 1/3[$\bar{1}\bar{1}$1]. Partial dislocations resolved at P are Shockley partials bounding the pure stacking faults. Diffraction contrast analyses of the dislocations and faults and determinations of the Burgers vectors of dislocations and the displacement vector of faults have been described in detail in [21].

Improvement of Ductility

The major mode of deformation of Al_3Ti has been demonstrated to be the ordered twinning of the type (111)[11$\bar{2}$]. However, there are only four independent ordered twinning systems. Therefore, in order to improve the ductility of polycrystalline Al_3Ti, in particular, at low temperatures, it is required not only to increase the activity of the ordered twinning but also to activate slip which is carried out by dislocations with Burgers vectors other than that of the type 1/6[11$\bar{2}$].

The Al_3X-type compounds of aluminum with elements of IV_A and V_A in the Periodic Table has been known to crystallize in the DO_{22}-type structure or structures closely related to the DO_{22}-type. Therefore, Al_3Ti is likely to form $Al_3(Ti,X)$-type ternary compounds with such elements. If the energy of the pure stacking fault led by the Shockley partials of the type 1/6[11$\bar{2}$] is lower in $Al_3(Ti,X)$ than in Al_3Ti, the addition of X to Al_3Ti would increase the activity of the ordered twinning in Al_3Ti and therefore improve the ductility of Al_3Ti. A search for elements which can

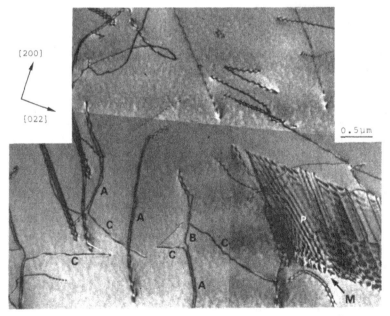

Fig.7 Deformation structure of Al_3Ti deformed at 760°C [21].

decrease significantly the energy of the pure stacking fault in Al_3Ti is now in progress.

If the major mode of deformation of an Al_3X compound with the DO_{22}-type structure is slip other than that by Shockley partials of the type $1/6[11\bar{2}]$, the addition of X to Al_3Ti may activate that slip system in Al_3Ti. Recently, Al_3V with the same structure as Al_3Ti has been found to deform via the motion of dislocations of the types $b=[110]$, $[100]$ and $[010]$ without twinning [22]. Therefore, vanadium additions are expected to increase the activity of these dislocations in Al_3Ti. Al_3Ti and Al_3V seem to show complete solid solubility as shown in Fig.8 [30] and $Al_3(Ti_{0.5}V_{0.5})$, for example, does show a deformation structure which is significantly different from that of Al_3Ti, namely, many dislocations rather than twins are observed. Figure 9 shows the deformation structure of this ternary compound deformed at $960^{\circ}C$. The strength of this ternary compound is much higher than that of Al_3Ti and it is brittle at room temperature. Deformation experiments are now in progress on $Al_3(Ti,V)$ compounds containing smaller amounts of vanadium in which both twinning and slip may be operative.

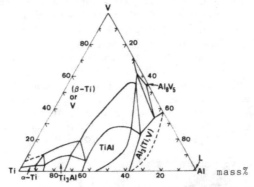

Fig.8 Ternary Ti-Al-V phase diagram ($800^{\circ}C$ section) [30].

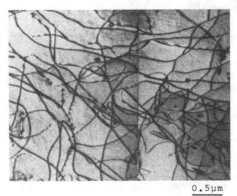

0.5μm

Fig.9 Deformation structure of $Al_3(Ti_{0.5}V_{0.5})$ at $960^{\circ}C$.

On the Al_3V side of the Al_3Ti-Al_3V quasibinary line, ductility has been found to be induced by the replacement of vanadium by titanium. Figure 10 shows stress–strain curves and deformation structures of Al_3V and $Al_3(V_{0.95}Ti_{0.05})$. As seen from the figure, Al_3V fractures with a very small fracture strain at $255°C$; however, $Al_3(V_{0.95}Ti_{0.05})$ shows a considerable compression ductility even at room temperature. In this ternary compound deformed at room temperature, not only dislocations but also numerous deformation twins and stacking faults are observed (see the corresponding deformation structures in Fig.10). Therefore, the improvement of ductility of Al_3V by the partial replacement of vanadium by titanium is obviously due to a substantial increase in the activity of the ordered twinning which is not operative in the binary compound Al_3V. This is an example which encourages us to pursue attempts to improve the low temperature ductility of Al_3Ti and Al_3Ti-base alloys.

Other possible alloying elements which may improve the ductility of Al_3Ti are zirconium and hafnium. Such elements may increase the ease of movement of dislocations of the type $\mathbf{b}=[110]$. Dislocations of this type are superlattice dislocations and they are expected to be dissociated into two superlattice partial dislocations with one-half that of the total $\mathbf{b}$ separated by a the ribbon of APB on (001) with the displacement vector of $1/2[110]$. Otherwise, the [110] superlattice dislocations may not be mobile because of the large magnitude of their Burgers vector. The dissociation is expected to occur on (001) because the energy of the APB on (001) is much lower than that on {111} since no contribution to the energy of the APB on (001) comes from the first-nearest-neighbor interactions. The DO_{22} structure can be derived from the $L1_2$ structure as noted earlier by introducing the $1/2[110]$-type APB on every (001) plane and by introducing it on every two (001) planes the DO_{23} structure can be derived. Therefore, introducing the $1/2[110]$-type APB on (001) plane in the DO_{22} structure

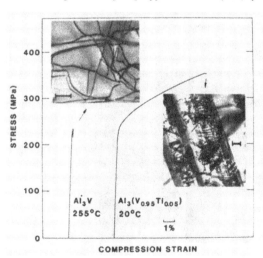

Fig.10
Stress–strain curves and deformation structures of Al_3V and $Al_3(V_{0.95}Ti_{0.05})$. Scales on photographs correspond to 0.25 μm.

is equivalent to creating four (001) layers of DO_{23}-type stacking in the DO_{22} structure. This suggests that the replacement of X in the Al_3X compounds with the DO_{22} structure by elements which form a stable DO_{23} phase with aluminum would decrease the energy of the 1/2[110]-type APB on (001) in the DO_{22} compounds. This would make the dissociation of the [110] superlattice dislocations on (001) in them well-defined. The addition of zirconium and/or hafnium which form an Al_3X compound with the DO_{23} structure is expected to increase the activity of the [110] superlattice dislocations in Al_3Ti and to improve its ductility particularly at high temperatures for the above mentioned reason. Lithium, which forms a metastable Al_3Li compound with the Ll_2 structure [35,36], may also be a possible alloying element to improve the high temperature ductility of Al_3Ti for the same reason. The fact that slip of the type (001)[110] is the major mode of deformation of the Ll_2 compounds at high temperatures also suggests great potential for slip of this type for improvement of the high temperature ductility of Al_3Ti.

DEFORMATION OF Al_3Ti-BASE TWO-PHASE ALLOYS

Al_3Ti-base two-phase alloys with average compositions of $Al_{3.22}Ti_{0.78}$, $Al_{3.07}Ti_{0.93}$ and $Al_{2.94}Ti_{1.06}$ have been deformed in compression at temperatures between 25 and 500°C and their stress-strain behavior and deformation structures have been studied [23]. The $Al_{2.94}Ti_{1.06}$ alloy contains Al_3Ti and TiAl and the other two alloys are made up of Al_3Ti and Al. The Al_3Ti+TiAl alloy is brittle at all temperatures investigated. Alloys containing Al as the dispersed secondary phase show a better compression ductility, but the fracture strain at room temperature is still less than 1%. Figure 11 shows a deformation structure observed on a specimen of $Al_{3.07}Ti_{0.93}$ alloy. At the beginning of deformation, crack initiation at grain boundaries is hindered by the dispersed phase of Al, but soon Al_3Ti grains start being broken (Fig.12). Refining of Al_3Ti grains would be required to achieve a higher ductility in such two-phase alloys consisting of Al_3Ti and Al. Yield strength of $Al_{3.07}Ti_{0.93}$ alloy was about 250 MPa at 25°C. The $Al_{3.22}Ti_{0.78}$ alloy

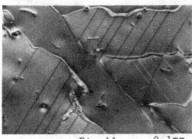

Fig.11   0.1mm   Fig.12   0.1mm

Fig.11 Deformation structure on $Al_{3.07}Ti_{0.93}$ (after yielding).
Fig.12 Deformation structure on $Al_{3.07}Ti_{0.93}$ (after fracture).

rolled at 600°C by about 50% reduction in a mild-steel sheath exhibited a yield stress of about 400 MPa at 25°C. The yield stress remains almost constant up to about 200°C and then starts decreasing rather quickly. A dispersed phase which is ductile and has a higher melting point is needed to achieve higher strength at higher temperatures.

OXIDATION RESISTANCE OF Al_3Ti

Al_3Ti is the only aluminide of titanium effective as a coating against oxidation since only Al_3Ti forms an oxide scale whose outermost layer is composed of Al_2O_3 which acts as a highly effective barrier against the diffusion of oxygen [31]. The external layer of oxide scale formed on Ti_3Al and TiAl is not Al_2O_3, but TiO_2 or a mixture of TiO_2 and Al_2O_3 [31,32]. Figures 13(a) ,for example, shows a cross-section of oxide scale formed on TiAl exposed to still air at 900°C for 100 h and Fig.13(b) shows a profile of concentration of titanium, aluminum and oxygen in the scale [32]. According to a measurement of weight gain during exposure of TiAl and Al_3Ti to still air at 900° for 200 h, the oxidation rate of Al_3Ti is slower than that of TiAl roughly by a factor of ten [33]. Therefore, Al_3Ti could be used as an oxidation-resistant coating material for not only titanium but also Ti_3Al and TiAl on which the alloy development efforts have been proceeding. Aluminizing of titanium and titanium-aluminum alloys can be accomplished by a variety of methods which include slurry spraying, molten dipping and pack cementation. In spite of this range of techniques, a well adhering layer of Al_3Ti is formed and no other aluminides arise because of the much faster diffusivity in Al_3Ti than in other aluminides of titanium [34]. It has been reported that the Al_3Ti coating is effective for protecting titanium against oxidation up to 850°C

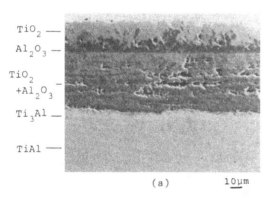

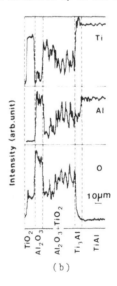

Fig.13
Scanning electron micrograph(a) and electron microprobe analysis profile (b) of a cross section of oxide scale on TiAl exposed to air at 900°C for 200 h [32].

above which the durability falls by the conversion of Al_3Ti into TiAl and Ti_3Al [31]. If the conversion of Al_3Ti into subaluminides is prevented, the Al_3Ti coating would be effective at higher temperatures. The problem with Al_3Ti coatings is that they show low ductility at room temperature, similar to other aluminides used for coating materials such as NiAl and FeAl. However, as mentioned in preceding sections, there seems to be a great hope that the low temperature ductility of Al_3Ti can be improved.

SUMMARY AND CONCLUSIONS

It has been demonstrated that the major mode of deformation of Al_3Ti is the ordered twinning of the type $(111)[11\bar{2}]$ which does not disturb the DO_{22} symmetry of the lattice. At higher temperatures, the four ordered twinning systems are augmented by slip of the types [110], [100] and [010]. At temperatures lower than 620^oC, Al_3Ti fractures with very small fracture strain, but at temperatures above 620^oC good compression ductility is observed. It has been suggested that the low temperature ductility of Al_3Ti can be improved by alloying processes. In fact, the room temperature ductility of Al_3V which is also a compound with the DO_{22}-type structure is improved substantially by the titanium additions. Vanadium, zirconium, hafnium and lithium have been suggested as possible alloying elements to improve the ductility of Al_3Ti. The low temperature ductility of Al_3Ti has not been improved yet to such an extent that the compound can be used as a structural material. However, since it shows good oxidation resistance, it could be used as a coating for titanium-aluminum alloys such as TiAl-base alloys on which development has been proceeding.

REFERENCES

1. J.B. McAndrew and H.D. Kessler, J. Metals 8, 1348 (1956).
2. D. Shechtman, M.J. Blackburn and H.A. Lipsitt,
   Met. Trans. 5, 1373 (1974).
3. H.A. Lipsitt, D. Shechtman and R.E. Schafrik, Met. Trans.
   6A, 1991 (1975).
4. H.A. Lipsitt, D. Shechtman and R.E. Schafrik, Met. Trans.
   11A, 1369 (1980).
5. S.M.L. Sastry and H.A. Lipsitt, Proceedings of the 4th International
   Conference on Titanium, Japan, p.1231, November 1980.
6. W.J.S. Yang, Met. Trans. 13A, 324 (1982).
7. W.J.S. Yang, J. Mats. Sci. Letts. 1, 199 (1982).
8. S.M.L. Sastry and H.A. Lipsitt, Met. Trans. 8A, 1543 (1977).
9. P.L. Martin, H.A. Lipsitt, N.T. Nuhfer and J.C. Williams, Proceedings
   of the 4th International Conference on Titanium, Japan, p. 1245,
   November 1980.
10. R.E. Schafrik, Met. Trans. 7B, 713 (1976).

11. M.G. Mendiratta and H.A.Lipsitt, J. Mater. Sci. 15, 2985 (1980).
12. P.L. Martin, M.G. Mendiratta and H.A. Lipsitt, Met. Trans. 14A, 2170 (1983).
13. S.M.L. Sastry and H.A. Lipsitt, Met. Trans. 8A, 299 (1977).
14. S.M.L. Sastry and H.A. Lipsitt, Acta Met. 25, 1279 (1977).
15. R.J. Kerans, Met. Trans. 15A, 1721 (1984).
16. T. Kawabata, T. Kanai and O. Izumi, Acta Met. 33, 1355 (1985).
17. M.J. Blackburn and M.P. Smith, U.S. Patent No. 4 294 615 (13 October 1981).
18. H.A. Lipsitt, in High Temperature Ordered Intermetallic Alloys, edited by C.C. Koch, C.T. Liu and N.S. Stoloff (MRS, 1985), p.351.
19. T. Tsujimoto and M. Adachi, J. Inst. Metals 94, 358 (1966); 95, 146, 221 (1967).
20. W.B. Pearson, in Lattice Spacings and Structure of Metals and Alloys, (Pergamon Press, Oxford, 1958).
21. M. Yamaguchi, Y. Umakoshi and T. Yamane, in press, Phil. Mag. 1986.
22. Y. Umakoshi, M. Yamaguchi and T. Yamane, Proceedings of the 1986 Nagoya Meeting of the Japan Institute of Metals, (The Japan Institute of Metals, Sendai, 1986), p.157.
23. M. Yamaguchi, Y. Umakoshi and T. Yamane, Proceedings of the 1986 Nagoya Meeting of the Japan Institute of Metals, (The Japan Institute of Metals, Sendai, 1986), p.158.
24. G. Vandershaeve, T. Sarrazin and B. Escaig, Acta Met. 27, 1251 (1979).
25. G. Vandershaeve and B. Escaig, Phil. Mag. 48, 265 (1983).
26. D.W. Pashley, J.L. Robertson and M.J. Stowell, Phil. Mag. 19, 83 (1969).
27. G. Vandershaeve and B. Escaig, Phys. Stat. Sol. 20(a), 309 (1978).
28. M. Yamaguchi, V. Vitek and D. P. Pope, Phil. Mag. 43,1027 (1981).
29. M. Yamaguchi, Y. Umakoshi and T. Yamane, in Dislocations in Solids, edited by H. Suzuki, T. Ninomiya, K. Sumino and S. Takeuchi (University of Tokyo Press, Tokyo, 1985) p.77.
30. K. Hashimoto, T. Tsujimoto and H. Doi, Japan J. Inst. Metals, 49, 410 (1985).
31. M. Kabbaj, A. Galerie and M. Caillet, J. Less-Common Metals, 108, 1 (1985).
32. K. Tsurumi, T. Miyashita, H. Hino, J. Fujioka and Y. Nishiyama, in Abstracts of Symposium on Plastic Deformation of Ordered Alloys and Intermetallic Compounds, Japan Institute of Metals, 16 July 1986 p. 13.
33. K. Tsurumi, T. Miyashita, H. Hino, J. Fujioka and Y. Nishiyama, (private communication).
34. F.J.J. van Loo and G.D. Rieck, Acta Met. 21, 61, 73 (1973).
35. J.M. Silcock, J. Inst. Met. 88, 357 (1959-60).
36. T. Yoshiyama, H. Hasebe and M. Mannami, J. Phys. Soc. Japan 25, 908 (1968).
37. D. Shechtman and L.A. Jacobson, Met. Trans. 6A, 1325 (1975).

Creep Deformation of Ta Modified
Gamma Prime Single Crystals

D.L. ANTON, D.D. PEARSON AND D.B. SNOW
United Technologies Research Center, East Hartford, CT 06108

ABSTRACT

The role of substitutional element alloying of single phase γ' has become of primary interest to alloy designers who would like to exploit its low density and excellent oxidation resistance. Current γ' alloys have not shown sufficient strength to be useful in a creep limited environment. In order to maximize the potential of single phase γ' alloys and to more fully understand the creep strengthening mechanisms in two phase Ni-base superalloys, it has become necessary to clarify the role of Al-substitution elements. Ta is a potent strengthening element in γ' as well as imparting beneficial surface stability to superalloys; its effect on the creep properties of Ni_3Al is the subject of this paper. The 1300°C isotherm of the Ni-Al-Ta system was determined in order to establish the γ' single phase field. Compositions were fabricated having chemistries which systematically varied both the Al:Ta ratio at Ni=75% and Ni:(Al+Ta) ratio at Ta=6%. Creep tests were conducted on <001> oriented single crystals at 760, 871 and 982°C. Electron microscopy was used to characterize the nature of slip deformation, confirm phase purity and to determine the existence of tetragonal distortions in these crystals. In this manner the strengthening due to Ta was examined in the absence of grain boundary effects. These γ' mono-crystals did not display classical creep response. Incubation creep was observed in all of the specimens tested. Surprisingly, the maximum incubation time was found to occur in the high ratio Ni:(Al+Ta) compounds, where less than 0.5% creep strain was obtained after 200 hours at stress. After incubation, either tertiary creep leading to failure, or apparently classic primary, secondary and tertiary creep ensued. In addition extremely long elongations, to 85%, were measured.

INTRODUCTION

Nickel-based superalloys have usable strength up to nearly 90% of their melting point. They derive their strength from the precipitation of the γ' phase on a fine scale. The fact that γ' is coherent with the matrix and the attendant slow coarsening rate, leads to exceptional creep strength. Factors which are known to increase creep strength include proper heat treatment to produce an initially small precipitate distribution, alloying to solid solution both the matrix and the γ' phase, and alloying to retard coarsening rates [1]. In addition to the beneficial effects of γ' on the mechanical properties of superalloys, its high Al content also enhances oxidation resistance. Recently, investigators have sought to make single phase γ' alloys to take advantage of the mechanical strength, oxidation resistance and low density that it offers [2,3]. Most of this work has been concentrated on developing alloys with a fine grain size and with controlled additions of boron to increase strength and ductility at room temperature. It is envisioned that these alloys would have a maximum use temperature of about 700°C at which point the fine grain size limits creep strength. For higher temperature use, elimination of grain boundaries becomes paramount and investigation of single crystal properties becomes essential. This single crystal approach has the further benefits of determining the intrinsic effect of alloy additions on creep strength aside from grain boundary related mechanisms as well as affording the opportunity to choose active slip systems.

While the anomalous increase in yield strength with temperature as well as the effect of alloying additions to γ' on yield properties are well known [4,5], little is known about these effects on elevated temperature creep strength. The need for a clearer understanding of the creep resistance of γ' alloys, before consideration of their use in high temperature applications, can not be overstated. The fact that not one paper was dedicated to creep in γ' alloys in the previous symposium proceedings, [6], points this out.

A recent study of creep deformation in Hf and B modified γ' showed that grain boundary mechanisms such as either Nabarro-Herring or Coble creep were dominant [7]. It is known that Ta is potent in increasing the yield strength of γ' [4,8] and may also improve its creep resistance. Since there would be no interfaces in a single phase γ' alloy to retard dislocation motion or lead as pathways for enhanced diffusion, creep strength would arise by retarding effects on the dislocations themselves through hardening.

This study was conceived with the intent of determining the role of Ta additions in stoichiometric and off stoichiometric single crystals of single phase γ'. The [001] crystal orientation was chosen to be characterized first to eliminate thermally activated cube slip as a possible strengthening mechanism. In addition, analysis of HOLZ lines through convergent beam diffraction was utilized in determining crystal symmetry changes brought about through Ta alloying.

EXPERIMENTAL METHODS

Phase Field Identification

The extent of Ta solubility in Ni_3Al was determined by preparing 100g arc melted buttons of Ni-Al-Ta alloys. Alloys were prepared of such chemistries so as to provide a thorough depiction of the Ta solubility in γ'. Homogenization of these alloys was conducted at 1300°C for 24 hours in a flowing Ar atmosphere. Subsequent optical metallography was conducted in order to determine the γ' phase field boundaries in the isothermal section of the ternary phase diagram.

Single Crystal Growth and Testing

Alloy chemistries were chosen so as to vary Ni:(Al+Ta) at constant 6% Ta, and to vary the Al:Ta at a constant Ni:(Al+Ta)=3.00 while maintaining single phase γ'.(Note: All chemistries given here are in atomic % unless otherwise noted.) These alloying guides are given as the solid lines in Fig. 1 with specific compositions shown as triangles and in Table 1. Chill cast bars 1.6 cm in diameter were fabricated using induction melting from commercial purity starting elements. These alloy chemistries will be referred to in the notation of Ni%-Al%-Ta% for brevity.

Oriented crystals 1.27 cm in diameter and approximately 15 cm long were grown using <100>

Table I

Gamma Prime Chemistries
(Atomic %)

Ni	Al	Ta	Ni:(Al+Ta)
75	23	2	3.00
75	19	6	3.00
75	17	8	3.00
77	17	6	3.35
74	20	6	2.85

aligned seeds by a modified Bridgman technique. Macro-etching followed by optical examination confirmed mono-crystal growth, while Laue back reflection techniques were used to insure crystal alignment to within 5° of <001>.

Homogenization for 24 hours at 1300°C in Ar followed by rapid air
quenching resulted in chemically homogeneous crystals. Tensile creep specimens
having a 2.54 cm long gage and 0.32 cm in diameter were machined. Tensile
creep tests were conducted in air using lever arm creep machines. Temperature
was controlled to within 3°C and elongation was measured vs. time with an LVDT
equipped extensometer clamped to ridges on the sample.

After rupture or test termination, final gage section minimum and maximum
diameters were measured and the degree of eccentricity, d_{min}/d_{max}, noted.
In addition, transverse section foil blanks were cut from selected specimens:
either from their gage section for dislocation observation or from their
respective grip regions for convergent beam diffraction pattern analysis. In
the latter case, the intention was to determine whether in some cases alloying
with Ta or deviation from stoichiometry had caused the crystal structure to
become non-cubic (i.e. tetragonal).

EXPERIMENTAL RESULTS

Phase Field Identification
Fig. 1 illustrates the 1300°C
isothermal section of the Ni-Al-Ta
ternary phase diagram in the
vicinity of the γ' single phase
region. The open symbols represent
phase pure γ' as determined
optically while the filled symbols
denote two or three phase alloys.
One notes that there is a distinct
bend in the γ' phase field at high
Ta concentrations; those greater
than 4%, towards the high Ni corner
of the diagram. Similar results
have since been reported in [13],
where it was noted that
considerable substitution of Al by
both Ni and Ta was found at 1250°C.
With the use of this isothermal
section, subsequent testing was

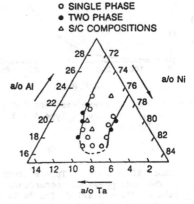

o SINGLE PHASE
● TWO PHASE
△ S/C COMPOSITIONS

Fig. 1 Ternary 1300°C isotherm for the
Ni-Al-Ta system.

conducted on alloys at the extremes of solid solubility and compared with
crystals having chemistries well within the phase boundaries.

Creep Deformation
Results of the creep testing are summarized in Table II where alloy
composition, testing parameters, incubation creep times, t_i, time to 2%
creep strain, $t_{2\%}$, time to rupture, t_r, strain to failure, ϵ_f and Minimum
creep rate are given. The incubation creep time is defined as the time at
temperature and stress required to obtain a minimum creep rate of $10^{-4} hr^{-1}$
while $t_{2\%}$, t, ϵ_f and ϵ_m were determined in the usual way.

These crystals did not display classical creep response with the typical
three stages of creep characterized by initial dislocation glide, steady state
creep and final necking instability and rupture. Instead, an incubation creep
at the onset of loading followed by either rapid elongation to failure or
classical three stage creep ensued. Typical elongation vs. time curves are
shown in Fig. 2 for both of these creep type responses. Fig. 2a illustrates
the former response with no minimum creep rate discernible for the 77-17-6
specimen tested at 760°C. This shall be designated as Type A creep for further

290

discussion here. The later mechanism of creep elongation is given in Fig. 2b
for a 75-19-6 specimen tested at 760°C, where incubation is followed by rapid
creep to a minimum creep rate which is in turn followed by necking instability
to failure. This shall be referred to as Type B creep. Only for crystals where
a valid minimum creep rate could be measured after the onset of rapid
straining is an entry made. Thus, Type B creep can be identified as occurring
in those specimens with a minimum creep rate reported. Additionally, a number
of specimens listed in Table I show no incubation creep time. One did occur in
these specimens, however the creep rate was greater than $10^{-4} sec^{-1}$.

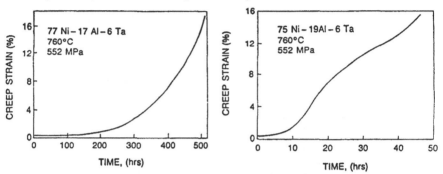

Fig. 2 Typical creep response for γ' single crystals (a) incubation followed
by rapid creep, Type A and (b) incubation followed by classical creep,
Type B.

Table II shows that Type B creep occurred only in the lower temperature
tests. The most prevalent occurrence of Type B creep is found in the 75-19-6
crystals at 760 and 871°C. The 74-20-6 crystal tested at 760°C and the 75-17-8
crystal tested at 871°C also displayed Type B creep. This last observation was
the one anomaly to the low temperature correlation of this data.

Fig. 3 illustrates the creep data for stoichiometric crystals (those
having 75% Ni) graphically as time to 2% creep strain, $t_{2\%}$, vs. Ta
concentration. For these alloys one notes that increasing Ta content is
beneficial to creep resistance only at 982°C. At the two lower temperatures,
increasing Ta reduced the creep resistance. This was most pronounced in the 2%
Ta case.

The 2% creep time data for the constant Ta concentration crystals is
given in Fig. 4. In general, there is little effect of Ni:(Al+Ta) on 2% creep
life above 760°C. At this temperature, however, creep resistance was greatly
enhanced for the Ni rich γ'.

The occurrence of incubation creep correlated very well with time to 2%
creep as would be expected. Those specimens displaying incubation creep led to
long lives to 2% creep strain.

Large strains to failures were exhibited by many of the specimens. In
general the 982°C tests resulted in strains of greater than 35% for all of the
chemistries tested. In addition, the 75-23-2 and 75-17-8 chemistries resulted
in strains of greater than 30%. The combination of these two chemistries are
unexpected since they are both of the same Ni:(Al+Ta) but of widely different
Ta concentrations.

A number of the specimens displayed eccentric gage cross sections at test
termination. These eccentricities were quantified by measuring the ratio of
minimum to maximum section dimensions. Large eccentricities result from

Burger's vectors limited to one plane and two directions (of the possible four planes, each with two directions). Thus large crystal rotations occurred and were responsible for much of the creep strain.

No correlation was found between the occurrence or magnitude of gage section eccentricity and type of creep, time to 2% strain, rupture or

Table II

Summary of Creep Results

Chemistry Ni-Al-Ta	Temp. (°C)	Stress (MPa)	t_i (hr)	$t_{2\%}$ (hr)	t_r (hr)	ϵ_f (%)	ϵ_m (sec^{-1})
75-23-2	760	552	26.0	72.0	290.0	28.2	
	871	414	0.9	7.8	25.3	49.6	
	982	207	0.0	1.3	8.4	80.1	
			0.0	0.2	1.5	85.0	
75-19-6	760	552	3.8	11.5	45.8	15.4	2.6×10^{-3}
			4.4	12.0	47.0	12.7	1.5×10^{-3}
	871	414	0.0	2.3	18.7	20.6	3.7×10^{-3}
	982	207	0.0	4.0	36.5	62.2	
75-17-8	760	552	1.2	7.5	12.7	44.9	
			3.5	7.3	13.6	38.9	
	871	414	0.0	1.8	7.0	43.4	1.1×10^{-2}
	982	207	6.9	35.0	135.1	38.9	
77-17-6	760	552	210.0	250.0	526.6	19.9	
			230.0	270.0	513.2	17.2	
	871	414	1.0	8.0	25.1	48.6	
	982	207	0.5	1.5	9.1	82.1	
74-20-6	760	552	1.3	5.5	36.0	26.3	2.6×10^{-3}
	871	414	0.0	3.5	37.0	28.6	
	982	207	0.0	2.5	19.0	60.3	

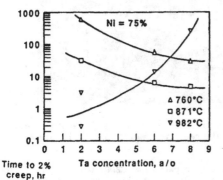

Time to 2% creep, hr

Ta concentration, a/o

Fig. 3 Creep life as a function of Ta concentration at stoichiometric Ni:(Al+Ta).

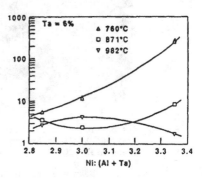

Ni:(Al + Ta)

Fig. 4 Creep life as a function of Ni:(Al+Ta) ratio at constant 6% Ta.

ductility. The only correlations that could be made were that only specimens having 75-17-8 or 74-20-6 chemistries displayed such behavior and that this only occurred at the two lower temperatures. Verification of creep specimen orientation techniques were made with no variation from the 5˚ limit on alignment found.

Microscopic Evaluation

Low-magnification bright field photomicrographs and ⟨100⟩ selected area diffraction patterns from thin foils of post-rupture gage sections were used to confirm that each specimen was single phase γ'. A survey of the dislocation distributions produced by varying amounts of creep strain has thus far not revealed any evidence of a dislocation cell structure. Fig. 5 shows the dislocation distribution in the gage section of specimen 77-17-6 after 17.2 % strain to failure, imaged under two-beam (g=022) bright field conditions with the foil plane close to (100), perpendicular to the tension axis. There are numerous screw dislocations on {111}, as well as small loops and tangles indicative of frequent intersection. The long lengths of screw or edge dislocations characteristic of slip on {100} [8] are absent, as expected due to the absence of resolved shear stress on this plane.

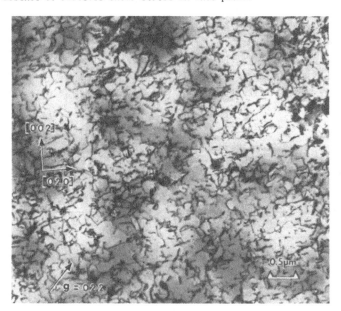

Fig. 5 Dislocation distribution in transverse section gage length, specimen 77-17-6, 17.2% strain to failure. Two-beam bright field; g=022, foil plane near (100).

Convergent-beam electron diffraction patterns (CBDP) were obtained from the undeformed threaded ends of several different alloy specimens. The symmetry of the higher-order Laue zone (HOLZ) lines within the central disk of

CBDP's provides a sensitive technique for determining whether the γ' unit cell
has deviated from cubic symmetry due to the presence of Ta and/or deviations
from stoichiometry [9]. In this investigation, CBDP's were obtained with the
incident beam parallel to one of the four <114> directions closest to the
[100] tension axis of the specimen. Although deviations from mirror plane
symmetry within the central disk of these patterns have not yet been
quantitatively evaluated, they permit the detection of tetragonality to a
degree of at least $c/a = 1.002$ [10,11]. In these specimens γ' is the only
phase present, so that the question of a contribution to the HOLZ line pattern
asymmetry from interphase elastic strains does not arise [11]. No deviation
from cubic symmetry was observed in any specimen of an alloy for which the
composition was stoichiometric (Al + Ta = 25 at%). However, the HOLZ lines
within <114> CBDP's from specimens 74-20-6 and 77-17-6 were asymmetric; i.e.,
they did not display mirror symmetry, as shown in Fig. 6. Thus far, these
patterns suggest that the degree of asymmetry increases with increasing
deviation from stoichiometry. In this example Fig. 6, it is greater for
specimen 77-17-6 (Al+Ta = 23 at%) than for specimen 74-20-6 (Al+Ta = 26 at%).

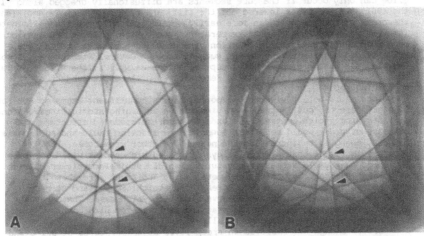

Fig. 6. Central disks of CBDP's with the incident beam parallel to <114>.
120 kV, beam diameter $\simeq$ 0.2 μm. Deviations of the HOLZ lines from
symmetry about a vertical mirror plane can be seen by comparing the
size of small triangles at the positions marked by arrows.
(A) Specimen 74-20-6. (B) Specimen 77-17-6.

DISCUSSION

The single phase γ' field determined here shows substantial substitution
of Ni on Al sites at Ta levels greater than 4%. Both the Ni rich and Ni poor
solubility limits of γ' curve towards the Ni rich corner of the ternary phase
diagram. Additions of Ta to γ' have previously been shown to linearly increase
its lattice parameter through 10 atomic per cent additions [1]. Thus above the
4% level of Ta in γ' the increased lattice parameter allows Ni to substitute
more readily on to Al sites than is normally possible. This is in disagreement
with [5] where it was shown that the Ni poor solubility limit did not shift to

the Ni corner of the phase diagram at 1250°C. This discrepancy may be accounted for by the 50°C difference in homogenizing temperature.

The strengthening effect of refractory metal substitutions in γ' is postulated to be due to increasing the APB energy on (111) planes while concurrently decreasing it on (001) planes [4]. Thus tantalum had been expected to significantly strengthen the creep resistance of γ' through solid solution strengthening and enhanced thermally activated (001) cross slip such as occurs in yield strength measurements. This has been shown not to be the case at the intermediate temperatures of 871°C and below, while at higher temperatures the potency of Ta was quite significant. Since creep and yield deformation have quite different time dependencies however their results must be approached independently.

Thin foil TEM observations have shown no evidence of cell formation therefore individual dislocation dynamics must explain these variations in creep response. At intermediate temperatures when a segment of the first of a pair of dissociated dislocations constricts and cross slips onto a cube plane due to thermal activation and resumes octahedral slip, further dislocation glide can only occur if the cube segments are diffusionally dragged along. If the (001) APB energy is lowered and the (111) APB raised through Ta additions the cross slipped segment can more easily spread thermally and thus eliminate itself as a barrier. At higher temperatures where cross slipped segment locks become less effective, solid solution strengthening dominates and Ta once again leads to creep strengthening such as through Cottrel atmospheres. A more detailed discussion of possible creep mechanisms as well as the effect of Ta an creep weakening at intermediate temperatures will be the subject of a succeeding paper.

Since γ' is not an electron compound, excess constituent atoms do not form Schottky defects, but instead simply reside on adjoining atomic sites. As one deviates from stoichiometry this leads to a more disordered structure . A decrease of order would result in lowering the stacking fault and anti—phase boundary energies by increasing the nearest neighbor mismatches. It was reported in [1,12] that Al—rich alloys displayed higher peak yield stresses and lower peak yield stress temperatures than did stoichiometric or Ni—rich materials. Furthermore, refractory metal alloy additions were shown to substantially increase the yield strength of stoichiometric and Al—rich γ' while little strengthening was observed in the Ni—rich case. These results were attributed to changes in the stacking fault energy with both stoichiometry and refractory metal alloying.

Little change was observed in creep lives at temperatures of 871°C and greater here. At 760°C, however, as Ni:(Al+Ta) increased, the creep resistance was enhanced considerably. Since moving away from stoichiometry is expected to lead to a less ordered alloy, one would at first expect the creep properties to be reduced for the nonstoichiometric crystals. Thus, the high temperature data again behaves as expected. Explanation of the intermediate temperature creep results, which show strengthening with increased Ni concentration, will need to await further TEM analysis.

The creep property considered here, $t_{2\%}$, was considerably effected by the incubation creep time, t_1, as well as the slow creep rate immediately following. The incubation creep is considered to be an artifact of the crystal growth process. During solidification, a nearly dislocation free crystals are fabricated. Johnston [14] showed that incubation times in creep of single crystal LiF were due to the generation of mobile dislocations. Since the creep data presented here does not readily agree with tensile data generated previously, it is suspected that Ta alloy additions may have a strong influence on the generation of dislocations in γ'.

REFERENCES

1. A.F. Giamei, D.D. Pearson and D.L. Anton, High Temperature Ordered Intermetallic Alloys , C.C. Koch, C.T. Liu and N.S. Stoloff eds., Materials Research Society, Pittsburgh, PA, 1985, p 293-308.

2. C.T. Liu and C.L. White, ibid, p 365-380.

3. A. Inoue, H. Tomioka and T. Masumoto, J. Mat. Sci. Letters, 1, 377-380, (1982).

4. R. Churwick, "Strengthening Mechanisms in Nickel-Base Superalloys", Ph.D. Dissertation, University of Minnesota, 1972.

5. B.H. Kear and D.D. Pope, 1984 ASM Conference on Refractory Alloying , ASM, Metals Park, Ohio, 1984, p 135-151.

6. High-Temperature Ordered Intermetallic Alloys , C. Koch, C. Liu and N. Stoloff eds., Materials Research Society, Pittsburgh, PA, 1985.

7. J.H. Schneibel, G.F. Petersen and C.T. Liu, J. Mater. Res., 1, 68-72, (1986).

8. P. H. Thornton, R. G. Davies, and T. L. Johnston, Metall. Trans., 1, 207-218, (1970).

9. A. J. Porter, M. P. Shaw, R. C. Ecob and B. Ralph, Phil. Mag. A, 44, 1135-1148, (1981).

10. H. L. Fraser, in Microbeam Analysis - 1982, K. F. J. Heinrich, ed., San Francisco Press, San Francisco, 1982, p. 54.

11. M. J. Kaufman, D. D. Pearson and H. L. Fraser, Phil. Mag. A 54, 79-92, (1986).

12. R.D. Rawlings and A.E. Staton-Bevan, J. Mat. Sci., 10, 505-514, (1975).

13. P. Willemin, O. Dugue, M. Durand-Charre and J.H. Davidson, Mat. Sci. and Tech., 2, 344-348, (1986).

14. W.G. Johnston, J. Appl. Phys., 33, 2716-2730, (1962).

CREEP CAVITATION IN A NICKEL ALUMINIDE

J. H. SCHNEIBEL* AND L. MARTINEZ**
*Metals and Ceramics Division, Oak Ridge National Laboratory, Oak Ridge, TN 37831-6116
**Instituto de Fisica, Universidad National Autonoma de Mexico (UNAM), Apdo. Postal 20-364, 01000 Mexico, D.F., Mexico

ABSTRACT

A nickel aluminide with the composition Ni-23.5 Al-0.5 Hf-0.2 B (at. %) was creep-tested in tension at constant load (initial stress 250 MPa, temperature 1033 K). The creep rate reaches a minimum at a strain of approximately 0.3%. Its increase at larger strains is partly caused by grain boundary cavitation. Cavity size distributions corresponding to different creep times and strains were determined metallographically and evaluated in order to obtain the cavity growth rate as a function of cavity size. Contrary to a newly developed model for crack-like diffusive growth the cavity tip velocity increases with cavity size. This discrepancy is attributed to cavity coalescence. Sintering at elevated temperatures (1473 K) reduces the diameter and increases the thickness of the disc-shaped cavities. Sintering of such cavities might be exploited to measure surface diffusivities.

INTRODUCTION

In recent years it has become possible to ductilize nickel aluminide alloys by appropriately deviating from the stoichiometric nickel to aluminum ratio, as well as by micro-alloying with boron [1,2,3]. Owing to their excellent high-temperature mechanical properties these alloys have received considerable attention. Much work has been devoted to understanding the exact mechanism by which they are ductilized [4,5,6] and the anomalous increase in yield stress with temperature which is characteristic of nickel aluminides and other ordered intermetallic compounds [7,8]. The mechanisms of creep and creep fracture of these alloys, however, have not been studied to a large extent, although they are very important in alloy design and reliability considerations.

The present work is primarily a contribution to the intergranular cavitation characteristics of these alloys. Like most metals and alloys under creep conditions, nickel aluminides fail by the nucleation, growth and coalescence of intergranular cavities. In the present work, we present measurements of the rate with which intergranular cavities grow in a nickel aluminide of composition Ni-23.5 Al-0.5 Hf-0.2 B (at. %). The measurements are not in complete agreement with a recently developed model for crack-like diffusive growth [9]. This inconsistency is most likely due to cavity coalescence.

EXPERIMENTAL PROCEDURE

An alloy with the nominal composition Ni-23.5 Al-0.5 Hf-0.2 B (at. %) was prepared from approximately 99.95% pure metals and a Ni-4 wt % B master alloy. The starting materials were arc-melted and drop-cast (in argon) into a copper chill mold (approximately 10 mm × 25 mm × 130 mm). From the cast slabs, 0.75 mm thick sheet material was fabricated by a sequence of rolling (typically 15% reduction in thickness per pass) and annealing (typically 3.6 ks at 1323 K). After rolling to the final thickness, tensile

creep specimens with a gage length of 26 mm and a cross-section of 3.2 mm ×
0.75 mm were machined and vacuum-annealed for 3.6 ks at 1273 K. The true
grain size (1.5 × mean intercept length) was approximately 80 μm.
Constant-load creep tests were performed in vacuum (10^{-4} Pa) at a temperature of 1033 K and an initial stress of 250 MPa. The specimen elongation was continuously monitored by means of an extensometer attached to the
specimen. Once a desired creep strain was reached, tests were interrupted
by cooling under load.
    The gage length of crept specimens was ground from the original
thickness (0.75 mm) to a thickness of 0.4 mm and then mechanically polished
to a mirror finish. The specimens were subsequently slightly etched in
order to improve grain boundary and cavity contrast during metallographic
examination. The maximum and minimum dimensions of the elongated cavities
were manually measured on the screen of a scanning electron microscope
(SEM), model JEOL 35 CF, at a magnification of 2000x. Approximately 300
cavities were evaluated per specimen. Depending on the specimen, this
required examination of an area between 1 and 4 mm².
    Using the Saltykov analysis [10], the measured data were converted into
cumulative true size distributions I(a,t) where a is the true radius
("half-size") of the assumed disc-shaped cavities (which are approximately
normal to the direction of the applied stress) and t is the creep time. In
order to minimize scatter, the cumulative distributions were numerically
smoothed. It should be pointed out that the definition of "cumulative"
employed in this work differs from the usual one in that I(a,t) refers to
the number of cavities with true half-size ⩾ a. In particular, the total
number of cavities per unit volume is given by I(0,t).

EXPERIMENTAL RESULTS

Creep Testing

    A typical creep curve corresponding to a test interrupted at ~5%
strain is shown in Fig. 1. The creep rate is minimum after a small
strain (~0.3%). At larger strains, the strain rate increases continuously.

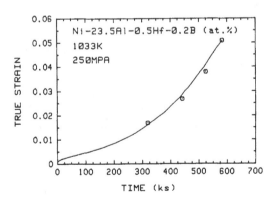

Fig. 1. Creep curve for Ni-23.5 Al-0.5 Hf-0.2 B (at. %) at a test temperature of 1033 K and an initial stress of 250 MPa. The open symbols
indicate creep times (strains) at which tests were interrupted for
metallographic examination.

The open circles in Fig. 1 correspond to tests which were interrupted by cooling under load. The corresponding specimens were metallographically analyzed.

Cavity Morphology

In Fig. 2 we give an example of the (usually crack-like) cavities observed on metallographic sections. Whereas the cavities in Fig. 2 exhibit smooth surfaces, some cavities were also irregularly shaped. This is due to cavity coalescence and/or matrix slip.

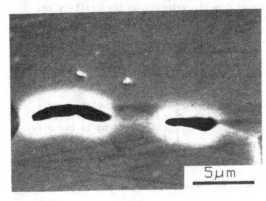

Fig. 2. SEM micrograph of creep cavities illustrating crack-like morphology (t = 320ks).

Influence of Cavitation on Creep Curve

Using relationships by Fullman [11], fractions of cavitated grain boundary area were evaluated from the cavity sizes measured on the metallographic sections, and the grain size. The stress increase due to cavitation was found to be comparable to that caused by the area reduction due to sample elongation. During creep to 5% strain, the effective stress (due to reduction in area caused by elongation and cavitation) increases from 250 MPa to ~305 MPa. Nevertheless, it would be difficult to explain the inverse creep transient observed for strains >0.3% by the increase in the effective stress alone, since one would have to assume a stress exponent significantly larger than the commonly accepted value of ~3 [12].

Cavity Growth Rates as a Function of Cavity Size

In Fig. 3 we show cumulative true cavity size distributions corresponding to two creep times t_1 = 320ks and t_2 = 524ks. The logarithm (base 10) of $I(a,t)$ is plotted versus the true cavity half-size, a. As discussed in previous work [13] the growth rate is obtained by solving

$$I (a_i, t_1) = I (a_j, t_2) \tag{3}$$

for a_i and a_j (and suitably chosen values of I). The tip velocity $v = \dot{a}$ of cavities of half-size

$$a = (a_i + a_j)/2 \tag{4}$$

is given by:

$$v = (a_j - a_i)/(t_2 - t_1) \ . \tag{5}$$

Tip velocities obtained from Eq. (3) are plotted in Fig. 4 as a function of half-size a. The scatter in the data points is a consequence of the statistical error in the cumulative distribution functions. Nevertheless, the experimental growth rates increase quite clearly with increasing cavity size, whereas the opposite is true for the theoretical line which will be discussed later. This result differs from previous experiments performed with an alloy of the same nominal composition [14]. That alloy did not exhibit a pronounced dependence of the tip velocity on the cavity size. The reasons for this discrepancy are presently not clear.

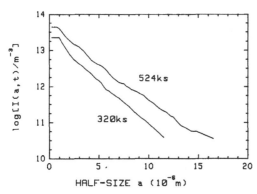

Fig. 3. Cumulative true cavity size distributions corresponding to two different creep times.

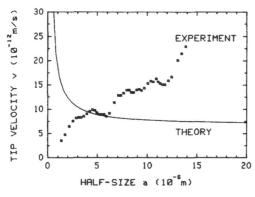

Fig. 4. Cavity tip velocities derived from the cumulative size distributions in Fig. 3.

Sintering Experiments

The diffusive growth of crack-like cavities like those observed in our work depends not only on the grain-boundary, but also on the surface diffusivity. If the characteristic time for shape change by surface diffusion is fast compared to that for cavity sintering by grain boundary diffusion, one should in principle be able to determine surface diffusivities by measuring the rate with which disc-shaped cavities evolve into quasi-equilibrium, spherically capped cavities. If, during sintering, the cavities become detached from the grain boundaries, grain boundary diffusion is no longer a factor, and the cavity morphology changes exclusively by surface diffusion. Annealing will then cause a decrease in diameter and an increase in thickness, but it will not change the cavity volume.

Unfortunately, sintering at 1033 K for periods up to 4.15 Ms (48 days) did not result in systematic changes in the size distributions. It appears that the changes due to sintering at this temperatures were too small to be resolved within the statistical error of our metallographic measurements. Sintering for 3.6ks at 1473 K, on the other hand, did result in observable changes. As expected, the diameters of the cavities tended to decrease, and the thicknesses of the larger cavities increased slightly.

DISCUSSION

We are now going to compare the measured growth rates with those theoretically anticipated. The cavity morphology suggests a model for crack-like diffusive growth. Previous models for this process are of limited use since they exhibit stress discontinuities which become increasingly severe as Λ/a decreases, where Λ is the characteristic diffusion length defined by Needleman and Rice [15]:

$$\Lambda = [(\delta_b D_b \Omega \sigma)/(kT\dot{\varepsilon})]^{1/3} . \tag{6}$$

The quantity $\delta_b D_b$ is the product of diffusional thickness of the grain boundaries and their diffusivity, $\Omega = 1.1 \cdot 10^{-29} m^3$ is the atomic volume, σ the applied stress, k is Boltzmann's constant, T the absolute temperature and $\dot{\varepsilon}$ the creep rate.

A model which avoids the discontinuity in the stress distribution has been recently developed by us [9]. The tip-velocity of crack-like cavities in this model is given by

$$v = \{H_2[(1 + H_1)^{1/2} - 1]\}^3 \tag{7}$$

where

$$H_1 = [2 \sigma \Lambda \delta_S D_S]/[\sin(\psi/2)\gamma_S \delta_b D_b Q(a/\Lambda)] , \tag{8}$$

$$H_2 = [(\delta_b D_b)/(2 \delta_S D_S \Lambda)] \times [(\delta_S D_S \Omega\gamma_S)/(kT)]^{1/3} \times Q(a/\Lambda) , \tag{9}$$

and, with $x = a/\Lambda$:

$$Q(x) = [1 - x^2/(1 + x)^2]/\{x[2\ln(1 + 1/x) + x^2/(1 + x)^2 - 1]\} . \tag{10}$$

Here, $\delta_S D_S$ is the product of diffusional thickness and diffusivity of the cavity surface, $\psi = 70°$ the dihedral angle at the cavity tip, and $\gamma_S = 2J/m^2$ the surface energy.

Employing a value of $\delta_b D_b = 10^{-22} m^3/s$ (which is only slightly higher than an upper bound value derived from diffusional creep measurements in this material [12]) and assuming $\delta_S D_S = 5 \times \delta_b D_b$ we obtain the solid line in Fig. 4. Although the magnitude of the observed tip velocities can be matched reasonably well by the model, the dependencies of the theoretical and experimental tip velocities on the cavity size are inconsistent. We suggest that cavity coalescence is responsible for this discrepancy. When small cavities coalesce, they disappear to form larger cavities. This means that the average growth rate of the smaller cavities will be reduced.

Similarly, the average tip velocity of larger cavities may be increased due to coalescence with smaller cavities. Thus, coalescence would tend to increase the slope of the theoretical line in Fig. 4. Since our analysis results in average growth rates, and since our metallographic observations provide qualitative evidence for coalescence, coalescence is an attractive explanation for the discrepancy between our model and the experimental results.

We have not yet attempted to infer surface diffusivities from our sintering experiments. However, such measurements appear to be a feasible technique to determine surface diffusivities, provided that sufficiently accurate size distributions can be obtained. A particular advantage might be that surface diffusivities (which may be sensitive to the chemical surface composition), could be measured "in situ," i.e., during annealing the surface composition would be thermodynamically determined by the bulk alloy composition. It would be difficult to achieve such conditions with diffusivity measurements on free surfaces.

CONCLUSIONS

Our results on creep cavity growth demonstrate the usefulness of our evaluation technique for cavity size distributions. Without this technique it would have been quite difficult, and far less obvious, to demonstrate the discrepancy between theoretical and experimental growth rates. It appears that coalescence is a significant factor in explaining this discrepancy. We also suggest that changes in cavity sizes and thicknesses during sintering may be exploited as a technique for surface diffusivity determinations.

ACKNOWLEDGMENTS

We would like to express our gratitude for financial support of this work by the Division of Materials Sciences, U.S. Department of Energy under contract DE-AC05-84OR21400 with Martin Marietta Energy Systems, Inc., and by Conacyt-Mexico, Grant PCCB-BNA-020348. We would also like to acknowledge the metallographic work of W. H. Farmer and the preparation of the manuscript by Gwen Sims.

REFERENCES

1. A. Aoki and O. Izumi, Nippon Kinzoku Gakkaishi 43, 1190 (1979).
2. C. T. Liu and J. O. Stiegler, Science 226, 636 (1984).
3. C. T. Liu, C. L. White and J. A. Horton, Acta Metall. 33, 213 (1985).
4. C. L. White, R. A. Padgett, C. T. Liu and S. M. Yalisove, Scr. Metall. 18, 1417 (1984).
5. M. K. Miller and J. A. Horton, Scr. Metall. 20, 789 (1986).
6. J. A. Horton and M. K. Miller, Acta Metall. 35, 133 (1987).
7. D. P. Pope and S. S. Ezz, Int. Metals Rev. 29, 136 (1984).
8. M. H. Yoo, Scr. Metall. 20, 915 (1986).
9. J. H. Schneibel and L. Martinez, submitted to Scr. Metall.
10. E. E. Underwood, in "Quantitative Microscopy," eds. T. DeHoff and F. N. Rhines, McGraw-Hill Book Company, New York, 1968, p. 149.
11. R. L. Fullman, Trans. AIME 197, 447 (1953).
12. J. H. Schneibel, G. F. Petersen and C. T. Liu, J. Mater. Res. 1, 68 (1986).
13. J. H. Schneibel and L. Martinez, Phil. Mag A, 54, 489 (1986).
14. J. H. Schneibel and L. Martinez, submitted to Met. Trans. A.
15. A. Needleman and J. R. Rice, Acta Metall. 28, 1315 (1980).

HIGH-TEMPERATURE FATIGUE CRACK PROPAGATION IN P/M Ni₃Al-B ALLOYS

K.-M. Chang, S.C. Huang, and A.I. Taub
General Electric Corporate Research and Development, Materials Laboratory
Schenectady, New York 12301

ABSTRACT

Ductile Ni_3Al-B type intermetallic alloys show a unique fatigue crack growth behavior at elevated temperatures. A crack propagation mechanism has been investigated in an experimental P/M alloy by testing the alloys with different fatigue frequencies at 400°C. The Ni_3Al-B intermetallic alloy shows a substantial time-dependence of fatigue crack growth rate when tested in air. Under a given cyclic stress intensity, an order of magnitude difference of crack growth rate was observed by decreasing the fatigue frequency. However, such a time-dependence did not occur when the alloy was tested in vacuum. It is concluded that "dynamic" embrittlement in an oxidation environment is the major factor controlling the fatigue crack growth in ordered Ni_3Al-B alloys at elevated temperatures.

INTRODUCTION

Boron-doped Ni_3Al intermetallic alloys have demonstrated a remarkable low-temperature ductility [1-3], which leads to the increasing interest in engineering structural applications. This ordered alloy has many unique mechanical properties, such as positive temperature-dependence of flow stress, high strain hardening rate at low temperatures, and good oxidization resistance at high temperatures. Recently, fatigue crack propagation at elevated temperatures has been evaluated in ductile Ni_3Al-B alloys [4,5].

Figure 1 shows the crack growth curve of a Ni_3Al-B based alloy measured at 400°C [4]. Also plotted in Figure 1 is the representing curve for high-strength superalloys. Fatigue crack growth was measured by using a 20 cpm (3 s per cycle) sinusoidal fatigue cycle. The ductile Ni_3Al-B based alloy has a significantly different behavior in fatigue crack propagation from those precipitation hardened superalloys. A very high value of threshold stress intensity, ΔK_{th}, at which fatigue crack starts to grow, was found in the Ni_3Al-B based alloy. The measured value was almost twice that of superalloys. On the other hand, stage II crack growth rate of this intermetallic alloy was found to be higher than that of superalloys by a factor of two.

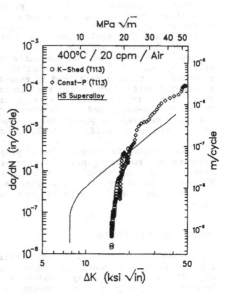

Figure 1. Crack growth curves.

This paper reports the results of the investigation aimed at understanding the mechanism of high-temperature fatigue crack propagation in ductile Ni₃Al-B alloys. The study is concentrated on the stage II crack propagation behavior. A time-dependence model based on the environmental embrittlement at crack tip is employed to explain the experimental observation.

EXPERIMENTAL

The nominal composition of the alloy studied is given in Table 1. The
equivalent aluminum concentration, defined as the total atomic percents of
Al and other elements that substitute for Al is designed to be 23.8%, since
the boron ductilization in Ni₃Al intermetallics occurs most effectively with
a substoichiometric composition [6]. The boron addition is set at 0.25
at.%.

Table 1   Chemical Composition of P/M T144 Alloy (wt.%)

Ni	Co	Al	Hf	Mo	Nb	Zr	B
bal.	11.5	11.5	1.75	1.00	0.50	0.05	0.05

One 8 kg heat was prepared by vacuum induction melting using high purity raw materials. The ingot was remelted and atomized into powders by a
gas atomizer. Screened powders of -140 mesh were canned, evacuated, and
sealed in vacuum. The consolidation was carried out by hot isostatically
pressing (HIP) at 1150°C/100 MPa for two hours. The alloy, designated as
T144, was then cold-rolled 20%, followed by an annealing at 1000°C for one
hour.

Subsized round tensile specimens with a gauge diameter of 2.54 mm were
employed to evaluate the tensile properties. Tensile tests were conducted
at a strain rate of 0.05 per minute. Fatigue crack propagation tests were
performed at 400°C. A single-edge-notched (SEN) sheet specimen was
employed, and the crack growth was monitored by the dc potential drop technique developed at our laboratory [7]. Normalized electric potential could
be converted into the corresponding crack length by a single analytical
relation, namely Johnson's equation [8]. A microprocessor continuously
recorded the testing data, and in real-time calculated the crack length, the
stress intensity factor, and the linear regression analysis of crack growth
rate. The servo-hydraulic test machine was modified to be controllable by
the microprocessor, and the testing mode could be automatically changed when
the crack reached the preset value. Some FCP tests were performed in a high
vacuum chamber to verify the influence of oxidation environment.

Metallographic samples were prepared using conventional mechanical
grinding and polishing procedures according to the standard laboratory process. An etching solution consisting of 10 ml HCl, 40 ml HNO₃, and 30 ml
H₂O was used to reveal microstructure. Fractography was examined under a
scanning electron microscope for each sample tested under every fatigue frequency.

RESULTS and DISCUSSION

Alloy T144 has a fully recrystallized, equiaxed grain structure after
the annealing treatment; the grain size is measured as ASTM 10 (10 to 15
μm), Figure 2. The alloy consists of the L1₂ type ordered phase predominantly, and only a small amount of secondary phase particles can be
detected.

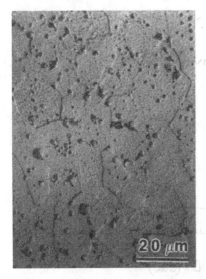

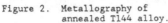

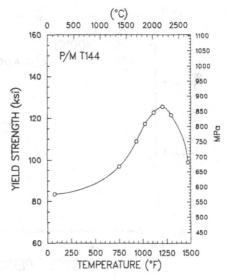

Figure 2.   Metallography of   Figure 3.   Relationship of yield
            annealed T144 alloy.   strength and temperature.

Figure 3 plots the measured yield strength as a function of temperature. The distinctive behavior of positive temperature-dependence of flow
stress in Ni$_3$Al-based intermetallics is evident. The yield strength
increases with the testing temperature from room temperature up to T$_p$, the
peak yield strength temperature, and then starts to decrease as the temperature is further increased. The T$_p$ temperature for P/M T144 under the
present strain rate is about 650°C.

Fatigue crack growth rate was measured at 400°C by using sinusoidal
waveforms with different frequencies: 10, 0.33, 0.05, and 0.01 Hz (0.1, 3,
20, and 100 s per cycle). The minimum-to-maximum load ratio was kept at R =
0.05 for all frequencies. The data are plotted in Figure 4, crack growth
rate under a given cyclic stress intensity, ΔK = 27.5 MPa m, as a fuction
of fatigue cycle period. The results of air tests show a strong time-
dependence; the crack accelerates substantially when the alloy is loaded by
slow fatigue cycles. The crack growth rate increases more than an order of
magnitude for the frequency range used in this study. In contrast, the
vacuum data indicated that the crack grows almost at a constant rate per
cycle in vacuum, independent of the fatigue frequency.

In addition to different frequency response of crack growth rate,
vacuum tests also change the fracture mode of fatigue cracking. As seen in
Figure 5, the fracture surface of air test under 0.1 s cycles consists of
mixing of transgranular and intergranular failure facets. When the fatigue
cycle period increases to 3 s, the fracture mode becomes completely intergranular. The same observation is found for 100 s cycles. On the other
hand, vacuum tests always show transgranular fracture facets with the conventional striation marks, even under the slow 100 s cycles (Figure 5(d)).
Therefore, the fast fatigue crack growth in ductile Ni$_3$Al-B intermetallic
alloys can be attributed to the environmental embrittlement. The oxygen in
hot air is believed to be the embrittling agent. Oxygen transports into the

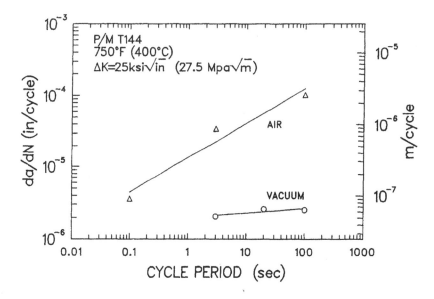

Figure 4. The relationship between crack growth rate and fatigue cycle period at 400°C for annealed T144 alloy.

crack tip under the cyclic loading, and embrittles the grain boundaries of the material in front of the crack. Such an embrittlement causes the crack to propagate along grain boundaries and accelerates the crack growth rate. A slow fatigue cycle provides a long period for oxygen transportation, and more oxygen can generate a larger and more brittle "embrittled" zone. As a result, the fatigue crack propagation shows a strong time-dependent behavior as seen in Figure 4.

Time-dependent fatigue crack propagation associated with environmental embrittlement has been reported in many other alloys, such as superalloys [9]. In general, fatigue crack growth can be characterized into two regimes. At a low temperature or under a fast frequency, the alloy shows a constant da/dN under a given ΔK. The crack growth is completely controlled by the cyclic stress intensity only. In this cycle-dependent regime, crack growth rate is believed to be insensitive to the variation of microstructure and alloy chemistry. As the temperature increases, the crack growth starts to accelerate substantially even if ΔK is kept at the same value. For a given temperature, da/dN increases dramatically with the cycle period. Such a time-dependent behavior is aggravated by increasing temperature. The fractography on the fracture surface suggests that the change of fracture mode corresponds to the transition from the cycle-dependent to the time-dependent regime.

Because of a high aluminum content, Ni_3Al-B intermetallic alloys have an excellent oxidation resistance at high temperatures. However, the kinetics of forming an Al_2O_3 protection layer at intermediate temperatures is too slow to prevent oxygen transport and embrittlement. As a result, the time-dependent fatigue crack propagation occurs at a much lower temperature in Ni_3Al-B alloys than in superalloys.

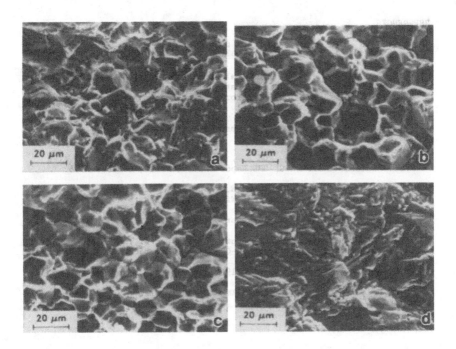

Figure 5.   Fatigue fracture mode: (a) 10 Hz, air; (b) 0.33 Hz, air;
             (c) 0.01 Hz, air; (d) 0.01 Hz, vacuum.

CONCLUSION

Fatigue crack propagation behavior of a ductile Ni_3Al-B intermetallic alloy, T144, has been studied at 400°C. Crack growth rate in air shows a strong time-dependence, i.e., da/dN increases with the fatigue cycle period for a given ΔK. The crack grows at a constant rate per cycle when tested in vacuum, independent of the fatigue frequency. The fracture mode is predominantly intergranular for air tests, while vacuum tests show completely transgranular failure. The environmental embrittlement mechanism is responsible for the fast crack growth in air; high aluminum content does not provide the alloy with adequate oxidation resistance at intermediate temperatures.

ACKNOWLEDGMENT

The author is indebted to E.H. Hearn and B.J. Drummond for providing direct assistant of experimental work. Many thanks are also due to L.J. Beha, C. Canestraro, D.A. Catharine, and H. Moran for their technical support in mechanical testing. Helpful discussions with Dr. M.G. Benz and Dr. L.A. Johnson are greatly appreciated.

308

REFERENCES

1. A.I. Taub, S.C. Huang, and K.-M. Chang: Met. Trans. A, 1984, vol. 15A, p. 339.

2. C.T. Liu and C.C. Koch: "Technical Aspects of Critical Materials Use by the Steel Industry, vol. IIB," National Bureau of Standards, NBSIR 83-2679-2, 1983, p. 42-1; also editor's note in Iron Age, September 24, 1982, p. 63.

3. K. Aoki and O. Izumi: J. Japan Inst. Met., 1979, vol. 43, p. 1190; ibid, 1979, vol. 43, p. 358.

4. K.-M. Chang, S.C. Huang, and A.I. Taub: "Rapidly Solidified Ni₃Al-B Based Intermetallics," GE-CRD Report No. 86CRD202, October 1986.

5. G.E. Fuchs, A.K. Kuruvilla, and N.S. Stoloff: "High Cycle Fatigue and Crack Growth in Polycrystalline Ni₃Al", submitted to Met. Trans. A.

6. C.T. Liu, C.L. White, and J.A. Horton: Acta Met., vol. 33, p. 213.

7. K.-M. Chang: "Improving Crack Growth Resistance of IN 718 Alloy Through Thermomechanical Processing," GE-CRD Report No. 85CRD187, October 1985.

8. H.H. Johnson: Mat. Res. Stand., 1965, vol. 5, no. 9, p. 442.

9. H.D. Solomon and L.F. Coffin: ASTM STP 520, p. 112, 1973.

STRUCTURE AND PROPERTIES OF POWDER METALLURGY Ni₃Al ALLOYS

D.D. Krueger, B.J. Marquardt, and R.D. Field
General Electric Company, Aircraft Engine Business Group, Evendale, OH 45215

ABSTRACT

A Ni_3Al + B base composition and derivatives containing Hf, Zr, Nb, or Nb + Hf substitutions for Al were processed by powder metallurgy, consolidated by compaction + extrusion, and evaluated for mechanical properties. The mechanical property evaluation consisted of tensile tests performed in air and vacuum, and fatigue crack growth tests performed in air. Microstructures, test specimen fracture morphologies, and tensile and fatigue deformation features were also studied. Some of the substitutions improved certain properties, but none improved ductility. The discussion considers property-fracture morphology-deformation structure relationships.

INTRODUCTION

The nickel aluminide Ni_3Al possesses certain mechanical and physical properties which encourage its use in structures requiring elevated temperature capability. Among these are the unusual flow behavior [1], good oxidation resistance [2], and low creep rates when alloyed with B and Hf [3]. For some nickel-base superalloy applications, Ni_3Al materials may offer improved specific strengths and reduced strategic element usage.

Until recently, interest in the development of polycrystalline Ni_3Al for structural use was tempered by its poor ductility. Brittle intergranular failure, now attributed to intrinsically weak grain boundaries [4,5], resulted in virtually no ductility at ambient temperatures. Aoki and Isumi [6], however, discovered that micro additions of B promoted transgranular fracture and significant improvements in ductility. This prompted a renewed interest in these materials, and considerable progress has subsequently been made in understanding the effects of stoichiometry [7-9], grain boundary chemistry [4,5,10], processing [11-13], microstructure [12-14], and solid solution ternary additions [3,7,15] on the behavior of polycrystalline Ni_3Al. These efforts provide a basis for the selection of Ni_3Al alloys suitable for bulk processing and initial engineering property evaluations. This paper describes the microstructures, tensile properties, fatigue crack growth rates, and deformation features observed for five Ni_3Al alloys.

EXPERIMENTAL

Table I lists the atomic formulae of the alloys selected for study. Each contains an 0.5 a/o B addition which is within the range shown to give a good balance of strength and ductility [4], and well below the solubility limit [6,16]. The base composition, Alloy 1, has a hypostoichiometric aluminum content (Ni:Al = 76:24). It has been shown that the ductilizing effect of boron is diminished in stoichiometric and Al-rich alloys [9]. This has been attributed to a more pronounced atomic defect structure, which limits boron transport to grain boundaries [17]. Relative to Alloy 1, Alloys 2-5 contain Hf, Zr, Nb, and Nb+Hf substitutions for Al. Previous studies have shown that these elements have a strengthening effect [3,7,15].

A vacuum induction melted ingot (11.4 kg) of each alloy was prepared for argon gas atomization. After atomization, -150 mesh powder was consolidated by compaction + extrusion at 1066 and 1093°C, respectively. An extrusion ratio of about 7:1 (area reduction) yielded billets approximately 2.54 cm in diameter. After extrusion, materials were given a 1093°C/1h anneal treatment.

Mat. Res. Soc. Symp. Proc. Vol. 81. 1987 Materials Research Society

Table I. Composition of Ni$_3$Al Alloys

Alloy 1 $(Ni_{0.76}Al_{0.24})_{99.5}B_{0.5}$

Alloy 2 $(Ni_{0.76}Al_{0.23}Hf_{0.01})_{99.5}B_{0.5}$

Alloy 3 $(Ni_{0.76}Al_{0.23}Zr_{0.01})_{99.5}B_{0.5}$

Alloy 4 $(Ni_{0.76}Al_{0.19}Nb_{0.05})_{99.5}B_{0.5}$

Alloy 5 $(Ni_{0.76}Al_{0.19}Hf_{0.01}Nb_{0.04})_{99.5}B_{0.5}$

Microstructures were studied using optical metallography and transmission electron microscopy (TEM) on electropolished foils. TEM was also used to study tensile and fatigue deformation structures. Test specimen fracture morphologies were examined using scanning electron microscopy (SEM).

Tensile tests were performed for all alloys in air and selected alloys in vacuum(10^{-3}Pa), using a strain rate of $8.3 \times 10^{-5}s^{-1}$ to plastic strains in excess of 0.2%. The test specimens were axisymmetric with nominal gage diameters of 3.56 (air) or 2.54 (vacuum) mm.

Fatigue crack growth (FCG) tests were performed for selected alloys in air. The test specimens had a single edge notch (SEN) geometry with a 2.54 by 10.16 mm rectangular gage. FCG cycling was performed at 0.33Hz using a min./max. stress ratio (R) of 0.05. Crack growth was monitored using a direct current potential drop technique [18], as adapted to the SEN geometry [19]. For the study of fatigue deformation, TEM foils were obtained from the gage section of specimens cycled at 0.33Hz (in air), using a plastic strain range ($\Delta\epsilon_p$) of 0.1% and alt./mean strain ratio (A_ϵ) of 1.0. These specimens had the same geometry as the tensile specimens tested in air. All mechanical test specimens were machined with their axis parallel to the extrusion axis.

RESULTS AND DISCUSSION

The nominal as-extruded grain size of each alloy was 10–15 μm. After annealing, the grains coarsened to produce a duplex size distribution with a nominal diameter of 20–25 μm. All alloys were single phase Ni$_3$Al, with occasional small oxide particles. Chemical analysis showed good agreement between aim and check composition. Bulk oxygen and sulphur levels ranged from 120–190 and 8–14 ppm, respectively.

Tensile test data are listed in Table II. All alloys were evaluated in

Table II. Air and Vacuum Tensile Properties

	Temperature (°C)	UTS (MPa)	0.2% YS (MPa)	Elong. (%)
Alloy 1	RT	1361	376	44.5
(base)	399	709	560	8.6
	538	707 (887)*	658 (632)*	2.5 (29.9)*
	649	561	546	1.0
	760	TF (494)*	TF (452)*	TF (1.7)*
Alloy 2	RT	1425	579	29.6
(Hf modified)	399	970	823	6.5
	538	958 (1113)*	901 (919)*	2.2 (26.6)*
	649	TF	TF	TF
	760	TF (817)*	TF (761)*	TF (2.7)*
Alloy 3	RT	1407	569	29.1
(Zr modified)	399	987	798	7.6
	538	1003 (1120)*	925 (917)*	2.6 (26.2)*
	649	TF	TF	TF
	760	TF (850)*	TF (785)*	TF (3.2)*
Alloy 4	RT	1003	832	4.1
(Nb modified)	399–760	TF	TF	TF
Alloy 5	RT	1091	976	3.0
(Nb + Hf modified)	399	1227	1119	2.5
	538–760	TF	TF	TF

* Vacuum data   TF = Thread failure in specimen grip end

air over the room temperature (RT) to 760°C range. In general, it was noted that loss of ductility occurred with increasing temperature and a transition from transgranular or mixed mode fracture to that of 100% intergranular fracture. During RT testing, Alloys 1-3 attained from 29.1 to 44.5% elongation (EL) and showed little or no intergranular fracture. At 538°C, they reached between 2 and 3% EL and showed intergranular fracture near the edge of the specimen and mixed mode fracture near the center. Similar transitions in fracture mode have been observed elsewhere [20]. At 649°C, Alloy 1 fractured in a 100% intergranular fashion and had 1.0% EL, whereas Alloys 2 and 3 fractured in the threaded grip end before reaching an 0.2% offset yield stress (YS). This type of failure was also observed for Alloy 4 at 399°C and above, and Alloy 5 at 538°C and above. These less ductile alloys also showed a large degree of intergranular fracture at RT.

Tensile tests on Alloys 1-3 were conducted in vacuum at 538 and 760°C. When tested at 538°C, each alloy fractured by a mixed mode across the entire fracture surface and achieved more than 25% EL. The fracture surface of Alloy 1 exhibited some intergranular microvoid coalescence (IMC) as shown in Figure 1a. TEM study of this specimen revealed a very high dislocation density, consistent with its considerable work hardening, and small grains, 0.1-0.2 μm in size, located at the original grain boundaries, Figure 1b. It is believed that the small grains formed as a result of dynamic recrystallization and aided ductility. This phenomenon was not observed in either the RT or 760°C specimens. Although Alloys 2 and 3 also displayed recrystallized grains at grain boundaries (and ductility similar to Alloy 1), IMC was not observed on the fracture surfaces. When tested at 760°C, each alloy exhibited 100% intergranular fracture (no IMC) and only 3.2 to 1.7% EL. As shown in Figure 2a, the dislocation density is drastically reduced with the increase in test temperature. The dislocations are <110> type, with a large portion in the pure screw orientation. The rough faceted nature of 760°C fractures, Figure 2b, can be understood in terms of the low dislocation density resulting in coarse slip and limited dislocation interactions. These results are consistent with the findings of previous investigators [21] who correlated the drop in YS at high temperatures with a change from <110>/{111} to <110>/{001} slip. A comparison of air and vacuum results indicates that environmentally assisted fracture and loss of ductility precludes a loss in ductility associated with a change in slip behavior.

Several significant variations in tensile properties were found as a result of variation in chemical composition. The strengthening potency of Hf and Zr was found to be nearly identical with respect to the base alloy. At RT, the YS of Alloys 2 and 3 was roughly 50-55% higher than Alloy 1. As test temperature increased, this strengthening effect gradually decreased to about

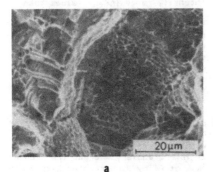 

a   b

Fig. 1. Alloy 1 (base) after 538°C vacuum tensile test; (a) SEM micrograph of fracture and (b) TEM micrograph of deformation structure.

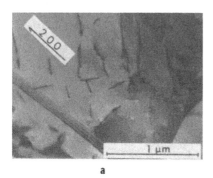 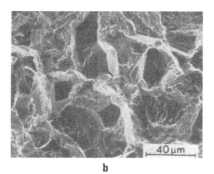

a   b

Fig. 2. Alloy 1 (base) after 760°C vacuum tensile test; (a) TEM micrograph
         of deformation structure and (b) SEM micrograph of fracture.

40% at 538°C. At 760°C, however, the YS of Alloys 2 and 3 was nearly 70%
higher than Alloy 1, indicating that Hf and Zr have a more pronounced
strengthening effect on the ⟨110⟩/{001} slip system or a more potent solid
solution strengthening effect at elevated temperatures. The Nb-containing
materials, Alloys 4 and 5, demonstrated even higher YS at RT, but lower
ultimate tensile strength (UTS) and EL despite the equivalent B level and
effective Al concentration. This suggests that Nb has a detrimental effect
on grain boundary strength. The partial replacement of Hf for Nb in Alloy 5
leads to higher UTS and increased temperature capability prior to grip end
failure.

      Fatigue crack growth tests were performed on Alloys 1-3 to determine
Region I (near-threshold) and Region II (Paris Law regime) rates at 399°C.
The Region I rates and threshold behavior were evaluated using load-shed
procedures. Tests on Alloys 2 and 3 were also performed to obtain Region II
rates at 538°C. The FCG test data are shown in Figure 3. Except for the
slight difference in Region I behavior for the two specimens of Alloy 2, agreement between duplicate tests is quite good.

      At 399°C, the Hf and Zr containing compositions (Alloys 2 and 3) have higher apparent threshold stress intensity range (ΔK_{th}) values than the base composition. In Region II, the relative performance of Alloy 1 reverses such that it exhibits the more desirable behavior and slower growth rates. At 538°C, the FCG rates for Alloys 2 and 3 nearly coincide, and are about an order of magnitude faster than those obtained for 399°C tests.

      SEM study of FCG specimen crack paths revealed that the lower ΔK_{th} of Alloy 1 at 399°C could be associated with fracture that was almost entirely intergranular, whereas the higher ΔK_{th} of Alloy 3 correlated with mixed mode fracture. The higher ΔK_{th} of Alloy 2, however, could not clearly be associated with mixed mode

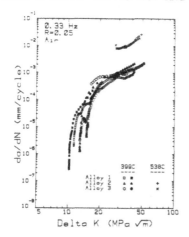

Fig. 3. Fatigue crack growth rates
         of Alloy 1 (base), Alloy 2
         (Hf modified), and Alloy 3
         (Zr modified).

fracture, as its morphology more closely resembled that of Alloy 1. In the
Region II regime at 399°C, the fracture mode of each alloy at low ΔK values
($\sim$20 MPa$\sqrt{m}$) was predominantly intergranular. At higher ΔK values, the amount
of transgranular cracking increased for the alloys showing lower growth
rates. Figure 4 shows the mixed mode morphology of Alloy 1 at a ΔK of
40 MPa$\sqrt{m}$. Faint slip traces were also observed on the grain facets in
intergranular regions. These features were consistent with the fatigue
deformation structure described later. The observations for 399°C tests
indicate that more favorable FCG behavior could typically be associated with
larger amounts of transgranular fracture. This would be expected since
transgranular fracture is generally associated with more cyclic plasticity or
damage accumulation than intergranular fracture. The fatigue fracture mode
for 538°C tests of Alloys 2 and 3 was similar, and intergranular over the
entire range in ΔK. Slip traces were again detected on the grain facets.
The more rapid growth rates at 538°C can be attributed to a lack of
transgranular fracture.

The TEM study of fatigue deformation was performed on Alloys 1-3 after
399°C cycling. Each alloy contained a low density of primarily screw
dislocations in the matrix, as previously observed in Ni$_3$Al materials
deformed at RT [22], and dislocations within the grain boundaries. The slip
traces observed on fracture surface grain boundaries thus resulted from the
relatively uninhibited slip of screw dislocations across the grains. A large
number of screw dislocation dipoles (identified through $\pm$g experiments) were
also observed. These have been associated with Kear-Wilsdorf pinning [23],
which prevents mutual annihilation by cross-slip. In the current study, weak
beam imaging revealed that the $\langle 110 \rangle$ super-dislocations were dissociated on
$\{100\}$ planes, with a seperation distance of approximately 5 nm, thus
supporting this hypothesis. A set of dipoles is shown in Figure 5, in which
the separation of two $\frac{1}{2}$ a$[10\bar{1}]$ superpartials is observed for a $[011]$ beam
direction, but not $[001]$, indicating dissociation on the (010) plane. This
result indicates that widespread pinning of dislocations by cross-slip onto
$\{100\}$ planes is occuring in this specimen. More work is needed to understand
the influence of deformation features, as well as environment on the FCG and
fatigue fracture behavior of Ni$_3$Al alloys.

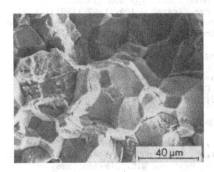

Fig. 4. SEM micrograph of crack path
at ΔK = 40 MPa$\sqrt{m}$, 399°C for
Alloy 1 (base).

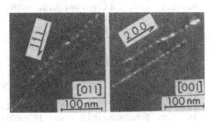

Fig. 5. Weak beam TEM micrographs of
screw dipoles (b = $[10\bar{1}]$) in
Alloy 1 (base) after 399°C
fatigue cycling ($\Delta\epsilon_p$ = 0.1%)

314

SUMMARY

The structures, tensile properties, fatigue crack growth rates, and deformation features of a Ni_3Al base alloy (Ni:Al = 76:24, B = 0.5 a/o) and four derivatives containing Hf, Zr, Nb, or Nb + Hf substitutions for Al were studied. Substituting for Al improved yield strength, but did not improve ductility. A comparison of air and vacuum tensile results indicated that environmentally assisted fracture and loss of ductility precludes a loss in ductility associated with a change in slip system. The Hf and Zr substitutions also improved threshold fatigue crack growth behavior at 399°C, but resulted in similar or faster Region II rates. Some aspects of the tensile and fatigue crack growth behavior were understood on the basis of fracture morphology and dislocation structure observations.

ACKNOWLEDGMENTS

The authors wish to thank L. Wojcik for producing the alloy powders, A. Taub for arranging the vacuum tests, and the Materials Laboratory at Wright-Patterson AFB for processing the extrusions. We also thank S.C. Huang for helpful discussions and J.R Groh and J.F. Wessels for technical assistance.

REFERENCES

1. R.W. Guard and J.H. Westbrook, Trans. AIME 215, 807 (1959).
2. E.A. Aitken, Intermetallic Compounds, J.H. Westbrook, ed., pp. 491-515, John Wiley and Sons, Inc., New York, N.Y. (1967).
3. C.T. Liu and C.L. White, Proc. MRS Symp. High-Temperature Ordered Intermetallic Alloys, C.C. Koch, C.T.Liu, and N.S. Stoloff, eds., vol. 39, pp. 365-380, Boston, MA. (1985).
4. C.T. Liu, C.L. White, and J.A. Horton, Acta Met. 33, 213 (1985).
5. T. Takasugi, E.P. George, D.P. Pope, and O. Izumi, Scripta Met. 19, 551 (1985).
6. K. Aoki and O. Izumi, J. Japan Inst. Metals 43, 1190 (1979).
7. R.D. Rawlings, A. Staton-BeVan, J. Mater. Sci. 10, 505 (1975).
8. O. Noguchi, Y. Oya, and T. Suzuki, Met. Trans. A 12A, 1647 (1981).
9. C.T. Liu, C.L. White, C.C. Koch, and E.H. Lee, Proc. Electrochem. Soc. on High Temperature Materials, M. Cubicciotti, ed., vol. 83-87, p. 32, Electrochem. Soc. Inc. (1983).
10. S.C. Huang, C.L. Briant, K.-M. Chang, A.I.Taub, and E.L. Hall, J. Mat. Res. 1, 60 (1986).
11. A.I. Taub, S.C. Huang, and K.-M. Chang, Proc. MRS Symp. High-Temperature Ordered Intermetallic Alloys, C.C. Koch, C.T. Liu, and N.S. Stoloff, eds., vol. 39, pp. 221-228, Boston, MA. (1985).
12. C.C. Koch, ibid., pp. 397-410.
13. K.-M. Chang, A.I. Taub., and S.C. Huang, ibid., pp. 335-342.
14. E.M. Schulson,ibid., pp. 193-204.
15. S.C. Huang, General Electric Company, unpublished research.
16. K.-M. Chang, S.C. Huang, and A.I. Taub, Proc. MRS Symp. Rapidly Solidified Metastable Materials, B.H. Kear and W.C. Giessen, eds., vol. 28, p. 401, North-Holland, N.Y. (1984).
17. A.D. Dasgupta, L.C. Smedskjaer, D.G. Legnini, and R.W. Siegel, Mat. Letters 3, 457 (1985).
18. R.P. Gangloff, Fat. Eng. Mat. and Structures 4, 15 (1981).
19. J.R. Wilcox and M.F. Henry, General Electric Co., unpublished research.
20. C.T.Liu and C.L. White, to be published in Acta Met.
21. P.H. Thorton, R.G. Davies, and T.L. Johnson, Met. Trans. 1, 207 (1970).
22. R.A. Mulford and D.P. Pope, Acta Met. 21, 1375 (1973).
23. N. Rao Bonda, D.P. Pope, and C. Laird, presented at the 1986 TMS Fall Meeting, Orlando, FL, Oct.7, 1986 (unpublished).

THE MICROSTRUCTURE AND TENSILE PROPERTIES OF EXTRUDED
MELT-SPUN RIBBONS OF IRON-RICH B2 FeAl

I. BAKER* AND D.J. GAYDOSH**
*Thayer School of Engineering, Dartmouth College, Hanover, NH 03755
**NASA-Lewis Research Center, Cleveland, Ohio 44135

ABSTRACT

The microstructure of extruded rods of iron-rich FeAl (B2-structure), as characterized by TEM, SEM, optical microscopy and x-ray diffractometry, consisted of elongated grains with a <111> fibre texture containing a high dislocation density. Numerous oxide particles were found, mostly in lines which reflected the matrix flow during extrusion. In addition, some large inclusions were present. Tensile testing of annealed, relatively dislocation-free specimens as a function of increasing temperature found increasing ductility up to 900K, above which a ductility drop occurred accompanied by a change in fracture mode, from transgranular cleavage to intergranular fracture. The yield strength, which was independent of temperature up to 800K (at ~500MPa), also decreased rapidly as diffusion became more important. The predominant slip vector changed from <111> to <100> around 700K

INTRODUCTION

Sainfort et al.[1] were the first to report the tensile properties of polycrystalline B2-structured FeAl. They found that the elongation to fracture of Fe-40Al specimens, of ~200μm grain size, (strained in air) increased monotonically as the temperature of straining increased, from ~8% at room temperature to greater than 40% at 1241K. The fracture mode was reported to be mainly intergranular at room temperature and transgranular cleavage at higher temperatures.
Mendiratta et al. [2] have examined the tensile properties of iron-rich Fe-35Al, of 30μm grain size, produced using rapid solidification followed by hot extrusion. The alloy showed 7% elongation at room temperature and the strain to fracture increased as the temperature of straining increased up to 773K, failure occurring transgranularly. However, at 873K a ductility drop was noted and fracture was intergranular.
This paper presents the microstructure and tensile properties of hot-extruded, fragmented, melt-spun ribbons of an iron-rich FeAl alloy and compares the results to the earlier works.

EXPERIMENTAL

Fragmented ribbons of FeAl, melt-spun under an argon atmosphere from cast material, were canned in an evacuated mild steel jacket and hot-extruded to rod at a 16:1 reduction ratio at 1250K. Transmission electron microscope (TEM) examination revealed that the as-extruded material had a residual dislocation density, while material annealed for one hour at 1273K was relatively dislocation-free. X-ray diffractometry of the extruded rod showed that it had a <111> fibre texture. Chemical analysis determined that it contained (in at.%) Fe-60.49, Al-36.88, Ni-2.22, Mo-0.23, Co-0.09 and Cr-0.09.
Optical microscopy revealed that the annealed material had a slightly elongated grain structure (23 μm grain dia. in longitudinal sections and 14 μm grain dia. in transverse sections). Lines of particles, presumably oxides, were observed. In longitudinal sections the oxide particles were mostly aligned along the extrusion direction, see fig.1a, whilst in transverse sections the oxides (and grain structure) had a swirly appearance, see fig.1b, reminiscent of the structure of hot-extruded NiAl [3]. These features presumably reflect the flow of the material during extrusion. An interesting feature of the oxide particles is that whilst in many places the line of oxides corresponds to a grain boundary, in some places the line of oxides passes through a grain, indicating that the oxides do not significantly impede grain growth, at least under the processing conditions used here. Grain boundaries that coincide with a line of oxide particles probably correspond to the former free surfaces of the unconsolidated melt-spun ribbon which must have become covered with oxide during the processing.

316

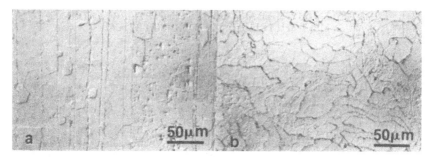

1. Optical micrographs of a) longitudinal and b) transverse sections of extruded rod.

In addition to the oxide particles, large inclusions (20-200 μm in length) were also present, see fig.2. These were presumably stronger than the matrix at the temperature at which extrusion took place because plastic flow of the FeAl occurred around them. Energy dispersive spectroscopy (EDS) of some of these inclusions during examination in a scanning electron microscope (SEM) revealed that they consisted of the impurities found in the chemical analysis. A separate TEM-EDS analysis showed that the Ni was present in the matrix also at about 1.5 at. % level. Thus the composition of the matrix is approximately 37 or 38 at.% Al and ~ 1.5 at.% Ni with the balance Fe.

Dumbbell-shaped tensile specimens (gauge diameter 3.1mm ; gauge length 33.0mm) were machined by centreless grinding from the annealed FeAl rod and electropolished to give a scratch-free surface. Specimens were strained to fracture under tension at a constant crosshead speed of 0.25 mm.min.$^{-1}$ (strain rate ~1 x 10^{-4} s^{-1}) in air at temperatures from 295K to 1100K and in vacuum (<5 x10^{-3} Pa.) above 500K. Fracture surfaces were examined in an SEM and polished and etched longitudinal sections of the fractured specimens were examined using both SEM and optical microscopy. Thin foils prepared from discs [4] both from the gauge of fractured specimens and from some lightly-strained specimens were examined in a 120 KeV TEM.

2. Optical micrograph of a large inclusion present in extruded rod.

RESULTS

The yield strength (YS), measured as 0.2 % proof stress, and ultimate tensile strength (UTS) of the alloy are shown as a function of temperature in figure 3a. Between 295K and 800K the yield strength (~500 MPa.) varied little; yielding (accompanied by a clearly audible clicking) was discontinuous and a period of work-hardening followed the Lüders regime. Above 800K the yield strength decreased rapidly, yielding occurred smoothly and the yield strength corresponded to the ultimate tensile strength. Sainfort et al. [1] similarly found little change in the yield strength (~270 MPa) of Fe-40Al as a function of temperature up to 923K but a rapid decrease in yield strength at higher temperatures. The Lüders strain increased up to 700K (0.38 at 300K, 0.56 at 500K, 0.84 at 700K) but decreased (to 0.67) at 800K.

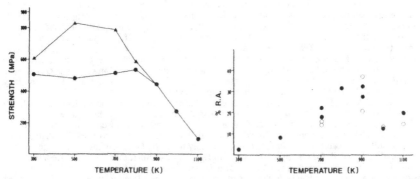

3. Graphs of a) YS (•) and UTS (▲), and b) %RA (○ vacuum, ● air), as a function of temperature.

Ductility , as measured by the percentage reduction in area (% R.A.), increased up to 900K then dropped slightly, fig. 3b. Elongation was not used as a measure of ductility since this largely reflected the length of gauge over which necking occurred and the extent of cracking, and showed considerable scatter. There was no clear evidence that straining FeAl in air affects ductility since testing in vacuum produced similar results. Fracture was wholly by transgranular cleavage up to 900K, fig 4a: this low energy fracture mode is probably the reason for the scatter in the measured R.A. Only one example was found of an inclusion on a fracture surface and since this was the specimen strained at 800K which showed considerable ductility (32% R.A.) the implication is that the inclusions do not prevent ductility in this material, although they may be expected to reduce it somewhat. At 1000K, the fracture mode changed to a mixture of transgranular cleavage, fig 4b, and intergranular cavitation, fig 4c, while at 1100K failure was wholly intergranular, fig. 4d. This fracture mode change and ductility drop are similar to that reported by Mendiratta et al. for iron-rich FeAl [2]. Fracture surfaces at all temperatures contained numerous cracks.

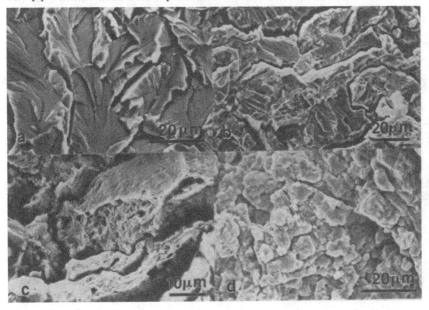

4. SEM fractographs of specimens strained at $10^{-4}s^{-1}$ at a) 900K; b,c) 1000K; d) 1100K.

Examination of polished longitudinal sections, fig 5, showed that delamination of grains had occurred along the lines of oxide particles. This delamination is the origin of the cracks observed on the fracture surfaces. At and above 700K considerable necking of individual grains occurred at the fracture surface. At higher temperatures (≥900K) rupture occurred along transverse boundaries. Whilst ruptured transverse boundaries were rare at 900K they were common at 1100K and cavities occurred at triple points where transverse grain boundaries met longitudinal grain boundaries, see fig 5d. Examination of the polished longitudinal sections also showed that the inclusions cracked during testing up to 900K but that above that temperature they remained intact and rupture occurred at the matrix/particle interface.

Up to 700K, TEM examination of tested specimens revealed, that most dislocations had a <111> Burgers' vector, with occasional <100> dislocations. By comparison, at 800K most dislocations had a <100> Burgers' vector but a few <111> dislocations were still present. Above 800K only <100> dislocations were found. At low dislocation densities most <100> dislocations were found to be close to edge character, as previously noted by Umakoshi and Yamaguchi [5] in FeAl single crystals deformed by <100> slip. In fig.6, for example, three sets of dislocations can be observed. Note that one set, arrowed, is almost out of contrast but can be observed due to the residual contrast arising because of the anisotropy of FeAl. Their line directions are aligned along the three <110> directions which are perpendicular to the <111> foil normal; that is, they are <100> edge dislocations gliding on {110} planes. Above 1000K dislocation networks were observed.

Fig 7 shows how yield strength decreases with decreasing strain rate at higher temperatures (≥ 900K); yielding occurred continuously at all strain rates. At 900K, increases in strain rate (to 10^{-2} s^{-1}) still resulted in transgranular cleavage but a decreased strain rate of 10^{-5}s^{-1} resulted in intergranular cavitation with some dimple-like features on the fracture surfaces, similar to features previously observed at 1000K at 10^{-4}s^{-1}, indicating local plastic deformation. Straining at this lower strain rate also resulted in more rupture of transverse grain boundaries. At 1000K, straining at 10^{-1}s^{-1} resulted in a change to almost completely transgranular cleavage with little intergranular failure whilst straining at 10^{-5}s^{-1} resulted in a wholly intergranular fracture mode. Longitudinal sections showed that transverse grain boundary rupture was absent after testing at 10^{-1}s^{-1} but extensive after testing at the slow strain rate (10^{-5}s^{-1}). Finally at 1100K, straining at 10^{-1}s^{-1} resulted in a mixture of transgranular cleavage and the previously observed (at 10^{-4}s^{-1}) rounded intergranular fracture.

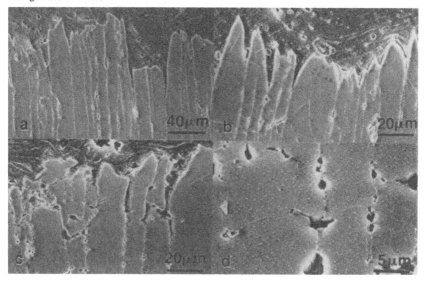

5. Longitudinal sections of specimens strained at 10^{-4}s^{-1} at a)700K, b) 900K and c,d) 1100K.

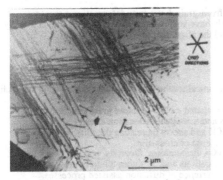

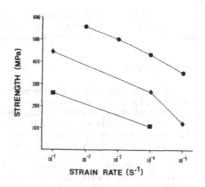

6. TEM image showing three sets of edge
<100> dislocations gliding on {110}
planes at 900K.

7. Graph of yield strength versus strain
rate. (●900K, ◆1000K, ■1100K)

To summarize, increased strain rate favoured transgranular cleavage and little rupture of
transverse grain boundaries, whilst decreasing strain rate favoured intergranular cavitation and
extensive rupture of transverse grain boundaries. The strain rate changes had little effect on
ductility except for straining at 900K at $10^{-5}s^{-1}$ where the R.A. fell to only 12%, a similar drop to
that previously observed (at $10^{-4}s^{-1}$) when testing at higher temperatures.

DISCUSSION

For FeAl <111> slip at low temperatures is now well-established [5-9]. In the present
case, <111> slip is predominant up to 700K, meaning that sufficient independent slip systems are
available [10] for uniform plastic flow in a polycrystal [11]. The poor low temperature ductility
could be due to the presence of the oxides and/or inclusions, however, it may be that the difficulty
of dislocation cross-slip also plays a role. It has been suggested that restricted cross-slip in ordered
alloys may lead to reduced ductility even if five independent slip systems are available [12]. The
increasing fracture strain with temperature (up to 700K) may, thus, be related to increasing
thermally-activated cross-slip.
Several workers [6,8,9] have noted that FeAl exhibits a change in slip direction with
increasing temperature from <111> to <100>, which means a reduction from five to only three
independent slip systems [10] which are insufficient for uniform plastic flow in a polycrystal [11].
At 800K <100> slip predominated but dislocations with <111> Burgers' vectors were found.
Presumably, enough of the latter were present, particularly near grain boundaries, to provide
sufficient slip systems to maintain the compatibility between grains.
Above 900K (~0.55T$_m$) recovery processes, particularly near grain boundaries, could have
provided the additional deformation modes [13] to allow general plastic flow. Evidence for these
diffusional processes is the lack of work hardening, the yield strength decrease and the strain rate
sensitivity of the yield strength, suggesting easier deformation (diffusion-assisted) mechanisms: the
onset of <100> slip does not alone account for the yield strength decrease since no decrease was
observed at 800K. At and above 1000K diffusion-assisted processes were also evident from the
presence of dislocation networks.
To explain the ductility drop observed here (and by Mendiratta et al. [2]), at 1000K we can
invoke the concept of the equicohesion temperature [14]. Below this temperature grain boundaries
are stronger than the grain interiors so that failure occurs transgranularly. Above this temperature,
the grain boundaries are weaker than the grain interiors, hence deformation is localized in the grain
boundaries, as evidenced by grain boundary sliding and intergranular cavitation. Because this
localized grain boundary deformation is diffusion-assisted, cavitation occurs more readily with
increasing temperature and decreasing strain rate. Presumably, then, the reason that a drop in
ductility occurred in both the RS material examined here and that examined by Mendiratta et al. [2]
but not in the conventionally processed FeAl examined by Sainfort et al. [1] is because of the finer
grain size of the RS material (20-30μm) compared to the conventionally processed material

(200μm) and hence the greater contribution from grain boundary deformation processes. The material tested here might also be expected to behave somewhat differently from conventionally-processed FeAl because of the strong texture and the alignment of oxides and grains.

CONCLUSIONS

Microstructural characterization and mechanical testing of a consolidated melt-spun iron-rich FeAl alloy led to the following conclusions:-
1. Consolidated melt-spun FeAl had a slightly elongated <111>-textured grain structure which contained some inclusions, numerous oxides and a residual dislocation density.
2. FeAl deforms mainly by <111> slip up to 700K and shows increasing ductility with increasing temperature, presumably due to the increasing ease of cross-slip.
3. Above 700K slip appears to be mainly by <100> dislocations. Ductility was still possible, at 800K due to the presence of a few <111> dislocations and, at higher temperatures (where no <111> dislocations were found) due to the occurrence of diffusion-assisted processes.
4. Above 900K, a ductility drop occurs, accompanied by a change in fracture mode from transgranular cleavage to intergranular cavitation. This is due to grain boundary weakening (as expressed by the concept of the equicohesion temperature).

ACKNOWLEDGEMENTS

The authors wish to thank M.V. Nathal, J.D. Whittenberger, J.R. Stephens and E.M. Schulson for helpful discussions. I. Baker gratefully acknowledges the support of a Case-NASA Co-operative R & D Aerospace Fellowship.

REFERENCES

1. G. Sainfort, P. Mouturat, P. Pepin, J. Petit, G. Cabane and M. Salesse, Mem. Sci. Rev. Met. 60 (1963) 125
2. M.G. Mendiratta, S.K. Ehlers, D.K. Chatterjee and H.A. Lipsett, in Rapid Solidification Processing-Materials and Technologies III, Ed. R. Mehrabian, NBS, Gaithersburg, MD,1982, p 240
3. I. Baker and E.M. Schulson, Met. Trans.A 15A (1984) 1129
4. I Baker and D J Gaydosh, Phys. Stat. Sol. (a) 96 (1986) 185
5. Y. Umakoshi and M. Yamaguchi, Phil. Mag. A 41 (1980) 573
6. T. Yamagata and H. Yoshida, Mat. Sci. Eng. 12 (1973) 95
7. T. Yamagata, Trans. Japan. Inst. Metals 18 (1977) 715
8. Y. Umakoshi and M. Yamaguchi, Phil. Mag. A 44 (1981) 711
9. M.G. Mendiratta, H.-M. Min and H.A. Lipsett, Metall. Trans. A 15A (1984) 395
10. G.W. Groves and A. Kelly, Phil. Mag. 8 (1963) 877
11. R. Von Mises, Z. Angew. Math. Mech. 8 (1928) 161
12. T.L. Johnston, R.G. Davies, and N.S. Stoloff, Phil. Mag. 12 (1965) 305
13. G.W. Groves and A. Kelly, Phil. Mag. 19 (1969) 977
14. G.E. Dieter, Mechanical Metallurgy, McGraw-Hill, New York, 1961, p 345

EFFECT OF ALUMINUM ADDITION ON DUCTILITY AND YIELD STRENGTH OF Fe$_3$Al ALLOYS WITH 0.5 wt % TiB$_2$*

C. G. McKAMEY, J. A. HORTON, AND C. T. LIU
Metals and Ceramics Division, Oak Ridge National Laboratory
Oak Ridge, TN 37831

ABSTRACT

Studies have been conducted of the mechanical properties of Fe$_3$Al alloys containing 24 to 30 at.% Al, to which 0.5 wt% TiB$_2$ was added for grain refinement. In tensile tests conducted at room temperature, it has been found that, as the aluminum content is increased, the yield strength decreases sharply from 760 to 310 MPa. The decrease in yield strength is accompanied by a four-fold increase in room-temperature ductility. Ordered iron aluminides (containing no disordered α phase) showed a clear increase in yield strength with temperature above 300°C. Their strength reached a maximum around 600°C, above which it decreased sharply. All these results will be discussed and correlated with stability of superlattice dislocations as a function of aluminum content.

INTRODUCTION

Alloys based on Fe$_3$Al are of interest because of their excellent oxidation resistance and low material cost. Unfortunately, low strength above 600°C and a lack of ductility at room temperature limit their use as structural materials. Recent studies [1] have shown that mechanical failure of pure Fe$_3$Al is transgranular, contrary to earlier results [2]. Nevertheless, improvement in ductility is still needed. No ternary additions were found which dramatically improved properties. In the current study, TiB$_2$ was added to control grain size and thereby give some property improvements. This also facilitated a more in depth study of the yield strength as a function of composition and temperature.

The tensile behavior of Fe$_3$Al-based alloys has a strong dependence on temperature, composition, and heat treatment [3-6]. The yield stress (σ_y) near the Fe$_3$Al composition increases with temperature above 300°C to a maximum value near 550°C and then decreases sharply. This temperature corresponds to the second order phase transformation temperature between the DO$_3$ and the B2 ordered structures [7,8]. Morgand et al. [9] showed that this peak in yield stress occurs clearly at compositions from about 23 to 32% Al which coincides with the composition range of the DO$_3$ phase. This type of yield behavior has been observed in many other ordered systems including Ni$_3$Al [10], FeCo [4], CuZn [11], and Ni$_3$Mn [12]. Reports of its occurrence in Fe$_3$Al, although numerous, have been conflicting with regard to the explanations for its occurrence. Although most reports are of a single peak at approximately 550°C, some researchers have not seen the peak at all at certain compositions [13,14]. Morgand et al. also reported the presence of two peaks in the yield stress in a sample of 23.9% Al [9]. Different mechanisms have been proposed to explain its presence as a function of temperature, including the cross slip model proposed by Kear and Wilsdorf [15] and Takeuchi and Kuramoto [16] and a change in dislocation configuration with degree of order proposed by Stoloff and Davies [17]. However, none of the proposed mechanisms appears to be entirely applicable to the Fe$_3$Al system.

*Research supported by the U.S. Department of Energy, Morgantown Energy Technology Center, Surface Gasification Materials Program under contract DE-AC05-840R21400 with Martin Marietta Energy Systems, Inc.

Within the DO_3 structure, the possible superlattice dislocations and their imperfect variants are shown in Fig. 1 [3,13,18]. Associated with the imperfect variants will be deformation-induced antiphase boundary (APB) trails (Fig. 1d-f). Since the motion of the imperfect types leads to the formation of nearest-neighbor and next-nearest-neighbor APB's (NNAPB's and NNNAPB's), these configurations are mobile only when the magnitude of the external stress is sufficient to allow for production of APB's. At compositions of 24-26% Al, ordinary single dislocations, sometimes with trailing APB's (similar to Fig. 1e and f), have been shown to be responsible for the deformation [5]. At higher aluminum levels, deformation is thought to occur by APB coupled dislocations [5]. The change in dislocation character results from a change in APB energy. As the aluminum content increases, the energy of the DO_3 APB (NNNAPB) decreases and the B2 APB (NNAPB) energy increases [13,18], resulting in two-fold superdislocations as shown in Fig. 1b. There is one report of the existence of four-fold superlattice dislocations (Fig. 1a), at compositions from 25-30% Al [19].

The purpose of this work is to clarify two points concerning the mechanical properties of Fe_3Al alloys containing 24 to 30% Al: (1) to explain the conflicting reports on the appearance of the yield behavior, and (2) to discuss the effect of aluminum content on the yield stress and ductility.

EXPERIMENTAL PROCEDURES

Alloys with compositions of Fe-24 to 30% Al, each containing 0.5 wt % TiB_2 added for grain refinement, were prepared by arc melting under argon and drop casting into water-cooled copper molds. After homogenizing for five hours at 1000°C, the alloys were hot-rolled to a thickness of approximately 0.9 mm, starting at 1000°C and finishing at 650°C. Final warm rolling to approximately 0.76 mm was done at 600°C.

Tensile samples with a gage section of 0.76 by 3.18 by 12.70 mm were punched from the rolled sheet. For tensile testing at various temperatures,

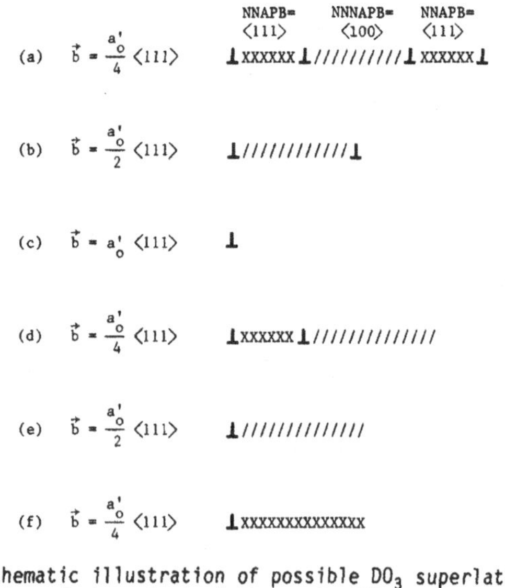

Fig. 1. Schematic illustration of possible DO_3 superlattice dislocations and their imperfect variants (after ref. 18).

samples were first given a standard heat treatment of one hour at 850°C (for
recrystallization) plus seven days at 500°C (for producing DO_3 order). All
tests were conducted in an Instron testing machine at a strain rate of 3.3 ×
$10^{-3}s^{-1}$. Temperatures of the tests varied between room temperature and
800°C. Samples to be tested at temperatures below 400°C were cleaned and
deburred by either electropolishing or vapor blasting.

RESULTS AND DISCUSSION

Figure 2 shows the room temperature tensile properties as a function of
aluminum concentration. The 0.2% yield stress was highest for the 24-26% Al
alloys (≈750 MPa) and then decreased rapidly to about 350 MPa for the 30% Al
alloy. The transition from high σ_y values at 26% Al to lower values at
27% Al coincides with the boundary between the $\alpha+DO_3$ and DO_3 phase fields at
≈500°C (the temperature used for our ordering heat treatment). Previous
studies have shown that compositions near 24% are somewhat affected by an
age-hardening reaction due to the precipitation of disordered α from the
ordered DO_3 phase [20]. Our heat treatments at 500°C were sufficient to
cause this reaction to occur, as indicated by the presence of the disordered
α phase between ordered thermal DO_3 domains in the TEM micrograph shown in
Fig. 3. Dark field images using <111> and <002> diffraction vectors show the
dark regions to be disordered and the bright regions ordered. The works of
Stoloff and Davies [4], Morgand et al. [9], and Saburi et al. [5] show that
the 24-26% Al composition is in the range where the dislocation mode changes
from the glide of single $\frac{1}{4}a_0'$<111> dislocations (where a_0' is the lattice
parameter for the DO_3 structure) associated with the α phase to glide of
$\frac{1}{4}a_0'$<111> dislocation pairs in the DO_3 superlattice. However, any possible
fault contrast associated with slip dislocations in the 24 and 25% Al samples
of this study were obscured by the scale of the ordered and disordered
regions. The higher strength of these alloys at room temperature may be
related to the presence of both disordered and ordered phases. Alloys of
26% Al and higher do not age-harden at 500°C because they lie outside of the
$\alpha+DO_3$ phase field at that temperature [20].

This figure also shows that the ductility exhibited a four-fold increase
from 1% at 24% Al to 5% at 30% Al. This increased ductility is apparently
due to the decreased yield stress with increasing aluminum content. Note

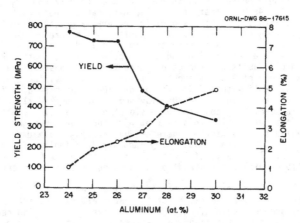

Fig. 2. Room temperature yield strength and elongation versus composition
of iron-aluminides.

324

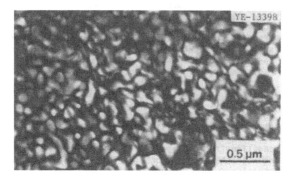

Fig. 3. A <111> dark field transmission electron micrograph of Fe-24% Al showing its two phase nature: bright regions are ordered DO$_3$ phase, dark regions are disordered α phase.

that all the alloys in this study exhibited essentially intergranular fracture [14], which was not affected by the increase in aluminum content.

Dislocation and APB types were studied by TEM as a function of composition. The different types of dislocations and APB's can be distinguished in TEM by using the appropriate diffraction conditions [21]. Imaging with a <111> superlattice diffraction vector will give rise to contrast from both the NNAPB and NNNAPB trails, while a superlattice diffraction vector of <002> or <222> will give rise to contrast only from NNAPB trails. Also, it has been shown by Crawford and Ray [19] that, as the aluminum content increases, the APB energy associated with the DO$_3$ order (NNNAPB) decreases while the APB energy associated with the B2 order (NNAPB) increases. Therefore, the number of superlattice dislocations with <111> fault vectors is expected to decrease with an increase in aluminum level, while those with <100> fault vectors should increase.

As noted above, the 24% Al alloy exhibited a two-phase structure and the dislocations were obscured by the scale of the APB's (Fig. 3). At 26% Al, which was single phase (DO$_3$), thermal APB's of both B2 and DO$_3$ ordering were observed (Fig. 4a,b). Most of the slip dislocations were coupled with a <111> fault with dislocation separation generally greater than 0.5 μm (Fig. 4c). Some of the curved <100> faults, which were originally thought to be thermally produced, were seen to terminate in dislocations (Fig. 4d), suggesting they were deformation induced. The presence of both kinds of deformation induced faults was expected, since it has been shown that the energies of the two types of APB are about equal at this composition [19].

Thermal APB's with a fault vector of <100>, which results from the DO$_3$ order, were present in the 28-30% aluminum alloys, but no <111> thermal faults were seen. Any coupling of dislocations by a <111> fault was generally not resolvable, in agreement with the expected higher APB energy of the <111> fault. Movement is therefore by one 1/2a$_0$'<111> dislocation instead of two 1/4a$_0$'<111> types. Also, many long straight faults of the <100> type were present. The energy of these APB faults is so low that they should not impede slip of dislocations and so no coupling of their dislocations was observed.

The high yield stress of the 24% Al alloy, which at first appears to result from the two-phase nature of the material, may be partially a result of the DO$_3$ ordered regions, which probably have a composition near 26% Al. At this composition, the high yield stress is caused by a low mobility of dislocations coupled loosely with APB's, the glide of which is expected to be subjected to a high lattice frictional stress [17]. Above 26% Al, the APB

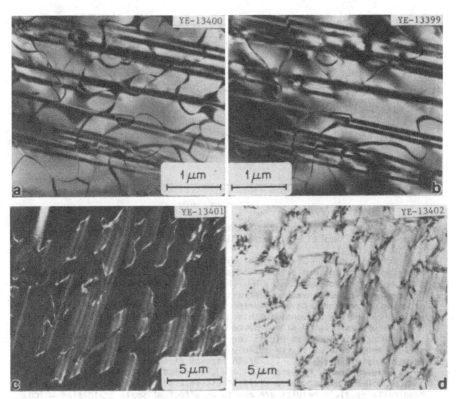

Fig. 4. Transmission electron micrographs of Fe-26% Al. (a) <111> dark field showing both DO$_3$ and B2 APB's. (b) <222> dark field showing only B2 APB's. (c) <200> dark field image showing <111> faults between dislocations. (d) Bright field showing curved APB ending in a dislocation.

energy of the <111> fault vector increases substantially with increasing aluminum content. The continuous drop in yield stress with aluminum concentration above 26% in Fig. 2 is thereby attributed to tightly coupled superlattice dislocations (paired 1/4a$_0$'<111> type), behaving like single 1/2a$_0$'<111> dislocations.

Figure 5 shows the 0.2% yield strength as a function of temperature for several of the compositions studied. For clarity only the data for the 24, 28, and 30% Al alloys are included. The curve for the 25% Al alloy was similar to the 24%; the curves of the 26 and 27% Al alloys were similar to those of 28 and 30%. Also included is the data of Mendiratta et al. [13] for an alloy of 35% Al prepared from consolidation of rapidly solidified powders. It is seen from this figure that, under the conditions of our test, the alloys containing 24 and 25% Al did not show the same yield behavior as was seen for the 26-30% Al alloys. As noted above, Inouye [20] reported that alloys of 24 and 25% Al are age-hardenable above 400°C due to the precipitation of α from the ordered DO$_3$ phase. The higher yield strength of these alloys at ambient temperatures is a consequence of that age-hardening reaction produced by our ordering heat treatment of seven days at 500°C. Inouye also showed that by slow cooling from above 550°C, with no aging, the anomalous yield stress peak could be produced at these compositions.

326

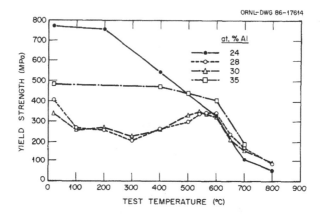

ORNL-DWG 86-17614

Fig. 5. Yield strength of iron-aluminides versus test temperature. (The data for 35% Al was taken from ref. 13).

Our data for the 26-30% Al alloys showed the expected anomalous yield behavior with a maximum between 550 and 600°C. This composition range coin-cides with the presence of the DO_3 phase field, as evidenced by the phase diagram. As noted above, similar maxima, lying at or near the critical ordering temperature, have been reported to occur in other superlattices [4,17]. Therefore we consider this phenomenon to be a general feature of Fe_3Al alloys associated with ordered DO_3 lattices.
The data included in Fig. 5 for the 35% Al alloy were taken from the work of Mendiratta et al. on rapidly solidified powders [13] and showed no anomalous yield behavior. The grain size of their samples was approximately 35 μm. In contrast, Morgand et al. [9], who studied samples with a grain size of 150-200 μm, showed a yield stress peak for this composition. Mendiratta et al.'s results are due to an effect of small grain size which hardens the alloy at low temperatures through grain boundary strengthening and weakens the alloy at high temperatures through grain boundary sliding [22].

ACKNOWLEDGMENTS

The authors would like to thank E. H. Lee for performing the tensile tests and Gwen Sims for preparation of the final manuscript.

REFERENCES

1. J. A. Horton, C. T. Liu, and C. C. Koch, in High-Temperature Alloys: Theory and Design, edited by J. O. Stiegler, (TMS-AIME, Warrendale, Pa., 1984), pp. 309-321.
2. M. J. Marcinkowski, M. E. Taylor, and F. X. Kayser, J. Mater. Sci. 10, 406 (1975).
3. M. J. Marcinkowski and N. Brown, Acta Metall. 9, 764 (1961).
4. N. S. Stoloff and R. G. Davies, Acta Metall. 12, 473 (1964).
5. T. Saburi, I. Yamauchi, and S. Nenno, J. Phys. Soc. Jpn. 32(3), 694 (1972).
6. S. K. Ehlers and M. G. Mendiratta, AFWAL-TR-82-4089, Wright-Patterson Air Force Base, Ohio, 1982.
7. H. Okamoto and P. A. Beck, Metall. Trans. 2, 569 (1971).

8. K. Oki, M. Hasaka, and T. Eguchi, Jap. J. Appl. Phys. 12(10), 1522 (1973).
9. P. Morgand, P. Mouturat, and G. Sainfort, Acta Metall. 16, 867 (1968).
10. C. T. Liu and J. O. Stiegler, Science 226, 636 (1984).
11. Y. Umakoshi, M. Yamaguchi, Y. Namba, and K. Murakami, Acta Metall. 24, 89 (1976).
12. M. J. Marcinkowski and D. S. Miller, Philos. Mag. 6, 871 (1961).
13. M. G. Mendiratta, S. K. Ehlers, and D. K. Chatterjee, in Proceedings of National Bureau of Standards Symposium on Rapid Solidification Processing, Principles and Technologies IV, (National Bureau of Standards, Washington, D. C., 1983) pp. 240-245.
14. C. G. McKamey, C. T. Liu, J. V. Cathcart, S. A. David, and E. H. Lee, ORNL/TM-10125, Oak Ridge National Laboratory, September 1986.
15. B. H. Kear and H. G. Wilsdorf, Trans. AIME 224, 382 (1962).
16. S. Takeuchi and E. Kuramoto, Acta Metall. 21, 415 (1973).
17. N. S. Stoloff and R. G. Davies, Prog. Mater. Sci. 13, 1 (1966).
18. H. J. Leamy and F. X. Kayser, Phys. Status Solidi 34, 765 (1969).
19. R. C. Crawford and I. L. F. Ray, Philos. Mag. 35(3), 549 (1977).
20. H. Inouye, in Materials Research Society Symposia Proceedings, vol. 39, High Temperature Ordered Intermetallic Alloys, ed. C. C. Koch, C. T. Liu, and N. S. Stoloff, (Materials Research Society, Pittsburgh, 1985), pp. 255-261.
21. M. J. Marcinkowski and N. Brown, J. Appl. Phys. 33(2), 537 (1962).
22. T. P. Weihs, V. Zinoviev, D. V. Viens, and E. M. Schulson, Acta Metall. (to be published).

EFFECTS OF Mo and Ti ADDITIONS ON THE HIGH TEMPERATURE COMPRESSIVE
PROPERTIES OF IRON ALUMINIDES NEAR Fe$_3$Al

R. S. DIEHM AND D. E. MIKKOLA
Department of Metallurgical Engineering
Michigan Technological University
Houghton, MI 49931

ABSTRACT

Hot compression testing has been used to examine the effects of Mo
and Ti additions on the yield strength and rate of work hardening for
cast alloys near Fe$_3$Al. A few powder processed materials have also been
studied. Significant improvements in high temperature compressive
properties on alloying have been related to increases in the DO$_3 \rightarrow$B2
transition temperature and the associated changes in the nature of the
dislocations involved in the deformation processes.

INTRODUCTION

Iron aluminides near Fe$_3$Al have good potential as low-cost
materials for use in high temperature environments causing oxidation,
sulfidation, and carburization. Because of the DO$_3 \rightarrow$B2 ordering
transition, binary alloys suffer a loss in strength at about 550°C;
however, recent work has shown that the transition temperature can be
raised by as much as 300°C by alloying with modest levels (< 10 at. pct.
total) of Mo and Ti [1]. The purpose of the current work has been to
evaluate the high temperature compressive properties of these modified
iron aluminides, with particular interest in developing alloys having
significant density-compensated strengths at or above 700°C.
    The currently accepted phase diagram for the region near Fe$_3$Al is
shown in Fig. 1 [2]. Of interest here is the second order transformation
DO$_3 \rightarrow$B2 involving the ordered structures described in Fig. 2. It is
important in consideration of deformation processes to note that the

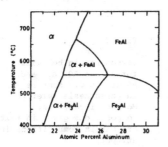

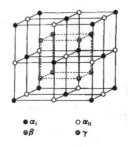

Fig. 1. Fe-Al phase diagram
         near Fe$_3$Al.

Fig. 2. DO$_3$ unit cell, Al occupies
         Y sites. For B2 order in
         Fe$_3$Al , Fe and Al occupy β
         and Y sites with equal
         probability.

Mat. Res. Soc. Symp. Proc. Vol. 81. ʲ1987 Materials Research Society

transformation to the B2 superlattice develops a simple cubic Bravais lattice from the fcc Bravais lattice of the DO_3 ordered structure. Also of importance is the development of two phase structures as a function of Al content and temperature.

It has been generally found that the yield strength of binary alloys increases with temperature to a peak value in the vicinity of the transformation temperature [3-5]. This behavior has been related to changes in the deformation mechanisms from superlattice dislocations to unit dislocations as the degree of order changes. In some instances the changes can also be attributed to the formation of a two phase structure as the temperature is increased [5]. The current study shows that the anomalous behavior also occurs in iron aluminides modified through alloying, despite the significantly higher temperatures and despite changes in the Burgers vectors of the dislocations carrying the deformation.

EXPERIMENTAL

The alloys studied were cast by Howmet Turbine Components Corp. (HTCC) using master alloys and/or virgin materials. Investment casting was used, with no mold insulation, and the castings were air cooled. Specimens (3x3x7 mm) for compression testing were cut from the castings using a wafering saw with flowing coolant. All surfaces were metallographically polished through 600 grit abrasive. Testing of the specimens was done in a resistance furnace, with the load transmitted through alumina rams. Initial strain rates were 5×10^{-5} (s^{-1}). Specimens for study with transmission electron microscopy were spark machined into 3 mm disks from thin slices that were ~ 0.25 mm thick. Thin foils were produced by electropolishing with 30 pct. HNO_3-70 pct. methanol at -35°C.

RESULTS

Typical stress-strain response behavior of the binary (A0, Fe-27.4 Al) and two ternary alloys (A2, Fe-27.2 Al-1.5 Mo and A5, Fe-25.8 Al-4.3 Ti) at 650°C is shown in Fig. 3. The critical ordering temperatures, T_c, for alloys A2 and A5 are 566°C and 662°C respectively, compared to 531°C for the binary alloy [1]. Parameters determined from curves of this type

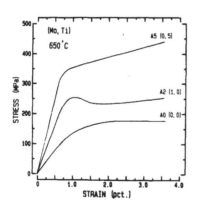

Fig. 3.

Stress-strain behavior for alloys A0, A2, and A5 at 650°C. Numbers in parenthesis refer to Mo and Ti contents. Compositions given in text are matrix compositions in at. pct.

that will be reported below include the 0.2 pct. offset yield stress and the work hardening rate taken from the slope of the linear region of the stress-strain curve beyond a few percent strain. A "yield drop" type behavior is observed in many instances where the test temperature is just above T_c, such as with A2 in Fig. 3. It appears that this is related to the premonitory local order associated with establishment of DO_3 long range order.

The variation of yield strength with test temperature for several alloys is shown in Fig. 4a. The general nature of the temperature dependence is similar for all the alloys, with evidence for an anomalous peak in strength of the type found for the binary alloy. However, the peak extends to higher temperatures with the more complex alloys. As can be noted, these increases in temperature are related to the values for T_c for each of the alloys. A similar temperature dependence occurs with the rate of work hardening (Fig. 4b), with the rate approaching zero at temperatures 75 to 125°C above T_c. Also, there is a regular increase in yield strength with solute content (combined Mo and Ti) up to $\sim$ 10 at. pct. where there is an apparent slight decrease, Fig. 5.

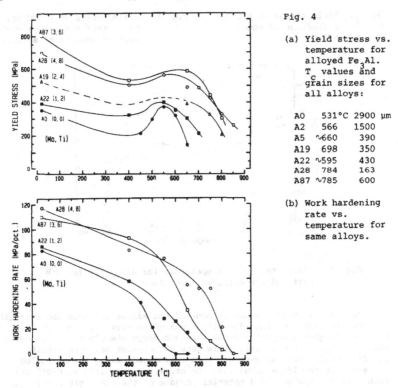

Fig. 4

(a) Yield stress vs. temperature for alloyed Fe_3Al. T_c values and grain sizes for all alloys:

A0	531°C	2900 μm
A2	566	1500
A5	$\sim$660	390
A19	698	350
A22	$\sim$595	430
A28	784	163
A87	$\sim$785	600

(b) Work hardening rate vs. temperature for same alloys.

The yield behavior also depends on Al content, as shown in Fig. 6, exhibiting a minimum in yield strength near 28 at. pct. Al for the various alloys examined. The increase in yield strength away from the minima appears asymmetric, being steeper at low Al levels. This has been interpreted in terms of the formation of two phase structures at lower levels of Al.

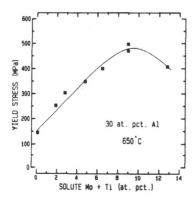

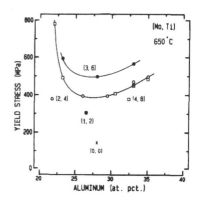

Fig. 5. Yield stress at 650°C as a function of total solute.

Fig. 6. Yield stress at 650°C vs. Al content for alloys with varying amounts of solute.

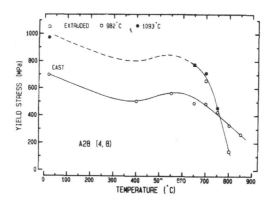

Fig. 7. Yield stress vs. temperature for alloy A28 in both cast and P/M extruded conditions.

Hot extrusion of powders of some of the alloys enhanced the strength at lower temperatures significantly compared to casting, e.g., see Fig. 7. However, at high temperatures the strength drops sharply to values below those for the cast material. This general behavior has been attributed to the fine grain size and residual work introduced during the extrusion process. It should be noted that the stress-rupture life at 650°C was much less for the extruded material because of the finer grain size (10 vs. 163 μm) [6].

Recent studies of the deformation behavior of Fe_3Al have shown that deformation at room temperature involves motion of $\frac{1}{2}$ <111> dislocations creating both near neighbor (NN) and next near neighbor (NNN) APB [7]. Increasing the temperature to 300°C causes slip to occur through motion of $\frac{1}{2}$ <111> dislocations with creation of NNN APB trails.

Examination of foils of A19 (Fe-25.2Al-1.9Mo-3.9Ti) deformed to about
five percent strain at room temperature, 650°C, and 800°C gave the
following results, Fig. 8. Room temperature deformation is with ½ <111>
dislocation pairs, which create NNNAPB. Increasing the temperature to
650°C causes deformation to occur by motion of ½ <111> unit dislocations.
Finally, at 800°C, which is in the B2 region for the alloy (T_c = 698°C),
the dislocations change to the ½ <100> type characteristic of the B2
structure, as expected. Repeating this series of observations with A82,
which has a greater amount of Al (Fe-33Al-2Mo-4Ti) gave interesting
results. Room temperature deformation occurs through the motion of <110>
pairs of dislocations. Changes to <110> Burgers vectors on alloying were
first observed by Longworth and Mikkola as discussed elsewhere [8].
Increasing the deformation temperature to 650°C causes the pairs to give
way to ½ <110> unit dislocations. As with alloy A19, ½ <100> dislocations
are again responsible for the deformation at 800°C. In order to evaluate
the effects of further stabilization of the DO_3 structure by alloying, the
behavior of A28 (Fe-22.8Al-3.9Mo-8.1Ti) at 650°C was studied. In this
case T_c (784°C) is considerably above the deformation temperature. As
expected, deformation occurs through <110> dislocation pairs.

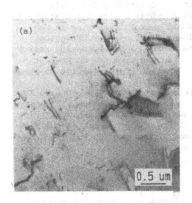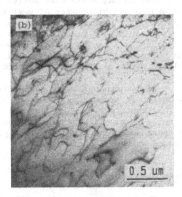

Fig. 8. Dislocations in alloy A19. (a) paired dislocations formed
at room temperature, and (b) unit dislocations at 650°C.

DISCUSSION

It is clear that the deformation behavior of modified iron aluminides
near Fe_3Al is determined largely by the stability of the DO_3 structure
relative to the B2 structure. In particular, the strength of the NN and
NNN interactions determines the type and nature of the dislocations, e.g.,
superlattice vs. unit dislocations, as well as the temperature at which
transitions in behavior take place. Detailed studies of the effects of
alloying with Mo and Ti on the stability of the DO_3 structure have been
completed recently [1]. These have shown that T_c increases with solute to
near 800°C at about 6-8 at. pct. solute. Up to that level of solute the
variation can be described approximately by the relation ΔT_c = 376 - 14.5
C_{Al} + 23 C_{Mo} + 29.7 C_{Ti}. Site occupancy studies have shown that the
substitution of Mo and Ti in the DO_3 structure occurs selectively by
replacing Fe on the β sites [1], Fig. 2. It has been suggested that
further stabilization of the DO_3 structure beyond 6-8 at. pct. solute does

334

not occur because at that point one β site is solute occupied (statistically) per DO_3 unit cell. Increasing solute beyond this point creates NN or NNN interactions between solute atoms because of the "excess" solute, and these interactions cause saturation of the stabilizing effects of the solute [1]. Solute additions can also force those Al atoms in excess of the capacity of the γ sites to occupy α sites, Fig. 2, thereby causing destabilization of the DO_3 structure.

The transformation from the B2 to the DO_3 superlattice causes the active Burgers vectors to change from <100> to <111>, or to <110> type if the DO_3 superlattice is highly stable. This latter change is not surprising because the Bravais lattice of the DO_3 structure is fcc. It may be that <110>-type dislocations can be found in many of the alloys, including certain binary alloys, at very low temperatures. At this time it is not clear whether the change from <111> to <110> slip has significant effects on mechanical properties of the DO_3 phase beyond those associated with APB energies. Certainly, the availability of different slip modes may enhance ductility and affect other responses such as creep resistance and strain rate sensitivity.

The temperature dependence of the deformation behavior is apparently related to the total active Burgers vector. At low temperatures dislocation pairs are most active leading to relatively low yield strengths and high rates of work hardening. Near T_c the dislocations move individually because of decreases in NNNAPB energy, resulting in relatively higher yield strengths and lower rates of work hardening. These changes can be correlated with the anomalous strengthening near T_c. As noted previously, the anomalous strengthening with low Al content alloys may also arise in part from the formation of the disordered phase near T_c.

Powder processing to refine the microstructure yields improved low temperature properties; however, high temperature properties, including stress-rupture life, are decreased because of the fine grain size which results in significant grain boundary sliding. Improvement of the high temperature properties of powder processed alloys through the use of second phases is currently being studied.

ACKNOWLEDGMENTS

The authors gratefully acknowledge the support of Howmet Turbine Components Corporation in funding this work as well as supplying the alloys.

REFERENCES

1. R. T. Fortnum and D. E. Mikkola, Mat. Sci. Eng., in press.
2. S. M. Allen and J. W. Cahn, Acta Metall. 23, 1017 (1975).
3. N. S. Stoloff and R. G. Davies, Acta Metall. 12, 473 (1964).
4. A. Lawley, A. E. Vidoz, and R. W. Chan, Acta Metall. 9, 287 (1961).
5. H. Inouye, in High-Temperature Ordered Intermetallic Alloys, edited by C. Koch, C. Liu, and N. Stoloff (Materials Research Society, Pittsburgh, PA, 1985) Symposia Proceedings, Vol. 39, pp. 255-261.
6. R. T. Fortnum, M.S. Thesis, Michigan Technological University, Houghton, MI 49931 (1985).
7. M. G. Mendiratta, S. K. Ehlers, D. K. Chatterjee, and H. Lipsitt, Metall. Trans. 17A (1986).
8. H. Longworth and D. E. Mikkola, J. Metals 37, 41A (1985) also to be published elsewhere.

HIGH TEMPERATURE STRENGTH OF ORDERED AND DISORDERED Ni₄Mo

H. P. Kao, C. R. Brooks and K. Vasudevan
Materials Science and Engineering Department
The University of Tennessee
Knoxville, TN 37996

ABSTRACT

The alloy Ni-20 at. % Mo is FCC and disordered at high temperatures, but below 868°C forms a superlattice of the D1a type. The disordered structure can be retained by quenching, and subsequent aging below 868°C can develop a variety of ordered domain structures. In the disordered condition, the yield stength at 25°C is typically 140-210 MPa (20,000-30,000 psi) and in this condition the alloy is quite ductile (e.g., 50% tensile elongation). In the ordered condition, the strength may double (depending on the aging treatment) but the alloy becomes very brittle (e.g., less than 1% elongation), fracturing along the former FCC, high angle boundaries. High temperature tensile mechanical properties have never been reported for the ordered condition, and in this paper we present the results of such measurements from 25 to 1000°C.

The tensile properties were measured for two ordered domain sizes: 24 nm and 3,000 nm. For the 24 nm domain size, the yield strength at 25°C was 862 MPa (125,000 psi). This decreased with increasing test temperature to about 690 MPa (100,000 psi) at 800°C, then decreased greatly to about 210 MPa (30,000 psi) at 1000°C. In the ordered condition, the alloy had a very low ductility (1% elongation or less) from 25°C to the disordering temperature of 868°C, above which it increased considerably (e.g., 30% elongation). The alloy with the 3,000 nm domain size showed similar behavior, although the strength was less. (The ordered alloy with 24 nm domain size had a yield strength-temperature curve similar to that of Waspaloy.) These results are correlated with fractography of the tensile samples.

INTRODUCTION

Some ordered alloys have sufficient strength at elevated temperatures that they are being considered for high-temperature, structural application [1]. A prominent example is Ni₃Al, which shows increasing strength up to about 800°C [2]. The alloy Ni-20 at . % Mo can be ordered by appropriate heat treated below the critical temperature of 868°C to obtain a yield strength at 25°C of about 830 MPa (120,000 psi). This contrasts to a typical value of 340 MPa (50,000 psi) for Ni₃Al at 25°C. The high strength of ordered Ni₄Mo provided the motivation for the determination of the temperature dependence of the strength, which is the subject of this paper.

The physical metallurgy and mechanical properties of Ni-Mo alloys has been reviewed recently [3]. For the Ni₄Mo composition, above the critical temperature of 868°C, the structure (α) is face-centered cubic (FCC), but is short-range ordered (SRO). This structure can be retained by cooling rapidly (e.g., water quenching). Upon aging, the α decomposes to the ordered structure (β), with an accompanying increase in strength which passes through a maximum with aging time. The alloy also becomes very brittle.

Apparently the high temperature strength of ordered Ni₄Mo has never been measured. In this paper we report tensile mechanical properties from 25 to 1000°C, and also fractographs of the tensile samples are presented.

Mat. Res. Soc. Symp. Proc. Vol. 81. ' 1987 Materials Research Society

EXPERIMENTAL PROCEDURE

The details of the alloy preparation and heat treatments have been reported [4]. The alloy was high purity, and chemical analysis showed that it contained 70.50 wt. % Ni. The tensile tests were conducted using a screw driven Instron machine equipped with a collar for induction heating in a vacuum system pumped by a turbo vacuum pump. Tensile samples conforming to ASTM E8-66 were punched from 0.51 mm thick sheet. The gage length was 12.7 mm. The samples were sealed in quartz tubes filled with argon, solution treated at 1000°C for one hour, then cooled rapidly by breaking the tubes in water. This retained the α phase. Two aging treatments were then used to develop two different domain sizes. Samples aged at 725°C for 1,000 min. had a domain size of 24 nm; samples aged for 10 days at 850°C had a domain size of 3000 nm. Previous studies [4] indicated that the domain size would change negligibly during tensile testing below 868°C. The long-range order (LRO) parameter was about 0.9 for the samples aged at 725°C. For the samples aged at 850°C, it was not measured. The equilibrium value at 850°C may not have been retained upon quenching. Below 800°C, the LRO parameter approached the equilibrium value of unity before the tensile test could begin. From about 800 to 868°C, the LRO parameter decreases smoothly to zero [5]. The thermal equilibration time prior to tensile testing above 800°C was probably sufficient to attain the equilibrium value of the LRO parameter.

For tensile testing, samples were heated from 25°C to the test temperature in about 5 min., held 10 min. to equilibrate the sample thermally, then tested. The strain rate was about 0.2 min-1, so that the testing time was from 1 to 5 min. From the load-strain diagrams, the yield strength was taken as that at 0.2 % strain, except when the strain at fracture was less than 0.2 %, in which case the yield strength was taken as the fracture stress.

RESULTS AND DISCUSSION

The data are tabulated in Table 1, and plotted in Figure 1. The strength decreases slightly from 25 to about 500°C, remains approximately constant up to about 800°C, then decreases rapidly. The ductility is quite low in the ordered condition, and only shows a significant increase above the critical temperature of 868°C.

At 25°C, the finer domain structure (24 nm) had a yield strength of 862 MPa (125,000 psi) and an elongation at fracture of about 1.4 %. The structure with a domain size of about 3000 nm had a yield strength of 630 MPa (91,400) psi and an elongation less than 0.1%. Other data [5] indicate that this lower strength is more strongly associated with a lower degree of LRO than the larger domain size. The temperature dependence of the yield stength for the large domain size structure is similar to that for the finer structure (Figure1), although the strength is about 20% lower, and the elongation is less.

The elongation upon testing samples at 1000°C, where the structure is α, is about 30%. However, if the sample is quenched from this temperature (retaining the α structure) then tested at 25°C, the elongation is about 80%. We comment on this difference in the discussion below on fractography.

Figure 2a shows the fracture surface topology of the sample tested at 25°C after aging at 725°C for 1000 min. This treatment produced a grain boundary reaction product whereby the high angle, former α boundaries (inside of which was a fine

Table 1
The 25°C and elevated temperature tensile mechanical properties of Ni$_4$Mo, for two different domain sizes prior to testing.

DOMAIN SIZE 24 nm			
Test Temp. (°C)	Yield Strength (MPa)	Tensile Strength (MPa)	Elongation (%)
25	863.3	1035.6	2.4
25	860.6	890.2	*
300	757.9	771.7	0.4
400	709.0	729.0	3.0
500	695.9	721.4	2.18
600	698.7	719.3	1.0
700	713.1	751.7	1.5
750	697.3	742.1	1.4
800	691.1	778.6	1.9
820	620.1	675.9	1.4
830	561.5	691.1	1.8
850	512.6	612.5	*
900	321.8	357.6	6.6
1000	194.3	197.1	37

*Broke at pin hole during tensile testing.
Heat treatment procedure: Step 1, Solution heat treated at 1000°C for 30 min, then water quenched. Step 2, Aged at 725°C for 1000 min, then water quenched (domain size about 24 nm).

DOMAIN SIZE 3000 nm			
Test Temp. (°C)	Yield Strength (MPa)	Tensile Strength (MPa)	Elongation (%)
25	629.7	629.7	Nil
600	523.6	557.4	Nil
750	486.4	511.9	Nil
800	472.0	483.7	Nil
850	441.0	472.7	*
900	364.5	372.1	Nil
1000	162.6	173.6	26.1

*Broke at pin hole.
Heat treatment procedure: Step 1, Solution heat treated at 1000°C for 30 min then water quenched. Step 2, Aged at 850°C for 14,400 min (10 d) to get large domain size (about 3 μm), then water quenched.

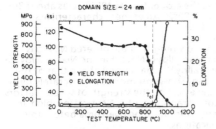

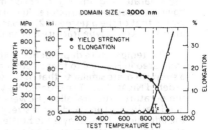

Figure 1. Yield strength and elongation as a function of test temperature for ordered Ni$_4$Mo, for two different domain sizes prior to testing. T$_c$ denotes the critical temperature of 868°C.

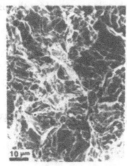

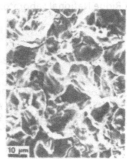

(a) Aged 725°C, 1000 min   (b) Aged 850°C, 10 d

Figure 2. Fractographs of the tensile samples tested at 25°C for two different domain sizes. (a) 24 nm (b) 3000 nm.

338

structure of β domains) have migrated, with a coarser domain structure formed behind them. This structure is brittle, with fracture occurring along these high angle, migrated boundaries [6]. This causes the fracture surface topology to be finer compared to that for shorter aging times, where fracture occurs along the high angle, former α boundaries. Figure 2b shows the topology of the tensile sample tested at 25°C after aging for 10 days at 850°C. The fracture topology is coarser because the grain boundary reaction product has encompassed completely the former grains, forming a coarse, patchy, β domain structure. Fracture is still occuring along the high angle boundaries. [5].

The fracture surfaces of samples aged at 725°C and tested at 700 and 800°C are shown in Figure 3. The topology of the samples tested below 800°C was similar to that of samples tested at 25°C (Figure 1a), and is typified by the example in Figure 3a. However, testing at 800°C gave the topology shown in Figure 3b. The fracture appears more classically intergranular, although the surfaces have a fine roughness to them. It appears that there is a fundamental change in the fracture mechanism, probably involving diffusional processes.

When samples of the alloy are quenched from above the critical temperature of 868°C, the α phase is retained . The ductility upon testing at 25°C is high, and the fracture mechanism is by void coalesence, as shown by the dimpled topology in Figure 4a. Samples tested at temperatures above 868°C, in the α region, also showed high ductility. However, the fracture topology was not dimpled, but had the morphology shown in Figure 4b-c. The fracture surface is rough, but not characteristic of fracture by void coalesence or cleavage. Numerous separations have formed along the gage length (Figure 5). At high magnification, the surface of the separated regions (Figure 4e) is similar to that of the fracture surface (Figure 4c), having a nodule-like structure. Recall that the elongation at 1000°C was about 30%, where that of the α tested at 25°C was about 80%. All of these characteristics indicate a creep-type fracture mechanism [6].

The temperature dependence of the yield strength of the ordered Ni_4Mo is compared in Figure 6 to that of other common alloys. Note that the Ni_4Mo alloy has a strength quite similar to that of the superalloy Waspaloy, and above that of the ordered Ni_3Al alloy. The ordered Ni_4Mo alloy shows promise as a moderately high temperature structural alloy. However, the effect on the strength of several factors need clarifying, such as domain size, prior α grains size, etc. A critical problem in the adoption of this alloy is the brittleness, which must be overcome before this alloy can be seriously considered for applications.

ACKNOWLEDGEMENTS

Dr. C. T. Liu of Oak Ridge National Laboratory kindly arranged for the use of some equipment.

REFERENCES

1. N. S. Stoloff, Int. Metals Rev. 29, 123 (1984).
2. See D. P. Pope and S. S. Ezz, Int. Metals Rev. 29, 136 (1984).
3. C. R. Brooks, J. E. Spruiell and E. E. Stansbury, Int. Metals Rev. 29, 210 (1984).
4. H. P. Kao and C. R. Brooks, Scripta Met., 20, 1561 (1986).
5. H. P. Kao, Ph.D. dissertion, The University of Tennessee (1986).
6. Metals Handbook, 8th edition, Vol. 9 Fractography and Atlas of Fractographs, American Society for Metals, Metals Park, Ohio (1974).

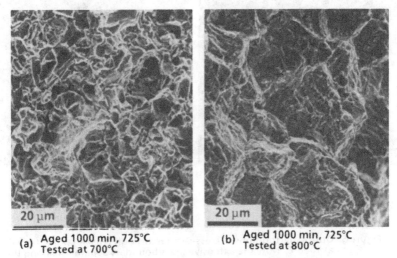

(a) Aged 1000 min, 725°C
    Tested at 700°C

(b) Aged 1000 min, 725°C
    Tested at 800°C

Figure 3. Fractographs of the tensile samples tested at 700 and 800°C.
The domain size was 24 nm prior to testing.

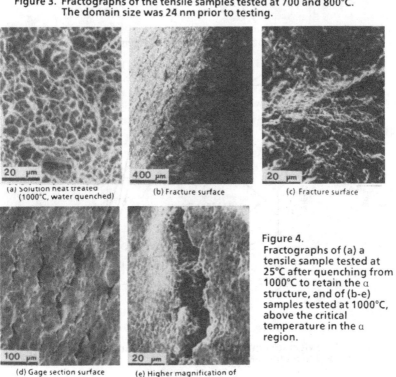

(a) Solution heat treated
    (1000°C, water quenched)

(b) Fracture surface

(c) Fracture surface

(d) Gage section surface

(e) Higher magnification of
    gage section surface

Figure 4.
Fractographs of (a) a
tensile sample tested at
25°C after quenching from
1000°C to retain the α
structure, and of (b-e)
samples tested at 1000°C,
above the critical
temperature in the α
region.

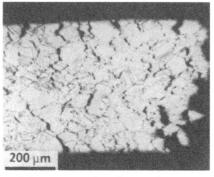 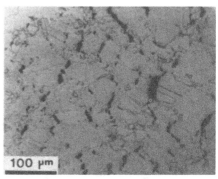

(a) Intergranular microcracks formed at grain boundaries

(b) High density of discontinuous microcracks or voids along grain boundaries. This region was 6 mm from fracture surface

Figure 5. Optical micrographs of cross-section through tensile sample tested at 1000°C. Note the extensive grain boundary separation along the gage length (compare to Figure 4 d).

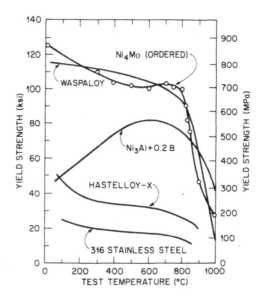

Figure 6. The yield strength as a function of temperature of ordered Ni₄Mo and four other alloys.

PART IV

Alloy Design and Microstructural Control

GRAIN BOUNDARY CHEMISTRY AND DUCTILITY IN NI-BASE Ll$_2$ INTERMETALLIC COMPOUNDS

A. I. Taub and C. L. Briant
General Electric Corporate Research and Development, PO Box 8, Schenectady, NY 12301 USA

ABSTRACT

The available experimental data for the Ni$_3$X, Ll$_2$ intermetallics indicate a strong effect of chemistry on the tendency for brittle intergranular fracture. Combining the data for the binary compounds with the effects of macroalloying with X substituents and microalloying with boron, the resistance to intergranular fracture varies as Fe ~ Mn > Al ~ Ga > Si > Ge. Two models have been proposed to explain these chemistry effects. The basis for both models is that electronic charge localization is responsible for the brittle behavior of the ordered intermetallics. The models differ in the measure of charge redistribution; one uses valency differences whereas the other uses electronegativity differences.

INTRODUCTION

Pure metals generally exhibit brittle intergranular fracture only when impurities such as S, P, Sn and Sb segregate to their grain boundaries and weaken them [1]. In contrast, many intermetallic compounds exhibit brittle intergranular fracture at ambient temperature even when prepared from high-purity metals [2,3]. For some classes of intermetallic compounds low ductilities are expected because of the absence of at least five independent, operable slip systems. However, the ordered FCC (Ll$_2$) intermetallic compounds slip on the {111}<110> system and therefore do not suffer from this restriction. For example, at room temperature single crystals of Ni$_3$Al have been shown to exhibit high ductilities in all crystal orientations [4] while the high-purity, polycrystalline alloy fractures intergranularly with negligible plastic strain to failure. The poor ductility is due to premature fracture along the grain boundaries which are not strong enough to withstand the stresses required for deformation of the matrix grains.

In this paper, we will review our current understanding of the chemical aspects controlling the tendency for grain boundary fracture of intermetallic compounds. The approach will be to first summarize the available experimental results and then present two of the models that have been proposed to explain the data. Since the majority of the work in this area has been conducted on Ni-base compounds, we will restrict the discussion to this alloy system.

EXPERIMENTAL DATA

There are six binary Ni$_3$X alloys that form the Ll$_2$ crystal structure (X=Fe,Mn,Al,Ga,Si,Ge). All six compounds are ductile as single crystals but only Ni$_3$Fe and Ni$_3$Mn are ductile when prepared in polycrystalline form. The remaining four compounds exhibit brittle intergranular fracture when deformed at room temperature.

The strong effects of chemistry on the ductility of these polycrystalline binary intermetallics has prompted alloying studies by several investigators. We will only examine the results of those studies which employed combinations of the six elements that make up the binary alloys (ie-Fe, Mn, Al, Ga, Si, Ge). In addition, the effects of doping with grain boundary active elements will also be summarized.

A. Effects of Alloying With Aluminum Substituents

Takasugi, Izumi and Masahashi [5] have examined the effect of many substitutional elements, including those of interest here, on the ductility of
Ni₃Al. They found that 5 atomic percent substitutions of Ga, Ge and Si for
Al resulted in brittle intergranular fracture while substitutions of Fe and
Mn resulted in substantial plasticity and predominantly transgranular fracture. Alloying with greater amounts of Fe and Mn (up to 15 atomic percent)
resulted in increased ductilities. However, the substitutions were done
entirely for aluminum while other studies have indicated that Fe and Mn substitute equally for Ni and Al [6]. Therefore, although the authors report
the structures to be single phase $L1_2$, for aluminum substitutions above
about 10 atomic percent the presence of a nickel-rich solid solution phase
is suspected.
Improvements in ductility have also been reported by Inoue, et. al.
[7], for melt quenched Ni₃Al alloyed with Fe and Mn. In that study the
improved ductility was associated with the refined antiphase domains
obtained by rapid solidification and the ductility was dramatically reduced
when the alloys were annealed.

B. Effects of Boron-Doping

The dramatic increase in ductility of Ni₃Al obtained by boron doping
was first reported by Aoki and Izumi [8] and later confirmed by several
investigators [9-12]. Plastic strains to failure greater than 50% have been
reported for the boron-doped aluminide compared to almost zero ductility for
the unmodified alloy processed under identical conditions. These improvements in ductility are associated with a transition from pure intergranular fracture to a completely transgranular fracture mode.
The degree to which boron improves the deformation behavior of the
nickel aluminide is controlled by both the boron and aluminum concentrations
[13]. As shown in figure 1, only nickel-rich compounds respond to the boron
doping, with the best results being obtained for Ni:Al ratios above 76:24.
For these Ni-rich deviations from stoichiometry, the improvement in ductility is observed for boron doping levels as low as 0.05 atomic percent. The
ductility drop observed for boron concentrations greater than about 2 atom
percent has been associated with the appearance of $M_{23}B_6$ borides at the
grain boundaries [10]. However, the correlation of boride precipitation and
ductility degradation has not been resolved [14].
The effects of boron doping on the other nickel-base intermetallics
that normally exhibit brittle intergranular fracture have also been reported
[15]. Ni₃Ga behaves in a similar manner to Ni₃Al with respect to boron
doping. The normally brittle binary alloy is made ductile with boron additions, but not for Ni-poor deviations from stoichiometry. Ni₃Si was also
made ductile by boron doping, but the fracture morphology was only partly
transgranular (figure 2b) compared to fully transgranular fracture for the
aluminum and gallium compounds (figure 2a). Ni₃Ge showed no improvement in
either ductility or fracture mode with boron doping (figure 2c).
Figure 3 shows the variation in tensile ductility of boron-doped Ni₃Al
as either Si or Ge is substituted for Al. As expected from the trends in
the binary alloys, both elemental substitutions lead to a degradation in
ductility. The reduced ductilities were accompanied by a transition to
intergranular fracture.
Studies of the boron ductilization effect have shown that the boron
segregates strongly to the grain boundaries [9,16,17]. Figure 4 shows that
for nickel aluminide, the boron concentration at the grain boundaries varies
almost linearly with the doping level. Converting the peak height ratios to
atomic percent according to the method described in the Handbook of Auger
Electron Spectroscopy provides a grain boundary concentration enhancement
of about 10 times the bulk concentration. The boron segregation in the gallium and silicon intermetallics appears to follow the trend observed for the
nickel aluminide. The enhanced segregation for the germanium compound may
be related to the very low solubility of boron (<0.1 atomic percent) [18].

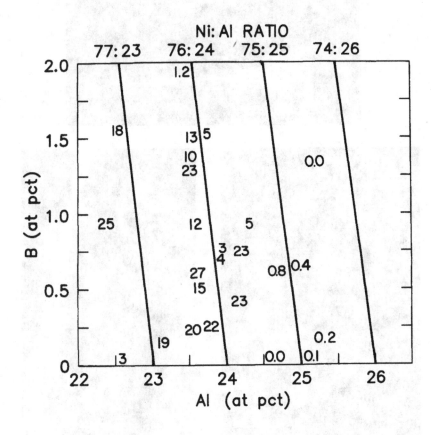

Figure 1- Plastic strain to fracture in a tensile test (see numbers) of melt spun Ni$_3$Al + B as a function of Al and B concentration. The solid lines indicate constant Ni:Al ratio. The ribbons were annealed at 1100C for 2 hours prior to tensile testing [13].

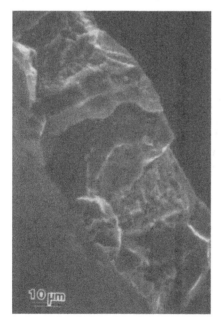
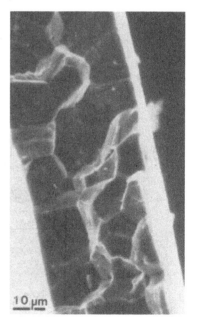

Figure 2- Fracture surfaces of boron-doped (a) $Ni_{bal}Ga_{24.8}B_{0.6}$, (b) $Ni_{bal}Si_{23.6}B_{0.5}$, and (c) $Ni_{bal}Ge_{21.9}B_{0.2}$.

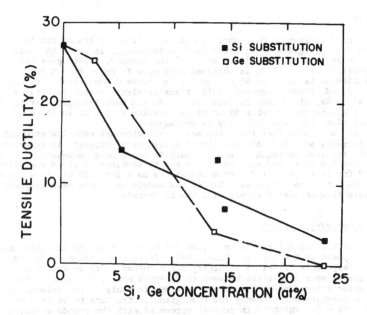

Figure 3- Variation of plastic strain to failure of boron-doped Ni₃Al with Si and Ge substitution for Al [14].

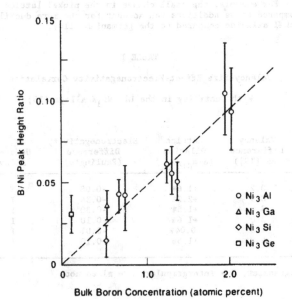

Bulk Boron Concentration (atomic percent)

Figure 4- B/Ni peak height ratios measured on the grain boundaries of the indicated Ni₃X intermetallics as a function of bulk boron concentration The boron 179 eV peak was normalized by the Ni 848 eV peak [18].

MODELS

The experimental data show strong and consistent trends for the tendency for intergranular fracture in Ni_3X intermetallics as X is varied between Fe, Mn, Al, Ga, Si and Ge. For the undoped, binary polycrystalline intermetallics ductility is obtained only when X= Fe or Mn. Moreover, substitution of Fe or Mn in Ni_3Al results in improved ductility. For the boron-doped, binary compounds fully transgranular fracture is obtained when X = Al or Ga, mixed mode failure for X = Si and pure intergranular for X = Ge. Correspondingly, when Si or Ge is substituted for Al in the boron-doped nickel aluminide, the ductility is decreased.

Let us assume that the boron provides additional cohesive strength to the boundary and that this additional increment of strength is superimposed on the inherent strength of the boundary in the binary compound. Then the data for the doped and undoped alloys can be combined to yield a consistent trend for intergranular fracture resistance as a function of the X species: Fe ~ Mn > Al ~ Ga > Si > Ge. Two of the models that have been proposed to explain this chemistry effect will now be described.

A. Valency/Size Effect

The valency difference criterion for brittle fracture in intermetallics was proposed by Takasugi and Izumi [19]. In an examination of the tendency for intergranular fracture in binary A_3B compounds they showed that the larger the valency difference between the A and B atoms, the greater the tendency for intergranular fracture. As shown in Table 1, the valency criterion predicts the tendency for transgranular fracture to be Fe > Mn >> Al ~ Ga > Si ~ Ge, which is in general agreement with the trends observed experimentally. To further differentiate between compounds of equal valency difference, a size effect criterion is used. As the size difference between the A and B atoms increases, the tendency for intergranular fracture increases. For example, the small change in the nickel lattice parameter with Si compared to Ge additions can account for the more ductile behavior of the nickel silicide compared to the germanium alloy.

TABLE 1

Valency-Size Effect-Electronegativity Correlation

With Ductility in the Ll_2 Ni_3X Alloys [15]

X Species	Valency Difference (Δz [19])	Lattice Dilation ($a\text{-}a_{Ni}/a_{Ni}$)	Electronegativity Difference (Pauling's)	Undoped Alloy	Boron-Doped Alloy
Fe	0.2	+1.0%	-0.08	T	-
Mn	0.9	+2.2%	-0.36	T	-
Al	3.0	+1.5%	-0.30	I	T
Ga	3.0	+1.6%	-0.10	I	.T
Si	4.0	-0.04%	-0.01	I	M
Ge	4.0	+1.5%	+0.10	I	I

T = transgranular, I = intergranular, M = mixed mode

B. Electronegativity Model

In pure metals, it has been shown that the embrittling potency of grain boundary segregants is related to the electronegativity of the segregating species [20]. The more electronegative the segregant atom, the greater is the tendency for it to pull charge out of the metal-metal bonds at the boundary, thereby reducing the cohesive strength and promoting intergranular fracture. The authors have proposed that the grain boundaries of the A_3B compounds can be examined in a similar manner. Geometric modeling of the A_3B compounds has shown that the A-A bonds dominate at the grain boundary [19] so the cohesive strength may be determined by the extent to which the B atom draws electronic charge out of the A-A bonds.

A standard measure of charge transfer between atoms is the electronegativity scale devised by Pauling. As shown in Table 1, this scale successfully correlates the ductility of the Ni_3X intermetallics where X is from group III (Al,Ga) or group IV (Si,Ge). In particular, Ge is the most electronegative X element and therefore the most likely to draw charge out of the Ni-Ni bonds. Correspondingly, Ni_3Ge shows the greatest tendency for intergranular fracture. However, this scaling fails to account for the ductile behavior of Ni_3Fe. This problem may result from the difficulties in comparing electronegativities of transition metals. Full-scale quantum mechanical calculations have shown that Pauling's electronegativity scale, which was based primarily on molecular data, does not work well for transition metal alloys [21,22].

Eberhart and Vvedensky [23] have recently suggested that a more appropriate charge transfer scale for intermetallic compounds is the s-orbital electronegativities obtained from calculations of ground-state atomic structures [24]. As shown in figure 5, this scale predicts the ductility variation to be $Ni_3Fe \sim Ni_3Mn \gg Ni_3Al > Ni_3Ga \gg Ni_3Si \sim Ni_3Ge$. The agreement with experiment is good except for the nearly equal ranking of Ge and Si, which Pauling's scale correctly ranks as Si > Ge. A third ranking based on the Mulliken electronegativity scale is also seen to incorrectly rank Ge > Si. In summary, the electronegativity difference provides a good first order prediction of the tendency for intergranular fracture, but the discrepancies in the rankings obtained with the different scales indicate that more detailed considerations of electronic bonding are required.

MODEL COMPARISONS

Both the valency/size effect and electronegativity models are based on the concept that localization of electronic charge is responsible for the brittle behavior of intermetallics. In the case of valency difference, the localization is associated with the expected degree of covalency of the bonds. The electronegativity arguments are more closely related to charge transfer away from the bonds of the majority atoms.

The model of choice is the one that provides a scaling which most satisfactorily explains the experimental data. For the binary alloys, we have seen that both models make the same ductility predictions. Figure 6 shows the results of a study in which the tendency for intergranular fracture was studied for boron-doped Ni_3X alloys where X was combinations of Al, Ga, Si and Ge. A ductility ranking based on the average valence of the X atoms produces two curves depending on whether the alloys contain Si or Ge. As was the case for the binary alloys, the size effect can be used to account for this difference. On the other hand, the average Pauling electronegativity of the X atoms provides a single parameter that successfully correlates the data. Studies are in progress with other substitutional elements to further test the two models.

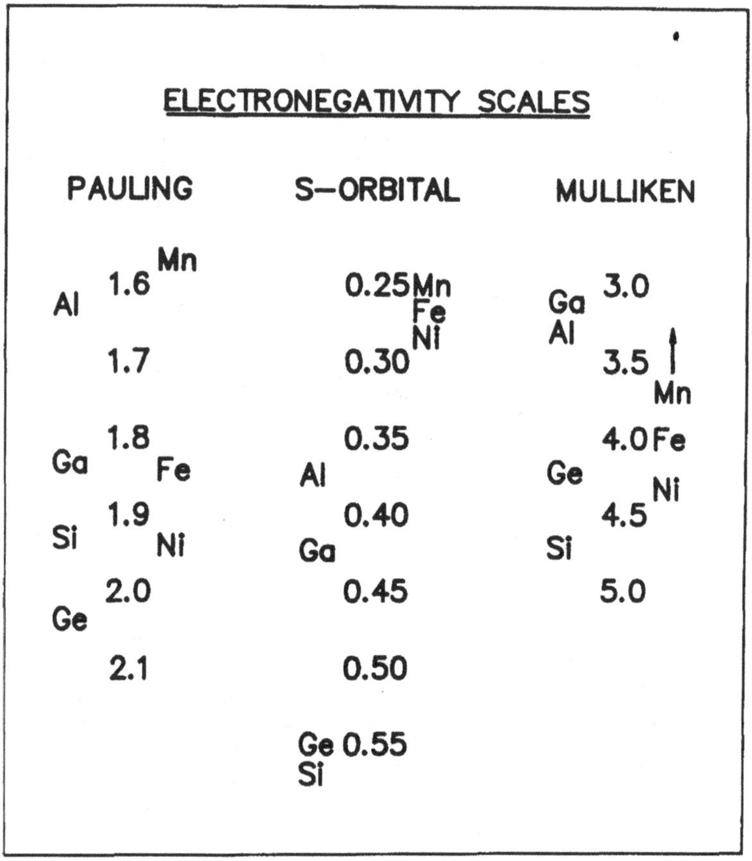

ELECTRONEGATIVITY SCALES

PAULING	S—ORBITAL	MULLIKEN
Al 1.6 Mn	0.25 Mn Fe Ni	Ga 3.0 Al
1.7	0.30	3.5 ↑ Mn
Ga 1.8 Fe	Al 0.35	Ge 4.0 Fe Ni
Si 1.9 Ni	Ga 0.40	Si 4.5
Ge 2.0	0.45	5.0
2.1	0.50	
	Ge 0.55 Si	

Figure 5- Electronegativities of the elements of interest as measured on the indicated scales. The s-orbital values are expressed in Hartrees after Eberhart and Vvedensky [23].

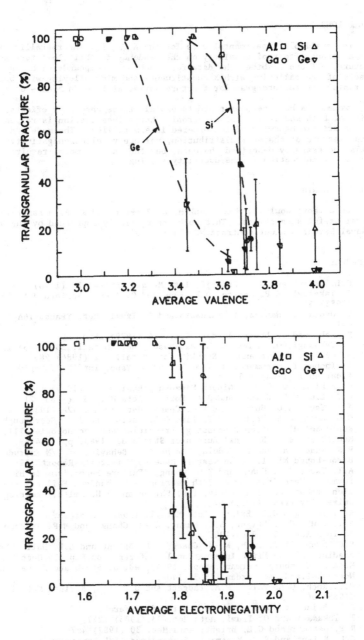

Figure 6- The percentage of the fracture surface that is transgranular for a series of boron-doped Ni$_3$X intermetallics where X is combinations of Al, Ga, Si and Ge as indicated. The data are plotted as a function of the weighted average (a) valency and (b) Pauling electronegativity of the X atoms [14].

CONCLUSIONS

The available experimental data for the Ni_3X, Ll_2 intermetallics indicate a strong effect of chemistry on the tendency for brittle intergranular fracture. If one combines the data for the binary compounds with the effects of macroalloying with X substituents and microalloying with boron, the resistance to intergranular fracture varies as Fe ~ Mn > Al ~ Ga > Si > Ge.

Two models have been proposed to explain these chemistry effects. The basis for both models is that electronic charge localization is responsible for the brittle behavior of the ordered intermetallics. The models differ in the measure of charge redistribution, valence vs electronegativity. Both models successfully described the data, but the valence model requires a size effect correction to the ductility ranking.

ACKNOWLEDGMENTS

The authors would like to thank M. R. Jackson and R. P. Messmer for several useful suggestions. This work was partially supported by the Office of Naval Research under contract N00014-86-C-0353.

REFERENCES

1. C.L. Briant and S.R. Banerji, Int. Metals Review, 23 (1978) 154.
2. T. Takasugi, E.P. George, D.P. Pope and O. Izumi, Scripta Met. 19 (1985) 551.
3. T. Ogura, S. Hanada, T. Masumoto and O. Izumi, Met. Trans. 16A (1985) 441.
4. K. Aoki and O. Izumi, J. Mat. Sci., 14 (1979) 1800.
5. T. Takasugi, O. Izumi and N. Masahashi, Acta Metall. 33 (1985) 1259.
6. S. Ochiai, Y. Oya and T. Suzuki, Acta Metall. 32 (1984) 289.
7. A. Inoue, T. Masumoto, H. Tomioka and N. Yano, Int'l. J. Rapid Solidification 1 (1984) 115.
8. K. Aoki and O. Izumi, Nippon Kinzoku Gakkaishi 43 (1979) 1190.
9. C.T. Liu, C.L. White, and J.A. Horton, Acta Met. 33 (1985) 213.
10. A.I. Taub, S.C. Huang, and K.M. Chang, Met. Trans. 42A (1984) 399.
11. C.C. Koch, J.A. Horton, C.T. Liu, O.B. Cavin, and J.O. Scarbrough, in Rapid Solidification Processing, Principles and Technologies III, R. Mehrabian, ed., National Bureau of Standards, 1983, pp. 264-69.
12. P.S. Khadkikar and K. Vedula, "Mechanical Behavior of P/M Extruded Boron-Doped Ni_3Al," Case Western Reserve University Report (1985).
13. A.I. Taub, S.C. Huang and K.M. Chang, "Failure Mechanisms in High Performance Materials, Proc. 39th Meeting of Mechanical Failures Prevention Group", ed. J.G. Early, T.R. Shives and J.H. Smith, Cambridge University Press (1985) 57.
14. A.I. Taub and C.L. Briant, Acta Metall. (1987) in press.
15. A.I. Taub, C.L. Briant, S.C. Huang, K.-M. Chang, and M.R. Jackson, Scripta Met. 20 (1986) 129.
16. S.C. Huang, A.I. Taub, K.-M. Chang, C.L. Briant and E.L. Hall, "Proceedings of the Fifth International Conference on Rapidly Quenched Metals, Wurzburg, Germany, Sept. 1984", ed. S. Steeb and H. Warlimont, North Holland, Amsterdam (1985) 1407.
17. C.C. Koch, C.L. White, R.A. Padgett and C.T. Liu, Scripta Met. 19 (1985) 963.
18. C.L. Briant and A.I. Taub, unpublished research.
19. T. Takasagi and O. Izumi, Acta Met. 33 (1985) 1247.
20. R.P. Messmer and C.L. Briant, Acta Met. 30 (1982) 457.
21. C.L. Briant and R.P. Messmer, "Proceedings AIME Symposium on Embrittlement by Liquid and Solid Metals," St. Louis, MO, Oct. 24-28, 1982, ed. M.H. Kamdar, TMS-AIME, Warrendale, PA (1983) 79.
22. R.E. Watson and L.H. Bennett, Phys. Rev. B, 18 (1978) 6439.

23. M.E. Eberhart and D.D. Vvedensky, Phil. Mag., in press.
24. E. Clement and C. Roetti, Atomic Data and Nuclear Data Tables 14
    (1974) 177.

DUCTILITY AND FRACTURE BEHAVIOR OF
POLYCRYSTALLINE Ni₃Al ALLOYS*

C. T. LIU
Metals and Ceramics Division, Oak Ridge National Laboratory, Oak Ridge,
TN 37831-6117

ABSTRACT

This paper provides a comprehensive review of the recent work on
tensile ductility and fracture behavior of Ni₃Al alloys tested at ambient
and elevated temperatures. Polycrystalline Ni₃Al is intrinsically
brittle along grain boundaries, and the brittleness has been attributed to
the large difference in valency, electronegativity, and atom size between
nickel and aluminum atoms. Alloying with B, Mn, Fe, and Be significantly
increases the ductility and reduces the propensity for intergranular fracture in Ni₃Al alloys. Boron is found to be most effective in improving
room-temperature ductility of Ni₃Al with <24.5 at. % Al.
The tensile ductility of Ni₃Al alloys depends strongly on test
environments at elevated temperatures, with much lower ductilities
observed in air than in vacuum. The loss in ductility is accompanied by
a change in fracture mode from transgranular to intergranular. This
embrittlement is due to a dynamic effect involving simultaneously high
localized stress, elevated temperature, and gaseous oxygen. The embrittlement can be alleviated by control of grain shape or alloying with chromium
additions. All the results are discussed in terms of localized stress
concentration and grain-boundary cohesive strength.

INTRODUCTION

Nickel aluminide, Ni₃Al, is an intermetallic compound that forms an
L1₂-type ordered crystal structure in the solid state. Its unit cell has
an ordered fcc lattice structure with nickel atoms occupying face-centered
sites and aluminum atoms occupying corner sites. The aluminide is of
interest for elevated-temperature structural use because its yield
stress shows an increase rather than a decrease with increasing temperature [1–3], reaching a maximum around 600–700°C. It is the most
important strengthening constituent (γ´ phase) of commercial nickel-base
superalloys, and it is responsible for their high-temperature strength and
creep resistance. In addition, Ni₃Al alloys are resistant to air oxidation because of their ability to form adherent oxide surface films that
protect the base metal from excessive attack [4,5].
Single crystals of Ni₃Al are highly ductile, whereas polycrystals are
very brittle at ambient temperatures [6–10]. The low ductility and brittle
intergranular fracture was the major obstacle that prohibited the development of polycrystalline Ni₃Al alloys as engineering materials.
Significant progress has been made recently in improving the ductility of
Ni₃Al polycrystals, starting with the pioneer work of Aoki and Izumi [11].
They found that microalloying with boron imparts substantial ductility
to Ni₃Al. The success of their effort has stimulated much interest in
study of grain-boundary fracture and ductility enhancement through control
of metallurgical variables in Ni₃Al alloys [5,12–28]. By careful control

*Research sponsored by the Division of Materials Sciences, U.S.
Department of Energy, under contract DE-AC05-84OR21400 with Martin Marietta
Energy Systems, Inc.

rapidly with increasing temperature, presumably due to approaching the $DO_3 \rightarrow B2$ transformation temperature of $\sim 540°C$. In contrast, the Fe-35 and -40Al alloys, which consist of the single phase B2 structure at all temperatures, were reasonably strong up to 600°C [yield strength ~ 414 MPa (60 ksi)].

Thin-foil transmission electron microscopy (TEM) of the deformed alloys indicated that in the Fe-25Al alloy, the room-temperature plastic flow is governed by the movement and cross-slip of ordinary dislocations creating anti-phase boundary (APB) bands. In contrast, with increasing Al content, the plastic flow occurs by the movement of superlattice dislocations. Alloys possessing some ductility at room-temperature failed in a transgranular cleavage mode, while the Fe-50Al alloy, with no ductility, exhibited intergranular fracture.

Traditionally, iron-aluminides are thought to be very brittle at room-temperature, with almost no ductility and an associated intergranular failure mode. The lack of ductility has been interpreted as a fundamental property of the weak "disordered" grain-boundaries in the ordered polycrystalline iron-aluminides [6]. However, the present results comprising of reasonable ductility values, extensive motion of ordinary and superlattice dislocations with Burgers vectors having <111> directions, and transgranular cleavage fracture modes provide a self-consistent tensile behavior and firmly establish that iron-aluminides are not inherently brittle at room temperature.

In addition to a comprehensive determination of tensile behavior of binary iron-aluminides, some work was done on the basic dislocation mechanisms. In particular, slip directions in the B2 Fe-Al alloys from Fe-35 to -50Al were determined by standard TEM $\vec{g}.\vec{b}$ contrast analysis in specimens deformed from room temperature to 700°C [7]. At low temperature, the predominant slip direction was found to be <111>, while at high temperature it was <100>. The transition temperature for the change in slip direction was composition dependent, it was $\approx 350°C$ for 50Al and increased with decreasing Al content. Attempts were made to predict theoretically the preferred slip system in the B2 iron aluminides [8]. Anisotropic elasticity theory was used to calculate the energies and mobilities of dislocations belonging to the slip systems {110} <111> and {110} <100> in the Fe-35, -40, and -50Al alloys. Based upon energy values alone, it appeared that the preferred room-temperature slip system should be {110} <100>, contrary to the experimentally observed system {110} <111>. However, when the mobility parameter, as modified by the consideration of atomic radii ratio, R_{Fe}/R_{Al}

of the aluminum concentration, boron content, and thermomechanical treatment, tensile elongations greater than 50% were achieved in Ni-24 at. % Al [5,12,15].

In contrast to room-temperature properties, only limited data on elevated-temperature properties are available for Ni_3Al alloys, and the reported results are quite conflicting. Liu et al. [29,30] tested in tension conventionally cast, cold fabricated, and recrystallized boron-doped Ni_3Al at temperatures up to 1200°C in vacuum (10^{-3} Pa). They reported tensile ductilities above 40% for temperatures up to 600°C with a ductility minimum of 5 to 10% in the temperature range between 800 and 1000°C. Taub and coworkers [13,27,28] tested melt-spun ribbons of boron-doped Ni_3Al at elevated temperatures in an argon atmosphere. Unlike Liu et al., they observed a severe loss in tensile ductility above 400°C. Furthermore, Hanada et al. [19] and Vedula et al. [31] observed a steady decrease in ductility with test temperature, and the ductility dropped to zero at 500 to 600°C when Ni_3Al alloys with or without boron additions were tested in air. The study of environmental effects indicates that tensile ductility of Ni_3Al alloys depends strongly on test environment, with much lower ductility observed in air than in vacuum [32,33].

This paper provides a comprehensive review of the ductility and fracture behavior of Ni_3Al alloys tested at room and elevated temperatures. Emphasis is placed on understanding metallurgical variables affecting the grain-boundary properties of Ni_3Al alloys and on improving the ductilities of the alloys by using physical metallurgy principles.

GRAIN-BOUNDARY FRACTURE AND ALLOYING EFFECTS

The brittleness of polycrystalline Ni_3Al originates at grain boundaries. In most metals and alloys, intergranular brittleness is associated with strong segregation of harmful impurities (e.g. S, P) to grain boundaries [34], causing embrittlement. Studies of fracture in high-purity polycrystalline Ni_3Al alloys using Auger electron spectroscopy, however, revealed brittle intergranular fracture without appreciable segregation of impurities at grain boundaries [5,15,20,21]. The grain boundary is, therefore, considered to be intrinsically weak in Ni_3Al. It is to be noted that the grain boundaries in Ni_3Al can be further embrittled by segregation of impurities. Sulfur was identified as a trace element that segregates to and embrittles grain boundaries in impure Ni_3Al [35].

In order to understand the nature of intrinsic grain-boundary weakness, Takasugi and Izumi [17,18], Taub et al. [36], and Farkas et al. [37] have recently conducted a systematic study of grain-boundary fracture in $L1_2$-ordered alloys Ni_3X, where X = Fe, Mn, Al, Ga, Si, or Ge. They attempted to correlate the tendency for grain boundary fracture with the valency difference [18] between Ni and X atoms, electronegativity difference [36], and size ratio [37] (defined as the ratio of the nearest-neighbor atomic distances between Ni-X and Ni-Ni atoms). As indicated in Table 1, the grain boundary cohesion generally decreases with increasing differences in valency, electronegativity, and atom size between Ni and X atoms. It appears that the correlation can be best established by a combined consideration of both electronic and size differences. An interesting way to verify this correlation was recently provided by Taub et al. [36], who microalloyed Ni_3X alloys with boron additions. The grain boundaries in Ni_3Al, Ni_3Ga, and Ni_3Si, which are marginally brittle as predicted from the correlation, can be ductilized by microalloying with boron, whereas the boundaries in Ni_3Ge, which are severely brittle, cannot be made ductile by boron additions. These observations are apparently in line with the predictions from the correlation.

The studies of grain-boundary cohesion have led to many metallurgical solutions to the brittle intergranular fracture problem in Ni_3Al, as shown

Table 1. Correlation of valency difference, electronegativity
difference, and size ratio with fracture
behavior in ordered $L1_2$ alloys Ni_3X

Alloy	Valency[a] difference (ΔZ)	Electronegativity[b] difference (Pauling's)	Size[c] ratio $\left(\dfrac{d_{Ni-X}}{d_{Ni-Ni}}\right)$	Fracture behavior[d] Undoped	B-doped
Ni_3Fe	0.2	-0.08[e]	1.02	D	—
Ni_3Mn	0.9	-0.36	1.01	D	—
Ni_3Al	3.0	-0.30	1.135	B	D
Ni_3Ga	3.0	-0.10	1.083	B	D
Ni_3Si	4.0	-0.01	1.089	B	D
Ni_3Ge	4.0	$+0.10$	1.132	B	B

[a]Data from Takasugi et al. (Ref. 18).
[b]Data from Taub et al. (Ref. 36).
[c]Data from Farkas (Ref. 37).
[d]Data from Taub et al. (Ref. 36).
 D = ductile, B = brittle
[e]Taub et al. (Ref. 36) suggest that the high value for Ni_3Fe may
be resulted from difficulties in comparing electronegativities
of transition metals.

in Table 2. Takasugi and Izumi [18] found that a partial replacement of
aluminum with iron or manganese in Ni_3Al reduces the valency and size differences between nickel and "modified aluminum" atoms, resulting in significantly improving the ductility and lowering the propensity for
grain-boundary fracture in Ni_3Al alloys. Microalloying with boron additions, which occupy interstitial sites in Ni_3Al, dramatically improves
the ductility and virtually completely suppresses intergranular fracture
in Ni_3Al [5,11,12,15] and Ni_3Al+Fe [38]. A peak ductility of 50%
at room temperature has been achieved in the $L1_2$ alloys Ni-24% Al and Ni-20%
Al-10% Fe doped with 0.5 and 0.2 at. % B. Beryllium, having an atom size
similar to boron, also improves the ductility of Ni_3Al, although its effect
is moderate [24]. On the other hand, the interstitial element carbon
does not produce any beneficial effect on Ni_3Al alloys [25].

The beneficial effect of B, Fe, and Mn has also been observed in
Ni_3Al alloys prepared by rapid solidification using melt-spinning [13,14,
39] and in-rotating-water methods [40,41]. Alloying with chromium, which
has a valency value (z = 1.2) and atom size (d = 12.8 nm) comparable to
Mn (z = 0.9 and d = 13.1 nm), produces no appreciable improvement in ductility of Ni_3Al prepared by conventional melting and casting [18], but it
leads to significant improvements in rapidly solidified Ni_3Al alloys [41,42].
The difference is presumably due to the fine grain size and low degree of
order attained in rapidly solidified materials containing a high density of
antiphase boundaries (APBs).

Figure 1 is a plot of room-temperature tensile ductility as a function of boron addition in Ni-24 at. % Al. Microalloying with boron
sharply increases the ductility and completely suppresses brittle
intergranular fracture. This striking effect of boron on the ductility
occurs over a wide range of boron concentration where boron is in solid
solution {the solubility of boron ≈ 1.5 at. % [15]}. To understand the
beneficial effect of boron, Auger electron spectroscopy has been used
extensively to study the segregation behavior of boron. White et al.
[15,22,23] have observed an unusual segregation behavior of boron in

Table 2. Effect of alloying addition on room-temperature
ductility and fracture behavior of Ni₃Al alloys
prepared by conventional melting and casting

Alloying element	Alloy composition (at. %)	Tensile ductility (%)	Fracture mode	Reference
	Ni₃Al	~0	Intergranular	15,24
B	Ni-24 Al-0.5 B	35–54	Transgranular	5,11,15
B,Fe	Ni-20 Al-10 Fe-0.2 B	50	Transgranular	38
Mn	Ni-16 Al-9 Mn	16	Transgranular	18
Fe	Ni-10 Al-15 Fe	8	Mixed	18
Be	Ni-24 Al-5.5 Be	6	Mixed	24

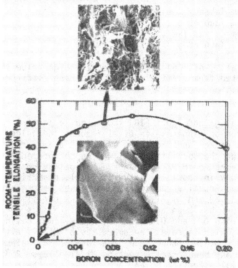

Fig. 1. Effect of boron additions on tensile elongation and fracture
behavior of Ni₃Al (24 at. % Al) tested at room temperature
(Ref. 15).

Ni₃Al. Boron tends to segregate strongly to grain boundaries in Ni₃Al but
not to free surfaces (Fig. 2). This relationship between grain boundary
and free surface segregation is in contrast to the well known behavior of
sulfur and other embrittling impurities, which tend to segregate more
strongly to free surfaces than to grain boundaries. Furthermore, the
study of the effect of thermal history on grain-boundary chemistry clearly
indicates the equilibrium nature of boron segregation to Ni₃Al grain boundaries [23]. All these observations are in agreement with the prediction
from a classical thermodynamic theory developed by Rice [43]. According
to this segregation theory, solutes that tend to segregate more strongly
to grain boundaries should increase the grain boundary cohesive energy,
whereas solutes segregating more strongly to free surfaces should lower
grain-boundary cohesion. The study of segregation of boron provides the
first direct confirmation of Rice's prediction for beneficial solutes.

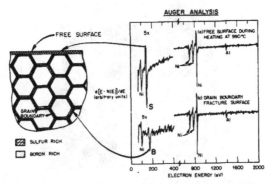

Fig. 2. Auger electron spectroscopic study showing the striking difference in interfacial segregation behavior of boron and sulfur
in Ni₃Al: while boron segregates strongly to grain boundaries
and not at all to free surfaces, the reverse is true for sulfur.

Recently, Sickafus and Sass [44] have observed that segregation of gold
changes grain-boundary dislocations in iron bi-crystals with twist boundaries. The implication of their work is that solute segregation possibly
enhances the mobility of dislocations and improves slip accommodation at
the grain boundary; both effects, in turn, lower the stress concentration
and the tendency for cracking along the boundary. By carefully measuring
the Hall-Petch parameter, K_y, in Ni₃Al alloys with and without boron,
Schulson et al. [45] and Vedula et al. [31] observed that microalloying
with boron lowers the effectiveness with which grain boundaries strengthen
the alloys. These results support the possibility that boron segregation
affects the grain-boundary dislocations and thereby the grain-boundary
ductility of Ni₃Al. Further studies are certainly required to understand
both the chemical and structural aspects of boron segregation in Ni₃Al.

BORON ADDITION AND STOICHIOMETRIC EFFECT

Alloy stoichiometry was found to have a strong effect on the ductility and fracture behavior of boron-doped Ni₃Al [15,46]. Boron
is most effective in improving the ductility and suppressing intergranular
fracture in Ni₃Al alloys containing <24 at. % Al. At higher aluminum concentrations, the ductility decreases sharply (Fig. 3), and the failure
mode changes from transgranular to mixed mode and then to completely
intergranular fracture. Auger studies of freshly fractured surfaces of
boron-doped samples indicate that the aluminum content has no observable
effect on carbon, oxygen, and sulfur segregation [15]. Instead, the
intensity of boron segregation decreases significantly and the grain-
boundary aluminum concentration increases moderately with increasing bulk
aluminum concentration (Fig. 4). These results simply suggest that
deviations from alloy stoichiometry influence grain-boundary chemistry,
which, in turn, affects grain-boundary cohesion and the overall ductility
of nickel aluminides.

Two models that have been suggested to explain the stoichiometry
effect are based on considerations of defect structure and valency difference between nickel and aluminum atoms. In order to understand how
alloy stoichiometry affects the boron segregation in Ni₃Al, the defect
structure (such as vacancies) in alloys containing 24 to 26 at. % Al and
0 to 500 wt ppm B (0 to 0.2 at. %) was studied using positron-lifetime and
Doppler-broadening analyses [47]. The measurement of the trapped-state

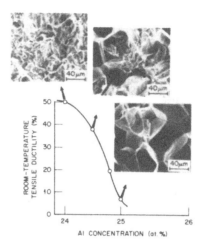

Fig. 3.   Effect of aluminum concentration on room-temperature ductility
and fracture behavior of Ni$_3$Al doped with 0.1 at. % B (Ref. 15).

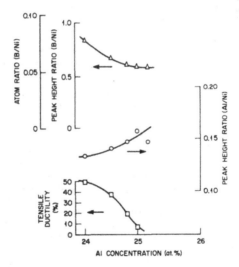

Fig. 4.   Effect of stoichiometry on grain-boundary segregation (in terms
of peak-height ratio) and room-temperature tensile ductility of
boron-doped Ni$_3$Al containing 24 to 25.2% Al (Ref. 15).

intensity evidenced that Ni$_3$Al with 24% Al trapped no positrons while
alloys with 25 and 26% Al trapped about 10% of the positrons (Fig. 5).
With the addition of boron to these alloys, the degree of positron trapping
increased in both the 25 and 26% Al alloys but remained unchanged in the
24% Al alloy.   An estimate based on the positron trapping gives a constitutional vacancy concentration of 10^{-6} to 10^{-5} in the 25 and 26% Al
alloys.   The inability of boron to ductilize polycrystalline Ni$_3$Al having

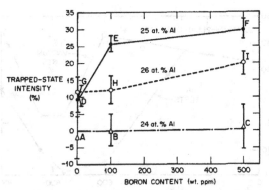

Fig. 5. A plot of the trapped-state intensity obtained from positron annihilation studies, as a function of boron concentration in 24, 25, and 26 at. % Al Ni_3Al alloys (Ref. 47).

stoichiometric (25 at. % Al) or hyperstoichiometric (>25% Al) compositions was attributed to boron clustering at constitutional vacancies, which reduces the level of boron available for segregation to the grain boundaries. The boundaries remain brittle when the amount of boron segregation is insufficient. In line with this thought, Choudhury et al. [23] have observed that Ni_3Al is more resistant to intergranular fracture when a higher level of boron is present at the boundaries. In an attempt to ductilize a hyperstoichiometric alloy, boron at levels as high as 1.0 at. % (or 0.2 wt %) was added to Ni-25.2 at. % Al; limited results indicate that the alloy remains brittle at room temperature [48].

Takasugi et al. [18] attempted to rationalize the stoichiometry effect by considering a valency difference between nickel and aluminum atoms occupying the sublattice sites in Ni_3Al. The occupation of aluminum sublattice sites with excess nickel atoms in hypostoichiometric alloys gives a valency difference of +3, resulting in a great enhancement of the grain-boundary cohesive strength of Ni_3Al. On the other hand, the occupation of nickel sublattice sites with excess aluminum atoms in hyperstoichiometric alloys gives a valency difference of −3, substantially weakening the grain-boundary cohesion. As a consequence, boron additions only effectively improve the ductility of Ni_3Al with excess nickel rather than excess aluminum. It should be pointed out that a successful explanation of the stoichiometry effect by considering the valency difference should also be able to explain the lower level of intergranular boron segregation in hyperstoichiometric Ni_3Al, as shown in Fig. 4.

Environmental Effect and Metallurgical Solutions

Although Ni_3Al alloys exhibit good oxidation resistance, their ductilities are found to be sensitive to test environments at elevated temperatures [32,33]. Figure 6 compares the tensile elongation of a nickel aluminide tested in air and in vacuum (10^{-3} Pa) as a function of test temperature. The alloy shows distinctly lower ductilities when tested in air than in vacuum at temperatures above 300°C, and the severest embrittlement occurs in the temperature range of 600 to 850°C. The embrittlement is due to the presence of gaseous oxygen, because an effect similar to that in air was observed in a pure oxygen atmosphere. Similar embrittlement has been observed in many ordered intermetallics, including boron-doped Ni_3Al containing up to 16 at. % Fe [49], Ni-15% Co-24% Al-1.0% B [50], Ni-10% Co-24% Al-0.25% B [50], and FeAl alloys [51].

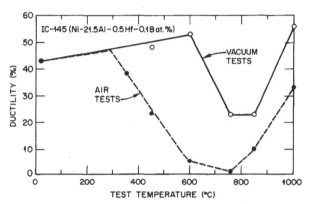

Fig. 6. Comparison of tensile elongation of Ni-21.5 Al-0.5 Hf-0.1 B (at. %) tested in vacuum and air (Ref. 53).

Figure 7 compares the tensile elongation of Ni_3Al alloys containing 0 or 0.5 at. % Hf as a function of (Al + Hf) concentration in air and in vacuum at 600°C [33]. All the alloys exhibit excellent ductility when tested in vacuum, and the ductility appears to increase somewhat with decreasing concentration. In contrast, the ductility of the alloys is dramatically lower when tested in air. The loss in ductility is accompanied by a change in fracture mode from transgranular to intergranular (Fig. 8). It is important to note that the alloys show a small but steady increase in ductility with decreasing aluminum concentration. In other words, Ni_3Al alloys are less embrittled by oxidizing environments when they contain lower levels of aluminum.

The environmental effect is clearly demonstrated in Fig. 9, where the 760°C ductility is plotted as a function of air (or oxygen) pressure [48]. The pressure was kept constant by leaking air into a dynamic vacuum system. The alloy, Ni-23% Al-0.5% Hf-0.1% B, exhibited about 1% elongation when tested in air at 760°C. With the decrease in air pressure, the ductility increases continuously and reaches as high as 26% in a vacuum of 10^{-5} Pa. The increase in ductility is accompanied with a change in fracture mode from brittle intergranular to ductile transgranular.

Although the embrittlement is caused by oxygen, experimental evidence indicates that it is not associated with oxidation per se (e.g. formation of oxide scales, grain-boundary oxidation, etc.). This deduction is based on several observations. First, preoxidation causes a slight increase rather than a decrease in ductility obtained in air (Fig. 7). Second, specimens that fractured in a brittle manner in air at elevated temperatures (e.g. 600–760°C) remained ductile and showed ductile transgranular fracture during subsequent bend-tests at room temperature. In other words, the air testing at elevated temperatures does not lead to any embrittlement at room temperature. Third, alloy specimens lost elevated-temperature ductility almost instantaneously when air was leaked into the vacuum system, indicating that diffusion is not required for embrittlement. All these observations suggest that the embrittlement is caused by a dynamic effect involving simultaneously high localized stresses, elevated temperature, and gaseous oxygen. According to this mode of embrittlement, the weakening of atomic bonds by chemisorption of oxygen plus stress concentration at grain boundaries opened to air lead to nucleation of microcracks along the boundaries. Premature fracture of Ni_3Al alloys with a low overall ductility is caused by continuously embrittling fresh crack tip and subsequently propagating the crack along the boundary.

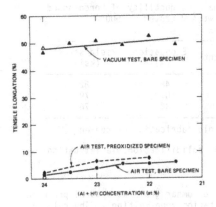

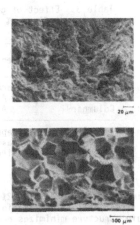

Fig. 7. Plot of tensile elongation of Ni₃Al alloys as a function of (Al + Hf) concentration for bare specimens tested at 600°C in vacuum or air, and for specimens preoxidized at 1100°C/2 h + 850°C/5 h and tested in air. Closed symbols for alloys with 0.5% Hf and open symbols for the alloy without hafnium (Ref. 32).

Fig. 8. Comparison of fracture mode of boron-doped Ni₃Al (24 at. % Al) tested in vacuum and air at 600°C. (a) Transgranular fracture, tested in vacuum. (b) Intergranular fracture, tested in air (Ref. 32).

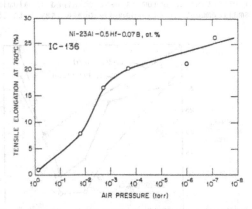

Fig. 9. A plot of tensile elongation at 760°C as a function of air pressure for Ni-23 Al-0.5 Hf-0.07 B (at. %), showing the environmental effect on ductility.

The dynamic embrittlement has to be overcome in order to use aluminide alloys in oxidizing environments. Two metallurgical solutions have proven effective in reducing the embrittlement. One is the control of grain shape in nickel aluminides through processing techniques [52]. As shown in Table 3, the dynamic embrittlement at 600°C is essentially

364

Table 3. Effect of grain shape on ductility of boron-doped
Ni-24 at. % Al-0.2 wt % B tested at 600°C

Grain shape	Test environment	Elongation (%)	Yield stress (ksi)
Equi-axed[a]	Vacuum	48	82
Equi-axed[a]	Air	0.2	79
Columnar[b]	Air	33	76

[a]Produced by repeated cold fabrication of conventional melted and cast ingot.
[b]Produced by directional solidification via levitation zone-melting.

eliminated in Ni-24% Al-0.2% B with columnar grain structures produced by directional solidification (levitation zone-melting). The columnar grain structure minimizes normal stresses across the grain boundary and thus suppresses brittle fracture along the boundary even though the boundary is weakened by chemisorption of oxygen.

An alternate solution to the problem of dynamic embrittlement is to add moderate amounts of chromium to nickel aluminides [53]. As shown in Fig. 10, alloying with 8% Cr increases the ductility at 600–850°C from 6 to 20%. A short-term air oxidation at 600°C indicates that the beneficial effect of chromium is to promote rapid formation of protective chromium oxide films that shut off gaseous oxygen from grain boundaries and base metal. The chromium-containing alloys showed, nevertheless, higher tensile ductilities in vacuum than in air. This indicates that chromium additions alleviate but do not completely eliminate the embrittlement. The beneficial effect of chromium is best obtained in aluminide alloys containing a small amount of disordered γ phase (say 5 to 15% γ).

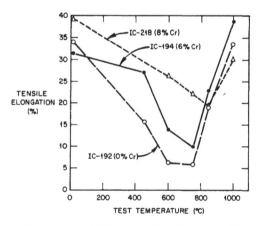

Fig. 10. Effect of chromium additions on tensile elongation of Ni$_3$Al alloys tested at room and elevated temperatures (Ref. 53).

It is important to realize that the ductility loss in nickel aluminides is a complicated phenomenon and is affected by many other metallurgical variables, in addition to test environment, aluminum level and grain shape. Limited results available at present indicate that the ductility and fracture behavior of Ni_3Al alloys are also influenced by grain size [54,55], boron level [28], thermomechanical treatment [48,50], and trace impurities {such as oxygen and sulfur [28]}. Taub et al. [50] have showed that boron-doped Ni-10% Co-24% Al exhibited a ductility of 9.9% in vacuum at 760°C for the heat prepared by conventional melting and casting but a ductility of 0.1% for the heat prepared by Osprey forming. Takeyama and Liu [54] have observed a decrease in ductility from 15% to 0 with an increase in grain size from 15 to 75 μm for tests done at 850°C in a high vacuum of 10^{-4} Pa. Additional studies are, of course, necessary for further characterizing these metallurgical variables, with special attention to their synergistic effects.

SUMMARY

(1) Grain boundaries are intrinsically brittle in high-purity Ni_3Al. A systematic study of grain-boundary fracture in $L1_2$-ordered alloys Ni_3X (X = Fe, Mn, Al, Ga, Si, and Ge) suggests that the large difference in valency, electronegativity, and atom size between nickel and aluminum atoms contributes to the intrinsic brittleness of grain boundaries in Ni_3Al.

(2) Alloying with B, Mn, Fe, and Be significantly increases the ductility and reduces the propensity for intergranular fracture in Ni_3Al alloys. Boron is found to be the most beneficial element, and a high ductility of about 50% at room temperature has been achieved in Ni-24 at. % Al and Ni-20% Al-10% Fe doped with 0.02 to 0.10 wt % B. In agreement with Rice's prediction for beneficial solutes, boron shows a unique segregation behavior, that is, it tends to segregate strongly to grain boundaries in Ni_3Al but not to free surfaces. This segregation behavior results in an effective increase in grain-boundary cohesive energy that suppresses fracture along the boundary.

(3) Doping with boron is ineffective in ductilizing Ni_3Al with >25 at. % Al. Auger analyses indicate that a deviation from alloy stoichiometry influences grain-boundary chemistry, thereby affecting the grain-boundary cohesion and overall ductility of Ni_3Al alloys.

(4) Tensile ductilities of Ni_3Al alloys depend strongly on test environments at elevated temperatures, with much lower ductilities observed in air than in vacuum. The loss in ductility is accompanied by a change in fracture mode from transgranular to intergranular. Detailed characterization indicates that the embrittlement is caused by a dynamic effect involving simultaneously high localized stresses, elevated temperature, and gaseous oxygen.

(5) Two metallurgical solutions have proven effective in reducing the embrittlement. One is the control of grain shape in Ni_3Al alloys. Columnar grain structures minimize normal stresses across the grain boundary and thus suppresses brittle intergranular fracture. The other solution is to add moderate amounts [6–10%] of chromium. The beneficial effect of chromium is to promote a rapid formation of protective oxide films that shut off gaseous oxygen from grain boundaries and base metal.

(6) In addition to test environment, the elevated-temperature ductility of Ni_3Al alloys appears to be affected by other metallurgical variables, including aluminum level, grain size, grain shape, trace impurities (such as oxygen and sulfur), boron level, and thermomechanical treatment. Additional studies are necessary for characterizing these variables, with special attention to their synergistic effects.

366

REFERENCES

[1] P. H. Thornton, R. G. Davies, and T. L. Johnston, Metall. Trans. 1, 207 (1970).
[2] P. A. Flinn, Trans. TMS-AIME 218, 145 (1960).
[3] R. G. Davies and N. S. Stoloff, Trans. TMS-AIME 233, 714 (1965).
[4] E. A. Aitken, Intermetallic Compounds, pp. 491–515, Wiley, New York (1967).
[5] C. T. Liu and C. C. Koch, Technical Aspects of Critical Materials Used by the Steel Industry, vol. 11B, p. 42, National Bureau of Standards (1983).
[6] E. M. Grala, Mechanical Properties of Intermetallic Compounds, p. 358, Wiley, New York (1960).
[7] R. Moskovich, J. Mater. Sci. 13, 1901 (1978).
[8] K. Aoki and O. Izumi, Trans. Japan Inst. Metals 19, 203 (1978).
[9] A. V. Seybolt and J. H. Westbrook, Acta Metall. 12, 449 (1964).
[10] K. Aoki and O. Izumi, Nippon Kinzoku Gakkaishi 41, 170 (1977)
[11] K. Aoki and O. Izumi, Nippon Kinzoku Gakkaishi 43, 1190 (1979).
[12] C. T. Liu, C. L. White, C. C. Koch, and E. H. Lee, Proc. Symp. High Temperature Materials Chemistry II, p. 32, Electrochem. Soc. Inc. (1983).
[13] A. I. Taub, S. C. Huang, and K. M. Chang, Metall. Trans. A 15A, 399 (1984).
[14] S. C. Huang, A. I. Taub, and K. M. Chang, Acta Metall. 32, 1703 (1984).
[15] C. T. Liu, C. L. White, and J. A. Horton, Acta Metall. 33, 213–219 (1985).
[16] C. C. Koch, J. A. Horton, C. T. Liu, O. B. Cavin, and J. O. Scarbrough, in Rapid Solidification Processing, Principles and Technologies IV, ed. R. Mehrabian, p. 264, National Bureau of Standards (1983).
[17] T. Takasugi and O. Izumi, Acta Metall. 33, 1247–1258 (1985).
[18] T. Takasugi, O. Izumi, and N. Masahashi, Acta Metall. 33, 1259 (1985).
[19] S. Hanada, S. Watanabe, and O. Izumi, J. Mater. Sci. 21, 203–210 (1985).
[20] T. Takasugi, E. P. George, D. P. Pope, and O. Izumi, Scr. Metall. 19, 551–556 (1985).
[21] T. Ogura, S. Hanada, T. Masumoto, and O. Izumi, Metall. Trans. A 16A, 441–443 (1985).
[22] C. L. White, R. A. Padgett, C. T. Liu, and S. M. Yalisove, Scr. Metall. 18, 1417–1420 (1984).
[23] A. Choudhury, C. L. White, and C. R. Brooks, Scr. Metall. 20, 1061 (1986).
[24] T. Takasugi, N. Masahashi, and O. Izumi, Scr. Metall. 20, 1317 (1986).
[25] S. C. Huang, C. L. Briant, K. M. Chang, A. I. Taub, E. L. Hall, J. Mater. Res. 1, 60–67 (1986).
[26] K. M. Chang, A. I. Taub, and S. C. Huang, Proc. MRS Symp. High-Temperature Ordered Intermetallic Alloys, vol. 39, p. 335, ed. C. C. Koch, C. T. Liu, and N. S. Stoloff, MRS Publication (1985).
[27] A. I. Taub, S. C. Huang, and K. M. Chang, Proc. MRS Symp. High-Temperature Ordered Intermetallic Alloys, vol. 39, p. 221, ed. C. C. Koch, C. T. Liu, and N. S. Stoloff, MRS Publication (1985).
[28] A. I. Taub, K. M. Chang, and S. C. Huang, Proc. ASM Int. Conf. on Rapidly Solidified Materials, San Diego, Calif., Feb. 3-5, 1986, p. 297, ed. P. W. Lee and R. S. Carbonara, American Soc. for Metals, 1985.
[29] C. T. Liu and C. L. White, Proc. MRS Symp. High-Temperature Ordered Intermetallic Alloys, vol. 39, p. 365, ed. C. C. Koch, C. T. Liu, and N. S. Stoloff, MRS Publication (1985).
[30] C. T. Liu, High-Temperature Alloys: Theory and Design, ed. J. O. Stiegler, pp. 289–308, Am. Inst. Mech. Engrs. (1984).
[31] K. Vedula, private communication (1986).
[32] C. T. Liu, C. L. White, and E. H. Lee, Scr. Metall. 19, 1247–1250 (1985).
[33] C. T. Liu and C. L. White, Acta Metall., accepted for publication (1986).
[34] D. F. Stein and L. A. Heldt, Interfacial Segregation, ed. W. C. Johnson and J. M. Blakely, pp. 239–260, ASM, Metals Park, Ohio (1977).
[35] C. L. White and D. F. Stein, Metall. Trans. A 9A, 13 (1978).
[36] A. I. Taub, C. L. Briant, S. C. Huang, K. M. Chang, and M. R. Jackson, Scr. Metall. 20, 129–134 (1986).

[37] D. Farkas, private communication (September 1985).
[38] J. A. Horton, C. T. Liu, and M. L. Santella, submitted for publication in Metall. Trans., 1986.
[39] K. M. Chang, S. C. Huang, and A. I. Taub, MRS Symp. Proc. 28, Elsevier Science Publishing, p. 401 (1984).
[40] A. Inouye, T. Masumoto, H. Tomioka, and N. Yano, Int. J. Rapid Solidification, vol. 1, pp. 115–142, 1984–85.
[41] A. Inouye, H. Tomioka, and T. Masumoto, Metall. Trans. A 14A, 1367 (1983).
[42] S. C. Huang, E. L. Hall, K. M. Chang, and R. P. Laforce, Metall. Trans. A 1685–1691 (1986).
[43] J. R. Rice, The Effect of Hydrogen on the Behavior of Metals, pp. 455–465, AIME Publication, New York, New York (1976).
[44] K. Sickafus and S. L. Sass, Scr. Metall. 18, 165–168 (1984).
[45] E. M. Schulson, T. P. Weihs, I. Baker, H. J. Frost, and J. A. Horton, Acta Metall. 34, 1395–1399 (1986).
[46] S. C. Huang, K. M. Chang, and A. I. Taub, Proc. ASM Int. Conf. on Rapidly Solidified Materials, San Diego, Calif., Feb. 3-5, 1986, p. 255, ed. P. W. Lee and R. S. Carbonara, American Soc. for Metals (1985).
[47] A. DasGupta, L. C. Smedskjaer, D. G. Legnini, and R. W. Siegel, Mater. Letters 3, 457–461 (1985).
[48] C. T. Liu, unpublished results (1986).
[49] C. T. Liu, to be published in Symp. Micon 1986, AIME Publication (1987).
[50] A. I. Taub, K. M. Chang, and C. T. Liu, accepted for publication in Scripta Metallurgica (1986).
[51] D. J. Gaydosh, paper presented at the 115th TMS/AIME Annual Meeting at New Orleans, La., March 2–6, 1986.
[52] C. T. Liu and B. F. Oliver, unpublished results (1986).
[53] C. T. Liu and V. K. Sikka, J. Metals 38, 19–21 (1986).
[54] M. Takeyama and C. T. Liu, unpublished results (1986).
[55] T. P. Weihs, V. Zimoviev, D. V. Viens, and E. M. Schulson, submitted for publication to Acta Metall (September 1986).

DISPERSOIDS IN INTERMETALLIC ALLOYS: A REVIEW

CARL C. KOCH
Materials Science & Engineering Department, North Carolina State University, Box 7907, Raleigh, NC 27695-7907

ABSTRACT

The existing work on structure and mechanical behavior of intermetallic alloys containing dispersed second phases — dispersoids — is reviewed. The dispersoids considered in the review are those given the conventional definition as inert, stable, insoluble phases in the matrix, such as oxides. The only detailed mechanistic studies of dispersoids in intermetallics have been carried out on the model material Cu_3Au, and these did not include elevated temperature, eg. creep, mechanical behavior. A number of investigations on dispersoids in the potentially important elevated temperature materials — the aluminides — are reviewed. The materials discussed include FeAl, Fe_3Al, NiAl, Ni_3Al, and Ti_3Al. Many of these studies are preliminary in nature and much more research and development is needed to allow for the design of intermetallic-dispersoid materials with improved fabricability and elevated temperature performance.

1.0 Introduction

The scope of this paper is focused on the existing literature dealing with dispersoids in intermetallic compounds. In keeping with the theme of the symposium it will stress the influence of dispersed phases on the mechanical properties of intermetallics — both those of potential use as structural materials such as the aluminides, and model materials such as Cu_3Au.

Most engineering alloys are multiphase. The structure of such alloys often consists of the major phase — the matrix — which surrounds the second (minor) phase. The second phase can be categorized as a "dispersoid" or a "precipitate". A "precipitate" phase belongs to the alloy system (binary or multicomponent) of the matrix phase while a dispersoid is introduced "synthetically" into the matrix and is relatively inert. Second phase particles can harden the matrix by obstructing dislocation motion. Dislocations may circumvent the obstacle by various alternatives such as by-passing, cross-slipping, climbing, shearing, etc. The synthetically introduced dispersoids, such as oxides or carbides, are usually non-shearable. It is these particles that this review is mainly concerned with.

First a brief review will be given of dispersion-strengthening in general, including a description of the microstructures of dispersed phase alloys, their tensile properties, and their high temperature behavior. A discussion of processing techniques for dispersed-phase alloys will follow, specifically applied to intermetallic-alloy-dispersoid systems. Then a review of the limited literature on specific systems will follow. The discussion will include studies where the dispersoid itself was the chief hardening agent and those in which the dispersoid was introduced as a tool for refinement of the matrix grain size.

2.0 Dispersion-Strengthening

The goal of dispersed-phase alloys is to strengthen the alloy matrix by impeding the motion of dislocations. The matrix remains the major load-bearing constituent. The metallic matrix will be strengthened in proportion to the effectiveness of the dispersion as a barrier to the motion of dislocations.

The first synthetic alloys consisting of an oxide dispersed into a ductile metallic matrix were the SAP (sintered aluminum powder) alloys produced by Lenel et al [1] who made bars by extrusion of a mixture of 1–10 vol.% of Al_2O_3 powder in pure Al powders. Most commercial materials contain 4–15 weight % oxide and are manufactured from aluminum powders ball-milled and simultaneously oxidized. The cold-compacted powder is vacuum-sintered and extruded into billets which are finally shaped by extrusion, rolling, forging, etc. The Al_2O_3 particles in SAP exhibit a wide distribution of sizes (~10 nm to 1 μm) and a lack of coherency with the matrix. The main advantage of dispersion-strengthened materials is the ability to maintain the yield-strength increase and creep resistance increase over a wide temperature range, up to T ≈ 0.8 T_M of the melting point of the metallic matrix. The effectiveness of the dispersoids depends on their insensitivity to high temperatures. This distinguishes dispersion-hardened materials from precipitation-hardened alloys, which soften with increasing temperature due to significant coarsening of the structure. The dispersoids may be oxides, carbides, silicides, nitrides, borides and refractory metal particles which are insoluble and incoherent with the matrix. Examples of other dispersion-strengthened systems include $Cu-SiO_2$ and $Cu-Al_2O_3$ (2), $W-ThO_2$ (3), $Ni-Al_2O_3$, $NiCrAlTi-Y_2O_3$, Cu-W and Ni-TiC(4).

The average size of the dispersoid particles is critical to the mechanical and metallurgical properties of dispersed-phase materials. Strudel [5] has described the effects of small particles (1-100 nm), medium-size particles (0.1–1 μm), and coarse particles (5-50 μm) on the mechanical behavior of dispersed-phase alloys. Small dispersoids have the strongest effect on the yield strength and tend to be spherical in shape to minimize surface energy. Medium-size dispersoids have a strong inhibiting effect on recrystallization and grain growth. Their strengthening effect is seen more on the high-temperature creep resistance than on the low-temperature yield strength. Coarse dispersoids create problems of compatibility of deformation and large stress gradients can be generated in their vicinity. They tend to be a source of weakness rather than strength.

The characteristic features of dispersoid containing alloys are: 1) their high yield strengths at relatively low temperatures (up to ~0.2 T_M), 2) their high rate of work hardening, and 3) their ability to maintain fairly high flow stresses up to temperatures close to their melting point (0.90 – 0.95 T_M). The proposed models for these experimental observations have been reviewed by several authors [5,6,7]. The higher yield stresses in dispersion-hardened alloys are explained by the Orowan mechanism. The Orowan mechanism [8] involves the need for an additional stress, $\Delta\tau$, to expand dislocations and enable them to bypass non-deformable particles of a second phase. $\Delta\tau_{Orowan} = 2T/bD_s$ where T = line tension of the dislocations, b = Burgers vector, and D_s = mean planar particle spacing. Subsequent refinements of Orowan's theory have led to expressions for $\Delta\tau$ which give excellent agreement with experimental results for the increased yield strengths in dispersoid containing materials. For example Brown and Ham (9) give

$$\Delta\tau_{Orowan} = 0.81 \frac{Gb}{2\pi} \frac{1}{(1-v)}^{\frac{1}{2}} \frac{1}{(D_s-d_s)} \ell n \left(\frac{d_s}{r_o}\right)$$

where v = Poisson's ratio, d_s = mean planar particle diameter, and r_o = inner cut-off radius of the dislocation line energy.

The quantitative understanding of the work-hardening of dispersion-strengthened alloys is a difficult task (as it is in single phase alloys). In dispersoid-containing alloys, contributions to work-hardening come from stress necessary for dislocations to by-pass the particles and to cut the high dislocation density created and stabilized in such materials. Ashby [10] has concluded that the useful work-hardening of dispersion-strengthened alloys stable to high temperatures reflects the ability of the particles to pin dislocation cell boundaries.

The creep resistance observed in dispersion-strengthened alloys at high temperatures (T ≥ 0.5 T_M) is affected by the dispersoids in two ways: 1) by dispersoids acting as barriers to moving dislocations, 2) by the influence the dispersoids have on creating the dislocation structure (cell boundaries etc.) during deformation, and the subsequent effect of this structure on dislocation mobility during creep.

3. Processing Methods for Dispersed-Phase Alloys

Several methods have been used for fabrication of dispersion-strengthened materials. These can be divided into four general groups: 1) mechanical mixing (including mechanical alloying) of powders followed by consolidation methods, 2) chemical reactions which include a variety of procedures such as internal oxidation of the matrix metal, surface oxidation of matrix metal, surface oxidation of matrix metal powder, preferential chemical reduction of mixed oxides, and co-precipitation of matrix and dispersed phase compounds followed by preferential chemical reduction of the matrix metal phase, 3) co-electrodeposition of metal and oxide particles, and 4) solidification from the liquid metal.

The dispersoid-intermetallic compound composites to be discussed in this paper have been prepared, in the main, by either mechanical mixing or by chemical reactions such as internal oxidation, sulfidation, etc. The preparation methods for each system to be reviewed will be described.

4. Dispersoids in Intermetallic Alloys - Specific Alloy Systems

The number of dispersoid-intermetallic alloy systems which have been studied is limited. They include the model ordered LI_2 system Cu_3Au as well as nickel-, iron-, and titanium aluminides which are of interest for elevated temperature structural materials.

4.1. Cu₃Au - Dispersoid Alloys. LI₂ Matrix

Sastry and Ramaswami [11] studied the plastic deformation of ordered and disordered Cu_3Au single crystals with and without a dispersion of Al_2O_3 particles. Cu_3Au-Al (Al ranging from 0.03 to 0.10 wt.%) single crystals were internally oxidized to form the Al_2O_3 dispersoids. The dispersoid size d_s and spacing D_s, as determined by transmission electron microscopy, ranged from $d_s = 45$ nm to $d_s = 79$ nm and $D_s = 592$ nm to $D_s = 359$ nm. The stress-strain curves of the ordered Cu_3Au and $Cu_3Au-Al_2O_3$ single crystals are illustrated in Figure 1. The tensile behavior of ordered Cu_3Au was characterized by a high initial work hardening rate. This linear work hardening extended up to approximately 70% shear strain. The Al_2O_3 dispersoids raised the yield stress of ordered Cu_3Au. The stress-strain curves of ordered $Cu_3Au-Al_2O_3$ showed an initial parabolic work-hardening region followed by a linear work hardening region, which, above 25% strain, was essentially parallel to that for Cu_3Au without the Al_2O_3 dispersoids. The microstructural observations indicated the superdislocations move as pairs of unit dislocations and, on bypassing the dispersoids, leave behind both glide or Orowan superdislocation loops and superdislocation prismatic loops. A planar, partly closed cell-like structure developed on the primary slip plane with increasing strain. The antiphase domain boundaries (APD) became elongated with increasing strain but the APD size remained unchanged. The yield stress was between $(\tau_1 + \tau_2)$ and $(\tau_1{}^2 + \tau_2{}^2)^{1/2}$ where τ_1 and τ_2 are the yield stress of the Cu_3Au matrix and the dispersoid bypassing stress for superdislocations respectively. The initial stages of deformation was dominated by the geometrically necessary dislocations, which arise from dislocation-particle interactions, giving rise to parabolic work hardening. At large strains the work hardening was controlled by the statistically stored dislocations, which are due to chance encounters and mutual trapping of dislocations. The results were consistent with Ashby's theory of work-hardening in two-phase systems [12].

Pattanaik and Ardell [13] determined the critical resolved shear stress (CRSS) of disordered and ordered Cu_3Au single crystals containing SiO_2 dispersoids. The

372

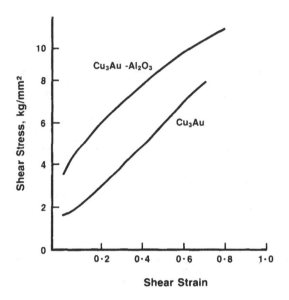

Figure 1. Shear stress vs. shear strain for Cu_3Au single crystals
with and without Al_2O_3 dispersoids (after Sastry and
Ramaswami, ref. 11)

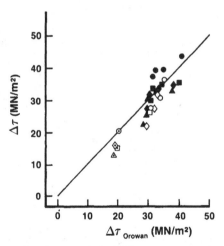

Figure 2. Increment in CRSS, $\Delta\tau$, measured experimentally vs $\Delta\tau$
calculated from modified Orowan Theory for Cu_3Au single
crystals with SiO_2 dispersoids (after Pattanaik and
Ardell, ref. 13)

SiO_2 particles were produced by internal oxidation. The spherical SiO_2 particles had mean diameters, d_s = 36 to 58 nm, and center-to-center particle spacing, Ds = 286 to 649 nm. The APD size was ≈ 43 nm, too large to contribute to the strengthening. TEM examination of (111) sections showed Orowan-type bowing of superdislocations between particles. However, superdislocation Orowan loops were not observed. Values of $\Delta\tau_{Orowan}$ were calculated using the expression for $\Delta\tau$ of Brown and Ham (9), except that the Burgers vector of the superdislocations was taken as b = a<110>. Figure 2 shows a plot of the experimentally measured increment in the CRSS $\Delta\tau$ versus the calculated $\Delta\tau_{Orowan}$ values. The experimentally observed values of $\Delta\tau$ agree reasonably well with the values of $\Delta\tau_{Orowan}$ suggesting that the dislocation pair constituting the superdislocation does by-pass the particles as a unit. Thus, the TEM observations of Orowan bowing of superdislocations between the dispersoids and the agreement between the experimentally observed increment in the CRSS and the theoretically predicted Orowan stress led to the conclusion that the yield stress in the ordered Cu_3Au-SiO_2 crystals is controlled by the Orowan process.

Ardell and Pattanaik [14] studied the work-hardening of dispersion-strengthened ordered Cu_3Au-SiO_2 single crystals. The SiO_2 particles had a definite effect on the early part of the stress-strain curves (Stage I) of ordered Cu_3Au-SiO_2 crystals. However, the Stage II slopes are very nearly the same for ordered Cu_3Au and Cu_3Au-SiO_2 and it was assumed that the same mechanism that controls Stage II hardening in ordered pure Cu_3Au also is operative in ordered Cu_3Au-SiO_2 Microstructural analysis revealed the following dislocation structures. At 295K after 10% strain a large number of long straight segments of superdislocations in screw orientation were observed. Short superdislocation dipoles of edge or near-edge orientation were observed at many of the SiO_2 particles. After 30% strain, long straight superdislocation screw segments had cross-slipped onto the most favorably oriented cube-type cross-slip plane (010) and were widely spaced because of the lower APB energy on {100}. Stage II work-hardening in both pure Cu_3Au and Cu_3Au-SiO_2 is controlled by the cross-slip of superdislocations onto (010). The interaction of glide dislocations with the SiO_2 particles and their associated plastic zones appears to explain Stage I work-hardening behavior of ordered Cu_3Au-SiO_2.

The fundamental studies of the effects of dispersoids on the mechanical behavior and dislocation structure of ordered Cu_3Au single crystals cited above were limited to temperatures of 373K or lower. Thus no "high" temperature studies or creep behavior were obtained for this model ordered alloy.

4.2. Ni-Al - Dispersoid Alloys. B2 Matrix

The study of oxide dispersion-strengthened NiAl by Seybolt [15] was probably the first study of an intermetallic alloy - dispersoid system for possible structural applications. It was undertaken as part of a program to determine the mechanical behavior of NiAl as a potential high temperature structural material. A Ni–49 at.% Al alloy was at first chosen for study since it was believed that the room temperature toughness was superior to the stoichiometric composition. However, the 50–50 alloy was also examined. The alloys were prepared by first mixing in a ball mill powders of Ni_2Al_3, Ni, and oxide. The oxide was added as μm size - Al_2O_3, ThO_2, or Y_2O_3. Analysis of the base $Ni_{51}Al_{49}$ showed about 0.3 wt.% O_2 after final ball milling which corresponds to approximately 1 vol % Al_2O_3. Thus, the "base" Ni-Al alloy contained unavoidable quantities of Al_2O_3 The powder mixtures were consolidated by hot extrusion. Chemical and microstructural analysis after consolidation revealed a variety of dispersoid volume fractions and particle sizes. The volume fractions ranged from 3 to 6% and the particle sizes from about 50 nm to 0.5 μm in diameter. One coarse particle dispersion with ~20 μm diameter Al_2O_3 particles (5 vol %) was prepared.

Mechanical properties were measured by bend, tensile, and stress rupture tests. Bend angle was used as a measure of ductility. Figure 3 shows bend angle

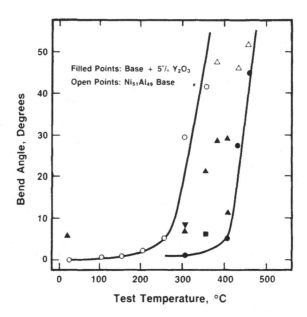

Figure 3.   Bend angle vs. test temperature for NiAl with and without
            a 5 vol.% Y₂O₅ dispersion (after Seybolt, ref. 15)

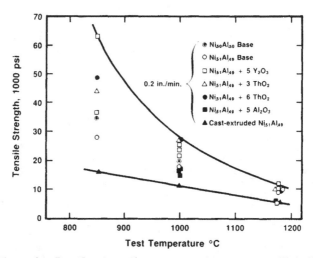

Figure 4.   Tensile strength vs. test temperature for NiAl disper-
            sions at a strain rate of 0.2 in/min (after Seybolt,
            ref. 15)

versus test temperature for NiAl with and without a 5 vol.% Y_2O_3 dispersion. The bend transition temperature was raised by about 150°C by the presence of Y_2O_3. That is, the Y_2O_3 dispersion embrittled NiAl. This result was unexpected since previously it had been found that dispersed oxides could lower the ductile-brittle transition temperature of metals such as tungsten [16]. The embrittling effect of dispersed oxides in NiAl was confirmed by high temperature tensile tests where the percent elongation was reduced by the presence of the dispersoids. Tensile tests were made at either 0.2 in/min. or 0.002 in/min. There was a large strain rate sensitivity. The tensile strength versus test temperature for a strain rate of 0.2 in/min. is shown in Figure 4. The lower curve was obtained from data on NiAl prepared by melting, casting, and extrusion and was essentially free from oxide dispersions. The oxide free material defines the bottom of a wide band of varying strength. The upper boundary of the band was determined by the NiAl + 5 vol.% Y_2O_3 or the NiAl + 3 vol.% ThO_2 alloys. While the strength differences are significant at 850°C, the strength increments decrease with temperature. Low strain rate had the same effect as higher temperature in wiping out the strength differences. At the fast strain rate the best Y_2O_3 dispersion was stronger than the much finer ThO_2 dispersion. However, the reverse was true at the slower strain rate and in stress-rupture tests. At a given value for the Larson-Miller parameter, $P = T (20 + \log t) (.001)$, the rupture stress for NiAl with oxide dispersions was twice that of NiAl without any dispersion. In general, the ThO_2 dispersions showed better stress-rupture behavior than the Y_2O_3 or Al_2O_3 dispersions. It was not known if this was due to its finer size, better distribution, or better stability. Of the two major mechanisms for retention of high-temperature strength by oxide dispersions in metals, 1) direct slip interference by the dispersoids and 2) particle-stabilization of the effects of cold work at high temperature, only mechanism 1) applies for brittle NiAl. Therefore, the increase in stress-rupture strength was not sufficient to be competitive with exisiting high temperature materals such as Inconel 200- and "TD" nickel.

Sherman and Vedula have studied "dispersion" strenghtening in NiAl containing 2.26 at.% Nb [17]. Homogenization of this alloy resulted in the formation of the intermediate phases NiAlNb and $NiAlNb_2$. High temperature strengthening of this alloy was apparently due to a dispersion of fine precipitates (~50 nm) of hexagonal NiAlNb. This should be considered a "precipitation" strengthening rather than dispersion strengthening system however, since the precipitates are soluble in the NiAl matrix. They coarsen significantly, even during elevated temperature testing, resulting in a drop in flow stress during deformation.

4.3. FeAl - Dispersoid Alloys. B2 Matrix

The B2 structure FeAl alloy was included in Seybolt's study of dispersion strengthening in NiAl [15]. Only one alloy was investigated, with 40 at.% Al and about 12 vol.% Al_2O_3. Only 5 vol.% Al_2O_3 was intended, but alloy oxidation during processing brought the total up to 12%. The Al_2O_3 dispersoids had average sizes of about 0.1 μm and ranged up to about 0.4 μm. The $Fe_{0.6}Al_{0.4}$ + Al_2O_3 alloy showed a four-fold increase in stress-rupture strength over that of a $Fe_{.6}Al_{.4}$ cast and wrought alloy reported by Sainfort et al [18].

Strothers et al (19) are presently studying oxide-dispersion strengthening of an Fe-40 at.% Al alloy containing 0.1 at.% Zr and 1000 wppm of boron.

4.4. Fe₃Al - Dispersoid Alloys. DO₃ Matrix

The work on dispersoids in Fe_3Al alloys is limited to studies of grain refinement due to additions of TiB_2 particles. Slaughter and Das [20] prepared stoichiometric Fe_3Al with 1.5 mole percent TiB_2. The alloy was made by first vacuum induction melting the components, then rapidly solidifying the remelted alloys to powder by centrifugal atomization and quenching in jets of helium. The calculated cooling rates were 10^5 to 10^6 degrees/s for -140 mesh particles. The −140 mesh powders were compacted by hot isostatic pressing (HIPing) or forging at various temperatures. The compacts were then thermomechanically processed. In the as-compacted powders all the TiB_2 particles were in the interdendritic or intercellular regions of the cast structure. Their average diameter was about 25.0 nm and the number density was about 2.6×10^{15} particles/cm^3. The particles could be uniformly dispersed by thermomechanical processing without rapid coarsening. Details of the thermomechanical processing were not given but final heat treatments involved anneals at 1040°C (2h) and 480°C (2h) followed by water quenching, which developed the equilibrium DO_3 structure. The TiB_2 dispersion was stable for temperatures up to 1040°C and stabilized a fine grain or subgrain size –1 to 2 μm diameter.

More recently, McKamey et al [21,22] have studied the mechanical properties of Fe_3Al alloys containing 24 to 30 at.% aluminum to which 0.5 wt.% TiB_2 was added for grain refinement. They found that grain size increased very little with increasing temperature above 700°C remaining at 40 to 50 μm in diameter to 1000°C. These observations indicate that the TiB_2 particles are effective in retarding grain growth in these alloys. Details of this work are described by McKamey et al [22] in this symposium.

4.5. Ti_3Al - Dispersoid Alloys. DO_{19} Matrix

The high strength to weight ratio of Ti_3Al makes it a particularly attractive material for applications such as aircraft engine components. However, it has very low ductility and toughness at room temperature. By alloying additions and microstructural modifications useful ductility has been developed in this material [23]. One approach has been to refine the slip length [24]. Konitzer and Fraser [25] have produced a refined dispersion of Er_2O_3 in Ti_3Al using rapid solidification. Er_2O_3 dispersions were first studied in Ti and by rapid solidification methods very fine dispersoids (~20 nm) were obtained by precipitation heat treatments [26]. However, on heating above the α (hcp)/β (bcc) transus, the precipitates coarsened significantly, presumably due to increased diffusion rate in bcc Ti. For Ti_3Al, the transformation on heating the α_2 DO_{19} phase is $\alpha_2 \rightarrow \alpha + \alpha_2$ which occurs at 1100°C. An alloy of Ti_3Al + 0.4 at.% Er was arc-melted. Oxygen was introduced as a major impurity in the Er and minor impurity in Ti and Al. The surface regions of the arc melted buttons were rapidly solidified by laser melting. The rapidly solidified material was annealed at various temperatures. The initial microstructure (~20 nm particles), produced by annealing at 973K, did not coarsen significantly after 10h anneals up to 1173K.

Rowe and coworkers [27,28,29] have carried out studies of dispersoids in Ti_3Al and Ti_3Al-Nb alloys prepared with 0, 5, 6, 7.5 and 10.5 at.% Nb and either 0.5 at.% Er or 0.5 at.% Ce with 0.2 at.% S. The alloys were rapidly solidified by melt spinning. The structures of the as-melt-spun ribbons varied from α_2 (hcp DO_{19}) for Ti_3Al to β + β_2 (bcc with B2 ordering) for Ti_3Al-Nb. Indications of the omega phase were also observed in the Ti_3Al-Nb alloys. After annealing at 750°C the $\beta + \beta_2$ (+ ω?) structure transformed to the equilibrium α_2 phase. Dispersoids of Er_2O_3, Ce_2S_3, or $Ce_4O_4S_3$ were identified, depending on the alloy additions. The as-melt-spun dispersoids were 10 to 50 nm in size. The ribbons were consolidated by hot isostatic pressing at 850°C to 950°C and extruded at 925°C. Alloys without Nb were extruded at a temperature of 1108°C. Consolidated alloys showed preferential dispersoid depletion and coarsening at grain boundaries. This appeared to be related to grain boundary migration and particle dragging along with the grain boundary. Alloys with the highest Nb content showed the least grain boundary coarsening. Extrusion caused much greater coarsening of the Er_2O_3 than the Ce_2S_3 or $Ce_4O_4S_3$ dispersoids. Bend tests on ribbon of the as-melt-spun and aged alloys indicated that the bend ductilities of alloys with dispersoids were lower than the same alloys without a dispersoid.

4.6. Ni$_3$Al - Dispersoid Alloys. Ll$_2$ Matrix

Dramatic improvements in the ductility of polycrystalline Ni$_3$Al resulting from small boron additions (~0.5 at.% B) were first reported by Aoki and Izumi [30]. Subsequently, Liu and coworkers [31] were able to achieve room temperature tensile elongations of over 50% and almost complete inhibition of intergranular fracture through control of boron content and Ni-Al stoichiometry. This development has stimulated much research aimed at synthesis of Ni$_3$Al-base elevated temperature structural materials - a significant fraction of this Symposium is devoted to this material. Included in these studies are preliminary investigations of Ni$_3$Al-dispersoid systems.

Baker et al [32,33] added titanium and boron to rapidly solidified Ni$_3$Al powders in the hope of stabilizing a fine grained structure, as Slaughter and Das [20] had accomplished for Fe$_3$Al. Ingots of Ni$_3$Al (24.4 at.% Al) containing 1.5 at.% titanium and 2.8 at.% boron and stoichiometric Ni$_3$Al were remelted and rapidly solidified by centrifugal atomization with droplet cooling by high velocity helium gas. Quench rates for the droplets (25 μm to 80 μm in size) were estimated to be about 10^5 K/s. Ni$_3$Al particles exhibited both dendritic and lamellar structures while Ni$_3$Al (B, Ti) showed only dendrites. The rapidly solidified powders were annealed at 800°C. After annealing the majority (about 80%) of the powders developed a grain structure, and both materials, i.e. with and without Ti, B additions, exhibited similar results. Only optical microscopy was carried out on sections of the powders, so the TiB$_2$ dispersions were not detected. After the 800°C anneal the Ni$_3$Al (B,Ti) powders which had crystallized had somewhat higher hardness values (413 VHN) than similar Ni$_3$Al samples (363 VHN). This was attributed to solute hardening, but again the internal structure is not known since TEM was not carried out. More recently, TEM was carried out on these powders in the as-rapidly solidified condition [34]. The focus of this study was the phases present, the observed martensitic structures, and compositional variations. The presence or absence of borides was not mentioned.

Huang et al [35] studied the effects of boron additions in rapidly solidified Ni$_3$Al which was prepared by melt spinning. Ingots of (Ni$_{0.75}$ Al$_{0.25}$)$_{100-x}$ B$_x$ containing up to 6 at.% boron as well as (Ni$_{75}$ Al$_{25-x}$ Ti$_x$)$_{0.99}$ B$_{0.01}$ alloys with Ti concentrations up to 15 at.% were prepared. Rapidly solidified ribbons approximately 6 mm wide by 35 μm thick were produced by melt spinning at an estimated cooling rate of 5 x 10^5 K/s. One of the goals was to determine the boron solubility limits under the conditions of rapid solidification. It was found, by x-ray and TEM measurements, that a limit of complete boron solubility of at least 1.5 at.% boron was obtained. With boron ≥ 1.5 at.%, M$_{23}$B$_6$ borides were observed at grain and cell boundaries. The presence of borides at the grain boundaries severely embrittled the ribbons. Additions of titanium in excess of ~3 at.% for an alloy containing 1 at.% B also reduced the bend fracture strain of the ribbons from 1.0 to < .05. The ductility reduction coincided with the appearance of M$_{23}$B$_6$ borides, from TEM studies. It appears that Ti promotes the formation of the M$_{23}$B$_6$ boride which has a marked embrittling effect in Ni$_3$Al.

Lauf and Walls [36] studied dispersoids in alloy powder of Ni$_3$Al + 0.5 at.% Hf + 500 wt. ppm B. The powder was milled to facilitate densification by hot pressing. One sample was made with the addition of 10 volume % SiC whiskers. After hot pressing in vacuum for 15 minutes at 1300°C under 4000 psi pressure, metallographic examination indicated that the fibers had reacted with the Ni$_3$Al forming one or more liquid Ni-Al-Si phases. Al$_2$O$_3$ dispersions were produced by mixing the Ni$_3$Al-base powder with 1 wt.% Al$_2$O$_3$ powder and mechanically alloying in a Spex mill for 10 h. Examination of the Ni$_3$Al-base particles after milling showed layers of Al$_2$O$_3$ powder folded into them. This powder was also hot pressed at 1300°C for 15 minutes in vacuum at 4000 psi. The resulting compact had theoretical density, a fine grain size, and a uniform dispersion of Al$_2$O$_3$ particles. The room-temperature microhardness was 488 ± 27 DPH compared to 225 ± 10 DPH for the Ni$_3$Al without the Al$_2$O$_3$, but processed the same. Hot microhardness tests showed that the hardness of the Ni$_3$Al + Al$_2$O$_3$ decreased more rapidly with increasing temperature than that of Ni$_3$Al such that a cross-over occurred with Ni$_3$Al

378

+ Al_2O_3 actually softer at the highest test temperature (1000°C). It was suggested that this may be due to the small grain size of the $Ni_3Al + Al_2O_3$ with resulting enhanced grain boundary sliding. More work is needed to clarify this observation.

Donnelly and Koch [37] have also used mechanical alloying to prepare oxide dispersions in Ni_3Al-base alloy powder. The powder used contained 0.5 at.% Hf and 500 wppm B. Oxide dispersoids of Al_2O_3, Y_2O_3, or ThO_2 were studied in concentrations from 0.5 to 3.5 volume %. In order to improve the powder yield (i.e. to prevent excessive cold-welding to the milling media and vial walls) a process control agent - 1 volume % of stearic acid - was added to some of the Ni_3Al-oxide powder mixtures. Mechanical alloying was carried out until a "steady state" was reached in terms of a stabilized hardness and particle morphology. This took about six hours of milling in a Spex mill at ambient temperature (about 60°C average powder temperature). The powders were canned in mild steel and hot isostatically pressed at 1300°C for 3h. Samples containing the stearic acid process control agent could not be hot (or cold) swaged after HIPing because they were brittle. Samples without the process control agent could be swaged. Room temperature microhardness indicated that the presence of the process control agent (PCA) — i.e. its decomposition products (carbides?) was a major factor in the hardness increases observed. Samples containing no oxides or PCA had a room temperature hardness of 325 DPH after HIPing. Samples containing both oxide dispersoids and the PCA had hardness values between 450 and 490 DPH, relatively insensitive to either type or volume fraction of oxide. Samples with no oxides but with the PCA had a hardness of 430 DPH compared to a sample with 2.5 volume % Al_2O_3 but no PCA and a hardness of 380 DPH. Work in progress includes microstructural examination by TEM and mechanical tests at elevated temperatures.

5.0. Summary

Inert dispersoids, such as stable oxides, have been shown for many years to be important strengtheners in metals and alloys, especially at temperatures at high fractions of the melting point. Studies of dispersoids in intermetallics are limited in number and scope as indicated in this review. The only detailed mechanistic studies have been carried out on the model material Cu_3Au. These have been limited to yield strength and work hardening effects at relatively low temperatures. No mechanistic studies of elevated temperature, e.g. creep phenomena, in intermetallic-dispersoid systems have been reported. The work on dispersoids in the aluminides has been mainly phenomenological and developmental in scope. However, work on such systems as $Ni_3Al + Al_2O_3$ and $Ti_3Al + Er_2O_3$ is in progress. It is expected more definitive mechanical property studies coupled with TEM observations will be forthcoming on such materials in the near future. This will allow for critical evaluations of the relative importance of the dispersoid effects on dislocation mobility, stabilization of dislocation structure, and grain structure. This, in turn, should provide the background for design of intermetallic-dispersoid materials with improved mechanical properties, both for fabricability and elevated temperature service.

ACKNOWLEDGEMENTS

The preparation of this paper was supported by the Department of Energy, Energy Conversion and Utilization Technology Office, on Subcontract Number 19X–43368C from Oak Ridge National Laboratory.

REFERENCES

1) F. V. Lenel, G. S. Ansell, and E. C. Nelson, Trans AIME, **209**, 117 (1957).

2) O. Preston and J. J. Grant, Trans AIME, **221**, 164 (1961).

3) H. Schreiner, Pulvermetallurgie elektrischen Kontakte, in Reine und angewandte Metallkunde in Einzeldarstellung vol. 20 ed. W. Köster (Springer, Berlin) 1964.

4) G. Frommeyer, "Metallic Composite Materials" in Physical Metallurgy, 3rd edition, ed. by R. W. Cahn and P. Haasen, North Holland 1983, p. 1860.

5) J.-L. Strudel, "Mechanical Properties of Multiphase Alloys" ibid pp. 1415-1416, and references cited therein.

6) G. S. Ansell, "The Mechanism of Dispersion Strengthening: A Review", in "Oxide Dispersion Strengthening" Proc. 2nd Bolton Landing Conference, ed. by G. S. Ansell, T. D. Cooper, and F. V. Lenel, Gordon and Breach, 1968, p. 61.

7) M. F. Ashby, "The Theory of the Critical Shear Stress and Work Hardening of Dispersion-Hardened Crystals", ibid, p. 143.

8) E. Orowan, **Symposium on Internal Stresses in Metals and Alloys,** Institute of Metals (1948), p. 451.

9) L. M. Brown and R. K. Ham, in "Strengthening Methods in Crystals", edited A. Kelly and R. B. Nicholson, Wiley, New York, 1971, p. 9.

10) M. F. Ashby, ibid, pp. 137-192.

11) S. M. L. Sastry and B. Ramaswami, Acta Metallurgica, **23**, 1517 (1975).

12) M. F. Ashby, Phil. Mag. **21**, 399 (1970).

13) S. Pattanaik and A. J. Ardell, Phil. Mag. A, **45** 1047 (1982).

14) A. J. Ardell and S. Pattanaik, Phil. Mag. A, **50** 361 (1984).

15) A. U. Seybolt, Trans. ASM, **59**, 860 (1966).

16) R. J. Jaffe, B. C. Allen, and D. J. Maykuth, Plansee Proceedings, 1961, Metallwerke Plansee-Springer, Vienna (1962).

17) M. Sherman and K. Vedula, J. Mater. Science, **21**, 1974, (1986).

18) G. Sainfort, P. Mouturat, P. Pepin, J. Petit, G. Cabane, and M. Soluse, Mem. Sci. Rev. Met. **60**, 125 (1963).

19) S. Strothers, N. Mantravadi, and K. Vedula, this sypmposium.

20) E. R. Slaughter and S. K. Das in "Proc. of the 2nd International Conf. on Rapid Solidification Processing", ed. R. Mehrabian, B. H. Kear, and M. Cohen, Claitor's Publishing Division, Baton Rouge, La., (1980) p. 354.

21) C. G. McKamey, C. T. Liu, J. V. Cathcart, S. A. David, and E. H. Lee, Evaluation of Mechanical and Metallurgical Properties of Fe$_3$Al-Based Aluminides, ORNL/TM-10125.

22) C. G. McKamey, C. T. Liu, and J. A. Horton, Effect of Aluminum Addition on Ductility and Yield Strength of Fe_3Al Alloys with 0.5 wt.% TiB_2, this symposium.

23) H. A. Lipsitt, High-Temperature Ordered Intermetallic Alloys, ed. by C. C. Koch, C. T. Liu, and N. S. Stoloff, MRS Symposia Proceedings **39** (1985) p. 351.

24) S. M. L. Sastry and H. A. Lipsitt, Acta Met. **25** 1279 (1977).

25) D. G. Konitzer and H. L. Fraser, Mat. Res. Soc. Symposium Proc., **39**, 437 (1985).

26) D. G. Konitzer, B. C. Muddle and H. L. Fraser, Scripta Met. **17**, 963 (1983).

27) J. A. Sutliff and R. G. Rowe, "Rare Earth Oxide Dispersoid Stability and Microstructural Effects in Rapidly Solidified Ti_3Al and Ti_3Al-Nb" in **Rapidly Solidified Alloys and Their Mechanical and Magnetic Propertis**, ed. B. C. Giessen, D. R. Polk, and A. I. Taub, Proc. Mat. Res. Soc. Symposium, **58**.

28) R. G. Rowe, J. A. Sutliff, and E. F. Koch, "Dispersoid Modification of Ti_3Al-Nb Alloys ibid.

29) R. G. Rowe, J. A. Sutliff, and E. F. Koch, "Comparison of Melt Spun and Consolidated Ti3Al-Nb Alloys With and Without a Dispersoid" in **Rapid Solidification Technology for Titanium Alloys** ed. F. H. Froes, D. Eylou, and S. M. L. Sastry, Proc. AIME Conf. New Orleans, LA, March 1986, TMS-AIME.

30) K. Aoki and O. Izumi, Nippon Kinzoku Gakkaishi **43**, 1190 (1979).

31) C. T. Liu, C. L. White, C. C. Koch, and E. H. Lee, Proc. High Temperature Materials Chemistry II, p. 32, Electrochemical Society, Inc. (1983).

32) I. Baker, F. S. Ichishita, V. A. Surprenant, and E. M. Schulson, Metallography 17, 299 (1984).

33) I. Baker, F. S. Ichishita, and E. M. Schulson, Mat. Res. Soc. Symp. Proc. **28**, 395 (1984).

34) I. Baker, J. A. Horton, and E. M. Schulson, Metallography, **19**, 63 (1986).

35) S. C. Huang, A. I. Taub, K. M. Chang, C. L. Briant, and E. L. Hall, Rapidly Quenched Metals (RQ5) vol. II, ed. S. Steele and H. Warlimont, Elsevier Science Publishers (1985) p. 1407.

36) R. J. Lauf and C. A. Walls, ECUT Quarterly Report, July 1–Sept. 30, 1985, ORNL p. 56.

37) S. G. Donnelly and C. C. Koch, ECUT Quarterly Reports, Sept. 1, 1983 to Sept. 1, 1986, North Carolina State University.

B2 ALUMINIDES FOR HIGH TEMPERATURE APPLICATIONS

K. VEDULA[*] AND J.R. STEPHENS[**]
* Associate Professor, Department of Metallurgy and Materials Science, Case Western Reserve University, Cleveland, Ohio, 44106.
** NASA Lewis Research Center, Cleveland, Ohio, 44135.

INTRODUCTION

The B2 aluminides are currently being investigated for potential high temperature structural applications. Although they are not being as actively pursued as the titanium aluminides or the Ll_2 nickel aluminide, the B2 aluminides are very attractive from density considerations. Several recent reviews of the potential for aluminides are available in literature [e.g Ref. 1,2]. Table I is a comparison of the titanium, nickel and iron aluminides of interest and shows that B2 NiAl and FeAl have the major advantage of lower densities than Ni_3Al and Fe_3Al. In addition, the melting point of NiAl is over 200K higher than conventional nickel based superalloys. Hence, although low density is the prime driving force, at least in NiAl a temperature advantage is also possible. Both of these aluminides have the advantage of containing very inexpensive elements. In fact, the thrust towards the B2 aluminides evolved from a program aimed at conserving strategic aerospace materials at NASA Lewis Research Center. A recent thrust at NASA Lewis Research Center has been to consider these aluminides as matrix materials for fiber reinforced composite systems.

Table I. Comparison of physical properties of aluminides.

Compound	Crystal Structure	Melting Temperature	Density g/cc
Ti_3Al	DO_{19}	$1600^\circ C$	4.2
TiAl	Ll_0	$1460^\circ C$	3.9
Fe_3Al	DO_3	$1540^\circ C$	6.7
FeAl	$B2$	$1330^\circ C$	5.6
Ni_3Al	Ll_2	$1390^\circ C$	7.5
NiAl	$B2$	$1640^\circ C$	5.9

The aluminides of interest for this paper are B2 FeAl and NiAl and the phase diagrams of interest are presented in Figure 1. The other B2 aluminide, CoAl, was initially explored at NASA Lewis Research Center, but was dropped due to extreme brittleness and inability to process suitable material. The Fe-Al phase diagram has been modified from the Metals Handbook [3], such that it is in general agreement with the investigations by Swann and coworkers [4,5,6] as well as Allen and Cahn [7,8,9]. The Ni-Al phase diagram has been modified to include the phase Ni_5Al_3 based on some recent studies [10,11,12,13]. The NiAl phase is a monotectic with about a 15 at% range in solid solubility and a congruent maximum in melting temperature of ~ 1913K ($1640^\circ C$). FeAl, on the other hand, does not possess a high melting temperature; rather the melting temperature continuously decreases with Al concentration to about 1523K ($1250^\circ C$) at 52 at% Al.

The B2 (CsCl) crystal structure, illustrated in Figure 2, is a body centered ordered structure. It is important to realize that the BCC ordered alloys behave quite differently from the FCC ordered alloys such as the Ll_2, Ni_3Al.

POINT DEFECTS

An understanding of the point defect behavior of these alloys is important because they form the basis for diffusion assisted processes. The wide range of deviations from stoichiometry, in the B2 phase field, is

382

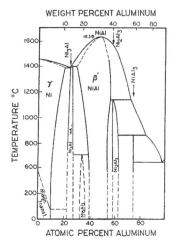
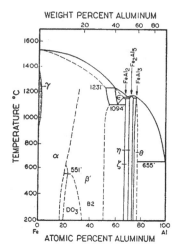

Figure 1. Phase Diagrams for Fe-Al and Ni-Al

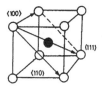

Figure 2.

The B2 Crystal
Structure

accomodated by the formation of defect structures. For Ni or Fe rich alloys, antistructure defects are predominant where excess Ni or Fe atoms fill Al sites [14,15,16]. For Al-rich alloys, excess Al atoms, generally, do not form antistructure defects; instead these atoms fill normal Al sites and the predominant point defects are vacant Ni sites [14,16,17]. However, the point defect structure of NiAl at stoichiometry is believed to be composed of triple defects [18,19]. consisting of two vacancies on Ni sites plus a Ni atom on an Al site. The defect structure in Al-rich and stoichiometric FeAl alloys is somewhat uncertain although combinations of antistructure defects and Fe vacancies combinations have been suggested, similar to the NiAl alloys [15,18,20,21,22].

Although the composition dependence of the type and number of point defects in these FeAl and NiAl alloys has been explored extensively, their temperature dependence has not been examined in any great detail. The limited information shows that in FeAl, even at 1073K (800°C) the concentration of vacancies is about 40 times greater than that found in a normal metal at its melting temperature [23]. Stoichiometric NiAl, on the other hand, appears to possess concentrations of vacancies at elevated temperatures, which are similar in magnitude to those in many metals [24,15].

Vacancies play a major role in high temperature mechanical deformation because of the importance of diffusion processes. Therefore, some connection between the point defect structure and high temperature mechanical properties of B2 alloys must exist. It is possible that certain point defects or combinations of defects decrease atomic mobility while others increase it. Furthermore, solid solution alloying will most certainly affect defect concentrations [25] and thereby play a key role in the mechanical behavior at high temperatures. These correlations need to be better understood before a proper basis for alloy design can be established.

SINGLE CRYSTALLINE MATERIAL

Single crystals of both FeAl and NiAl are ductile at room temperature. The details of operative slip systems have been studied to some extent and the behavior is shown to be a sensitive function of orientation and temperature.

There is general agreement that FeAl single crystals deform in the slip direction <111>, although there is some disagreement about the slip plane. Yamagata and Yoshida [26] report slip basically on {211} with {101} becoming active at higher test temperatures while Crawford [27] reports slip only on {101} at room temperature. Umakoshi and Yamaguchi [28,29] have observed a transition in active slip systems from {110}<111> to {110}<100> and or {100}<100> as the test temperature increased in the range 470 to 1000 K for B2 FeAl single crystals in vacuum. This transition temperature was found to change with deviations from stoichiometry.

The <111> dislocations require the formation of an antiphase boundary (APB) as can been seen from the crystal structure. Ray, Crawford and Cockayne [30] examined thin foils cut from a room temperature compression tested Fe-35at%Al single crystal and found the total Burger's vector to be a<111> type which was split into two a/2<111> partial dislocations separated by an antiphase boundary (APB). They also measured the APB energies and found that the APB energy increased linearly with increasing Al content. The trend shown by these experimental values has been extrapolated to higher aluminum contents by Crimp and Vedula [31] using calculations based on Marcinkowski and Brown [32]. Yamagata [33] has observed double images when viewing the a<111> type dislocations in deformed Fe-48at% Al single crystals, which he ascribes to the anisotropy of FeAl rather than to the partial dislocation-APB geometry. This interpretation is consistent with a large value of APB in higher Al alloys, making it difficult to resolve the individual partial dislocations.

In polycrystalline FeAl alloys a similar change in slip direction is observed, as a function of temperature and aluminum content by Mendiratta et.al. [34], although there is disagreement as to whether the transition temperature increases or decreases with aluminum content. Crimp and Vedula [31] have observed the dislocations in hot extruded polycrystalline FeAl alloys and find that Fe-50at%Al contains essentially all <100> dislocations whereas the Fe-40at%Al alloy contains some <111> dislocations along with the <100> type. Subsequent plastic deformation of the Fe-40at%Al alloy at room temperature clearly shows an increased number of <111> dislocations.

A number of studies of near-stoichiometric NiAl single crystals have been conducted. The <100> oriented crystals are more difficult to deform than other orientations making this alloy very anisotropic. In compression all orientations possess some ductility irrespective of the test temperature [35,36,37]. However, due to the anisotropy of slip, the non-<100> crystals are much more ductile at lower temperatures than the <100> oriented crystals. At 673K and above, all orientations possess considerable ductility even when tested in tension. There is general agreement that for non-<100> oriented crystals, deformation involves {100}<001>, and {110}<001> slip systems at all temperatures. The behavior of <100> oriented crystals is less clear and conflicting data exist; at or below room temperature <111> slip direction with {123} [37,38] and {112} slip planes [39] have been observed; while high temperature deformation has been reported to be the result of slip in <111> [37], <110> [37,39,40] and <100> [36,41,42] directions.

Results for low temperature deformation of NiAl indicates that only <100> slip is operative (except for <100> orientations), which is insufficient for general deformation in a polycrystalline alloy [35,43,44]. In addition, neither antiphase boundaries nor superdislocations are found in annealed or deformed material, since <100> dislocations do not create APBs. The dislocation configurations in an extruded, stoichiometric NiAl alloy, however, show evidence for the existence of glissile dislocations with Burger's vectors of <100>, <110> and <111> [45]. The additional two vectors provide enough independant slip systems at high temperatures, to satisfy Von Mises criterion for general plastic flow in a polycrystalline aggregate. These additional slip systems coupled with climb mechanisms which play an important role at elevated temperatures, make NiAl ductile at elevated temperatures.

POLYCRYSTALLINE MATERIAL

One of the earliest attempts to judge the potential of intermetallic compounds for use as high temperature structural materials was conducted by Westbrook [46]. Through microhardness testing he was able to demonstrate that the strengths of FeAl and NiAl are dependent on both composition and temperature. Typical hardness-composition curves at low temperature are reproduced in Figure 3, which reveals that a minimum in hardness at nominally 50 at% Al exists at low temperatures in NiAl alloys. Westbrook interpreted this behavior to mean that vacant atom sites (Al-rich alloys) as well as substitutional atoms (Al-poor alloys) provide strengthening effects. In contrast, the FeAl system did not possess any minimum in hardness-composition profiles. No explanation of this behavior was put forth. It is likely that this difference is due to the different slip systems which are active in each of the two alloys at low temperatures i.e. the <100> type in NiAl and the <111> type in FeAl. The <100> dislocations do not require the formation of antiphase boundaries, whereas the <111> type need to form antiphase boundaries and deviations from stoichiometry can affect the APB energy quite significantly.

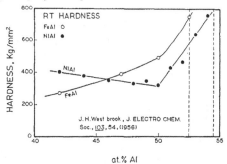

Figure 3. Room Temperature Hardness as a of Composition in FeAl and NiAl Alloys [46].

A few tensile tests on polycrystalline FeAl, containing nominally 40 at% Al, have been conducted by Sainfort and co-workers [47,48] as a function of temperature. These tests indicate tha FeAl has some room temperature ductility. Chatterjea and Mendiratta [49,50] have examined alloys containing 35 and 50 at% Al and find that about 7% tensile ductility can be obtained at room temperature for the Fe-35at%Al alloy, but the Fe-50at%Al alloy fractures in a brittle manner without any ductility.

More recently, the low temperature tensile properties of FeAl alloys have been studied by Crimp and Vedula [51,52]. Details of some of their recent results are presented elsewhere in these proceedings. Crimp and Vedula have confirmed that iron rich deviations from stoichiometry can result in appreciable tensile ductilities at room temperature. This has been attributed to the lower antiphase boundary energy with lower aluminum content, making it easier to move <111> dislocations and allowing for plastic yielding to take place before intergranular failure. Small amounts of boron additions have been found to result in some improvement in the room temperature tensile ductility and this is believed to be due to strengthening of the grain boundaries in a manner similar to the boron effect in Ni_3Al. In addition, the rate of cooling from an annealing temperature of 1273K (1000°C) has been

TEMPERATURE DEPENDANCE OF STRENGTH

The strengths of the two B2 aluminides, NiAl and FeAl, decrease with increasing temperature, as do regular metals. The unique positive temperature dependence of strength observed in Ni$_3$Al is peculiar to the Ll$_2$ crystal structure because of the anisotropy of stacking fault and antiphase boundary energies [57]. (In B2 alloys, such a temperature dependence has been observed in β brass). A drastic decrease in strength accompanied by a sudden increase in the ductility is observed in both FeAl and NiAl alloys at about 0.4 T$_m$ [58,59], and is illustrated in Figure 4 for FeAl alloys [59] and Figure 5 for NiAl alloys [35]. The reasons for this brittle to ductile transition are not entirely clear. The temperature dependence of slip behavior and the onset of diffusion assisted processes must definitely play a role in this transition behavior.

Another interesting feature is the intermediate temperature embrittlement reported by Gaydosh in FeAl alloys [59] and seen in Figure 4 as well. The reasons for this are not yet clear, but the behavior resembles a

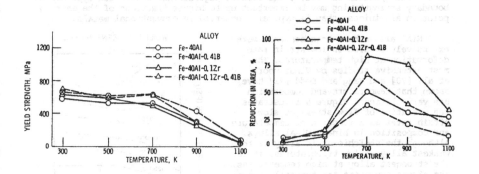

Figure 4: Temperature dependence of strength and ductility of FeAl alloys [59]

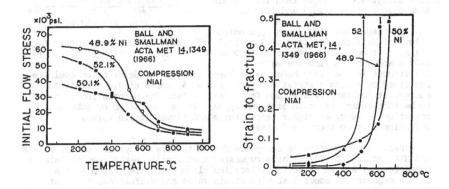

Figure 5. Temperature dependance of strength and ductility of NiAl alloys [35]

similar phenomenon observed in boron—doped Ni_3Al tested in air [60].

Elastic modulus is a key factor in the design of high temperature structural parts and some effort is being devoted to quantifying modulii of NiAl and FeAl as functions of temperature [61]. These measurements show high values of modulus at low temperatures with a gradual decrease as temperature increases.

HIGH TEMPERATURE CREEP DEFORMATION

The high temperature deformation mechanisms of these aluminides have not been extensively studied. Recent studies by Whittenberger [62,63,64] and Vedula et.al. [65,66,67,68] have been restricted to compression creep deformation. Earlier studies of creep in Fe—Al alloys have been limited to alloys containing up to 28 at% Al [69]. Recently, Whittenberger [63] has shown that FeAl (in the range of 41.7 to 48.7 at% Al) deforms by two independant deformation mechanisms with the same activation energy in the temperature range 1100 to 1400 K. Both mechanisms are dependent on grain size; however, for the high stress exponent regime (n~5), strength increases with a decreasing grain size, while for the lower stress exponent regime (n~1), the opposite is true. These results suggest that Hall—Petch type grain boundary strengthening may be important up to higher fractions of the melting point in aluminides, than is typically observed in conventional metals.

NiAl alloys have been somewhat more extensively studied in low strain rate deformation at high temperatures. The most extensive studies by Vandervoort et al. [70] and Yang and Dodd [71] have shown that temperature and composition are very critical. Figure 6 illustrates the variation of flow stress at slow strain rates as a function of temperature and composition in binary Ni-Al alloys. Although the stoichiometric alloy is the weakest alloy at low temperatures, it is the strongest alloy at high temperatures. The stress exponents are typically between 3 and 5, suggesting combinations of dislocation climb controlled creep and viscous glide controlled creep. These studies also indicate that fine grain size alloys are stronger than large grain size alloys between 1073 and 1273 K.

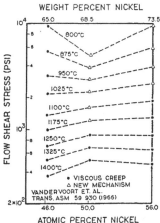

Figure 6. High temperature flow behavior of NiAl alloys [70]

Whittenberger [64] recently studied alloys of NiAl ranging from 43.9 to 52.7 at% Al in compressive creep from 1200 to 1400 K and did not find any major differences in the creep behavior of these alloys. The deformation could be described by a single stress exponent (around 5) and a single activation energy, at low temperatures and fast strain rates. The deformation is found to be independant of grain size and appears to be controlled by subgrain formation. However at higher temperatures and slower strain rates, diffusional creep seems to dominate.

Results of compressive creep studies of ternary alloys of FeAl at 1100 K and NiAl at 1300 K [72,73] have suggested that similar mechanisms operate in these ternary alloys as well. Dislocation climb controlled creep with a stress exponent of about 5 at high strain rates and diffusional creep at lower strain rates with a stress exponent closer to 1 are typically observed. Ternary NiAl alloys with second phase strengthening, demonstrate a stress exponent closer to 3, possibly indicating viscous glide control.

GRAIN BOUNDARY PROPERTIES

Very few studies have addressed the critical issue of grain boundaries in these alloys. Although addition of boron has slightly improved the room temperature ductility of FeAl alloys, it is not clear if boron segregation to grain boundaries is responsible. In NiAl alloys, addition of boron has not resulted in a any significant improvement in room temperature ductility, nor has it caused a change in the fracture mode.

Grain growth studies [73,74,75] in NiAl and FeAl alloys indicate that grain boundary mobility is fairly high in these ordered microstructures. In powder processed alloys of NiAl and FeAl, grain growth is very slow at low temperatures but takes off very rapidly at high temperatures. This is believed to be due to the presence of the prior particle oxides. A wide prior particle size distribution in extruded microstructures can lead to abnormal grain growth at high temperatures. Another interesting observation is the evidence of very rapid grain growth in samples compression tested at about 1273K (1000°C), suggesting a significant stress assisted increase in the mobility of grain boundaries.

An interesting phenomenon related to grain boundaries in NiAl is the observation of dynamic recrystallization at temperatures of the order of 1073K (800°C) [76]. Figure 7 illustrates the effect of dynamic recrystallization on the tensile stress–strain behavior of powder processed, hot extruded Ni-45at%Al at 1073K (800°C). This is interesting because it suggests that recovery mechanisms are slow in these ordered stuctures and this would impact strongly on their creep behavior.

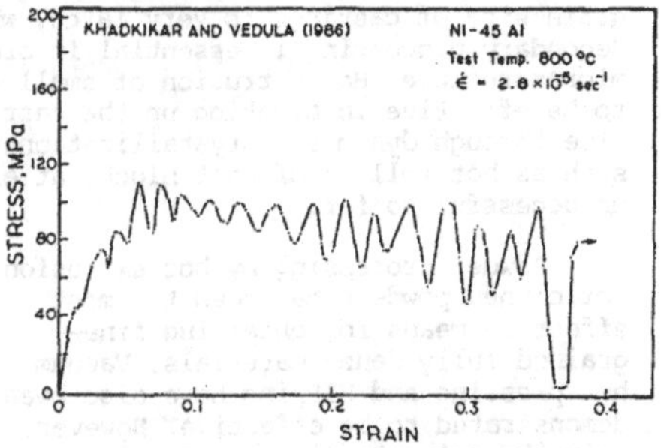

Figure 7. Dynamic recrystallization during hot tensile testing of NiAl [76]

ALLOYING ADDITIONS

In view of the inadequate strengths of the binary alloys at temperatures of interest, alloying additions are needed to modify properties. Information on the phase diagrams for ternary additions to NiAl and FeAl is scarce [72,77,78].

Ternary additions result in substantial improvements in the high temperature compressive creep of both NiAl and FeAl alloys [66,67]. Several elements have been studied, including those which have a large solubilities (Cr, Mn, Ti, Co) and those which have limited solubilities (Nb, Ta, Hf, Mo, W, Zr, B) in the binary alloys. Boron has been found to have a significant impact on the creep behavior in addition to its effect on room temperature ductility of FeAl alloys. An alloy containing 0.1 at% Zr and 0.2 at% B has been found to be the best among the alloys investigated at 1100 K (Figure 8) [72]. The activation energy for the deformation of this alloy at 1100 K

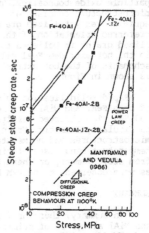

Figure 8. Effect of alloying on creep of FeAl alloys [53]

has been shown to be about 50% higher than that of the binary alloy, suggesting that diffusion processes have been made more difficult with addition of Zr and B.

In the case of NiAl, ternary additions which result in the precipitation of second phases have shown the best creep resistance at 1300 K. In particular, small amounts of additions of Ta and Nb have been shown to cause substantial improvements in compressive creep behavior. Precipitates of ternary intermetallic phases have been observed and are believed to cause strengthening by dislocation-particle interactions. These ternary intermetallics have been carefully characterized and are typically the Heusler type compounds (Ni$_2$AlTa, Ni$_2$AlNb) and the hexagonal compounds (NiAlTa, NiAlNb) [79]. The phases found in the microstructure are in agreement with the limited amount of ternary phase diagram information available [80].

PROCESSING CONSIDERATIONS

Conventional casting of these alloys has proven to be difficult. Segregation results in regions of very high aluminum contents which are brittle and cause microcracking during cooling or subsequent processing. The grain size of castings is very large, which is an added disadvantage. Secondary processing is essential in order to obtain any reasonable microstructure. Hot extrusion of small castings in containers has been shown to be effective in breaking up the cast structure and in refining the grain size through dynamic recrystallization [81,82]. Standard secondary operations such as hot rolling of cast blocks at elevated temperatures have been unsuccessful so far.

Powder processing by hot extrusion of canned powders has been the most effective means for obtaining fine-grained fully dense materials. Vacuum hot pressing and HIPping have also been demonstrated to be effective. However, only the hot extrusion process is effective in breaking up the prior particle oxide boundaries. The resultant microstructure consists of fine equiaxed recrystallized grains. Some preferred orientation of the recrystallized grains is believed to be present in the hot extruded FeAl alloys. A schematic of the hot extrusion process is shown in Figure 9.

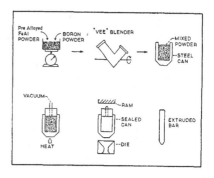

Rapid solidification by melt spinning of both FeAl and NiAl have been attempted [83,84]. The observed results are not

Figure 9. Schematic of the P/M Hot Extrusion Process

particularly dramatic. Grain sizes are not much finer than hot extruded microstructures and no metstable microstructures are observed. Room temperature bend tests of ribbons show that similar effects of deviations from stoichiometry and heat treatment are observed in the melt spun materials as were observed in the P/M hot extruded material.

SUMMARY

In summary, therefore, the room temperature brittleness of these B2 FeAl and NiAl alloys is of primary concern. In the case of FeAl, significant tensile ductilities of up to 8% have been obtained at room temperature in iron-rich alloys. The major reason for this appears to be that <111> dislocation are mobile and, hence, it is possible to satisfy Von Mises criterion for yielding without fracturing. Intermediate temperature embrittlement will, however, be of concern for thermomechanical processing of these alloys into any suitable wrought product.

The development of NiAl, however, is seriously hampered by room temperature brittleness. Although some earlier reports have indicated compressive ductilities and even some tensile ductilities at room temperature in polycrystalline NiAl alloys, more recently, the attempts at making polycrystalline NiAl alloys ductile have not been successful. The most likely composition to possess adequate ductility is the stoichiometric composition. Careful control of oxygen contamination could be another critical issue. But the most central aspect appears to be the problem of adequate number of mobile dislocations on sufficient number of slip systems. If somehow additional slip systems in hard orientations can be activated, a solution to the problem may be possible. One approach to this has been suggested by the work of Gibala on oxide film softening in BCC metals [84]. Softening of BCC metals and single crystal NiAl is observed when a thin oxide film is present on the surface. The explanation for this is the activation of sources for mobile dislocations resulting from stresses at the interface. It may be possible to exploit this mechanism internally at interfaces of oxide dispersions in polycrystalline materials.

Yet another approach which has been suggested and attempted to a limited extent is the possibility of alloying to modify the slip direction from <100> to <111>. Empirical parameters such as the electron-to-atom (e/a) ratio and radius ratio have a strong influence on deformation behavior of intermetallic compounds. Alloying additions can, hence, be made which may induce changes in these parameters in the direction of improved ductility. Some B2 compounds are ductile at room temperature and hence, correlations between the empirical parameters and ductility in several B2 compounds may provide clues for improving the room temperature ductility of NiAl. The e/a approach appears to have some relevance in B2 TiX alloys [85] which shows how the plastic deflection of TiX compounds correlates with the electron-to-atom ratios in TiX compounds. The radius ratio approach must also be feasible, since the bonds in intermetallic compounds can be rated in terms of their deviations from metallic nature [86] and appropriate alloying additions could be made to improve the metallic nature. The more metallic the bond, the greater the chances of activating <111> dislocations as in BCC metals. Approaches based on such empirical correlations have, in fact, been demonstrated to have some relevance as illustrated in the (Fe,Ni,Co)$_3$V system [87].

Grain refinement is yet another avenue as suggested by Schulson since this will encourage crack deflection. Oxide dispersion strengthening appears to show promise of grain refinement and enhanced room temperature ductility in a preliminary study of FeAl alloys [88]. However oxygen content could be of concern in P/M materials, and in particular in ODS materials.

In terms of processing opportunities, powder processing and hot extrusion in containers, appears to be the most appropriate technique until the hot workability of these alloys can be improved. Injection molding may offer some interesting possibilities of circumventing the difficulties of conventional techniques in obtaining near-net shapes of such low ductility materials.

REFERENCES

1. N.S. Stoloff, MRS Symposia Proceedings, High Temperature Ordered Intermetallic Alloys, Vol. 39, 1984, p.1.
2. 'Structural Uses of Ordered Intermetallic Alloys' Report by National Materials Advisory Board, NMAB-419, National Academy Press, Washington, D.C. 1984.

3. American Society for Metals, Metals Handbook, 8th ed. Vol. 8., Metals Park, Ohio, 1973.
4. P.R. Swann, W.R. Duff and R.M. Fisher, Met. Trans. Vol. 3. p. 409, 1972.
5. P.R. Swann, W.R. Duff and R.M. Fisher, Trans. Met. Soc. AIME, V. 245, p.851, (1969).
6. P.R. Swann, W.R. Duff and R.M. Fisher, Phys. Stat. Sol. Vol.37, p. 577 (1970).
7. S.M. Allen and J.W. Cahn, Acta Met. Vol. 23, p. 1017, 1975.
8. S.M. Allen and J.W. Cahn, Acta Met. Vol. 24, p. 425, 1976a.
9. S.M. Allen and J.W. Cahn, Acta Met. Vol. 10, p. 451, 1976b.
10. K. Enami and S. Nenno, Trans. Jpn. Inst. Met. Vol. 19., p.571, 1978.
11. I.M. Robertson and C.M. Wayman, Metallography, Vol. 17, p. 43, 1984.
12. M.F. Singleton, J. Murray and P. Nash, submitted to Bull. Alloy Phase Diagrams.
13. P.S. Khadkikar and K. Vedula, submitted to J. of Materials Research.
14. N. Ridley, J. Inst. Metals, Vol. 94, p. 255, 1966.
15. A.J. Bradley and A. Taylor, Proc. Roy. Soc. A136, p. 210, 1932.
16. A.J. Bradley and A. Taylor, Proc. Roy. Soc. A159, p. 56, 1937.
17. W.J. Wang, F. Lin and R.A. Dodd, Scripta Met. Vol. 12, p. 237, 1978.
18. J.P. Neumann, Acta Met. Vol. 28, p. 1165, 1980.
19. R.J. Wasilewski, J. Phys. Chem. Solids, Vol. 29, p. 39, 1968.
20. A. Taylor and R.M. Jones, J. Phys. Chem. Solids, Vol. 6, p.16, 1958.
21. J.P. Neumann, Y.A. Chang and C,M, Lee, Acta Met. Vol. 24, p. 593, 1976.
22. D. Paris, P. Lesbats and J. Levy, Scripta Met, Vol. 9, p. 1373, 1975.
23. K. Ho and R.A. Dodd, Scripta Met., Vol.12, p. 1055, 1978.
24. G.G. Libowitz, Met.Trans, Vol.2, p. 85, 1971.
25. H. Jacobi and H.J. Engell, Acta Met., Vol.19, p. 701, 1971.
26. T. Yamagata and H. Yoshida, Mat. Sci. and Engg. Vol. 12, p. 95, 1973.
27. R.C. Crawford, Phil. Mag. Vol. 33, p. 529, 1976.
28. Y. Umakoshi and M. Yamaguchi, Phil. Mag. A, Vol. 41, p. 573, 1980.
29. Y. Umakoshi and M. Yamaguchi, Phil. Mag. A, Vol. 44, p. 711, 1981.
30. I.L.F. Ray, R.C. Crawford and D.J.H. Cockayne, Phil. Mag. Vol.21, p. 1027, 1970.
31. M.A. Crimp and K. Vedula, unpublished results.
32. M.J. Marcinkowski and N. Brown, Acta Met. Vol. 9, p. 764, 1961.
33. T. Yamagata, Trans. JIM, Vol. 18, p. 715, 1977.
34. M. Mendiratta, H.M. Kim and H.A. Lipsitt, Met. Trans. 15A, p. 395, 1984.
35. A. Ball and R.E. Smallman, Acta Met. Vol. 14, p.1349, 1966.
36. R.J. Wasilewski, S.R. Butler and J.E. Hanlon, Trans. Met.Soc. AIME, Vol. 239, p.1357, 1967.
37. R.T. Pascoe and C.W.A. Newey, Met. Sci.J. Vol.2, p. 138, 1968.
38. R.T. Pascoe and C.W.A. Newey, Phys. Stat. Sol. Vol. 29, p.357, 1968.
39. M.H. Loretto and R.J. Wasilewski, Phil. Mag. Vol. 23, p. 1311, 1971.
40. J. Bevk, R.A. Dodd and P.R. Strutt, Met. Trans. Vol.4, p. 159, 1973.
41. H.L. Fraser, R.E. Smallman and M.H. Loretto, Phil. Mag. Vol.28, p. 651, 1973.
42. H.L. Fraser, M.H. Loretto and R.E. Smallman, Phil. Mag. Vol.28, p. 667, 1973.
43. A. Ball and R.E. Smallman, Acta Met, Vol.14, p.1517, 1966.
44. E.P. Lautenschlager, T.C. Tisone and J.O. Brittain, Phys. Stat. Sol. Vol.20, p.443, 1967.
45. C.H. Lloyd and H.M. Loretto, Phys. Stat. Sol. Vol. 39, p. 163, 1970.
46. J.H. Westbrook, J. Electrochem. Soc. Vol. 103, p. 54, 1956.
47. G. Sainfort, P. Mouturat, Mme P. Pepin, J. Petit, G. Cabane and M. Salesse, Mem. Sc. Revde Met., Vol. 40, p. 125, 1963.
48. G. Sainfort, J. Gregoire, P. Mouturat and M. Romeggio in Fragilite et Effets de L'irradiation, eds. M. Salesse and M. Caudron, p. 187, Presses Universitaires de France, Paris, France, 1971.
49. D.K. Chatterjee and M.G. Mendiratta, J. Metals, Vol.33, no.12, p.5, 1981.
50. D.K. Chatterjee and M.G. Mendiratta, J. Metals, Vol.33, no.12, p.6, 1981.
51. M.A. Crimp, K. Vedula and D.J. Gaydosh, this proceedings.
52. M.A. Crimp and K. Vedula, Mat. Sci and Eng.,Vol. 78, p. 193, 1986.

53. N. Mantravadi, Ph.D. thesis, Case Western Reserve University, 1986.
54. A.G. Rozner and R.J. Wasilewski, J. Inst. Metals, Vol. 94, p. 169, 1966.
55. E.M. Schulson and D.R. Barker, Scripta Met., Vol. 17, p. 519, 1983.
56. E.M. Schulson, COSAM Program Overview, NASA TM. 830006, NASA Lewis Research Center, 1982, p. 175.
57. B.H. Kear and H.G.F. Wilsdorf, Trans. TMS-AIME Vol. 224, p. 382, 1962.
58. M.G. Mendiratta et. al. Technical Reports, AFWAL/MLLM, Contract F33615-81C-5059, 1982-1984.
59. D.J. Gaydosh, NASA Lewis Research Center, unpublished results.
60. C.T. Liu, C.L. White and E.H. Lee, Scripta Met. Vol. 19, p. 1247, 1985.
61. A. Wolfenden, NASA report, No. WAG3-314, Dec, 1985.
62. J.D. Whittenberger, Mat. Sci and Eng., Vol. 57, p. 77, 1983.
63. J.D. Whittenberger, Mat. Sci. and Engg., Vol 77, p. 103, 1986.
64 J.D. Whittenberger, paper accepted for publication in J. Mat. Sci.
65. K. Vedula et.al., Modern Developments in Powder Metallurgy, Vol 16, p. 727, 1984.
66. R.H. Titran, K. Vedula and G. Anderson, MRS Symposia Proceedings, High Temperature Ordered Intermetallic Alloys, Vol. 39, p. 319, 1984.
67. K. Vedula et. al., MRS Symposia Proceedings, High Temperature Ordered Intermetallic Alloys, Vol. 39, p. 411, 1984.
68. M. Sherman and K.Vedula, J. Materials Science, Vol. 21, p. 1974, 1986.
69. A. Lawley, J.A. Coll and R.W. Cahn, Trans. Met. Soc. AIME, Vol. 218, p.166, 1960.
70. R.R. Vandervoort, A.K. Mukherjee and J.E. Dorn, Trans. ASM, Vol. 59, p. 930, 1966.
71. W. Yang, R.A. Dodd and P.R. Strutt, Met. trans. Vol. 3, p. 2049, 1972.
72. N. Mantravadi and K. Vedula unpublished results.
73. V.M. Pathare and K. Vedula, unpublished results.
74. I. Aslanidis, M.S. Thesis, Case Western Reserve University, 1985.
75. M.A. Crimp and K. Vedula, unpublished results.
76. P.S. Khadkikar and K. Vedula, unpublished results.
77 V.M. Pathare, M.S. Thesis, Case Western Reserve University, 1984.
78 P. Nash, MRS Symposia Proceedings, High Temperature Ordered Alloys, Vol. 39, p.423, 1984.
79. V.M. Pathare, K. Vedula and G. Michal, paper submitted to Scripta Met.
80. K. Vedula et.al. unpublished work in progress.
81. E.M. Schulson and D.R. Barker, Scripta Met., Vol. 17, 1983, p. 519.
82 D.J. Gaydosh and M.A. Crimp, MRS Symposia Proceedings, High Temperature Ordered Alloys, Vol. 39, p. 429, 1984.
83. C.C. Law and M.J. Blackburn, United Technologies Corporation, Pratt & Whitney Group, Interim Reports, AFWAL/MLLM, Contract No.F33615-84-C-5067.
84. R. Noebe and R. Gibala, MRS Symposia Proceedings, High Temperature Ordered Intermetallic Alloys, Vol. 39, p.319, 1984.
85. R. Scholl, D.J. Larson and E.J. Friese, J. App. Phy., Vol. 39, No.5, p. 2186, 1968.
86. E.P. Lautenschlager et. al. Acta Met., Vol. 15, p. 1347, 1967.
87. C.T. Liu, Int. Met. Reviews, Vol. 29, p. 168, 1084.
88. S.D. Strothers and K. Vedula, unpublished results.

A REVIEW OF RECENT DEVELOPMENTS IN IRON ALUMINIDES

MADAN G. MENDIRATTA, SANDRA K. EHLERS*, DENNIS M. DIMIDUK, WILLIAM R. KERR, SIAMACK MAZDIYASNI,** AND HARRY A. LIPSITT***
*Universal Energy Systems, Inc., Dayton, OH 45432-1894
**AFWAL/MLLM, Wright-Patterson AFB, OH 45433-6533
***Wright State University, Dayton, OH 45435-0001

ABSTRACT

This paper presents a review of recent research conducted at the Air Force Wright Aeronautical Laboratories (AFWAL)/Materials Laboratory to develop iron-aluminides as elevated temperature structural materials. The research consisted of investigations on the microstructure, tensile behavior, deformation, and fracture mechanisms of the DO_3 and B2 iron-aluminides. Binary Fe-Al alloys with a wide range of Al contents, solid-solution ternary alloys and precipitation- and dispersion-strengthened two-phase alloys have been investigated. It is shown that iron-aluminides have a potential to be structural materials at least up to 650°C.

INTRODUCTION

It has been known for quite some time that iron-rich iron aluminides are highly oxidation and sulfidation resistant at elevated temperatures [1,2]. This resistance stems from the formation of a highly protective and adherent Al_2O_3 scale. The two types of aluminides have compositions in the vicinity of Fe-25 at.% Al (DO_3 crystal structure) and Fe-35 to -50 at.% Al (B2 crystal structure). The densities of these intermetallics range from 5.49-6.68 gm/cm^3 which are ∿ 30% lower than those of commercial nickel-base superalloys. Also, the ordered DO_3 (Fe_3Al) and B2 (FeAl) crystal structures are body centered cubic derivative structures. These structures do not, a priori, exclude the possibility of extensive dislocation motion and plasticity at room temperature. In the past, however, iron-aluminides were thought to be very brittle.

Based upon the reasons enumerated above, the AFWAL Materials Laboratory decided about six years ago to investigate and develop iron-aluminides as elevated temperature structural materials. The approach was to determine first the tensile behavior of the binary alloys with a wide composition range, and then, based upon the baseline data and understanding of the flow and fracture behavior, to explore solid-solution and two-phase alloys. This paper presents a review of the results of these efforts. Most of the results have been either published or the papers have been accepted for publication and, therefore, this review only highlights the most important aspects of the results.

Concurrent to the in-house work at the AFWAL Materials Laboratory, two contractual efforts involving Pratt & Whitney and TRW were also initiated. The results of those efforts have been reviewed recently by Kerr [3], with an emphasis on processing/fabrication and mechanical properties. The present paper will mainly emphasize the microstructure-property relations.

BINARY Fe-Al ALLOYS

The initial focus of research on iron-aluminides was the tensile behavior of the binary alloys encompassing both the DO_3 and B2 phase-fields. Specifically, the strength-ductility-temperature relations were determined for Fe-25, -31, -35, -40, and -50Al alloys (all compositions are expressed in atomic percent in this paper) and the deformation and fracture behavior was characterized in an effort to understand the basic mechanisms governing the tensile behavior. It was considered important to obtain these data to serve as a baseline for ternary and quaternary alloying modifications. The detailed results have been published elsewhere [4,5], a brief summary of important findings is given below.

All alloys, except the Fe-31Al composition, were prepared using appropriate blends of rapid-solidification-rate (RSR) powders having two base compositions, Fe-25Al and Fe-50Al. The Fe-25Al powders were consolidated by extrusion at 900°C with an extrusion ratio of 16:1, while the Fe-35, -40, and -50Al compositions were consolidated by extrusion at 1000°C using the same extrusion ratio. The recrystallized grain-size was 50 μm for Fe-25Al and 30 μm for the rest of the alloys. The Fe-31Al alloy was cast as a 3-inch billet and then extruded at 900°C with a 16:1 ratio resulting in a grain size of ∿ 150 μm.

The results of tensile tests showed that the Fe-25 to -40Al alloys were ductile at room temperature. Maximum ductility and strength were exhibited by Fe-25Al, i.e., 8% elongation to fracture and a yield strength of ∿ 710 MPa (∿ 103 ksi). The Fe-31Al alloy with a large grain size also exhibited a room-temperature ductility of ∿ 5.6% elongation to fracture. The room-temperature ductility decreased with increasing Al content, e.g., 2.5% for the Fe-40Al alloy and premature, totally brittle failure for the Fe-50Al alloy. The latter alloy exhibited a ductile-to-brittle transition at ∿ 450°C. The room-temperature strain-hardening rate increased with increasing Al content. The yield strength of the Fe-25Al alloy dropped

was taken into account, the operative slip system could be predicted correctly.

Limited work was also carried out on the room-temperature fatigue behavior of Fe-25Al [9] and elevated temperature creep of Fe-25Al and Fe-35Al [10]. In a tension-tension cyclic mode (R = 0.1), Fe-25Al was found to have a reasonable fatigue strength of $\sim$ 620 MPa (90 ksi) corresponding to 10^6 cycles with a ratio of fatigue strength-to-tensile fracture stress of 0.6. The creep resistance of the binary aluminides was low even at moderate temperatures, e.g., the Fe-25Al alloy exhibited a minimum creep rate of $\sim 10^{-2}\%$ per hour at 276 MPa (40 ksi) at 450°C and Fe-35Al a rate of 5 x $10^{-1}\%$ per hour at 570°C at the same stress level. These high rates may be associated with the relatively open structure of the ordered b.c.c. lattice.

SOLID SOLUTION ALLOYS

A large number of ternary DO_3 and B2 alloys were investigated in an effort to determine the effect of solid-solution additions on the tensile behavior of iron-aluminides. It has been mentioned that the yield strength of the Fe-25Al alloy decreased rapidly with temperatures above $\sim$ 400°C, due to an approaching $DO_3 \rightarrow$ B2 phase transformation, therefore, the approach to solid solution strengthening consisted of exploring those alloying additions which will raise the transformation temperature, T_c, significantly. It was hoped that increasing the temperature to which the DO_3 phase is stable would also increase the elevated-temperature strengths. The Fe-35 and -40Al alloys, on the other hand, are presumed to consist of a single phase B2 structure up to their melting temperatures. The Fe-35Al alloy was chosen as the base alloy for solid-solution alloying, since this alloy exhibited a reasonable room-temperature ductility ($\sim$ 6%), and the yield strength did not drop significantly with temperatures up to 600°C.

The ternary element additions in the Fe-25Al included Ti, V, Cr, Mn, Ni, Nb, Mo, Ta, Cu, and Si. By examining in the TEM the size and morphology of the thermal anti-phase domains (APDs) of the DO_3 structure, it was found [11] that Cr, Ni, and Mo additions increased T_c moderately, while Ti and Si additions increased T_c significantly. More extensive work was accomplished on the latter two ternaries, important results are reviewed below.

The Fe-20Al-5Si alloy was annealed at various temperatures to vary the DO_3 thermal APD size. TEM examination of this variation revealed that the $DO_3 \rightarrow$ B2 transition occurs between 725-750°C. The compound Fe_3Si is isostructural with Fe_3Al and possesses the DO_3 order up to its melting temperature of $\sim$ 1250°C. The ternary Si atoms substitute at the Al sublattice sites in the DO_3 structure. This substitution results in a gradual increase in T_c with increasing Si content.

Tensile tests were carried out on an Fe-20Al-5Si alloy which was prepared by blending and extruding argon atomized, prealloyed powders of the compositions Fe-25Al and Fe-25Si [12]. The test results are summarized in Table I which also includes corresponding results on Fe-25Al for comparison. It can be noted that the yield strengths of the Fe-20Al-5Si alloy are significantly higher at 600 and 700°C than those for Fe-25Al, although the Si addition severely embrittles the Fe-25Al. Thus, an increase in T_c correlates well with the significant increase in elevated temperature strength values. Further work has shown that the addition of very fine TiB_2 dispersoids in Fe-20Al-5Si, through rapid solidification techniques, produces a highly stable substructure with a grain size of 1-2 μm and a room-temperature elongation to fracture between 3-5 percent [9].

TABLE I. Tensile Properties of Fe-25Al and Fe-20Al-5Si Alloys (P/M + Hot Extrusion).

Fe-25Al: T_c ≈ 540°C
Fe-20Al-5Si: T_c = 725-750°C

Test Temperature	σ_y, ksi		ϵ_f,%	
	Fe-25Al	Fe-20Al-5Si	Fe-25Al	Fe-20Al-5Si
R.T.	101	118	8	0.6
600°C	20	85	30	5
700°C	-	50	32	3.5

In contrast to Si, Ti substitutes preferentially at specific Fe sub-lattice sites in the DO_3 structure. In a Fe-25Al-5Ti alloy it was found that the DO_3 domain boundaries lie preferentially on {100} planes and that the DO_3 → B2 transition occurs between 800-825°C [13]. Preliminary bend tests indicated that this alloy was totally brittle at room-temperature. No tests have been carried out at elevated temperatures.

In the B2 Fe-35Al, the following ternary elements were added to explore the solid-solution effects: 1, 2, and 4 atomic percent of Ti, Cr, V, and Ni and 1 and 2 atomic percent of Nb, Mo, Ta, W, and Si. The alloys were made by melting and casting of 150 gm buttons which were homogenized at 1000°C for 168 hr. and isothermally forged at 900°C to one-half the original button thickness. Four-point bend tests were carried out to determine the elastic limit stress at 600 and 700°C and to obtain an approximate idea of the ductility at room temperature. A summary of the best strength values are given in Table II. By comparing the strength values of the base Fe-35Al alloy with those of ternary alloys, it is concluded that all ternary additions investigated produced significant solid-solution strengthening at 600 and 700°C. All alloys exhibited some measure of room-temperature ductility. The Si containing alloys were the most brittle.

TABLE II. Elevated Temperature Strength of Fe-35 Al-X Alloys

Alloys	Elastic Bend Strength, MPa (ksi)	
	Test Temperatures	
	600°C	700°C
Fe-35 Al	296 (43)	124 (18)
-4Ti	538 (78)	331 (48)
-2C	414 (60)	365.4 (53)
-2V	469 (68)	345 (50)
-2Ni	538 (78)	296 (43)
-1Nb	503 (73)	345 (50)
-1Mo	517 (75)	379 (55)
-2Ta	538 (78)	324 (47)
-1W	476 (69)	455 (66)
-2Si	724 (105)	345 (50)

TWO-PHASE IRON-ALUMINIDES

In order to explore which ternary additions might provide two-phase alloys, it was decided to determine the solubility limits of various elements in Fe-35Al and the composition of precipitated phases when the solubility is exceeded. A rapid approximate method for obtaining these data involved preparation of psuedo diffusion couples with a piece of element X being surrounded entirely by Fe-35Al powder, hot-pressing these compacts, and subsequent, long-time annealing at high temperatures. Electron probe microanalysis and scanning electron microscopy (SEM) were carried out to characterize microstructures in these diffusion couples. The details of the experimental technique and results are given elsewhere [9].

Based upon diffusion-couple data and further exploratory research, a number of two-phase iron-aluminide systems were identified. The types of microstructures and corresponding compositions are given in Table III. Extensive research was carried out on the thermal stability and strengthening potential of these two-phase iron aluminides. A brief review of the results is given in the following paragraphs.

TABLE III. Two-Phase Microstructures and Alloys

Microstructures	Alloys
DO_3 Matrix + Coherent $L2_1$ Precipitates	Fe-25Al-2Nb-X Fe-25Al-2Ta-X
DO_3 Matrix + TiB_2 Dispersoids DO_3 Matrix + Zr-Rich Dispersoids B2 Matrix + TiB_2 Dispersoids B2 Matrix + Hf-Rich Dispersoids	Fe-25Al + TiB_2 Fe-25Al-1Zr Fe-35Al + TiB_2 Fe-35Al + 1Hf
α Matrix + Coherent DO_3 Precipitates DO_3 Matrix + Coherent α Precipitates	Fe-XAl-5Ti where X = 15-25

DO_3 Matrix + $L2_1$ Coherent Particles

In Fe-25Al alloy, 2 at.% Nb could be totally dissolved at temperatures ≥ 1250°C. When an alloy with the composition Fe-25Al-2Nb was water-quenched from above 1250°C and aged at temperatures between 600-750°C, a metastable $L2_1$ phase, in the form of uniformly distributed coherent particles, precipitated in the DO_3 matrix. Figure 1 shows the distribution of this phase for a given aging treatment. The identification of this phase was carried out using convergent beam electron diffraction, scanning transmission electron microscopy (STEM), and iterative structure factor calculations [14]. When specimens were aged for longer times, and/or higher temperatures, large, incoherent particles began to appear in the microstructure at the expense of the coherent $L2_1$ phase. Diffraction patterns obtained from the large particles could be self-consistently indexed based upon a hexagonal lattice isostructural with a known compound Fe_2Nb ($MgZn_2$, C-14 structure). This latter phase was the stable phase in the Fe-25Al-2Nb alloy below ∿ 1250°C.

Preliminary tensile properties of the Fe-25Al-2Nb alloy containing coherent precipitates were measured by four-point bend testing. Figure 2 shows the proportional-limit strength as a function of test temperature (solid circles) for bend bars subjected to a heat treatment of 1350°C/2 hr. → water quench plus 700°C/2 hr. → air cool. The strength at room temperature is ∿ 1379 MPa (200 ksi), at 600°C is ∿ 827 MPa (120 ksi), and at 700°C is ∿ 483 MPa (70 ksi). The total (elastic + plastic) elongation at room temperature was determined to be 2.8%. The strength of the as-extruded alloy containing very large equilibrium-phase particles, is also shown in Fig. 2, as well as that of unalloyed Fe_3Al. These data indicate that this alloy exhibits significant precipitation strengthening. At room temperature $\Delta\sigma_{ppt}$ is ∿ 483 MPa (70 ksi) and at 600°C it is ∿ 345 MPa (50 ksi).

400

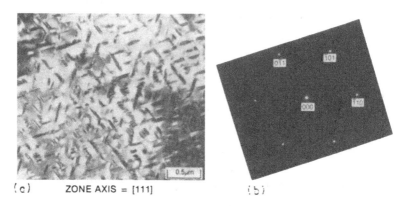

(c)   ZONE AXIS = [111]   (b)

Figure 1.   Bright-Field Electron Micrograph and Corresponding Diffraction
Pattern Showing Distribution of Coherent L2$_1$ Phase Particles in
Fe-25Al-2Nb. Heat treatment: 1350°C/2 hr. → water quench +
700°C/2 hr. age.

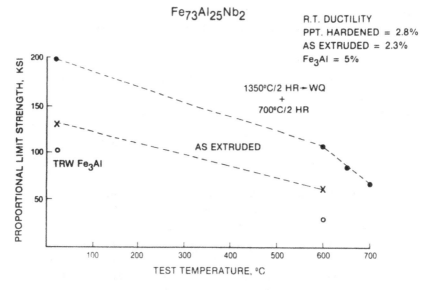

Figure 2.   Proportional Limit Bend Strength of Fe-25Al-2Nb Alloy Containing
Coherent L2$_1$ Particles as a Function of Test Temperature. For
comparison purposes data are also shown for the as-extruded alloy
and for Fe$_3$Al.

Even though the Fe-25Al-2Nb exhibited a potential for precipitation-
strengthening, the thermal stability of the coherent L2$_1$ phase was not
encouraging [15,16], especially for long-term elevated-temperature structural applications. A number of quaternary element additions to the base

Fe-25Al-2Nb were explored to determine whether the thermal stability could be improved. It was found that the addition of 2% Ti enhanced the thermal stability significantly, e.g., at 750°C and 800°C the coherent phase in the base alloy was stable for times less than 16 hr. and 1 hr., respectively. The corresponding times for the Ti containing alloy were > 100 hr. and 4 hr., respectively.

Iron Aluminides + Dispersoids

In both Fe-25Al and Fe-35Al, very fine dispersoids were introduced through powder metallurgy techniques. The different dispersoids were: TiB_2, Hf-rich, and Zr-rich particles. The prealloyed powders having the compositions indicated in Table III, were consolidated by hot-extrusion. The research on dispersoid-containing iron aluminides included microstructural characterization, investigation of thermal stability and determination of tensile behavior [16].

In general, the three types of dispersoid containing alloys had an as-extruded microstructure consisting of extremely fine grains, $\sim$ 1 μm in size, a high density of dislocations and subgrains, and fine dispersion of incoherent particles. It was found that the microstructures were highly stable with temperature, i.e., even after 500 hr. at 900°C the changes in microstructures were minimal. A summary of grain size, subgrain size, and particle size as a function of thermal exposure for the alloys investigated are given in Table IV.

TABLE IV. Summary of Size of Microstructural Features

Alloy	H. T. Condition	Average G.S. (μm)	Average Subgrain Size (μm)	Average Dispersoid Size (μm)
Fe-35Al-1Hf	As Ext.	1.5	-	0.2
	800°C/500 hrs.	1.3	-	0.2
	900°C/500 hrs.	2.7	-	0.2
Fe-35Al + TiB_2	As Ext.	1.0	0.3	0.04
	800°C/500 hrs.	1.0	0.3	0.04
	900°C/500 hrs.	1.0	0.3	0.1

Figure 3 compares the tensile properties of the Fe-35Al + TiB_2 and Fe-35Al-1Hf as a function of test temperature. Both alloys exhibited $\sim$ 8% elongation-to-fracture at room temperature and reasonably high yield strengths up to 600°C. Above 600°C, the strength dropped rapidly, such

402

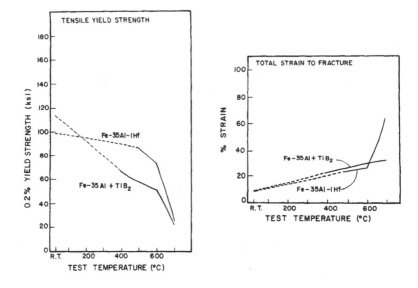

Figure 3.   Comparison of Tensile Properties of Two Dispersion-Hardened B2
Iron-Aluminides as a Function of Test Temperature.

behavior was also observed in the Fe-35Al alloy without dispersoids. It
appears that the dispersoids provide substructural and grain boundary
strengthening at low-temperatures but above 600°C, this strengthening
mechanism is not operative.

α Matrix + DO₃ Coherent Particles

It has been mentioned earlier that partial substitution of Fe with Ti
in Fe_3Al raises the $DO_3 \rightarrow B2$ transition temperature appreciably. In addition, as shown in Fig. 4, the Ti substitution expands the $(\alpha + DO_3)$ binary
phase-field to a much lower Al content and much higher temperature [13]. In
this figure the solid lines represent phase boundaries for the Fe-XAl-5Ti
compositions, while the dotted lines represent the binary Fe-Al phase diagram. The expanded phase diagram provided an opportunity to produce
α matrix + DO_3 coherent precipitate microstructures which were stable to
much higher temperatures than those possible in the binary Fe-Al alloys. A
number of heat treatments were carried out [13] which varied the volume
fraction and size of the coherent particles. Thus, the Fe-XAl-5Ti system
provided a set of alloys with a potential for precipitation strengthening.

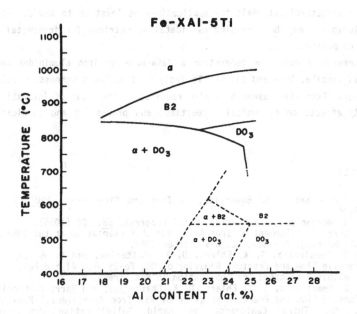

Figure 4. Phase Relations (Solid Curves) for Fe-Al Alloys Containing 5 at.% Ti. For comparison purposes, the binary Fe-Al phase diagram (dotted lines) is also shown.

Preliminary bend tests revealed that the Ti addition embrittles Fe_3Al considerably at room temperature and produces precipitation-strengthening at elevated temperatures; however, extensive tensile testing has not been carried out.

SUMMARY AND FUTURE DIRECTIONS

An important result of the research reviewed in this paper is that the DO_3 Fe-25Al and B2 Fe-35Al alloys are not inherently brittle and do possess reasonable ductility at room-temperature. It is shown that a number of alloying additions provide significant solid-solution strengthening up to 700°C in iron-aluminides. The strength values compare well with some of the available high-temperature stainless steels. Also, a number of ternary alloying elements have been identified which provide a range of two-phase microstructures. There is a potential for both precipitation-strengthening and dispersion-strengthening in iron-aluminides. Based upon the results available at present, we believe that iron-aluminides can be developed as

404

"useful" structural materials for applications at least up to 650°C. Also, iron-aluminides may be promising candidates as matrices for intermetallic-matrix composites.

There is a need for generating a data-base on iron aluminides which includes tensile, low- and high-cycle fatigue, fracture toughness, modulus, and creep. Extensive research is also required in the areas of interstitial impurity effects on mechanical properties, and processing and fabrication methods.

REFERENCES

1. C. Sykes and J. W. Bampfylde, J. Iron and Steel Inst. 130(II), 389 (1934).
2. E. R. Morgan and V. F. Zackay, Metal Progress, 68, 126 (1955).
3. W. Kerr, "Development of Iron Aluminides," Presentation at the 1986 ASM Fall Meeting, Orlando, FL, 6-9 October 1986.
4. M. G. Mendiratta, S. K. Ehlers, D. K. Chatterjee, and H. A. Lipsitt, "Tensile Flow and Fracture Behavior of DO_3 Fe-25 at.% Al and Fe-31 at.% Al Alloys," accepted for publication in Met. Trans. A.
5. M. G. Mendiratta, S. K. Ehlers, D. K. Chatterjee, and Harry A. Lipsitt, "Tensile Flow and Fracture Behavior of RSR Iron Aluminides," Presented at the Third Conference on Rapid Solidification Processing, Gaithersburg, MD, 6-9 December 1982, published in the conference proceedings.
6. M. J. Marcinkowski, M. E. Taylor, and F. X. Kayser, J. Mater. Sci., 10, 406 (1975).
7. M. G. Mendiratta, H. M. Kim, and H. A. Lipsitt, Met. Trans. A, 15, 395 (1984).
8. M. G. Mendiratta and C. C. Law, "Dislocation Energies and Mobilities in B2-Ordered Fe-Al Alloys," accepted for publication in the J. Mater. Sci.
9. M. G. Mendiratta, T. Mah, and S. K. Ehlers, Interim Technical Report, October 1982, under Air Force Contract F33615-81-C-5059 (Systems Research Laboratories, Dayton, OH).
10. S. Mazdiyasni, Unpublished research, AFWAL Materials Laboratory, Wright-Patterson AFB, OH.
11. M. G. Mendiratta and H. A. Lipsitt, Materials Research Society Symposium Proceedings, 39, 155 (1985).
12. S. K. Ehlers and M. G. Mendiratta, J. Mater. Sci., 19, 2203 (1984).
13. M. G. Mendiratta, S. K. Ehlers, and H. A. Lipsitt, "DO_3-B2-α Phase Relations in Fe-Al-Ti Alloys," accepted for publication by Met. Trans. A.
14. D. M. Dimiduk, M. G. Mendiratta, D. Banerjee, and H. A. Lipsitt, "A Structural Study of Ordered Precipitates in an Ordered Matrix Within the Fe-Al-Nb System," paper in preparation.
15. M. G. Mendiratta, T. Mah, and S. K. Ehlers, "Mechanisms of Ductility and Fracture in Complex High-Temperature Materials," AFWAL-TR-85-4061, Final Technical Report under Air Force Contract F33615-81-C-5059 (Systems Research Laboratories, Dayton, OH).
16. M. G. Mendiratta, T. Mah, and S. K. Ehlers, "Mechanisms of Ductility, Toughness and Fracture in Complex High-Temperature Materials," Interim Technical Reports (April 1985, November 1985, and May 1986), under Air Force Contract F33615-84-C-5071 (Universal Energy Systems, Inc., Dayton, OH).

SELECTION OF INTERMETALLIC COMPOUNDS FOR HIGH-TEMPERATURE MECHANICAL USE

ROBERT L. FLEISCHER
General Electric Research and Development Center, Schenectady, New York

ABSTRACT

In the widespread search for structural materials for high temperatures, it is useful to first apply basic physical considerations to eliminate the least promising candidates. Certain basic quantities are indicators of some of the several properties that are demanded of high-temperature materials. Those properties are high strength and specific strength, high stiffness and specific stiffness, and low creep and thermal expansion coefficient. The basic quantities (structure-insensitive in that they vary little with processing history and consequent microstructural changes) which are chosen to measure these properties are melting temperature T_m, specific gravity ρ, and elastic moduli C_{ij}. Fortunately the most readily available of these structure-insensitive properties, T_m and ρ, are the most useful. As T_m increases, the general trend is for stiffness and strength to increase and thermal expansion and creep to decrease. For space and aircraft applications, knowledge of ρ allows relative values of specific stiffness and specific strength to be estimated from the ratio T_m/ρ. Data for more than 280 intermetallic compounds have been collected and subdivided by crystal class so as to allow the most promising compounds and groups of compounds to be selected. The Ll_2 structures tend to lie in the least desirable regions of T_m and ρ and therefore are not the first candidates for maximizing high-strength at elevated temperatures. At present we have no simple structure-independent indicators of ductility and oxidation resistance--two other properties that are needed in high-temperature structural materials.

BACKGROUND AND RATIONALE

Materials are needed for high temperature use, spurred particularly by the drive to produce higher performance jet engines [1,2]. Studies over the past several years have demonstrated the potential of intermetallic compounds for use as elevated-temperature high-performance materials. Limits of temperature capability for materials such as the aluminides that are currently under development, however, do not meet the temperature requirements of more advanced engines. A simple, basic screening process would be helpful in identifying materials for subsequent experimental evaluation.

Much can be done toward preliminary screening of all compounds using basic structure-insensitive properties such as melting temperature T_m, specific gravity ρ, and elastic modulus E (Young's modulus) [3]. Because these properties are not significantly influenced by processing variables and the resultant microstructure and only slightly altered by minor variation in alloying elements, they are useful indicators of several vital properties. T_m is by far the most useful structure-insensitive property. First, it specifies the temperature interval over which materials are solid. Second, its magnitude is roughly proportional to the stiffness of a material, since the elastic moduli correlate strongly with melting temperature [4]. In addition, all models of strengthening give values of flow stress that increase with the magnitude of the elastic constants, which in turn increase with T_m. Fourth, expansion coefficients (which for convenience should be small) vary inversely with T_m. Finally, the creep rates define a maximum operating temperature that increases with T_m. Approximate limits on operating temperatures for single-phase materials are estimated to lie between $T_m/2$ and $2T_m/3$, with $T_m/2$ being more typical [5]. A simple zero-order rule-of-thumb is that, if the melting temperature is expressed in degrees centigrade, the operating temperature is roughly the same number in degrees Fahrenheit, i.e. a material that melts at 3400°C might be engineered to operate at 3400°F. Thus five separate considerations imply that melting temperature is a figure of merit for high-temperature materials.

Above the earth and in rotating parts, high specific strength (strength per unit density) and specific stiffness are important. It is, therefore, necessary to know the elastic moduli (as measures of stiffness and as structure-insensitive indicators of strength) and the specific gravity. Unfortunately, for intermetallic compounds data on elastic moduli are limited: We could locate measurements of E for only 24 of ~280 intermetallics that melt at or above 1500°C. Although E/ρ is the specific stiffness (which we want to know) and the specific strength σ/ρ is roughly proportional to E/ρ, in most cases we must presently reason more indirectly from T_m/ρ as roughly proportional to E/ρ, and in turn still more roughly proportional to σ/ρ.

DATA

Data were collected from sources referenced earlier [3]; more details will be presented separately [6]. Figures 1-4 give the T_m and ρ values for intermetallics, subdivided into specific crystal structure types. Figure 5 presents all of the intermetallic compounds that were identified. The curve is a rough envelope for the data. In general for aerospace uses one

407

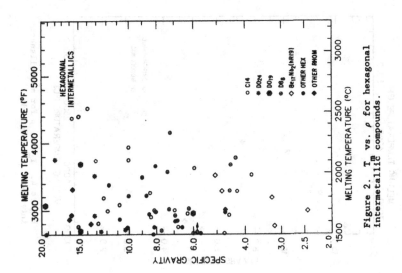

Figure 2. T_m vs. ρ for hexagonal intermetallic compounds.

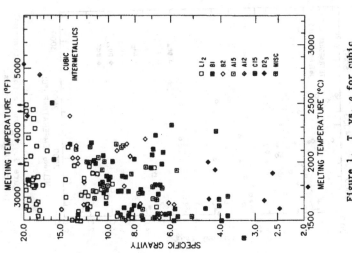

Figure 1. T_m vs. ρ for cubic intermetallic compounds.

408

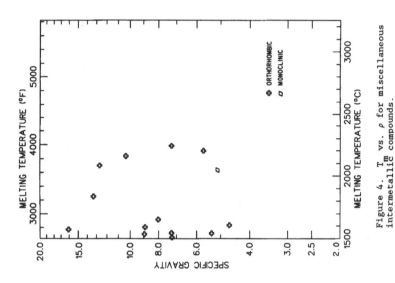

Figure 4. T_m vs. ρ for miscellaneous intermetallic compounds.

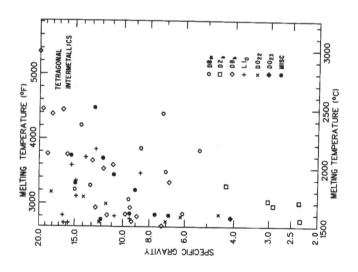

Figure 3. T_m vs. ρ for tetragonal intermetallic compounds.

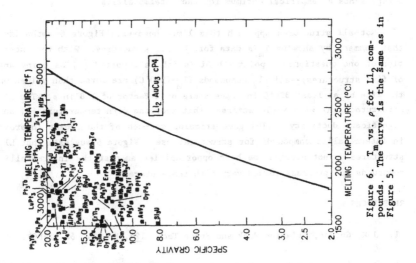

Figure 6. T_m vs. ρ for $L1_2$ compounds. The curve is the same as in Figure 5.

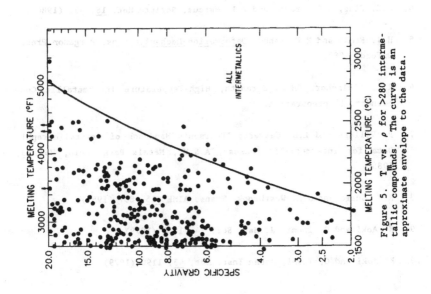

Figure 5. T_m vs. ρ for >280 intermetallic compounds. The curve is an approximate envelope to the data.

wants compounds that are of low ρ and high T_m, so that the curve in Figure 5 represents an empirical optimum for the intermetallics.

Not all structures approach this limit however. Figure 6 emphasizes that comment by showing T_m-ρ data for the Ll_2 structures. With the exception of one questionable point--Nb_3Si (which is reported [7] as being any of five structures)--all Ll_2 compounds (T_m>1500°C) are worse than values on the line by at least 320°C in temperature or a factor of 1.8 in ρ. The Ll_2 structure Ni_3Al with yield stresses that increase with temperature [8] and significant ductility [9,10] gave stimulus to much of the recent interest in intermetallic compounds for structural use. Figure 6 suggests that Ll_2 structures do not provide the best opportunities amongst the intermetallic compounds for progress toward very high temperatures.

REFERENCES

1. J.K. Gordon, Aviation Week and Space Tech. 124, No. 24, 63 (1986).

2. J.C. Lowndes, Aviation Week and Space Tech. 124, No. 25, 167 (1986).

3. R.L. Fleischer, J. Met. 37, 16 (1985).

4. M.E. Fine, L.D. Brown, and H.L. Marcus, Scripta Met. 18, 951 (1984).

5. H.J. Frost and M.F. Ashby, Deformation-Mechanism Maps, Pergamon Press, Oxford, 1982.

6. R.L. Fleischer, "High Strength, High-Temperature Intermetallic Compounds", in preparation.

7. P. Villars and L.D. Calvert, "Pearson's Handbook of Crystallographic Data for Intermetallic Phases", A.S.M., Metals Park, Ohio, Vol. 3, 1985.

8. R.W. Guard and J.H. Westbrook, Trans. AIME 215, 807 (1959).

9. K. Aoki and O. Izumi, J. Mat. Sci. 14, 1800 (1979).

10. K. Aoki and O. Izumi, Japan Inst. Met. 43, 1190 (1979).

EVALUATION OF MULTICOMPONENT NICKEL BASE L1$_2$ AND OTHER INTERMETALLIC ALLOYS
AS HIGH TEMPERATURE STRUCTURAL MATERIALS

D. M. SHAH AND D. N. DUHL
Materials Engineering and Research Laboratory
Pratt & Whitney
400 Main Street, East Hartford,CT 06108

ABSTRACT

Multicomponent nickel base intermetallics with the L1$_2$ structure were
evaluated as high temperature structural materials. The compounds were based
on the γ' composition of PWA 1480, a high strength single crystal nickel
base superalloy. The best balance of properties in the compound was achieved
with <111> oriented single crystals but no significant advantage could be
demonstrated over the precipitation hardened superalloys. Insufficient impact resistance was a major deficiency of the L1$_2$ compounds. Other nickel
base intermetallics were also evaluated but showed little advantage over
superalloys.

INTRODUCTION

The high temperature strength of nickel-base superalloys used in gas
turbine engines is primarily attributed to the γ' precipitates. It is
therefore logical to consider single phase γ'(L1$_2$) as the basis for a class
of high temperature structural alloys to replace the two phase γ/γ' superalloys. This paper addresses the possibility of developing a single phase γ'
aluminide with properties significantly superior to those of a two phase
superalloy for high temperature applications in a gas turbine engine.

APPROACH AND EXPERIMENTAL PROCEDURE

To assess the development potential of γ' it was imperative to evaluate
multicomponent γ' alloys so that synergistic effect of alloying additions
could be evaluated. This was achieved by evaluating alloys based on the γ'
composition of PWA 1480, a high strength single crystal superalloy currently
employed by Pratt & Whitney for turbine airfoils in several advanced gas
turbine engines [1,2]. A typical γ' composition analysed from a particular
heat of PWA 1480 is presented in Table-I along with its bulk and γ-matrix
compositions. The mass balance between the phases was reconciled assuming
approximately 65 volume per cent γ'.

Over 25 alloys were cast as arc melted buttons and homogenized for at
least 24 hours at 1177°C(2150°F) and furnace cooled. Their compositions and
a brief description of microstructure is presented in Table-II along with
the data for the binary and the ternary alloys studied previously[3,4]. The
presence of a widely dispersed, coarse, second phase was considered to have
little influence on the high temperature creep behavior. Three of the alloys 3-LA, 7, and 14 were subsequently cast as single crystals and the alloy

Table-I Analysis of PWA 1480 and Its Constituent Phases in Atom Percent

	Ni	Co	Cr	W	Ta	Al	Ti
PWA 1480	Bal.	4.93	11.89	1.29	3.95	11.23	1.76
γ' Precipitate	Bal.	2.73	1.94	0.89	5.19	14.40	2.52
γ-Matrix	Bal.	9.05	33.49	1.67	2.19	1.06	0.25

3-LA was evaluated in greater detail. All compression tests were carried out in air using 5mm x 5 mm x 15 mm parallelopiped specimens and tensile tests were conducted using standard tensile specimens with a 10 mm gauge length. Notched fatigue testing was done using notched specimens with K_T=2.

RESULTS AND DISCUSSION

The multicomponent alloys were hard and brittle but the necessary specimens could usually be machined with little difficulty as long as impact type loading was avoided. Cracking occurred sporadically in some single crystals but could not be associated with any defects.

Evaluation of Creep Behavior

All the alloys were evaluated at 982°C(1800°F) in compression at stress ranging from 103 MPa(15 Ksi) to 221 MPa(32 Ksi). The minimum creep rate for each alloy is presented in Table-II. The multicomponent alloys are divided into four groups and their creep behavior is compared with PWA 1480 and binary and ternary γ' alloys in Figure-1 on a plot of stress versus minimum creep rate. Creep rate can be reduced by an order of magnitude with tantalum additions to binary Ni₃Al. Additional improvements can be achieved with the Group A alloys, based on the γ' composition of PWA 1480, that contain tungsten in addition to the same level of tantalum as the ternary Ni₃(Al,Ta). Further improvements in creep resistance were achieved with the

Table-II Composition, Microstructure and Creep Results at 982°C(1800°F) for Binary, Ternary and Multicomponent γ' Alloys.

Alloy	Al	Ta	Ti	Co	Cr	W	Mo	Re	Hf	Cb	Microstructure	MPa	1/10⁵ hr
Ni3Al	23.5	-	-	-	-	-	-	-	-	-	Ni rich in dendritic core	103	* 470
Ni3(Al,Ta)	20.3	5.0	-	-	-	-	-	-	-	-	Interdendritic second phase	103	* 33
Group A - PWA 1480 'Based Alloys													
1	14.5	5.5	2.5	2.5	2.5	1.5	-	-	-	-	5% second phase	103	3.2
1-LA	13.8	5.5	2.5	2.5	2.5	1.5	-	-	-	-	2% second phase	103	2.6
7	16.0	5.5	-	2.5	2.5	1.5	-	-	-	-	Similar to binary alloy	221	120
10	16.0	5.5	-	5.0	2.5	1.5	-	-	-	-	Similar to binary alloy	221	77
11	15.0	5.5	-	10.0	2.5	1.5	-	-	-	-	Similar to binary alloy	221	55
12	15.0	5.5	-	5.0	2.5	3.0	-	-	-	-	Heavy dendritic segregation	221	51
13	14.0	5.5	-	10.0	2.5	3.0	-	-	-	-	Heavy dendritic segregation	221	56
14	16.0	5.5	-	10.0	2.5	1.5	-	-	-	-	Almost single phase	221	59
15	17.0	5.5	-	10.0	2.5	1.5	-	-	-	-	Almost single phase	221	148
16	15.0	5.5	-	10.0	5.0	1.5	-	-	-	-	Heavy dendritic segregation	221	53
17	15.0	5.5	-	15.0	2.5	1.5	-	-	-	-	Second phase in dendrite core	221	280
Group B - Refractory Element Substitution													
2	18.5	1.0	-	-	2.5	-	3.5	-	0.1	-	Acicular second phase	103	82
2-LA	17.5	1.0	-	-	2.5	-	3.5	-	0.1	-	γ/γ' eutectic	103	89
20	16.0	5.5	-	5.0	2.5	-	-	-	1.5	-	Mostly single phase	103	12
21	16.0	4.0	-	5.0	2.5	-	-	-	1.5	-	Mostly single phase	103	12
18	16.0	-	-	5.0	2.5	1.5	-	-	-	5.5	Single phase	221	800
19	16.0	-	-	5.0	2.5	-	-	-	1.5	5.5	Single phase	172	400
Group C - Multiple Refractory Elements													
8	16.0	4.5	-	2.5	2.0	1.5	1.5	-	-	-	10% interdendritic second phase	221	76
9	16.0	4.5	-	4.0	2.0	1.5	1.5	-	-	-		221	96
4	16.0	4.5	-	2.0	2.0	1.5	1.5	0.5	.05	-	Re, rich acicular and Cr	207	7.3
5	16.0	4.0	-	2.0	2.0	2.0	2.0	0.5	.05	-	rich script phase	207	95
6	16.0	3.5	-	2.0	3.0	2.0	2.0	1.0	.05	-		207	131
3	14.5	4.5	2.5	2.0	2.0	1.5	1.5	0.5	.10	-	Re rich acicular phase	103	3.3
3-LA	13.8	4.5	2.5	2.0	2.0	1.5	1.5	0.5	.10	-	10-15% coarse interdendritic phase	221	* 10.5
Group D - High Tantalum Alloys													
37	14.0	6.5	-	10.0	2.5	1.5	-	-	-	-	10-15 % interdendritic phase	221	95
38	13.0	7.5	-	10.0	2.5	1.5	-	-	-	-		221	55
39	11.0	7.5	-	10.0	2.5	1.5	-	-	-	2.0		221	255
40	13.0	8.5	-	10.0	2.5	1.5	-	-	-	-		221	100

*Data for <111> orientated single crystal

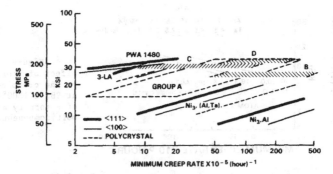

Figure-1 Compressive creep study of binary, ternary, and multicomponent γ' alloys (Group A,B,C and D) at 982°C(1800°F). See Table-II for compositions.

Group C alloys where refractory element content was increased further. Neither the substitution of refractory elements in Group B nor the increase in total tantalum content in Group D help reduce the creep rate.

A more detailed consideration of alloys from Group A in Table-II show 10 a/0 cobalt to be optimum (alloys 7,10,11 and17) and insensitivity of the creep rate to increased chromium and tungsten levels (alloys 15 and 16 and 11 and 13, respectively). Among the group B alloys substitution of refractory elements such as tungsten with molybdenum (alloys 2 and 2-LA) , tungsten with hafnium (alloys 20 and 21) or tantalum with columbium (alloys 18 and 19) is regressive. Improvement in creep resistance is best achieved with microstructurally unstable alloys 4, 8 and 9 from Group C but the creep resistance of PWA 1480 is barely exceeded.

Effect of Single Crystal Processing: A comparison of the most creep resistant alloys from Group C in polycrystalline form with PWA 1480 as a single crystal is not ideally right. The creep resistance of polycrystalline alloy 3-LA is improved at least two fold as a <111> oriented crystal. A similar trend is evident for the ternary Ni_3(Al,Ta) and binary Ni_3Al as well. In all cases the <111> orientation is more creep resistant than the <100> orientation as well as the polycrystalline material. Thus some improvement over PWA 1480 can be expected with refractory element rich Group C alloys but not with Group A alloys with comparable compositions.

Effect of Stoichiometry: To determine the role of stoichiometry the aluminum content was varied by 1.5 a/o for the ternary Ni_3 (Al,Ta) composition. For multicomponent alloy 4 both the total (Al+Ta) content and the ratio of Al to Ta were varied. The results for both the <100> oriented single crystal Ni_3(Al,Ta) and polycrystalline alloy 4 are presented in Figure-2. It is evident that the stoichiometry has a significant effect. However, it is important to recognize that the effect appears large in as much as it occurs with a minor variation in composition. The increase in creep rate, in absolute terms, is no more than a factor of two per atom percent. This is an important distinction central to the feasibility of developing a single phase γ' material with creep strength superior to a nickel base superalloy.

Stoichiometry is perhaps the strongest reason for expecting an ordered γ' to be more creep resistant than a two phase superalloy. Consider the schematic representation of a multicomponent γ/γ' alloy as presented in Figure-3. We expect a unique composition , such as S, within the γ' phase field with the lowest free energy and with the lowest concentration of "defects" to be "stoichiometric" and hence the most creep resistant. By the

414

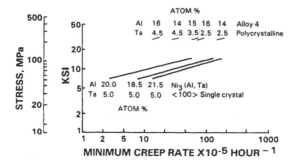

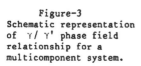

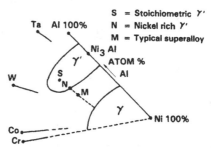

Figure-2
Effect of stoicheometry
on the creep resistance
at 982°C(1800°F)
of a ternary and a
multicomponent γ' alloy.

Figure-3
Schematic representation
of γ/ γ' phase field
relationship for a
multicomponent system.

term "defect" we mean the vacancy concentration or an Al atom at a Ni site
or vise versa but also, for example, a Ta atom on a wrong sub-lattice site
within the Al sites in a Ni_3(Al,Ta) ternary. Now consider a typical multicomponent superalloy composition M with its γ' precipitates having the
nickel rich composition N. The critical question is whether the difference
in the creep resistance of composition S and composition N is significant
enough to make development of single phase γ' alloys worthwhile. It may be
possible to estimate this theoretically at least for a ternary alloy with
kinetic considerations in conjunction with the cluster variation
technique[5]. However, given that the width of the γ' phase field for aluminum is no more than 5 atom percent, the experimental results presented
here suggest no more than half an order of magnitude increase in creep resistance. This is not withstanding the contribution of precipitation hardening in the case of superalloys. The contribution of this mechanism to
diffusion controlled high temperature creep behavior may not be significant
but its effectiveness in enhancing the intermediate temperature creep and
yield strength cannot be overlooked. The very slight difference in melting
temperature between the two phase alloy M and and single phase alloy S and
their high temperature order/disorder behavior are not significant factors
to be considered. In conclusion, based on the experimental evidence the development of a single phase γ' alloy (composition S) with significantly
better creep strength than its conjugate superalloy (composition M) is unlikely.

Evaluation of Flow and Fatigue Behavior

The ductilizing effect of boron addition to polycrystalline γ' [6,7],
the yield strength anomaly of single crystal γ' [8,9] and the effect of
stoichiometry on the strength of γ' [10,11] have been widely studied. However except for one study of binary Ni_3Al [12] the nature of the intrinsic
ductility of single crystals and its dependence on temperature, orientation

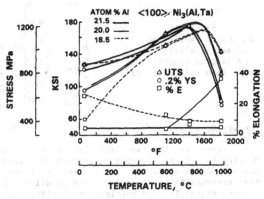

and stoichiometry has not been investigated. The tensile flow stress results for <100> oriented ternary Ni₃(Al,Ta) is presented in Figure-4 at three aluminum concentrations. The nickel-rich composition is most ductile at room temperature but has the lowest yield strength, however at 982°C(1800°F) it is the strongest. The aluminum rich composition is the weakest at high temperature and strongest at room temperature with approximately 5% ductility at all temperatures. The behavior of the stoichiometric composition is naturally between the two extremes.

Figure-4
Effect of stoicheometry on the tensile behavior of <100> oriented single crystal Ni₃(Al,Ta) ternary alloys.

Even though all compositions showed reasonable tensile elongations, the aluminum rich composition behaved in an extremely brittle manner under impact loading. The low ductility and impact strength could not be conclusively associated with the presence of the β-NiAl phase. The typical room temperature impact strength of 80-100 ft-lb for a superalloy compares with 85 ft-lb for the nickel rich, 40 ft-lb for the stoichiometric and 4 ft-lb for the aluminum rich compositions respectively.

The lack of impact strength became even more severe for multicomponent γ' alloys, typically being less than 4 ft-lb. However, the low temperature yield strength did not show an anomalous increase as has been observed in Pt₃Al type Ll₂ compounds[8]. Thus the brittle behavior could not be simply associated with behavior typical of body centered cubic alloys. The tensile ductility was significantly affected by the crystal orientation. A comparison of the 593°C(1100°F) tensile properties of <100> and <111> oriented single crystal alloy 3-LA with PWA 1480 is presented in Figure-5. Almost a ten fold improvement in ductility was obtained for the <111> orientation, albeit at the loss of strength. This is of little use in an application where impact loading can occur in an unpredictable direction.

The multicomponent γ' alloys were so notch sensitive that notched fatigue testing could not be carried out at meaningful stresses. Among the ternary Ni₃(Al,Ta) alloys the aluminum rich composition behaved in the same fashion. With the stoichiometric composition, a reasonable level of fatigue life could be achieved and with the nickel rich ternary Ni₃(Al,Ta), fatigue life comparable to PWA 1480 could be easily attained.

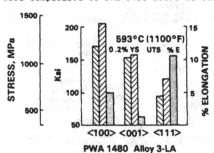

Figure-5
Comparision of 593°C(1100°F) tensile properties of single crystal γ' alloy 3-LA with PWA 1480.

416

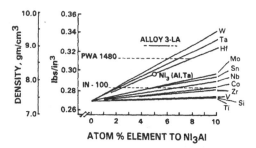

Figure-6
Estimated density of
ternary γ' alloys using
the lattice expansion
data from reference [13].

Density Considerations

Based on the available lattice expansion rate (da/dc) data [13] the density for eleven elements as ternary additions to Ni_3Al up to 10 atom % has been calculated. The results are presented in Figure-6. For comparison the experimental values for the superalloys IN-100 and PWA 1480 as well as ternary $Ni_3(Al,Ta)$ and multicomponent alloy 3-LA are also given. IN-100 represents a superalloy strengthened with Mo,Ti, and V, the elements belonging to the lower cluster of lines in Figure-6. These light elements however, do not contribute significantly to high temperature creep strength which can only be achieved by the addition of the denser refractory elements. Clearly there is little justification for developing single phase γ' alloys on the basis of low density.

OTHER NICKEL BASE INTERMETALLICS

For the sake of completeness, besides single phase $L1_2$ alloys other nickel base intermetallics were also explored. A pseudo-ternary phase diagram approach was taken. On a (Ni,Co)-(Cr,Mo)-(Si,Al) phase diagram four additional intermetallic phases with sufficiently wide composition ranges for developing useful structural alloys were identified as shown in Figure-7. These include β-NiAl, Laves, sigma and silicide. The β-NiAl alloys have been the subject of many investigations in recent years[14,15] and are known to be brittle at room temperature and offer no significant advantage in creep strength over conventional superalloys. The last three phase fields were explored with at least two multicomponent compositions with and without aluminum. The silicide alloys were too brittle to be evaluated. Laves and sigma type alloys were creep tested in compression. The creep resistance of one of the best sigma type alloys (Ni-14.5Co-29Cr-28Mo-9.5Si atom %) at 982°C(1800°F) was a factor of two better than ternary $Ni_3(Al,Ta)$. Laves type alloys were somewhat weaker yet better than the ternary γ'. However, with melting points comparable to a typical superalloy and with the lack of room temperature ductility these intermetallic alloys appear to offer little advantage.

Figure-7
(Ni,Co)-(Cr,Mo)-(Al,Si)
psudo-ternary phase diagram
showing major intermetallic
phase fields.

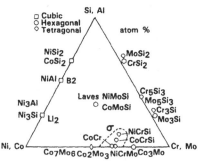

CONCLUSIONS

(1) Creep rate of multicomponent γ' alloys is almost two orders of
magnitude lower than that of binary Ni₃Al.

(2) For a given γ' alloy the creep resistance is highest for the
stoichiometric composition and can be improved by a factor of
two with <111> oriented single crystals.

(3) Reasonable tensile ductilities can be obtained for all γ' alloys
in single crystal form and can be significantly higher in
the <111> orientation.

(4) Tensile ductility and especially impact strength are significantly
lowered for aluminum rich ternary alloy and multicomponent γ' alloys.

(5) The creep resistance of the multicomponent γ' alloys are not
significantly better than two phase superalloys such as PWA 1480.

(6) The γ' alloys with creep resistance comparable to superalloys
offer no density advantage.

(7) Considering the balance of properties needed for gas turbine
applications none of the multicomponent γ' alloys as well as
other nickel base intermetallics appear to offer any real
benefit over existing advanced nickel base superalloys.

(8) A better understanding of the various aspects of ordering
affecting both high temperature diffusion versus low temperature
dislocation glide processes is necessary for a successful high
temperature structural application of intermetallic alloys.

ACKNOWLEDGEMENT

Technical assistance of Mr. C. L. Calverley and Mr. R. E. Doiron as
well as other support groups within Pratt & Whitney is appreciated.

REFERENCES

1. M. Gell, D. N. Duhl and A. F. Giamei, Superalloys 1980 , Edited by
J. K. Tien et al. (American Society for Metals 1980) pp.205-214
2. M. Gell and D. N. Duhl, Advanced High Temperature Alloys:
Processing and Properties - Nicholus J. Grant Symposium , Edited by
S. M. Allen et al.(American Society for Metals 1986) pp.41-49
3. D. M. Shah, Scripta Met. 17 , 997 (1983)
4. D.M. Shah, J. of Metals 11 (8), (1985)
5. J.K. Tien, Proceeding of Japan - U.S. Seminar on Superalloys ,
Edited by R. Tanaka, et al.(Japan Institute of Metals 1984) p.25
6. K. Aoki and O. Izumi, J. Japan Inst. Met., 43 , 1190-1196 (1979)
7. C.T. Liu, C.L. White and E.H. Lee, Scr. Met. 19 , 1247-1250 (1985)
8. D.P. Pope and S.S. Ezz, Int. Met. Rev. 29 , 136-167 (1984)
9. Y. Mishima, et.al., High Temperature Ordered Intermetallic Alloys-
MRS Symposia Proceedings-V39 ,Edited by C.C. Koch, C.T. Liu and
N.S. Stoloff (Material Research Society 1985) pp.264-277
10. J.A. Lopez and G.F. Hancock, Phys. Stat. Sol. (a), 2 , 469-474 (1970)
11. O. Noguochi, Y. Oya and T. Suzuki, Met. Trans. A, 12 A , 1647(1981)
12. S.M. Copley and B.H. Kear, Trans. TMS-AIME 239 , 977-984(1967)
13. Y. Mishiona and T. Suzuki, Proceedings of Japan-U.S. Seminar on
Superalloys , Edited by R. Tanaka et.al.(Japan Institute of Metals
1984) p. 25)
14. K. Vedula et al, High Temperature Ordered Intermetallic Alloys-
MRS Symposia Proceedings-V39 , Edited by C.C. Koch, C.T. Liu and
N.S. Stoloff (Material Research Society 1985) pp. 411-421
15. U.S. Airforce Contract No.F33615-84-5067.

SiC Reinforced Aluminide Composites

Pamela K. Brindley
NASA, Lewis Research Center, 21000 BrookPark Rd., Cleveland, OH 44136

ABSTRACT

The tensile properties of SiC fiber, Ti_3Al+Nb and SiC/Ti_3Al+Nb composite have been determined from 300 to 1365K. The composite results compared favorably to rule-of-mixtures (ROM) predictions in the intermediate temperature regime of 475 to 700K. Deviations from ROM are discussed. Composite tensile results were compared on a strength/density basis to wrought superalloys and found to be superior. Fiber-matrix compatibility was characterized for the composite at 1250 and 1365K for 1 to 100 hours.

INTRODUCTION

There is an ever increasing need to develop low density materials that can maintain their high strength and stiffness properties at elevated temperatures. Intermetallics have long been investigated with this high temperature, strength/density characteristic in mind and are likely candidates to replace superalloys in this regard. Some of the intermetallic materials currently being examined at this laboratory are aluminides based on Fe, Ni, and Ti[1]. The most suitable aluminides for elevated temperature application from a density viewpoint are based on Ti. In general, two classes of titanium aluminide have received attention[2]: those based on the α_2 composition (Ti_3Al) and those based on the γ composition (TiAl). However, the inherent brittle nature of intermetallics has inhibited their widespread application. Therefore, efforts are being made to enhance intermetallic ductility, primarily through alloying additions[3,4] and rapid solidification technology[5].

This investigation of composites of SiC reinforced aluminides (SiC/aluminide) was undertaken with the goal of examining materials with the potential of exhibiting both higher strength and lower density than monolithic aluminides. The composite of interest is based on an SiC/Ti_3Al+Nb structure. Nb addition to the Ti_3Al was chosen since it has been shown[3,4,6] to enhance ductility. Emphasis in characterizing this composite was placed on four areas. First, the feasibilty of fabricating the structure was investigated. Second, the tensile strength over a wide range of temperatures was determined and compared to strengths predicted by the ROM. Third, the fiber/matrix compatibility after fabrication and upon exposure to elevated temperature was examined in a reaction study. Measurements of reaction zone growth at 1250 and 1475K are presented. Finally, a strength/density(σ/ρ) comparison was made between the SiC/Ti_3Al+Nb composite, the Ti_3Al+Nb matrix, and typical wrought superalloys.

EXPERIMENTAL PROCEDURE

The SiC/Ti_3Al composite was a powder metallurgy product. The matrix material was produced by the Plasma Rotating Electrode Process (PREP). Continuous 140μm diameter SiC (SCS-6) fiber was utilized. The powder and fibers were consolidated by hot pressing to obtain 5cm x 15.25cm x 0.075cm composite plates with 40 volume percent fiber. Matrix-only plates were consolidated in similar fashion. The SiC fiber, matrix-only material, and the SiC/Ti_3Al+Nb composite were tested in tension in air from 300 to 1365K. Gauge lengths of 1.25cm and a strain rate of $1 \times 10^{-4}s^{-1}$ were employed.

SiC/Ti_3Al+Nb coupons were annealed in vacuum at 1250 and 1365K for 1 to 100 hours to determine the extent of fiber/matrix reaction zone growth.

420

The coupons were sectioned, metallographically prepared and etched. Measurements of the fiber/matrix reaction zone were made for both the annealed and the as-fabricated conditions at 1000X.

RESULTS AND DISCUSSION

Attempts to produce a fully consolidated SiC/Ti$_3$Al+Nb composite were successful as can be seen by examining the cross-section of the as-fabricated structure in figure 1a. and b. A two-phase matrix is evident as is good bonding between the fiber and matrix. A 3µm dark ring is evident around each fiber which is suspected to be a combination of reaction zone formed during fabrication and the 2µm Si/C surface coating present on the SCS-6 fibers before fabrication[7]. Also distinguishable around each fiber is a zone approximately 3.5µm thick which appears to be depleted of second phase.

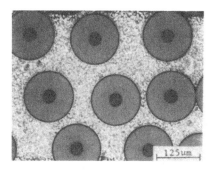 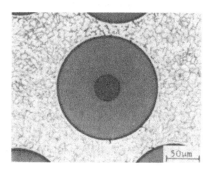

(a)   (b)

Figure 1. Cross-section of the as-fabricated SiC/Ti$_3$Al+Nb composite with (a) 40 volume % fibers. Note the reaction and depleted zones in (b).

Figure 2 contains the tensile strength data of the SiC fiber, the matrix-only material and the 40 volume % SiC/Ti$_3$Al+Nb composite. The ROM prediction based on the average fiber strength and the matrix strength is also included for comparison. In general, the actual composite strengths attained compared favorably to the ROM prediction, especially since the ROM based on average fiber strength does not account for bonding or interaction effects between the fiber and matrix. The closest match of ROM prediction and actual composite values occurred in the intermediate temperature regime, at 475 and 700K. At both the room temperature and elevated temperature extremes, however, the composite strengths were significantly less than those predicted; 60% and 40% less at 300K and 1365K, respectively.

The less-than-predicted strengths at room temperature can probably be attributed to the limited ductility of both the fiber and matrix. The SiC fiber has been shown to have a strain-to-failure of approximately 1%[7] at room temperature. Studies[3,4] of α_2 and α_2 alloys containing Nb report elongations which are immeasurable for Ti$_3$Al and range from 0.4 to 1.6% for α_2 alloys with Nb at room temperature. In addition, examination of the room temperature fracture surfaces of this SiC/Ti$_3$Al+Nb composite by scanning electron microscope (SEM), shown in figure 3, revealed no evidence of ductility. Further SEM examination showed minimal fiber pullout occurred throughout the fracture surface, suggesting matrix-controlled fracture with good fiber-matrix bonding.

At 475 and 700K the fracture surfaces contained a limited amount of dimpled regions within the matrix and fiber pullout was observed.

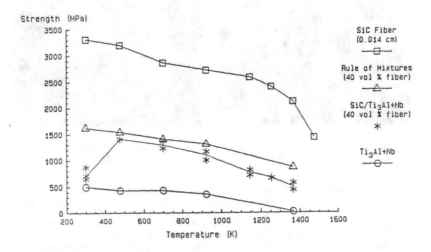

Figure 2. Tensile strength versus temperature for SiC, Ti$_3$Al+Nb, SiC/Ti$_3$Al+Nb, and ROM prediction.

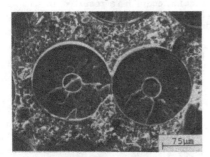

Figure 3. Room temperature tensile fracture surface of SiC/ Ti$_3$Al+Nb.

The fracture surfaces of the samples tested at 1150 to 1365K exhibited increased amounts of dimpled regions on the matrix surfaces as test temperatures were increased, indicating an increase in matrix ductility with temperature. Again, this is consistent with the increase in ductility with temperature reported for Ti$_3$Al alloys containing Nb[3,4,6], where elongations of 4.4 to 17% have been observed. Furthermore, an increased amount of fiber pullout was present, as evidenced in figure 4, as well as an increased amount of debonding between the matrix and the fiber/matrix reaction zone. At all temperatures where fiber pullout was observed, no matrix material appeared to remain on the fibers. However, the fiber and the reaction zone generally maintained a good bond. This is illustrated in figure 5 where the reaction zone appears to be bonded to the fiber and the matrix appears to be debonded from the reaction zone.

There are several possible explanations for the departures from predicted strengths of figure 2. One is the reaction zone degrades the strength of the fiber and causes it to fracture prematurely. However, it is questionable

422

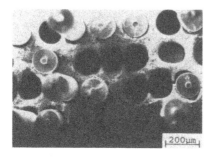 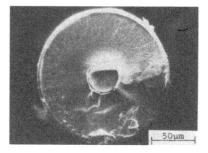

Figure 4. 1150K fracture surface of SiC/Ti$_3$Al+Nb.

Figure 5. SiC fiber debonded from matrix at 1365K.

whether the reaction zone could degrade the strength of the fibers at elevated temperatures in these tests since, in the intermediate temperature regime, full fiber strength was obtained, as evidenced by its comparison to the ROM. In addition, the SEM investigation of the tensile fracture surfaces did not reveal reaction zone growth during elevated temperature testing.

Another possible contributor to a departure from predicted strengths is debonding, caused perhaps by thermal expansion mismatch between the fiber and matrix. Assuming a reaction zone exists, this debonding could occur at the fiber-reaction zone interface and/or the reaction zone-matrix interface. In these tests, debonding of the latter was observed. It is possible that regions of debonded fibers throughout the test section could have resulted in an increased "ineffective" length of the fibers and reduced fiber bundle strength following the reasoning of Rosen[8] and ultimately resulting in less-than-predicted composite strengths.

In spite of the fact that the reaction zone does not appear to contribute to mechanical property degradation in the tests performed to date, it is expected that reaction zones can contribute significantly to strength

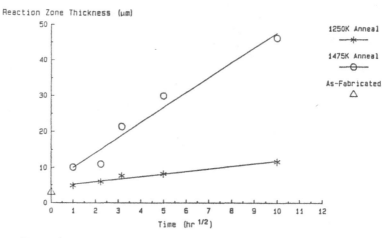

Figure 6. Reaction zone growth as a function of square root time.

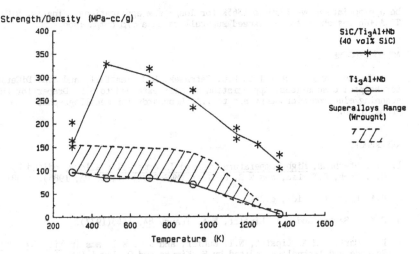

Figure 7. Comparison of σ/ρ values of SiC/Ti₃Al+Nb, Ti₃Al+Nb and wrought superalloys for a range of temperatures.

degradation during long-term elevated temperature testing. With this in mind, the rate of reaction of the SiC/Ti₃Al+Nb composite was determined at 1250 and 1475K. As can be seen in figure 6, the reaction products exhibited parabolic growth, with the rate of growth at 1475K approximately 5.5 times greater than that at 1250K. The rapid growth of reaction zone at 1475K indicates this composite will not be appropriate above 1250 to 1365K for long-term exposures.

Also included in this figure for comparison is the as-fabricated reaction thickness of 3μm, shown here at zero time. The effect of the reaction zone thickness on mechanical property degradation of the composite is the subject of continuing research. Lower fabrication temperatures and pressures are also being studied to decrease the initial quantity of reaction zone present after fabrication.

The SiC/Ti₃Al+Nb composite is most attractive in a σ/ρ comparison. In figure 7, the composite σ/ρ is plotted as a function of temperature and compared to a range of wrought superalloys. The superalloys chosen to represent this range of values were Rene 41 and Hastelloy X as the strongest and weakest, respectively. Even in this first attempt to fabricate and test SiC/ Ti₃Al+Nb, the composite σ/ρ exceeds that of superalloys.

CONCLUSIONS

The feasibility of fabricating a fully consolidated 40 volume % SiC/ Ti₃Al+Nb composite has been demonstrated. Tensile strengths for the composite were measured and were comparable to the ROM predictions at intermediate temperatures, but below the ROM predictions at room temperature and at 1150 to 1365K. From these preliminary data, the probable causes for the reduction in strengths were attributed to limited matrix ductility at room temperature and to reaction zone-matrix debonding at elevated temperatures. The fiber-matrix reaction zone did not appear to degrade the strength of the fibers in these tests. Much more investigation needs to be done to discern the role(s) of the debonding and the reaction zone during fracture. Reaction zone growth was measured and determined to be 5.5 times greater at 1475K than at 1250K. The rapid growth of reaction zone at 1475K indicates this composite will not be appropriate above 1250 to 1365K for long-term applications. Finally, SiC/Ti_3Al+Nb was shown to far exceed superalloys in a σ/ρ comparison.

ACKNOWLEDGEMENTS

The author is grateful to D.W. Petrasek, L.J. Westfall, and J.A. DiCarlo for helpful discussions. Appreciation is also extended to S.L. Draper for the scanning electron microscopy and to T.A. Leonhardt for metallographic preparation.

REFERENCES

1. J.R. Stephens, High Temperature Ordered Intermetallic Alloys, edited by C.C. Koch, C.T. Liu, and N.S. Stoloff (MRS Publication, PA, 1985), p.381.

2. H.A. Lipsitt, ibid., p.351.

3. S.M.L. Sastry and H.A. Lipsitt, Met. Trans. 8A, 1543(1977).

4. P.L. Martin, H.A. Lipsitt, N.T. Nuhfer, and J.C. Williams in Titanium '80 Science and Technology, edited by H. Kimura and O. Izumi (TMS-AIME Publication, PA, 1980), p.1245.

5. D.J. Gaydosh and M.A. Crimp, High Temperature Ordered Intermetallic Alloys, edited by C.C. Koch, C.T. Liu, and N.S. Stoloff (MRS Publication, PA, 1985), p.381.

6. C.F. Yolton, T. Lizzi, V.K. Chandhok, and J.H. Moll, presented at the 1986 Annual Powder Metallurgy Conference, Boston, MA, 1986 (in press).

7. J.A. Cornie, R.J. Suplinkas, and H. Debolt, ONR Contract N00014-79-0691, 1981.

8. B.W. Rosen, AIAA Journal 2, 1985(1964).

PART V

Metallurgical Properties

GRAIN BOUNDARY AND SURFACE SEGREGATION
IN BORON DOPED Ni₃Al

CALVIN L. WHITE(*) and ASHOK CHOUDHURY(**)
(*)Department of Metallurgical Engineering, Michigan Technological
University, Houghton, MI 49931
(**)Department of Materials Science and Engineering, University of
Tennessee, Knoxville, TN 37996-2200

ABSTRACT

Grain boundary segregation, and its effects on strength and ductility,
play an especially important role in the performance of intermetallic alloys
based on Ni_3Al. The effect of a given segregant on grain boundary strength
appears to be determined to some extent by its tendency to segregate to free
surfaces. In this paper, we briefly review the available information on
grain boundary and surface segregation in boron doped Ni_3Al.

INTRODUCTION

The intermetallic alloy Ni_3Al adopts the ordered Ll_2 crystal structure
at temperatures up to its solidus. This structure is a variant of the face
centered cubic structure, in which the Ni occupies the face centered sites,
and the Al occupies cube corner sites. In the ideal Ni_3Al lattice, all
aluminum atoms are surrounded by nickel atoms, and there are no Al-Al near
neighbors. The strong ordering tendency of Ni_3Al indicates that Al-Al
nearest neighbors are high energy defects in this ordered structure.

The flow stress of Ni_3Al, as well as several other strongly ordered Ll_2
alloys, increases as a function of temperature over a fairly broad range of
temperatures up to around 700 C [1,2]. This "anomalous temperature
dependence of the flow stress", plus its high aluminum content, which
imparts good oxidation and corrosion resistance, has generated considerable
interest in the development of alloys based on Ni_3Al for high temperature
structural applications.

While these attractive aspects of Ni_3Al have been known for some time,
practical use of alloys where Ni_3Al is the primary constituent was stifled
for many years by the extreme intergranular brittleness of the polycrystalline alloy. While single crystals of Ni_3Al exhibit excellent ductility at
room temperature, the same material exhibits very little ductility and fails
100% intergranularly in its polycrystalline forms. Early researchers
suggested that this brittleness might be related to impurity segregation
(possibly oxygen or sulfur) at grain boundaries [3,4]. Direct analysis of
grain boundaries using Auger electron spectroscopy (AES) did in fact confirm

strong intergranular sulfur segregation to Ni_3Al grain boundaries, and qualitatively confirmed that those alloys having the highest level of sulfur segregation were also the most brittle [5]. Research by Liu and co-workers [6,7], however, indicated that even high purity alloys exhibiting uncontaminated grain boundaries failed in a brittle intergranular fashion. These results indicate, that while impurities such as sulfur may be detrimental to the grain boundaries in Ni_3Al, this intermetallic alloy is intrinsically brittle and tends to fail intergranularly even in the absence of segregated impurities. Some other strongly ordered alloys also tend to fail intergranularly, however the intrinsic (as opposed to impurity induced) nature of their brittleness is not as well established. While we often assume that high purity metals and alloys will be ductile in the absence of impurity segregation, we should remember that intergranular failure is quite common in ceramics, where the tendencies for ordering are even more pronounced than for intermetallic compounds.

A beneficial effect of boron additions in Ni_3Al was first discovered by Aoki and Izumi [8], and somewhat later by Liu and co-workers [6,7]. The latter group also showed that the beneficial effect of boron additions is associated with boron segregation to grain boundaries in the ductilized alloys. Boron is apparently different from many other grain boundary segregants in that it strengthens rather than weakens the boundaries. Surface segregation studies by White et al. [8] and theoretical studies by Rice et al. [9-11] indicate that the unusual effect of boron segregation may be related to the unusual relationship between grain boundary and free surface segregation in this system. In this paper, we briefly review grain boundary and surface segregation in boron doped Ni_3Al, and comment on the mechanisms by which the unusual effects of boron segregation may occur.

SEGREGATION BEHAVIOR AND INTERGRANULAR FRACTURE

Equilibrium grain boundary segregation is the enrichment of a solute species in the vicinity of the grain boundary resulting from an energetic attraction (binding energy) of the solute species for sites at the boundary. Segregation is distinct from precipitation of a second phase at the boundary; segregation can and does occur in alloys where solute contents are well below the solubility limit, and for which no bulk second phase is stable. The binding energy of a solute for a grain boundary can result from chemical and structural factors (e.g. nearest neighbor distances, arrangements, etc.) that make solute occupation of sites at a grain boundary more energetically favorable than the sites it would otherwise occupy in the solvent lattice. Studies of disordered alloys often indicate solute binding energies on the order of 0.5 eV, which can result in solute enrichments on the order of 10^3 to 10^4 in a thin layer 2-3 atoms thick at the boundary. The thinness of this segregated region results from the short range nature of the distortions associated with most interfaces, and the fact that perfect lattice periodicity is restored upon moving a few atomic distances from the boundary plane. We should note that the effective repeat distance for long range ordered lattices is generally at least twice that for corresponding disordered alloys, hence we might expect that the thickness of solute enriched regions at grain boundaries in these alloys to be somewhat greater than in disordered alloys as well.

A detailed discussion of interfacial segregation is beyond the scope of this paper, however several well accepted points bear repeating [5,12,13,14]:

1. Except for extreme cases involving very fine grained materials and extremely strong grain boundary segregation, the fraction of the total solute content associated with the segregated region at an interface is negligible compared with the fraction in solid solution in the bulk lattice. This is primarily due to the extreme thinness of the interfacial region compared to the grain size.

2. The extent of segregation is expected to decrease as the segregation temperature is increased. This results from thermal activation of solute atoms out of the few energetically attractive intefacial sites into the far more numerous sites in the bulk lattice (a configurational entropy effect). Providing the kinetics of segregation are not too fast, it should be possible to quench in the level of segregation associated with various temperatures.

3. The level of segregation at any temperature should increase with overall solute content. This increase will not be strictly linear, however, due to saturation effects as the available interfacial sites become filled.

4. Segregation that is equilibrium in nature will necessarily decrease the interfacial energy. This is a direct consequence of the Gibbs adsorption equation, and is independent of any detailed assumptions regarding the nature of the interface-solute interaction.

Given that overall solute contents of a few atomic parts per million (appm) can result in grain boundary solute concentrations of several tens of atomic percent (at%), it is not surprising that segregation of this type can strongly affect the properties of grain boundaries, including their cohesion. The relevance of surface segregation behavior to intergranular fracture is somewhat less obvious, and is related to the effects of solute segregation on the ideal energy of grain boundary separation as developed by Rice and others [9-11]. Briefly, this approach considers the (mechanically) reversible separation of two grains normal to their common boundary. In applying the first law of thermodynamics, the work done in separating the two crystals (ω) is related to the positive sum of the interfacial energies of the two free surfaces created by the separation (γ_s), minus the interfacial energy of the grain boundary (γ_b, prior to separation).

$$\omega = 2\gamma_s - \gamma_b \qquad [1]$$

Qualitatively stated, the predictions of this treatment are as follows:

1. Reduction of grain boundary energy, γ_b, will increase cohesive energy.

2. Reduction of the interfacial energy of the surfaces created by fracture, γ_s, will decrease cohesive energy.

The converse of both predictions are also true. If grain boundary segregation alters either the energy of the grain boundary, or of the free surfaces, then its cohesive energy should be altered accordingly. We further suppose that any change in cohesive energy resulting from grain boundary segregation will be reflected in the effects of segregation on the macroscopic strength and ductility of the polycrystalline alloy.

In the absence of any effect on surface energy, therefore, equilibrium grain boundary segregation will decrease grain boundary energy and tend to enhance grain boundary cohesion. Since experience tells us that many (if not most) grain boundary segregants are embrittlers rather than strengtheners, we clearly must look beyond the effect of segregation on grain boundary energy alone, and consider its effect on the energy of the free surfaces created during grain boundary separation.

The fracture processes of interest here occur under conditions of low temperature and high strain rate which preclude compositional equilibration between the bulk crystals and the newly created intergranular fracture surfaces during the course of separation. Specifically, the compositions of those free surfaces will be "inherited" from the grain boundary, and are not determined by the tendency of the solute to segregate to the free surface. If the solute in question tends to segregate strongly to free surfaces as well as to grain boundaries, then the Gibbs adsorption equation predicts that the inherited solute should lower the free surface energy as well as the grain boundary energy. In this case, both relevant interfacial energies are decreased, and will tend to have opposing effects on grain boundary cohesion. The net effect of grain boundary segregation on cohesion will be determined by the relative magnitudes of its effects on these two opposing interfacial energy terms. Although extensive data are not available, embrittling grain boundary segregants are indeed generally observed to segregate even more strongly to free surfaces.

If, on the other hand, the segregant inherited by the fracture surface does not tend to segregate to free surfaces, then the Gibbs adsorption equation indicates that its presence should increase the energies of the surfaces created during separation. In this case, the segregant's effect on both the grain boundary and the free surface energies would tend to increase the cohesive energy of the boundary. Rice's initial paper on this subject acknowledged the possibility of such behavior, but it also pointed out that all known segregation behavior (at that time) was in fact characteristic of embrittling segregants.

It appears, therefore, that the difference in the tendency of a solute to segregate to free surfaces may be relevant to its tendency to embrittle or strengthen grain boundaries. Grain boundary segregants that also segregate strongly to free surfaces should be embrittlers, while grain boundary segregants that do not tend to segregate to free surfaces under equilibrium conditions should strengthen boundaries. Therein lies the basis for reviewing the surface segregation behavior in boron doped Ni_3Al alloys.

GRAIN BOUNDARY SEGREGATION IN Ni_3Al

Given the small concentration of boron required to ductilize Ni_3Al, and the fact that the ductilizing effect is clearly directed at strengthening the grain boundaries, there was ample cause to suspect that the effect is associated with boron segregation to grain boundaries. Unfortunately, direct observation of grain boundary composition in boron doped Ni_3Al was complicated by the fact that most methods of analysis required exposure of the boundaries by fracture. By achieving high ductility and inhibiting intergranular fracture, boron additions had also made study of grain boundary compositions in boron doped Ni_3Al much more difficult.

By notching specimens, and cooling them prior to in-situ fracture, Liu et al. [6,7] were able to obtain small regions of intergranular fracture in the boron doped Ni_3Al. Then using high spatial resolution AES to examine these small regions of intergranular fracture, it was possible to show that they were enriched in boron. Alternate AES analysis and inert ion sputtering indicated that this boron enrichment extended only few atomic distances below the fracture surface (i.e. only a few atomic distances away from the grain boundaries). This observation indicated that boron segregated strongly to grain boundaries, suggesting that the ductilizing effect is associated with such segregation.

If the ductilization is in fact associated with the segregation per se, and not just the overall boron content of the alloy, then variations in thermal history that alter the level of segregation in a given alloy should also alter its fracture behavior. This was in fact shown to be true in some work by Choudhury et al. [15], where Ni-24at%Al-300wppmB quenched from 1050 C had less boron segregation and more intergranular fracture than the same alloy that had been slow cooled from 1050 C. The scanning electron micrographs in Fig. 1-a and -b show a similar effect in Ni-24at%Al-100wppmB alloys [16]. The completely intergranular failure mode in the quenched material (Fig. 1-a) is transformed to mostly transgranular failure in the slow cooled one (Fig. 1-b). Such effects of thermal history were shown to be reversible, thus discounting the possibility that boron segregates to grain boundaries in Ni$_3$Al via non-equilibrium mechanisms [17].

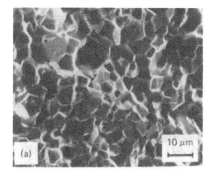

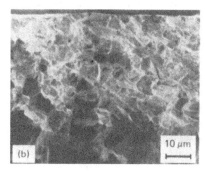

Fig. 1 Scanning electron micrographs of Ni-24at%Al-100wppmB specimens that were; (a) held at 1050 C for 30 min, then water quenched, and (b) held at 1050 C for 30 min and slow cooled (1C/min).

AES analysis of many (≅25) grain boundaries on the intergranular fracture in Fig. 1-a indicated that they exhibit a range of boron levels, as indicated by the histogram in Fig. 2-a. The percentage of the total grain boundaries analyzed having B/Ni atomic ratios (*100 and expressed as percent) is plotted as a function of that B/Ni ratio. The details of these analyses will be reported elsewhere, however we should note that quantitative analysis of Auger spectra were carried out using reference spectra of elements and compounds generated specifically for this study, and not using published handbook spectra. The distribution of boron levels observed in Fig. 2-a can arise from several causes, but the most obvious is the distribution of grain boundary structures expected in a random polycrystalline aggregate.

Although only a few grain boundaries were exposed in the slow cooled specimen, a similar histogram for this specimen is presented in Fig. 2-b. The median B/Ni ratio for this alloy is only slightly higher than for the quenched one, but only a few of the weakest boundaries are included in this analysis. Choudhury et al. also developed a procedure for hydrogen charging the ductilized alloys, and fracturing them intergranularly inside the AES analysis chamber [15]. Analysis of intergranular fractures in slow cooled specimens that had been hydrogen charged indicated significantly higher B/Ni ratios (Fig. 2-c). In comparing the distributions from Figs. 2-b and -c, it is important to keep in mind that the former is a sampling of the weakest boundaries, while the latter is a more representative sampling of all boundaries.

For a given thermal history, the bulk boron level is known to exert a strong effect on ductility in Ni_3Al [6]. According to the previous discussion of grain boundary segregation, we expect segregation to increase with bulk solute content, thus affecting grain boundary cohesion. Figure 2-d shows a B/Ni distribution for a Ni-24at%Al-1000wppmB alloy that was slow cooled and hydroged charged for AES analysis. Except for boron content, the specimens of Fig. 2-c and 2-d are identical. While it is clear that the increase from 100wppmB to 1000wppmB results in a significant increase in segregation level, it is also clear that the increase is not nearly the tenfold increase in the bulk. This can be attributed to a saturation effect that is predicted by statistical thermodynamic models of grain boundary segregation [5,12,13].

One of the more interesting aspects of the boron effect in Ni_3Al is its dependence on slight deviations from stoichiometry. Liu

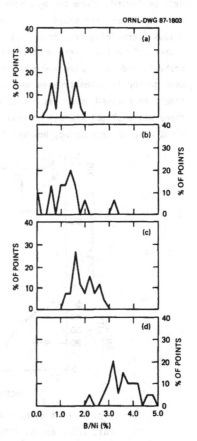

ORNL-DWG 87-1803

Fig. 2 Histograms showing the distributions of B/Ni ratios observed on grain boundaries in boron doped Ni-24at%Al alloys. See text for bulk boron levels and thermal histories.

et al. showed that the beneficial effect of boron is most pronounced for alloys containing less than 24.5 at% Al (bottom curve in Fig. 3) and decreases steadily as the stoichiometric composition (25 at% Al) is approached [6]. AES studies on these same alloys showed that the extent of boron segregation at the grain boundaries also decreased as the aluminum content increased above 25 at% (top curve in Fig. 3). These AES analyses were performed before the hydrogen charging technique used by Choudhury had been developed; therefore, only the weakest grain boundaries in the ductile alloys (containing less than 24.5 at% Al) could be analyzed. For this reason, the actual average boron levels in the substoichiometric alloys were probably underestimated and the effect of aluminum content on boron segregation is probably even more pronounced than originally thought, and should be revised upward as indicated by the arrows. In this same study, there was a slight indication that aluminum concentration at the boundaries might also be increasing, however the trend in those data was weak.

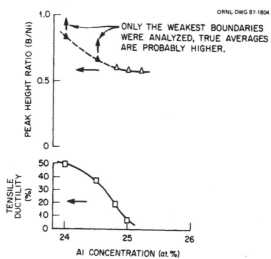

Fig. 3 Plot of B/Ni Auger peak height ratio (Ni, 102eV and B, 179eV, in top portion) and tensile ductility (bottom portion) versus %Al in Ni_3Al alloys containing 500wppmB. Filled symbols indicate AES analyses on grain boundaries in predominantly transgranular fractures, hence overall boron levels were probably underestimated as indicated by the vertical arrows.

In all of the previously mentioned studies where AES analyses of grain boundaries were reported (except those of White and Stein [5]), segregation of solutes other than boron was below the detection limit for AES. This is apparently due to the fact that these alloys were prepared from high purity melt stock, and the bulk sulfur levels were typically less than 3 wppm. At such low bulk concentrations, sulfur concentrations at the grain boundaries are apparently quite low, even though sulfur has a very high binding energy for grain boundary sites.

Because sulfur is a common impurity in nickel base alloys, and because it is known to embrittle Ni_3Al and other nickel base alloys, it would be useful to have information regarding the combined effects of sulfur and boron segregation in Ni_3Al. Very little has been done toward this end; we can, however, provide at least one relevant observation. Figure 4-a shows a scanning electron micrograph of a room temperature fracture in a

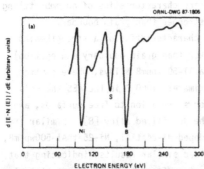

Ni-24 at%Al-300wppmB alloy. This alloy was cooled at a relatively slow rate from 1050 C, and ductile transgranular fracture was expected, however brittle intergranular fracture was observed instead. Subsequent chemical analysis indicated that this alloy contained 30 wppm sulfur. AES spectra from this intergranular surface indicated that the level of boron segregation was roughly as expected, but the sulfur segregation was much greater than observed in any of the ductile alloys (Fig. 4-b). The sulfur segregation in this alloy clearly had a profound embrittling effect on the grain boundaries. It seems unlikely that high purity alloys such as those used for much of the research described here (generally <3wppm S) will be routinely available for commercial production of Ni_3Al based alloys. Hence, a better under-standing of sulfur segregation behavior is required to enable alloy

Fig. 4 Auger spectrum (a) and scanning electron micrograph (b) of a Ni-24at%Al-500wppmB alloy contaminated with 30wppmS.

designers to "manage" any potential problems associated with sulfur
segregation.

FREE SURFACE SEGREGATION IN Ni_3Al

The first hint that the free surface segregation behavior of boron in
Ni_3Al might be different than for sulfur and other embrittling solutes was
obtained quite by accident during the coarse of AES analysis of small pores
on the intergranular fracture surface of a Ni-25.2at%Al-500wppmB alloy [6].
In spite of its strong segregation to the grain boundaries, boron was not
present at the surfaces of these pores. Conversely, while the bulk
concentration of sulfur was sufficiently low to decrease its grain boundary
segregation level below the detectability limit for AES, it was still high
enough to segregate strongly to the pores surfaces. Considered in the light
of our earlier discussion of segregation behavior and grain boundary
cohesion, sulfur exhibits the segregation characteristics of an embrittling
species (surface segregation is much stronger than grain boundary
segregation) while boron exhibits the characteristics of a strengthening
species (surface segregation much weaker than grain boundary segregation).

Subsequent AES studies on Ni-24%atAl-500wppmB alloys, where sputter
cleaned surfaces were heated to approximately 1000 C in the AES analysis
chamber, confirmed the absence of boron segregation to free surfaces, and
the strong segregation of sulfur in the ductilized alloy [8]. Similar free
surface segregation experiments on undoped Ni-24at%Al, Ni-25at%Al-500wppmB,
and Ni-26at%Al-500wppmB at 1000 C yielded similar results, indicating that
the absence of boron segregation to free surfaces in Ni_3Al is not strongly
dependent on alloy stoichiometry [18].

As with grain boundary segregation, we expect free surface segregation
to depend upon temperature, bulk solute content, and interfacial structure.
The results of unpublished research by White and co-workers on free surface
segregation in these alloys in the temperature range between 600 and 800C
indicates a most interesting phenomenon which we briefly describe here [18].

Using boron doped Ni_3Al that had been repeatedly heated to 1000 C then
sputter cleaned, White et al. were able to deplete the surface region of
sulfur, and anneal specimens at 1000 C without appreciable sulfur
segregation (Fig. 5-a). When these clean well annealed specimens were held
at temperatures between 600 and 1000 C, no boron segregation was observed
(in agreement with previous observations). Figure 5-b shows the Auger
spectrum from such a specimen after 90 min. at 760 C.

If these clean well annealed specimens were given a short sputtering treatment at room temperature prior to heating, however, boron segregation was observed at temperatures between 600 and 800 C. Figure 5-c shows the Auger spectrum from such a sputtered surface after heating for 90 min. at 700 C. As these sputtered specimens were heated above 800 C, the boron disappeared, in agreement with earlier surface segregation studies at 1000 C.

These observations confirm that boron does indeed segregate more strongly to grain boundaries than to well annealed free surfaces, but that boron will segregate to sputtered surfaces below 800 C. It appears that the sputtering process introduces some kind of damage that is not annealed-out until the specimen is heated above 800 C, and that this damage significantly affects the tendency for boron to segregate to free surfaces. DasGupta and Liu [19] have observed that long range order in $(Ni,Fe)_3Al$ single crystals (as indicated by low energy electron diffraction, LEED) is destroyed by

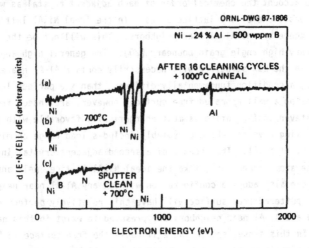

ORNL-DWG 87-1806

Fig. 5 Auger spectra from the surface of a Ni-24at%Al-500wppmB alloy after: (a) heating to 1000 C and sputter cleaning 16 times to deplete the near surface region in sulfur, then given a final 1000 C anneal without subsequent sputtering; (b) annealing the sample in (a) for 90 min at 700 C, and; (c) sputtering the sample in (b), then annealing for 90 min at 700 C.

sputtering, and does not return until the specimens are heated to nearly 1000 C. Several types of structural defects are probably introduced by the sputtering process, including interstitials, vacancies, and antisite defects (aluminum atoms on nickel sites, and vice-versa). It is also possible that sputtering alters the Al/Ni ratio of the near surface region by preferentially removing one or the other species, although no clear evidence for preferential sputtering is available.

In offering one possible explanation for the differences between surface and grain boundary segregation, and the effect of sputtering on surface segregation, we note that one manner in which a free surface differs from a grain boundary is the absence of a second crystal lattice to which the interfacial atoms must interact. Atoms at grain boundaries must adopt a configuration that will be a compromise between optima for the two adjacent crystals. In the case of ordered crystals, this optimum configuration must take into account the chemical order of each adjacent crystal as well as the physical positions of the lattice sites. In the ideal Ni_3Al lattice for example, there are no Al-Al near neighbors. This will not be the case for an arbitrary high angle grain boundary [20]. The general high angle grain boundary in the $L1_2$ structure will necessarily contain Al-Al nearest neighbors, a condition fundamentally different than the perfect lattice.

Atoms on a well annealed free surface, however, are indeed free to adopt whatever configuration is most energetically favorable with respect to the underlying crystal, which presumably includes long range chemical order in the case of Ni_3Al. The absence of a second adjacent crystal in the case of surface atoms means that, like the ideal Ni_3Al lattice, well annealed surfaces can also adopt a configuration without any Al-Al near neighbors.

The sputtered free surface will have this relatively perfect order destroyed and Al-Al near neighbors are presumed to exist in this damaged region. In this sense, sputtering will cause the free surface to "look" more like a grain boundary to the boron atoms. If boron atoms are somehow attracted to regions containing Al-Al near neighbors, then the observed grain boundary and free surface segregation behavior can be rationalized.

A second, though less strongly supported, explanation focuses on the possibility that the primary effect of sputtering might be to cause a compositional change in the near surface region via preferential sputtering. To date, AES results exhibit no clear evidence of such preferential sputtering effects, but further investigation is warranted. Both theoretical [21] and atom-probe field ion microscopy results [22] suggest that grain boundaries in Ni_3Al may be Ni-rich, with the attendant possibility that this Ni-enrichment may be related to boron segregation. If sputtering were shown to cause surface enrichment in nickel (e.g. by preferential removal of aluminum), then an explanation for the various observed segregation behaviors could be constructed on the basis of an attractive interaction between boron and nickel. One strength of such an explanation would be its ability to rationalize the effects of alloy stoichiometry on boron segregation behavior.

SUMMARY AND CONCLUSIONS

Boron segregation to grain boundaries in Ni-24at%Al alloys, and its effect on grain boundary cohesion, are a most interesting segregation phenomenon. The ductilizing effect of boron is clearly related to its segregation to grain boundaries, and not just its mere presence in the alloy. The extent of segregation is dependent on bulk boron level and thermal history in a manner that is consistent with an equilibrium type interaction with grain boundaries.

Unlike embrittling segregants, boron does not exhibit a strong tendency to segregate to well annealed free surfaces; although it does segregate to sputtered surfaces that have not been annealed. The unusual relationship between surface and grain boundary segregation in boron doped Ni-24at%Al is consistent with the classical thermodynamic view of grain boundary cohesion as proposed by Rice and Co-workers. The tendency of boron to segregate to grain boundaries and disordered free surfaces, but not to ordered free surfaces, suggests that the presence of Al-Al near neighbors in the former regions may be an important factor in the boron segregation. This conclusion is not consistent, however, with the observation that boron segregation decreases with increasing aluminum content (and presumably increasing Al-Al near neighbors at the grain boundaries). The possibility that the effect of sputtering versus annealing on free surface segregation in Ni_3Al may involve sputtering induced changes in surface composition should be investigated.

ACKNOWLEDGEMENTS

The authors gratefully acknowledge the financial support of the Division of Materials Sciences, U. S. Department of Energy under Contract No. DE-AC05-84OR21400 with Martin Marietta Energy Systems, Inc. We also acknowledge the encouragement, support, and collaboration of Dr. C. T. Liu during the coarse of much of the research cited here.

440

REFERENCES

1. J. H. Westbrook, Trans. TMS-AIME, 209, 898 (1959).

2. D. P. Pope and V. Vitek, in High Temperature Ordered Intermetallic Alloys, (MRS Symp. Proc. Vol. 59, edited by C. C. Koch et al., 1985) pp. 183-192.

3. A. U. Seybolt and J. H. Westbrook, Acta Metall., 12, 449 (1964).

4. J. H. Westbrook and D. L. Wood, J. Inst. Metals, 91, 174 (1962-63).

5. C. L. White and D. F. Stein, Metall. Trans. A, 9A, 13 (1978).

6. C. T. Liu, C. L. White, and J. A. Horton, Acta Metall., 33, 213 (1985).

7. C. T. Liu, C. L. White, C. C. Koch, and E. H. Lee, Proc. Symp. High Temperature Materials Chemistry II, ed. Munir et al., The Electrochemical Society, Inc. (1983).

8. K. Aoki and O. Izumi, Nippon Kinzaku Gakkaishi, 43, 1190 (1979).

9. C. L. White, R. A. Padgett, C. T. Liu, and S. M. Yalisov, Scripta Metall., 18, 1417 (1985).

10. J. R. Rice, in Effect of Hydrogen on Behavior of Materials, edited by A. W. Thompson and I. M. Bernstein (AIME New York, 1976), pp. 455-466.

11. J. P. Hirth, Phil. Trans. R. Soc. Lond. A, 295, 139 (1980).

12. J. P. Hirth and J. R. Rice, Metall. Trans. A, 11A, 1501 (1980).

13. D. McLean, in Grain Boundaries in Metals, (Oxford, Clarendon Press, 1957), pp. 116-150.

14. C. L. White and W. A. Coghlan, Metall. Trans. A, 8A, 1403 (1977).

15. C. L. White, J. Vac. Sci. Technol. A, 4, 1633 (1986).

16. A. Choudhury, C. L. White, and C. R. Brooks, Scripta Metall., 20, 1061 (1986).

17. A. Choudhury et al., unpublished research, Oak Ridge National Laboratory (1986).

18. H. Wiedersich and P. R. Okamoto in Interfacial Segregation, edited by W. C. Johnson and J. M. Blakely, ASM (1979) pp. 405-432.

19. C. L. White, C. T. Liu, and R. A. Padgett, unpublished research, Oak Ridge National Laboratory (1986).

20. A. DasGupta and C. T. Liu, Private communication, Oak Ridge National Laboratory (1986).

21. T. Takasugi and O. Izumi, Acta Metall., **31**, 1187 (1983).

22. S. M. Foiles, "Calculation of the Defect and Interface Properties of Ni_3Al", paper in this symposium.

23. D. D. Sieloff, S. S. Brenner, and M. G. Burke, "FIM/Atom Probe Studies of B-Doped and Alloyed Ni_3Al", paper in this symposium.

High Temperature Oxidation and Corrosion of
Metal-Silicides

G.H. Meier
University of Pittsburgh
Pittsburgh, PA 15261

The high temperature oxidation of metal silicides is briefly reviewed followed by a description of recent experiments on the oxidation and mixed oxidant corrosion of Ni- and Fe-silicides. The oxidation kinetics are related to the structure and composition of the oxide films formed under various exposure conditions. Finally, the effects of pre-formed SiO_2 films on minimizing the corrosion of Ni- and Fe-base alloys in sulfur-bearing gases are presented.

INTRODUCTION

Alloys and coatings for high temperature applications generally obtain oxidation resistance by the selective oxidation of Cr or Al to form protective surface films of Cr_2O_3 or Al_2O_3. As a result the formation of Cr_2O_3 and Al_2O_3 have been extensively studied [1,2]. The selective oxidation of Si to form SiO_2 films has not been investigated in great detail but appears to offer great potential as another means of protecting high temperature alloys and coatings. In this section the oxidation behavior of metal silicides will be briefly reviewed.

Refractory Metal Silicides

Monolithic $MoSi_2$ heating elements initially form SiO_2 and Mo-oxides but the latter evaporate leaving a protective SiO_2 film. This film results in usable lives in excess of 2000hr at 1650°C [3]. Similarly, disilicide coatings formed on refractory metals such as Mo can provide improved oxidation resistance due to the formation of a SiO_2 film. In this case, however, removal of Si to form the oxide results in conversion of the disilicide to a lower silicide immediately below the oxide. For example, Mo_5Si_3 forms between a SiO_2 layer and $MoSi_2$ [3] and $(Nb,W)Si_2$ in complex silicide coatings on Nb-alloys is converted to $(Nb,W)_5Si_3$ during oxidation [4]. The lower silicides of the refractory metals are generally unable to maintain protective SiO_2 films, presumably due to the reduced activity of Si.

The refractory metals silicides are also subject to degradation by silicide "pest" attack [3]. This form of degradation, which is associated with the penetration of oxidant into the silicide, becomes severe at intermediate temperatures on the order of 800°C [5].

Silicides of Iron and Nickel

The preparation of silicide coatings on Ni- or Fe-base alloys is difficult. However, $CrSi_2$-$NiSi_2$ coatings have been successfully applied to Ni-base superalloys by a slurry fusion technique. [6] Such coatings have exhibited slow oxidation rates at 1000°C.

Published oxidation studies on the Ni-Si and Fe-Si systems have concentrated on dilute solid solution alloys rather than the silicides. Cocking et al [7] studied the oxidation of Ni-Si alloys (0-5.2 wt pct Si) in air at temperatures between 900 and 1200°C. The oxidation rates decreased with increasing Si concentration. The oxidation products were observed.

to contain NiO, SiO_2, and Ni_2SiO_4. Kerr and Simkovich [8] studied the oxidation of Ni-Si alloys (0-10.1 wt pct Si) at 1000° in one atm. of O_2. Small additions of Si up to 1 wt pct were observed to accelerate the oxidation rate which was attributed to a doping effect of Si in the NiO scale. Oxidation rates continually decreased with increasing Si concentration above 2.5 wt pct Si. The oxidation products responsible for the slower oxidation rates were not clearly identified.

The oxidation behavior of solid solution Fe-Si alloys has been the subject of numerous investigations [9-13]. In cases where the additions of Si were sufficient to form an external film of SiO_2 extremely slow oxidation rates were observed. Atkinson [14] has analyzed the selective oxidation behavior of Fe-Si alloys and predicts that the critical atom fractions of Si required for external SiO_2 formation are 0.05 and 0.055 at 1000 and $900^\circ C$, respectively. Additions of Si to ferrous alloys have also been observed to increase their resistance to carburization [15].

Thin Silicide Films in Microelectronic Devices

Thin silicide layers are being used as contacts and interconnections in silicon integrated circuit technology because they have lower electrical resistivity than polycrystalline silicon and are compatible with the silicon substrates. Also, the passivation of these films by oxidation to form SiO_2 films has resulted in renewed interest in the oxidation properties of silicides [16]. The oxidation behavior of the following silicides formed on Si has recently been reviewed [17]:

$TiSi_2$	WSi_2	$IrSi_3$
$ZrSi_2$	$FeSi_2$	$NiSi$
$HfSi_2$	$RuSi_3$	$NiSi_2$
$NbSi_2$	$CoSi_2$	Pd_2Si
$TaSi_2$	$RhSi$	$PdSi$
$CrSi_2$	Rh_3Si_4	$PtSi$
$MoSi_2$	Ir_3Si_4	

Oxidation of silicide films on Si generally results in formation of SiO_2 while the silicide layer is preserved by diffusion of Si from the underlying substrate. Only for silicides for which the metal forms oxides considerably more stable than SiO_2, e.g. $ZrSi_2$ and $HfSi_2$, are metal oxides formed on the films. Ternary phase diagrams describing the oxidation of the silicides of Mo,W,Ta, and Ti have recently been calculated [18]. The oxidation data for the disilicides of these metals have been fitted to a linear-parabolic kinetic expression [19] similar to that derived by Deal and Grove [20] for pure Si

$$x^2 + Ax = B (t + \tau)$$

where x is the oxide thickness, B is the parabolic rate constant, B/A is the linear rate constant, and τ is a fitting parameter to account for initial oxide growth. The parabolic rate constants for the silicides were equal to those for pure Si while the linear rate constants were significantly higher for the silicides.

The oxidation of silicide films on SiO_2 substrates leads to depletion of Si from these films followed by the appearance of metal oxides in the films as the Si activity decreases [17].

Background of Current Studies

The majority of previous studies on silicide oxidation have concentrated on the disilicides. However, from the standpoint of possible structural alloys or coatings for ferrous or nickel-base alloys the disilicides are not acceptable because of poor mechanical properties and low melting point eutectics in the event of decreased Si content either due to oxidation or interdiffusion with an alloy substrate. In the Ni-Si system [21] the lowest eutectic occurs at T=964°C and X_{Si}=0.46, while in the Fe-Si system [21] it occurs at T=1200°C and X_{Si}=0.34. Therefore, silicides with lower Si content must be considered and the minimum Si concentration required to form a continuous SiO_2 film and the stability of this film become major concerns. The oxidation and mixed oxidant corrosion behavior of alloys in the Ni-Si and Fe-Si system have been under study at the University of Pittsburgh for several years. The following summarizes a portion of this work.

EXPERIMENTAL

Nickel-silicon and iron-silicon alloys were arc melted under a purified argon atmosphere and drop cast into a water cooled copper chill. The nominal and analyzed compositions of the alloys are given in Table 1.

TABLE 1 COMPOSITION OF ALLOYS STUDIED

Nominal Composition (Wt%)	Analyzed Composition (Wt%)
Fe-5Si	Fe-5.34Si
Fe-10Si	Fe-10.83Si
Fe-14Si	Fe-13.43Si
Fe-20Si	Fe-19.79Si
Ni-5Si	Ni-4.93Si
Ni-10Si	Ni-9.51Si
Ni-12.5Si	Ni-12.82Si
Ni-15Si	Ni-14.28Si
Ni-17.5Si	Ni-16.97Si
Ni-20Si	Ni-17.83Si
Ni-22.5Si	Ni-19.52Si

The ingots were homogenized in purified argon for 150 hrs at 1100°C. Phase determination by x-ray diffraction (XRD) was essentially consistent with published phase diagrams for the Fe-Si [21] and Ni-Si [21,22] systems presented in Figure 1. The Ni-Si diagram was constructed by combining the Ni-rich segment, recently re-determined [22], with the older diagram. Note that the latter portion is also in need of re-determination since it contains questionable and, in some cases, impossible constructions.

Specimens of dimensions 11x8x2mm for isothermal oxidation and 22x8x2mm for cyclic oxidation were cut from the ingots, polished through 600 grit SiC, and cleaned in alcohol and acetone prior to oxidation.

Isothermal oxidation tests were run in air and argon ($pO_2 \sim 10^{-4}$atm) for times as long as one week at 900, 1000, and 1100°C. The weight change of the specimens was measured continuously using a Cahn 2000 microbalance. Cyclic oxidation tests were run in air for up to 1075 cycles at 900, 1000, 1100°C. Each cycle consisted of 45 min. at temperature and 15 min. cooling at room temperature.

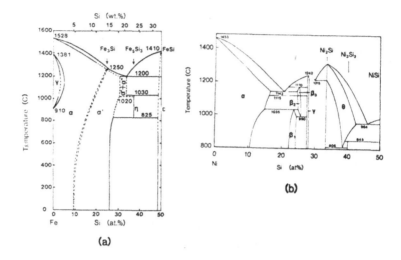

(a)

(b)

Figure 1 Metal-rich portions of (a) and Fe-Si and (b) the Ni-Si binary
phase diagrams.

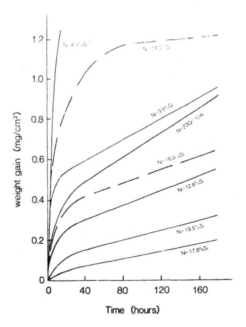

Figure 2 Isothermal oxidation rates (weight change vs time) for Ni-Si
alloys in air at 1100°C.

Specimens of Ni-20 wt pct Si were also exposed in high sulfur pressure gases (H_2-H_2 S-H_2O) to evaluate the ability of SiO_2 films to resist sulfidation.

The phases present in the corrosion products were studied using x-ray diffraction (XRD) and selected area electron diffraction (SAD). Their morphology and composition were studied using scanning electron microscopy (SEM), energy dispersive x-ray (EDX) analysis, and transmission electron microscopy (TEM).

RESULTS AND DISCUSSION

Isothermal Oxidation

The oxidation rates for Ni-Si alloys at 1100°C in air are illustrated in Figure 2. The significant features of these weight change curves are: [23]

(i)   The data for Ni-5Si are in good agreement with those of Cocking et. al.[7].

(ii)   The oxidation rates generally decrease with increasing Si concentration except for Ni-15Si, which oxidizes at a rate greater than that of Ni-10Si, and Ni-22.5Si which oxidizes slightly faster than Ni-20Si.

(iii)   The rate of oxidation of Ni-20Si is comparable to or slower than that for current Al_2O_3-forming alloys under similar conditions (Rate data for Ni-20Cr-10Al are plotted in Fig. 2 for comparison).

(iv)   Alloys containing greater than 10 wt% Si exhibit linear oxidation rates after a short initial period.

The corresponding oxidation rates for Fe-Si alloys at 1100°C in air are presented in Figure 3 [24]. The addition of 5 wt% Si decreased the oxidation rate by two orders of magnitude compared with that for pure Fe and 10 or more wt% Si resulted in oxidation rates comparable to that for NiCrAl. The weight change vs time plots are, however linear for these alloys, as was observed for Ni-Si. The oxidation behavior of both Ni-Si and Fe-Si alloys was qualitatively the same at 1000 and 900°C.

The exception of Ni-15Si to the trend of decreasing weight gains with increased Si content has not been completely explained. However, the error bars in Figure 2, which represent the data variation for multiple runs, indicate that this result is not due to experimental scatter. The shape of the weight gain curves in Figure 2 indicate that Ni-15Si undergoes more transient oxidation before a protective Si-rich scale can form. This may result from a decreased alloy interdiffusion coefficient in this alloy, which contains a significant fraction of the Ll_2-structure β -phase, but diffusion data are not adequate to test this hypothesis.

The oxide film responsible for the low oxidation rates of Ni-Si and Fe-Si alloys containing 10 wt%Si or more has been identified as SiO_2. For lower temperatures and shorter times TEM and SAD have indicated this film is vitreous (Figure 4). For long time exposures at temperatures of 900-1100 the films were indicated by XRD to be crystalline. Therefore, at lower temperatures e.g. 900-950°C the film formed as a vitreous layer and then devitrified. It has not been determined if the same sequence occurs at higher temperature, because TEM studies were not conducted, but this is considered likely. The identification of oxides formed on Fe-Si alloys at 1100°C is summarized in Table 2.

448

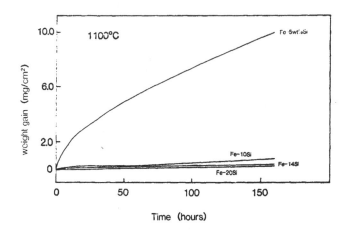

Figure 3   Isothermal oxidation rates (weight change vs time) for Fe-Si
            alloys in air at 1100°C.

TABLE 2 OXIDE SCALE FORMED ON Fe-Si ALLOYS AT 1100°C

Time	Oxidation/Si Content			
	5wt%	10wt%	14wt%	20wt%
1 hr	Fe_2O_3	$Fe_2O_3+SiO_2$[*1]	SiO_2[*1]	SiO_2[*1]
4 hrs	Fe_2O_3	$Fe_2O_3+Cristo$	$Cristo$[*2]	$Fe_2O_3+Cristo$
24 hr	Fe_2O_3	$Fe_2O_3+Tridy$[*3]	$Fe_2O_3+Cristo$	$Fe_2O_3+Cristo$
1 week	$Fe_2O_3+Fe_2SiO_4$	$Fe_2O_3+Fe_3O_4+Tridy$	$Fe_2O_3+Cristo$	$Fe_2O_3+Cristo$

*1 The structure of SiO_2 is not known, because the thickness of the
   scale was too thin to be detected by XRD.

*2 Cristo: Cristobalite

*3 Tridy: Tridymite

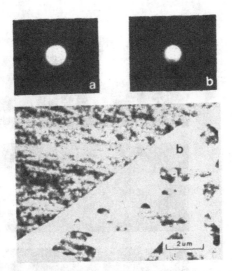

Figure 4   Transmission electron micrograph of the oxide film formed on
            Ni-20Si after one hour in air at 950°C and the selected area
            diffraction patterns of (a) the dark transient oxide (Ni_2SiO_4)
            and (b) the thin light areas (SiO_2).

Figure 5 shows the surface and cross-section of three Fe-Si alloys
oxidized for 1 week at 1100°C. The scale on Fe-5Si consists of a thick
outer layer of Fe_2O_3 and an inner layer of Fe_2SiO_4 dispersed in Fe_2O_3.
Figure 5 shows that the scale on Fe-14Si has completely spalled on cooling
but examination of the sapled oxide indicated a morphology similar to
Fe-20Si. An additional feature, observed only for Fe-14Si, was the presence
of facetted voids in the alloy substrate with filaments of SiO_2 growing
out from their base.

The linear oxidation rate observed for alloys with more than 10
wt% Si is surprising for alloys which oxidize as slowly as those in the
Fe-Si and Ni-Si systems. The slopes of the weight change vs time plots
for Ni-Si give linear rate constants in the range $1 - 2 \times 10^{-3} mg/cm^2 hr$ at
1100°C. Published data for the linear rate constant (Eqn. 1) are 0.1 mg/
cm^2/hr for oxidation of silicides such as $MoSi_2$ and WSi_2 and 0.02 mg/cm^2hr
for (100) Si [19]. (These conversions were made from oxide thickness
measurements using a density of SiO_2 of 2.27 g/cm^3 [25]). Therefore, it
is concluded that the linear rates observed are not associated with the
interface reaction of SiO_2 growth. The linear rate can, however, be
understood on the basis of Ni or Fe diffusion through a SiO_2 layer. If
this diffusion is rapid relative to the growth of the SiO_2 layer the rate
will be linear with time rather than parabolic. It was observed that
the scales on both Ni-Si and Fe-Si were initially rather pure SiO_2 e.g.
Fig. 4 and at longer times were covered with crystals of NiO or Fe_2O_3 e.g.
Fig. 5 and Table 2. In order to further investigate this phenomenon
experiments involving preoxidation at low p_{O_2} were performed.

450

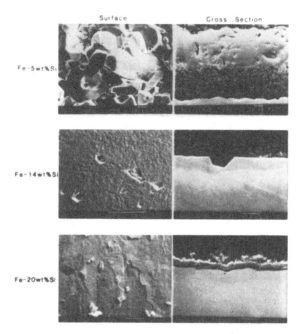

Surface   Cross Section

Fe-5wt%Si

Fe-14wt%Si

Fe-20wt%Si

Figure 5 Scale/gas interface and cross-section for three Fe-Si alloys
oxidized in air for one week at 1100°C.

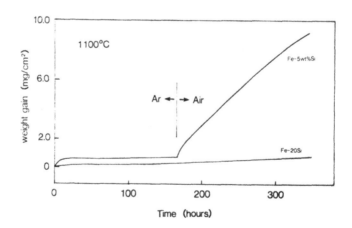

Figure 6 Isothermal oxidation rates for Fe-5Si and Fe-20Si in tank argon
($P_{O_2} = 10^{-4}$ atm) followed by air at 1100°C.

Figure 6 shows weight change vs. time data for Fe-5Si and Fe-20Si oxidized for 1 week in tank argon (residual $p_{O_2} \sim 10^{-4}$ atm) following which the atmosphere was changed to air without cooling the specimens. The oxidation rates for both alloys were quite low during the argon exposure but increased almost immediately when the atmosphere was changed to air. (The increase is not obvious for Fe-20Si because of the scale used in plotting Figure 6). The morphologies developed in Ar and Ar→air exposures for Fe-5Si are presented in Figure 7. During the argon exposure both alloys developed a continuous SiO_2 (cristobalite) film (dark phase) overlaid with small amounts of Fe_2O_3. Upon switching to air significant amounts of Fe_2O_3 began to form at the scale/gas interface, particularly for Fe-5Si. These data indicate that SiO_2 is permeable to Fe and that the permeation is more rapid at high p_{O_2}. Furthermore, it appears that the permeation is associated with diffusion of un-ionized Fe through the SiO_2 since the measured diffusivity of Fe^{3+} in amorphous SiO_2 is comparable to that for O_2 in this temperature range [26] and, thus, too low to account for the rapid growth of Fe_2O_3. Unfortunately the diffusivity of Fe^{3+} in cristobalite is not available. Acoustic emission experiments indicated that the Fe transport was not associated with cracking of the SiO_2 film.

in Ar   in Ar → Air

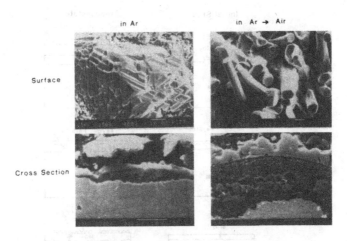

Surface

Cross Section

Figure 7   Scale/gas interface and cross-section of Fe-5Si oxidized in argon for one week followed by air for one week at 1100°C.

The oxidation mechanisms in air at 1100°C of the four Fe-Si alloys are summarized in the schematic diagrams of Figure 8. The initial stage of oxidation of Fe-5Si (N_{Si}=0.095) involves areas covered with a layered scale with an outer Fe_2O_3 and inner Fe_2SiO_4 + Fe_2O_3 layer separated by thin areas of SiO_2. However, the Si content of the alloy is too low to maintain the SiO_2 film and at steady state the two-layered scale covers the entire surface. This observation is not in agreement with the predictions of Atkinson [14] that N_{Si}=0.05 should be sufficient in this temperature range to develop and maintain a continuous layer of SiO_2. The reason for this is that Atkinsons's predictions, made primarily for applications in low p_{O_2} $CO-CO_2$ atmospheres, do not account for the growth rate of the transient Fe-oxides. It has recently been shown that the critical solute concentration for the transition from internal to external oxidation of an element such as Si can increase markedly with increased growth rate of the transient oxides. In the present case continuous SiO_2 films did form on Fe-5Si in argon ($p_{O_2} \sim 10^{-4}$atm) where the transient oxidation rate was reduced. For Fe-10Si the initial scale consists of larger areas of SiO_2 between islands of the two-layered scale which grow with continued exposure at 1100°C. At lower temperatures the thin SiO_2 layer covers most of the surface for times greater than 1 week. At 1100°C the SiO_2 layer, which may have nucleated as vitreous oxide, was observed to be cristobalite after 4 hrs and to transform to tridymite between 4 and 24 hours. This transformation was not observed in weight change vs time plots. However, a small effect would have been masked because of the significant contribution of Fe_2O_3 growth to the overall weight change. The scales formed initially on Fe-14Si and Fe-20Si were essentially pure SiO_2 which was identified as cristobalite for times from 4 hr to 1 week at 1100°C. With continued exposure Fe_2O_3 formed at the scale/gas interface due to the outward diffusion of Fe, probably metallic, through the SiO_2 layer. This resulted in essentially linear oxidation kinetics as the Fe was being transported through a SiO_2 layer of almost constant thickness. Qualitatively similar oxidation mechanisms were observed for the Ni-Si system.

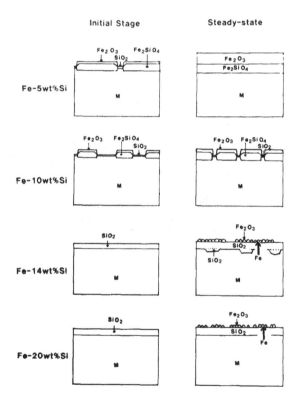

Figure 8   Schematic diagram of the oxidation mechanisms of Fe-Si alloys in air at 1100°C.

The effect of low p_{O_2} on the oxidation of Fe-Si and Ni-Si and the possible existence of an active-passive transition [27-29] were also investigated. No features associated with SiO volatilization were observed when the p_{O_2} was decreased to 10^{-4} atm (argon) even at the highest temperatures studied. However, lowering the p_{O_2} to 10^{-15} atm (H_2-H_2O) at 1150°C resulted in porosity in the alloy due to the selective volatilization of Si as SiO. This is illustrated in Figure 9.

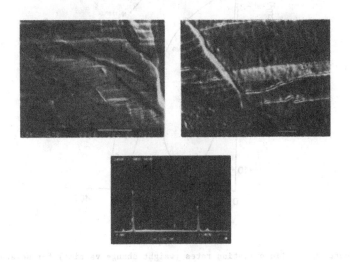

Figure 9   Surface of Ni-20Si after 30 min. oxidation in H_2-H_2O (p_{O_2}-10^{-15} atm) at 1150°C.

CYCLIC OXIDATION

The air cyclic oxidation of the Fe-Si alloys at 900, 1000, and 1100°C and that of four Ni-Si alloys at 1000 and 1100°C was investigated. The results for Ni-Si are presented in Figure 10. All four alloys showed weight losses after a small number of cycles with Ni-12.5Si exhibiting considerably larger losses than the other alloys. Data for Ni-20Cr-10Al are also plotted in Figure 10 for comparison. The Ni-22.5Si alloy was observed to change shape and dimensions (as did Fe-20Si) during cyclic oxidation exposure. The cause of this is believed to be the eutectoid reaction at 800°C (825°C for Fe-Si) indicated in Figure 1 but the reaction has not been fully characterized. The cyclic oxidation at lower temperatures (900°C) was extremely slow and the SiO_2 films were very adherent. This difference is due, in part, to the greater thermal stresses generated for the higher temperatures. However, an additional factor is believed to be that the crystalline SiO_2 formed at higher temperatures is more likely to crack on cooling due to displacive transformations, e.g. upper tridymite to lower tridymite.

454

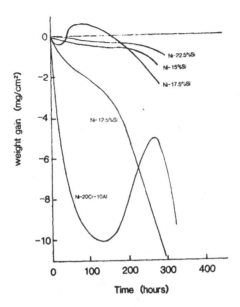

Figure 10   Cyclic oxidation rates (weight change vs time) for several Ni-
Si alloys at 1100°C (One cycle per hour).

SULFIDATION BEHAVIOR

As extensive investigation of the relative resistance to sulfidation/
oxidation of Cr_2O_3-, Al_2O_3-, and SiO_2-forming alloys has recently been
completed [30]. Figure 11 compares the corrosion behavior of Ni-20Si with
Ni-30Cr and Fe-18Cr-6Al in a gas consisting essentially of H_2 and H_2S
($p_{O_2} \sim 10^{-28}$ atm, $p_{S_2} \sim 10^{-6}$ atm) at 950°C. The Ni-30Cr alloy was almost
entirely converted to sulfide while the Fe-18Cr-6Al showed a thick sulfide
scale, which spalled on cooling, and significant attack of the underlying
alloy. The Ni-Si alloy appears rather unattacked at low magnification.
However,higher magnification shows porosity penetrating into the alloy Figure 12.
This porosity is the result of selective Si removal as a vapor species,
inferred from thermodynamic calculations to be SiS, which is analogous to the
SiO responsible for the porosity in Figure 9. Nevertheless, no condensed
sulfide was formed and when a continuous SiO_2 surface film was formed by
preoxidation in air it was protective against the formation of SiS and
resulted in extremely protective behavior of Ni-20Si in H_2-H_2S (Figure 13).
Preformed Cr_2O_3 and Al_2O_3 films were observed to be rapidly broken down in
this environment leading to catastrophic corrosion of the underlying alloy.
Thus, SiO_2 appears to be the best barrier to sulfur transport of the three
commonly employed protective oxides.

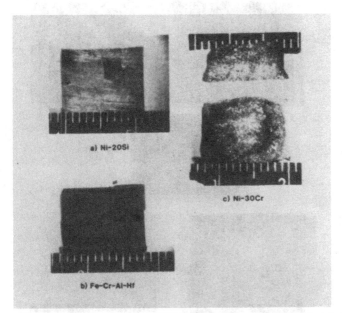

Figure 11   Photographs of (a) Ni-20Si, (b) Fe=18Cr-6Al-1Hf and (c) Ni-30Cr
exposed for 4 hours at 950°C in H_2-H_2S (p_{O_2}=10^{-28}, p_{S_2}=10^{-6}atm)

SUMMARY

The air oxidation rates of Fe-Si and Ni-Si alloys decreased with
increasing Si content above 5 wt% (with exception of Ni-15Si) at 900, 1000,
and 1100°C. Alloys containing 10 wt% Si or more oxidized at rates
comparable to or slower than Ni-20Cr-10Al due to the formation of an SiO_2
film. The oxidation kinetics were essentially linear after an initial
period due to the outward permeation of Fe-or Ni through a diffusion barrier
(SiO_2) of nearly constant thickness to form Fe_2O_3 or NiO at the scale/gas
interface. This phenomenon was not related to cracking of the SiO_2 layer.
The cyclic oxidation resistance of alloys with Si content on the order of
20 wt% was at least comparable to that of Ni-Cr-Al and SiO_2 films on
Ni-Si were observed to be much more resistant to sulfidation degradation
than either Cr_2O_3 or Al_2O_3 films.

ACKNOWLEDGMENT

The author gratefully acknowledges the collaboration of Professors
E. A. Gulbransen and F. S. Pettit, Dr. G. M. Kim, Mr. B. Warnes, and Mr.
T. Adachi in the research described above. The author also acknolwedges
the experimental contributions of Mr. K. F. Andrew and Mr. E. Hewitt and
the financial support of the Air Force Office of Scientific Research
under AFOSR Contract No. 80-0089 and the Department of Energy under
Martin Marietta Subcontract No. 19X-43346-C.

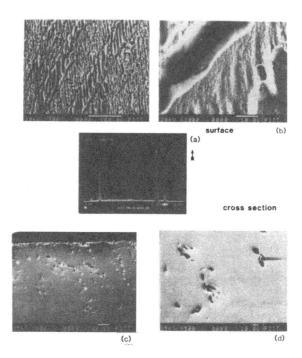

surface   (b)
(a)

cross section

(c)   (d)

Figure 12   Higher magnification micrographs of the surface (a,b) and
cross-section (c,d) of the Ni-20Si specimen in Figure 11.

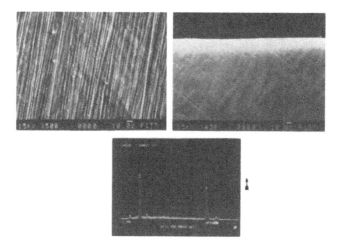

Figure 13   Surface and cross-section of Ni-20Si preoxidized 24 hours
in air at 950°C and then exposed to H_2-H_2S for 4 hours at
950°C.

REFERENCES

1. D. P. Whittle and H. Hindam, in Corrosion-Erosion-Wear of Materials in Emerging Fossil Energy Systems, edited by A. V. Levy, (NACE, 1983), p. 54.

2. G. C. Wood and F. S. Stott, in High Temperature Corrosion, edited by R. A. Rapp (NACE, 1983), p. 227.

3. L. L. Seigle, in High Temperature Oxidation-Resistant Coatings (NMAB Report, 1970), p. 20.

4. M. Levy, J. J. Falco, and R. B. Herring J. Less Comm. Metals, 34, 321 (1974).

5. J. J. Falco and M. Levy, J. Less Comm. Metals, 20, 291 (1970).

6. E. Fitzer, H-J. Maurer, W. Nowak and J. Schlichting, Thin Solid Films, 64, 305 (1979).

7. J. L. Cocking, L. D. Palmer, N. A. Burley and G. R. Johnston, to be published in Oxid. of Metals.

8. J. W. Kerr and G. Simkovich, in Properties of High Temperature Alloys, edited by Z. A. Foroulis and F. S. Pettit (The Electrochem. Soc., 1976), p. 675.

9. T. Nakayama and M. Sugiyama, J. Japan. Int. Metals, 23, 534 (1959).

10. C. W. Tuck, Corr. Sci., 5, 631 (1965).

11. R. Logani and W. W. Smeltzer, Oxid. Metals, 1, 3 (1969); 3, 15 (1971); 3, 279 (1971).

12. I. Svedung and N. G. Vannerberg, Corr. Sci., 14, 391 (1974).

13. P. T. Moseley, G. Tappin, and J. C. Riviere, Corr. Sci., 22, 87 (1982).

14. A. Atkinson, Corr. Sci., 22, 87 (1982).

15. A.Schnaas and H. J. Grabke, Oxid. Metals, 12, 387 (1978).

16. S. P. Murarka, Silicides for VLSI Applications, (Academic Press, New York, 1983), p. 134.

17. H. Jiang, C. S. Petersson, and M-A Nicolet, Thin Solid Films, 140, 115 (1986).

18. R. Beyers, J. Appl. Phys., 56, 147 (1984).

19. L. N. Lie, W. A. Tiller, and K. C. Saraswat, J. Appl. Phys. 56, 2127 (1984).

458

20. B. E. Deal and A. S. Grove, J. Appl. Phys., 36, 3770 (1965).

21. M. Hansen and K. Anderko, Constitution of Binary Alloys 2nd ed. (McGraw-Hill, New York, 1958).

22. Y. Oya and T. Suzuki, Z fur Metallk., 74, 21 (1983).

23. A. Ashary, G. H. Meier, and F. S. Pettit, "Advanced High Temperature Coatings Ssytem Beyond Curent State-of-the-Art Systems," Univ. of Pittsburgh, Final Report on AFOSR Contract No. 80-0089, April, 1986.

24. T. Adachi and G. H. Meier, "Oxidation of Iron-Silicon Alloys", submitted to Oxid. Metals.

25. M. Bartur and M. A. Nicolet, J. Electronic Mater., 13, 81 (1984).

26. A. Atkinson and J. W. Gardner, Corr. Sci., 21, 49 (1981).

27. C. Wagner, J. Appl. Phys., 29, 1295 (1958).

28. C. Wagner, Corr. Sci., 5, 751 (1965).

29. E. A. Gulbransen, K. F. Andrew, and F. A. Brassart, J. Electrochem. Sco., 113, 834, 1966.

30. G. M. Kim, Ph.D. Thesis, University of Pittsburgh, 1986.

KINETICS OF OXIDATION OF Ni ALUMINIDE EXPOSED TO OXYGEN-SULFUR ATMOSPHERES

K. NATESAN
Argonne National Laboratory, 9700 S. Cass Ave., Argonne, IL 60439

ABSTRACT

As part of a development effort on nickel aluminides based on Ni_3Al as structural materials for fossil energy applications, oxidation/sulfidation studies are being conducted at Argonne National Laboratory on materials that are being developed at Oak Ridge National Laboratory. Sheet samples of nickel aluminide, containing 23.5 at. % Al, 0.5 at. % Hf, and 0.2 at. % B were tested in an annealed condition and after preoxidation treatments. Continuous weight-change measurements were made at 875°C by a thermogravimetric technique in exposure atmospheres of air, a low-p_{O_2} gas mixture, and low-p_{O_2} gas mixtures with several levels of sulfur. Detailed analyses of the corrosion product scale layers were conducted using a scanning electron microscope equipped with an energy dispersive x-ray analyzer and an electron microprobe. The air-exposed specimens developed predominantly nickel oxide; the specimen exposed to a low-p_{O_2} environment developed an aluminum oxide scale. As the sulfur content of the gas mixture increased, the alumina scale exhibited spallation and the alloy tended to form nickel sulfide as the reaction phase. The results indicated that the sulfidation reaction in nickel aluminide specimens (both bare and preoxidized) was determined by the rate of transport of nickel from the substrate through the scale to the gas/alumina scale interface, mechanical integrity of the scale, and the H_2S concentration in the exposure environment.

INTRODUCTION

New structural materials based on nickel- and nickel-iron aluminides are being developed by Oak Ridge National Laboratory for applications in fossil energy systems such as coal gasifiers and fluidized-bed combustors. The nickel aluminide, Ni_3Al, is an intermetallic compound and has been shown to have superior strength properties for application at elevated temperatures [1]. Ni_3Al or γ' phase is also the major strengthening constituent in commercial nickel-base superalloys that are used in gas turbine applications. The high concentration of aluminum leads to the formation of aluminum-rich oxides when the material is exposed to environments with a wide range in oxygen partial pressures. The major difficulty with Ni_3Al as an engineering material, however, was its tendency for low ductility and brittle fracture in polycrystalline forms [2-4]. White and Stein [5] attributed the brittleness of polycrystalline Ni_3Al to segregation of sulfur at the grain boundaries. Aoki and Izumi [6] and Liu et al. [7-9] used microalloying processes to add small concentrations of dopants such as boron, carbon, titanium, cerium, etc., to Ni_3Al, thereby controlling the chemistry and cohesion of the grain boundaries in the material. Boron was identified to be the most effective in improving the tensile ductility at room temperature [6,7]. Liu et al. [9] have made a detailed study of the effects of boron concentration, alloy stoichiometry, and microstructure on the strength and ductility properties of Ni_3Al.

Even though the mechanical property studies on Ni_3Al (with and without varying alloying element additions) have been conducted over a wide temperature range, very limited work has been performed to evaluate the oxidation and sulfidation resistance of these materials. However, the oxidation studies conducted on a wide variety of alumina-forming alloys have shown that spallation of the oxide scale is a major drawback and that the adherence of protective oxide scales has to be improved via additions of microalloy (e.g., Y, Hf, La) constituents.

The purpose of the present work is to determine the kinetics of oxidation of nickel aluminide material and establish the composition and micro-structures of oxide scales. In addition, the material (in both bare and preoxidized conditions) was subjected to sulfur-containing environments to examine the sulfidation resistance of in-situ or preformed oxide scales on the material.

MATERIALS AND EXPERIMENTAL PROCEDURE

Nickel aluminide sheet material was obtained in a cold-worked condition from Oak Ridge National Laboratory. The material, identified as heat IC-50, had a composition of Ni-23.5 at. % Al-0.5 at. % Hf-0.2 at. % B (Ni-12.2 wt % Al-1.7 wt % Hf-0.05 wt % B). Coupon specimens of ~10 x 10 x 0.75 mm were cut from the sheet material and were given an annealing treatment by enclosing them under vacuum in vycor capsules and exposing them at 1000°C for 1 h. Based on the phase stability diagram for the Ni-Al system, the composition of the alloy falls in the region of the Ni_3Al phase. Ni_3Al has an ordered fcc structure that is isotopic with $AuCu_3$. In contrast, the NiAl phase has an ordered bcc structure isotopic with $CsCl$.

The oxidation experiments were performed using an electrobalance, made by CAHN Instruments Inc., that had a sample capacity of 2.5 g with a sensitivity of 0.1 µg. The weighing mechanism was enclosed in glass assemblies with a capability to isolate the reaction gas mixture from the balance mechanism. Helium gas was flowed through the balance assembly countercurrent to the reaction gas mixture, which flowed in the furnace chamber. The reaction chamber was connected to the balance assembly by an ~300 mm-long pyrex tube. The upper portion of the reaction chamber was baffled to minimize the heat losses and also to reduce contamination of the balance system by the reaction gas mixture (especially when it contains sulfur). The bottom portion of the reaction chamber was filled with glass balls which acted to mix and preheat the incoming reaction gas mixture prior to exposure of the specimens. The furnace consisted of three-zone Kanthal heating elements with a temperature capability of 1100°C in continuous operation.

The test specimens were suspended from the balance with an ~200-µm-diameter platinum wire. The tests were started by setting up the specimen at room temperature and purging the system with the appropriate reaction gas mixture. The specimen was heated in the reaction gas to the desired test temperature. The heating time was normally less than 1200 s. Upon completion of the test, the furnace was opened and the specimen was cooled rapidly in the reaction gas environment. The cooling time to reach ~100°C was generally less than 600 s.

Three types of pretreatments (namely, annealed and two preoxidation conditions) were used in the test program. Specimens in the annealed condition were surface-ground using diamond polish and cleaned with water and ethyl alcohol. One of the preoxidation treatments involved oxidation of the specimens at 875°C in a low-p_{O_2} atmosphere for 100 h. The second preoxidation treatment, performed by Oak Ridge National Laboratory, involved oxidation of the specimens at 1100°C for 188 h in wet hydrogen followed by a 188-h exposure in pure oxygen. Subsequent to thermogravimetric tests, the corrosion product scales were analyzed using several electron optical techniques. X-ray diffractometer analyses were made on "in-situ" scales using a GE XRD5 diffractometer with nickel filtered copper radiation. In addition, a Debye Scherrer camera with a 114.6-mm diameter was used with CuKα radiation for analysis of scales that were scraped of the surface of the exposed specimens. In addition, a scanning electron microscope (SEM) equipped with an energy dispersive x-ray analyzer and an electron microprobe in conjunction with a Kevex solid-state energy spectrometer were used to determine quantitatively the distribution of elements in scale and substrate.

RESULTS AND DISCUSSION

Oxidation in Air and Low-p_{O_2} Atmospheres

Figure 1 shows the thermogravimetric test data for the oxidation of Ni_3Al at 1000°C in air and three different low-p_{O_2} atmospheres. The weight gain data did not follow parabolic kinetics in an air environment, but an instantaneous rate constant of 1.9×10^{-12} g^2 cm^{-4} s^{-1} was obtained. In low p_{O_2}, the rate constant decreased to 1.3×10^{-13} g^2 cm^{-4} s^{-1}. Extensive analysis of the scale layers using electron optical techniques and an x-ray diffractometer showed that the alloy formed NiO in air and the HfO_2/α-alumina in the low-p_{O_2} atmospheres. Similar experiments were conducted at 875°C; oxidation rates decreased by a factor of 10 in air and by factors between 2 and 4 in low-p_{O_2} environments as the test temperature was decreased from 1000 to 875°C. Table I lists the phases identified using x-ray diffractometer analysis of specimens after exposure in several gas atmospheres.

Sulfidation in H_2-H_2S Atmospheres

Figure 2 shows the weight-change data generated for the sulfidation of Ni_3Al at 875°C in several H_2-H_2S gas mixtures containing H_2S concentrations in a range 0.38 to 4.7 vol %. Also shown in the figure (for comparison) are the weight-change data obtained for the material at 875°C in air. At H_2S levels below ~0.5 vol %, the weight-change data were essentially unaffected compared to those obtained in air or low-p_{O_2} atmospheres; at an H_2S level of 0.75 vol %, the reaction rate increased, but the data exhibited a parabolic behavior. On the other hand, at H_2S levels of 1.5 vol % or higher, the alloy exhibited accelerated corrosion as evidenced by the weight-change data. The results also indicate that the corrosion rate for the alloy under a sulfidation mode of attack is much larger than that acceptable for use of the alloy in practical applications. As a result, to be viable in fossil energy applications, the alloy must develop protective oxide scales in the exposure environment (which generally contains sulfur).

Oxidation-sulfidation in Mixed-gas Environments

Figure 3 shows the weight-change data for the Ni_3Al material that was preoxidized in the absence of sulfur at 875 or 1100°C and subsequently exposed at 875°C to several H_2-H_2S atmospheres. The samples that were preoxidized at 1100°C were prepared at Oak Ridge National Laboratory and, thus, had to be reheated during sulfur exposure at Argonne National Laboratory; the preoxidation and subsequent sulfur exposure at 875°C, conducted at Argonne National Laboratory, were performed isothermally without cooling the specimens. The data in Fig. 3 show that the preoxidation treatment (at 1100°C) resulted in a lower corrosion rate at a 1.5 vol % H_2S level than that observed on bare alloy (see Fig. 2). At a sulfur level of 4.7 vol %, the weight-change data for preoxidized specimens were somewhat lower than those for the bare alloy, but after ~100 h of exposure, the specimens exhibited accelerated corrosion. Similar behavior has been observed in other experiments conducted at 1000°C. Table I shows the x-ray diffractometer analysis of phases that formed on specimens that were preoxidized at 1100°C and sulfur-exposed at 1000°C. The preoxidation at low p_{O_2} enables formation of α-Al_2O_3, but subsequent exposure to a high sulfur level of 4.7 vol % H_2S resulted in the formation of Ni_3S_2 and HfS_2.

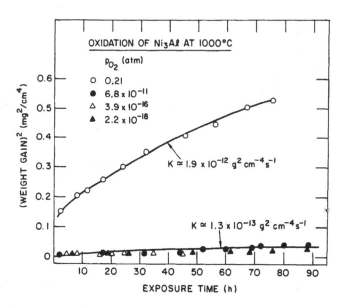

Fig. 1. Thermogravimetric test data for the oxidation of Ni$_3$Al at 1000°C.

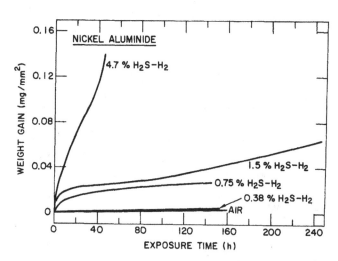

Fig. 2. Thermogravimetric test data for the sulfidation of Ni$_3$Al at 875°C.

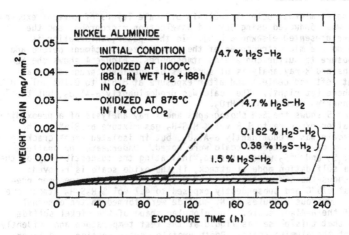

Fig. 3. Thermogravimetric test data for the sulfidation at 875°C of initially preoxidized Ni_3Al.

TABLE I. X-ray Diffractometer Analysis of Ni_3Al (IC-50) Specimens after Oxidation/Sulfidation

Temp. (°C)	Exposure Oxygen Partial Pressure (atm)	Reaction Phases Identified
1000	0.21	NiO, HfO_2
1000	6.8×10^{-11}	$\alpha\text{-}Al_2O_3$, HfO_2
1000	3.9×10^{-16}	$\alpha\text{-}Al_2O_3$, HfO_2
1000	2.2×10^{-18}	$\alpha\text{-}Al_2O_3$, HfO_2
875	0.21	NiO
875	1.2×10^{-18}	$\alpha\text{-}Al_2O_3$, HfO_2
875	6.5×10^{-21}	$\alpha\text{-}Al_2O_3$, HfO_2
1000	a	$\alpha\text{-}Al_2O_3$, HfO_2, NiO
1000	b	$\alpha\text{-}Al_2O_3$, HfO_2
1000	c	Ni_3S_2, $\alpha\text{-}Al_2O_3$, HfO_2, HfS_2

aPreoxidized at 1100°C for 376 h and subsequently exposed to 0.009 vol % H_2S/H_2.
bPreoxidized at 1100°C for 376 h and subsequently exposed to 0.162 vol % H_2S/H_2.
cPreoxidized at 1100°C for 376 h and subsequently exposed to 4.7 vol % H_2S/H_2.

Microstructural Observations

Extensive microstructural analysis of scale layers of several exposed specimens was conducted using SEM and an electron microprobe, and the results are presented elsewhere [10]. In this paper, information is presented on the microstructures of the preoxidized specimens before and after exposure to sulfur-containing atmospheres. Figure 4 shows the SEM photographs and x-ray analysis of surfaces of specimens preoxidized at 1100°C (at left and center) and after exposure at 875°C to 0.162 vol % H_2S-H_2 atmosphere (at right). The scale was predominantly α-Al_2O_3, but the grain boundaries were rich in HfO_2.

Figure 5 shows the SEM photographs and x-ray analysis of a preoxidized specimen after exposure to 1.5 vol % H_2S-H_2 gas mixture at 875°C. The external scale was predominantly α-Al_2O_3, but in isolated areas cracking and spalling of the external scale were noted. Underneath the spalled region, Ni_3S_2 and HfS_2 were observed, indicating the susceptibility of the alloy to a sulfidation mode of attack, if the oxide scale is removed. Figure 6 shows the SEM photographs and x-ray analysis of a specimen pre-oxidized at 875°C and subsequently exposed to 4.7 vol % H_2S-H_2 atmosphere at 875°C. Numerous globules of Ni sulfide were formed on the external surface of the α-Al_2O_3 scale. The globular shape of the nickel sulfide indicates that this phase was liquid at the test temperature and evidently did not wet the alumina scale. Depth profile, cross section, and taper-mount analyses of several preoxidized/sulfur-exposed specimens showed the thickness of oxide scales to be 4 to 8 µm. Nickel from the base alloy migrated through the oxide scale to the gas/scale interface and subsequently reacted with sulfur in the gas environment. The accelerated formation of the globules of sulfide on the sample surface led to increased weight change and eventually to catastrophic corrosion of the material. The microstructural analysis also showed voids in the alloy substrate (underneath the scale) and particles of (Ni,Hf) sulfides in the grain boundary regions to significant depths in the alloy which may be of concern with regard to the structural integrity of the material.

SUMMARY

A number of thermogravimetric tests were conducted to evaluate the oxidation kinetics and sulfur resistance of oxide scales developed on nickel aluminide material. In an air environment, the alloy developed predominantly nickel oxide. At low p_{O_2}, aluminum oxide formed but the grain boundary oxide phase was enriched in hafnium. The sulfur resistance of alumina scales was good up to 1.5 vol % H_2S in the gas mixture. Above this level, nickel-rich sulfide globules formed on the scale surface. The scale was susceptible to cracking and spallation. Preoxidation at 875°C for 100 h in low p_{O_2} did not improve sulfur resistance of the alloy significantly. Preoxidation at 1100°C for 376 h delayed the sulfidation attack at a 1.5 vol % H_2S level; however, if sulfidation reaction was initiated, the attack accelerated. The results indicate that the rate of sulfidation is determined by the rate of transport of nickel from the substrate to the gas/alumina interface, mechanical integrity of the scale, and the H_2S concentration in the exposure environment.

ACKNOWLEDGMENTS

This work was supported by the U. S. Department of Energy, Advanced Research and Technology Development Fossil Energy Materials Program, Work Breakdown Structure ANL-3(A), under contract W-31-109-Eng-38. The author is grateful to C. T. Liu and J. Cathcart of Oak Ridge National Laboratory

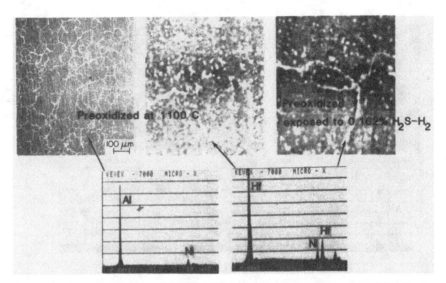

Fig. 4.   SEM photographs and EDAX analysis of Ni$_3$Al specimens preoxidized at 1100°C and exposed at 875°C to 0.162 vol % H$_2$S-H$_2$.

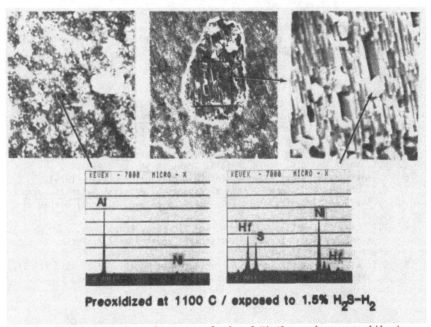

Fig. 5.   SEM photographs and x-ray analysis of Ni$_3$Al specimen preoxidized at 1100°C and exposed at 875°C to 1.5 vol % H$_2$S-H$_2$.

466

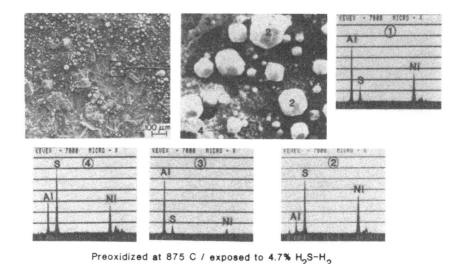

Preoxidized at 875 C / exposed to 4.7% H_2S-H_2

Fig. 6.   SEM photographs and x-ray analysis of Ni_3Al specimen preoxidized
at 875°C and exposed at 875°C to 4.7 vol % H_2S-H_2.

for supplying the material and for performing one of the preoxidation
treatments. D. L. Rink and R. W. Puccetti assisted with the experimental
program conducted at Argonne National Laboratory.

REFERENCES

1. D.P. Pope and S.S. Ezz, Intl. Metals Reviews 29(3), 136 (1984).
2. E.M. Grala, in Mechanical Properties of Intermetallic Compounds,
    edited by J.H. Westbrook (John Wiley, New York, 1960), p. 358.
3. R. Moskovich, J. Mater. Sci. 13, 1901 (1978).
4. K. Aoki and O. Izumi, Trans. Japan Inst. Metals 19, 203 (1978).
5. C.L. White and D.F. Stein, Metall. Trans. A 9A, 13 (1978).
6. K. Aoki and O. Izumi, Nippon Kinzaku Gakkaishi, 43, 1190 (1979).
7. C.T. Liu and C.C. Koch, Technical Aspects of Critical Materials Used
    by the Steel Industry, Vol. IIB, NBSIR 83-2679-9, National Bureau of
    Standards (1983).
8. C.T. Liu, C.L. White, C.L. Koch, and E.H. Lee, Proc. Symp.
    High-Temperature Materials Chemistry II, edited by Z.A. Munir,
    Electrochemical Society (1983).
9. C.T. Liu, C.L. White, and J.A. Horton, Acta Metall. 33(2), 213 (1985).
10. K. Natesan, Argonne National Laboratory report ANL/FE-87-1, to be
    published.

REACTION CHEMISTRY AND THERMODYNAMICS
OF THE Ni-Al AND Fe-Al SYSTEMS

ROBERT P. SANTANDREA, ROBERT G. BEHRENS, AND MARY A. KING
Materials Chemistry Group, Materials Science and Technology Division
Los Alamos National Laboratory, Los Alamos, NM 87545

ABSTRACT

The reaction chemistry of the Fe-Al and Ni-Al systems was studied using differential scanning calorimetry (DSC). Experimentally measured heats of reaction and compositions of reaction products were characterized in terms of sample heating rates, reaction temperature, shape and size of aluminum powder, and sample preparation. Standard enthalpies of formation were derived from DSC results. Reaction products were identified by x-ray powder diffraction techniques.

INTRODUCTION

The work reported here had two objectives. First, we sought to identify experimental conditions under which single phase material could be produced by condensed phase reactions occurring between aluminum and either nickel or iron. In an attempt to optimize the extent of such reactions, we used different types of metal powders as starting materials and, in some cases, decreased sample porosity by pressing samples into small pellets. Secondly, we attempted to determine standard enthalpies of formation of compounds in the Ni-Al and Fe-Al systems.

The condensed-phase reactions occurring between the elements were monitored using differential scanning calorimetry (DSC), and reaction products were identified by means of x-ray powder diffraction techniques. The effects of the type of aluminum powder used, sample preparation, and sample heating rate upon observed heats of reaction, product composition, and reaction temperature were examined.

A small amount of thermodynamic data is available for the Fe-Al and Ni-Al systems. Kubaschewski [1], using a calorimetric technique, measured heats of formation of nickel-aluminum alloys in the composition range 80-25 at.% Ni. Sandakov et al. [2] calculated an enthalpy of formation for NiAl(s) from calorimetry data. Biltz [3] reported a heat of formation for $FeAl_3(s)$, whereas Kubaschewski and Dench [4] calorimetrically measured heats of formation of various iron-aluminum alloys in the composition range 70-30 at.% Fe.

EXPERIMENTAL

The appropriate elements, in powder form, were blended together in the desired proportions. In the nickel-aluminum system, mixtures were prepared containing 75, 50, 40, and 25 at.% Ni, respectively. Similarly, in the iron-aluminum system, starting mixtures were prepared containing 14.3, 25, 28.6, 33.3, 50, and 75 at.% Fe. Metal powders used in these studies consisted of 3-7 μm Ni or spherical Ni, flake Al or spherical Al, and -325 mesh Fe.

Reactions occurring in the powder mixtures were monitored by a differential scanning calorimeter (Setaram DSC 111). Samples ranging in weight from 15 to 65 mg were loaded into open alumina boats and, under an argon atmosphere, heated in the DSC to a maximum temperature of 1100 K. Heating rates were varied from 5 to 30 K/min, depending on the experiment.

Reaction products were identified using Debye-Scherer x-ray powder diffraction (CuKα radiation, 114.59 mm diameter camera) photographs.

RESULTS

Ni-Al System

Three series of experiments were performed on this chemical system. In the first study, compositions ranging between 25 to 75 at.% Ni were prepared from 3-7μm nickel powder and flake aluminum and heated at 30 K/min. Identical compositions were then prepared from spherical nickel and spherical aluminum powders for use in the second set of experiments, in which samples were heated at 30 K/min. Heating rates were varied from 25 to 5 K/min. and only equimolar mixtures of spherical powders were used in the final set of experiments.

Table I summarizes the results obtained by heating mixtures of 3-7 μm nickel powder and aluminum flake in the DSC. Compositions studied are listed in the first column. The temperature range over which the onset of reaction occurs for a specific composition is given in the second column. Average standard enthalpies of reaction at 298.15 K, extrapolated from enthalpies of reaction measured at experimental temperatures, are listed in the third column. Phases present in the reaction product, as determined by x-ray diffraction, are listed in the fourth column. For compositions ranging from 25 to 50 at.% Ni, a single sharp exothermic reaction was observed. Reaction of starting materials containing 75 at.% Ni occurred in two overlapping steps; a slightly exothermic process beginning at 782 K and continuing to 1080 K, and a sharp exothermic reaction beginning between 872 and 897 K. This effect appeared to be analogous to that observed upon heating of amorphous alloys [5]. With the exception of one sample, all reaction products contained multiple phases.

Results obtained for experiments in which compositions prepared from spherical nickel and aluminum powders were heated at 30 K/min are summarized in Table II. All samples prepared from these powders reacted in a single, sharp exothermic process. With the exception of the equimolar composition of nickel and aluminum, which produced a mixture of Ni_2Al_3 and NiAl, reaction products were single phase.

Table I. Results of DSC experiments in which starting materials were prepared from 3-7 μm nickel powder and aluminum flake.

Atom % Ni	Reaction Temperature (K)	$-\Delta H^{o}_{298}$ (kJ/g-atom)	Phases Present
25	902-933	28.0	Ni_2Al_3, $NiAl_3$, NiAl, Al, Ni
40	928-933	44.4	Ni_2Al_3, NiAl
50	892-932	38.1	NiAl, Ni_2Al_3, Ni_3Al^*, $NiAl_3^*$, Al*
75	first exotherm: 782	25.4**	Ni_3Al, NiAl, Ni
	second exotherm: 872-897		

* not detected in all reaction products of samples studied with 50 at.% Ni
** combined heat of reaction of both processes

Table II. Results of DSC experiments in which starting materials were spherical nickel and spherical aluminum powders, and were heated at 30 K/min.

Atom % Ni	Reaction Temperature (K)	$-\Delta H^{o}_{298}$ (kJ/g-atom)	Phases Present
25	928-932	35.4	$NiAl_3$
40	932	48.1	Ni_2Al_3
50	928-932	50.4	NiAl, Ni_2Al_3
75	877-912	29.2	Ni_3Al

Table III summarizes results of experiments in which equimolar mixtures of spherical powders were heated at different rates. As the sample heating rate decreases, the temperature at which the exothermic reaction takes place gradually decreases. This effect may be due to a decreasing lag between the measured temperature and actual sample temperature as the heating rate is slowed. No systematic variation in reaction enthalpy was observed when the sample heating rate was changed. The average enthalpy obtained from these experiments is -51.7 ± 7.1 kJ/g-atom, which is only slightly greater than that obtained at a heating rate of 30 K/min. As with experiments in which this composition was heated at 30 K/min, reaction products obtained in this study were not single phase.

Table III. Results of DSC experiments in which the sample heating rate was varied for equimolar mixtures of spherical nickel and aluminum powders.

Heating Rate (K/min)	Reaction Temperature (K)	$-\Delta H^o_{298}$ (kJ/g-atom)	Phases Present
25	929-933	54.5	NiAl, Ni_2Al_3
20	925	45.2	NiAl, Ni_2Al_3
15	921	56.55	NiAl, Ni_2Al_3
10	918	59.22	NiAl, Ni_2Al_3, Al, $NiAl_3$
5	915	43.1	NiAl, Ni_2Al_3

Experimental results indicate that single phase nickel-aluminum alloys can be prepared from mixtures consisting of spherical powders, whereas corresponding mixtures of 3-7 μm Ni powder and Al flake yield multiphase reaction products. Possible differences in porosity of the two types of powder mixtures may account for the observed difference in reactivity.

With the exception of samples containing 75 at.% Ni, the temperature at which reaction occurs does not appear to be dependent upon composition. Reaction appears to occur just below the melting point of aluminum (933 K). The reaction temperature for samples consisting of spherical powders is reproducible; that is, little variation in reaction temperature from sample to sample is observed. The observed temperature range over which mixtures of nickel powder and aluminum flake react is generally larger than that observed for mixtures of spherical powders.

Standard enthalpies of reaction expressed in terms of kJ/g-atom are plotted as a function of composition in Fig. 1. Standard enthalpies of formation reported by

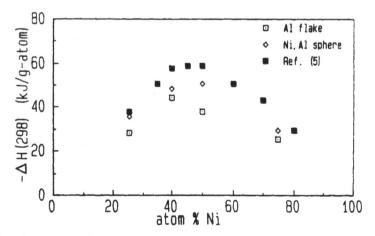

Figure 1. Enthalpies of Reaction for the Ni-Al System, Plotted as a Function of At.% Ni.

Kubaschewski [1] are also plotted in Fig. 1 for comparison. Experiments in which 3-7 μm Ni powder/Al flake mixtures were heated produced the least exothermic results. Incomplete reaction, as evidenced by the presence of several phases in the reaction products, accounts for the difference between these enthalpies and the other data. The fact that several phases are observed in all reaction products in this series of experiments precludes the use of these enthalpies as standard reference data.

Samples prepared from spherical powder yielded more exothermic results than those mentioned above and, with the exception of samples containing 50 at.% Ni, produced single phase materials. The enthalpies obtained with the spherical powders are less exothermic than those reported by Kubaschewski [1], with differences for results obtained at 40 and 50 at.% Ni being the largest. These discrepancies may arise in part from uncertainties in the enthalpy for 40 at.% Ni, as reported in Ref. [1], the incomplete reaction which we observed in experiments on 50 at.% Ni, and the fact that standard enthalpies in Ref. [1] are reported for an average temperature of 548 K.

Our studies show that the reaction chemistry of the Ni-Al system depends in part upon the type of metal powder used in sample preparation. Since the samples employed in previous work cannot be duplicated--no description of starting materials was given in Ref. [1]--exact reproduction of earlier results may not be possible.

Fe-Al System

Samples ranging in composition from 14.3 to 75 at.% Fe were prepared from spherical aluminum powder and -325 mesh iron powder and heated in the DSC at a rate of 15 K/min. In some cases, starting material was pressed into a cylindrical shape (2 mm diameter and 4 mm length) to quantitatively reduce sample porosity and promote reactivity.

Table IV summarizes the results of experiments carried out on the Fe-Al system. All samples appeared to react in a fast, exothermic manner. Samples pressed into pellets reacted

TABLE IV. Results of DSC experiments on the Fe-Al system.

Atom % Fe	Reaction Temperature (K)	$-\Delta H^o_{298}$ (kJ/g-atom)	Phases Present
14.3	897-927	13.8	Al_3Fe, Al
25	894-937	25.6	Al_3Fe, Al_5Fe_2, Al
28.6	904-929	25.4	Al_5Fe_2
33.3	901-932	23.2	Al_5Fe_2, Fe
50	911	19.6	AlFe, Al_5Fe_2, Fe
75	916-934	7.69	Fe_3Al, Fe

472

at a lower temperatures than those samples consisting of loose powders. No significant difference in heats of reaction measured for pellets and loose powders was observed. Many of the reaction products obtained in this study were multiphase, although some samples containing 50 at.% Fe and 28.6 at.% Fe did yield single phase AlFe and Al_5Fe_2, respectively.

Standard enthalpies of reaction, calculated from experimental results, are plotted as a function of iron concentration in Figure 2. The data reported by Kubaschewski and Dench[4] are also plotted for comparison. The results reported in this work are less exothermic than those given in Ref. [4], apparently because samples did not completely react.

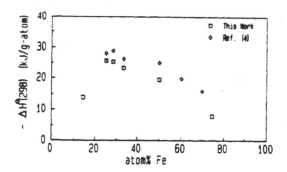

Figure 2. Enthalpies of Reaction for the Fe-Al System, Plotted as a Function At.% Fe.

REFERENCES

1. O. Kubaschewski, Far. Soc., Trans. 54, 814(1958).
2. V. M. Sandakov, Y. O. Esin, and P. V. Gel'd, Russ. J. Phys. Chem. 46, 897(1972).
3. W. Biltz, Z. Metallk. 29, 73(1937).
4. O. Kubaschewski and W. A. Dench, Acta Met. 3, 339(1955).
5. R. B. Schwarz, R. R. Petrich, and C. K. Saw, J. Non-Crystalline Solids 76, 281(1985).

TEMPERATURE AND ORIENTATION DEPENDENCE OF SURFACE
FILM EFFECTS IN SINGLE CRYSTAL NiAl

R.D. NOEBE, J.T. KIM AND R. GIBALA
Department of Materials Science and Engineering, The University of Michigan,
Ann Arbor, MI 48109.

ABSTRACT

During deformation of bcc metals and bcc-based ordered alloys, conditions of elastic and plastic constraint associated with the presence
of thin adherent surface films can be responsible for introducing increased
densities of mobile dislocations in the metal, resulting in enhanced
ductility and reduced yield and flow stresses of the film-coated materials.
In the present paper, surface film effects were investigated as a function
of temperature and crystallographic orientation for single crystal β-NiAl.
Appreciable temperature-dependent and orientation-dependent surface film
effects were observed, as were significant effects of film adherence
on the observation of surface film softening.

INTRODUCTION

Recent research has demonstrated that the presence of surface oxide
films on single crystals of β-NiAl can result in a substantial reduction
in flow stress and an increase in ductility of the base material at room
temperature [1-3]. This work was motivated by previous research on effects
of surface films on the mechanical behavior of body-centered cubic metals
at low homologous temperatures. In principle, surface film softening
can occur in any "dislocation-starved" material. The material may simply
have an insufficient density of dislocation sources and/or mobile dislocations to afford macroscopic plastic flow at lower possible flow stresses.
In this case, film-substrate compatibility interactions can then act
as a "pump" for the additional density of mobile dislocations necessary
for enhanced plasticity, as demonstrated in experiments by Ruddle and
Wilsdorf on nickel-plated copper single crystals [4]. When the substrate
material is additionally characterized by glissile dislocation species
of significantly disparate mobilities, e.g. $\langle 111 \rangle$ edge and screw dislocations in bcc metals, selective pumping of the high mobility dislocation
can result in especially large softening effects [5].

Examination of the fundamental dislocation dynamics of ordered intermetallic alloys suggests several possible manifestations of surface film
softening effects in these materials [3]. In the present paper we consider
the two different slip systems operative in single crystal NiAl (depending
upon orientation) for which surface film softening is expected to occur
by mechanisms different in detail and different in their effects on the
temperature dependence of the flow stress:

1.) $\langle 123 \rangle$-oriented single crystal NiAl deforms by motion of $\langle 001 \rangle$
dislocations on $\langle 001 \rangle \{110\}$ slip systems. Eshelby-type mobility
calculations [3,6] and computer simulations of dislocation core structures
[7] indicate that intrinsic mobilities of $\langle 001 \rangle$ edge and screw dislocations
are not significantly different. The effect of a surface film during
deformation of NiAl of this orientation is anticipated to be mainly one
of increasing the overall mobile dislocation density of the material
[3,4]. The expected effect of a surface coating on the flow stress of
such a material is portrayed in Figure 1(a), where the flow stress is
decreased in essentially equal amounts at all temperatures for which
the film-substrate interface can act as an efficient pump for additional
mobile dislocations.

Mat. Res. Soc. Symp. Proc. Vol. 81. ⸱ 1987 Materials Research Society

2.) <001>-oriented single crystal NiAl deforms by motion of <111> dislocations on {011}, {112}, or {213} slip planes. Mobility calculations and core simulation studies suggest that <111> edge and screw dislocations have significantly different intrinsic mobilities, similar to <111> dislocations in bcc metals [1-3,6,7]. In this case, the effect of a surface film may still be anticipated to increase the overall mobile dislocation density during deformation, but perhaps more importantly to bias the material with enhanced densities of highly mobile edge dislocations. With an increased density of high mobility edge dislocations, both the magnitude of the flow stress and its temperature dependence are expected to be reduced to that of a material for which the high-mobility dislocation is the principal carrier of plastic flow, as proposed in Figure 1(b).

The purpose of the present paper has been to investigate the extent to which the anticipated effects of surface films on mechanical behavior of ordered intermetallic alloys portrayed in Figure 1 can be realized experimentally. During the investigation of these temperature and orientation effects, experiments on different single crystal castings has additionally demonstrated the very significant role of film adherence during deformation, as well as possible effects of material variability, on the observation of surface film softening.

EXPERIMENTAL

The as-received NiAl single crystals were prepared by TRW Inc., Cleveland, Ohio at different periods of time during this ongoing investigation. One casting, designated A in Table I, was obtained from Dr. J.K. Doychak and is part of the same lot utilized in Doychak's oxidation studies of NiAl [8,9] and our own initial investigations of film softening in NiAl [1-3]. A second casting, designated B in Table I, was obtained directly from TRW. The primary difference in composition between the two materials is that zirconium was intentionally added to material A to improve thermal oxide adherence. Details concerning materials preparation, mechanical testing and oxide coating are presented elsewhere [1-3,10,11]. Mechanical testing was performed on 4mm x 2mm as-electropolished (uncoated) or oxide coated specimens in compression at an initial strain rate of $2x10^{-4}$/s. Oxide coatings of thicknesses in the range of 40-700 nm were formed thermally in air at 1000°C for times up to four hours. Under these conditions, the oxide formed is thought to be either delta-Al_2O_3 or twinned theta-Al_2O_3 [8,9,12]. Separate experiments involving oxide films grown at 800 and 1200°C and their effects on mechanical behavior have been reported previously [1-3,10].

TABLE I
Compositions of NiAl (at.%)

Material	Ni	Al	Si	Fe	Cr	Zr	Cu	O
NiAl (A)	52.10	47.13	0.52	0.09	0.05	0.03	0.03	0.03
NiAl (B)	51.72	47.52	0.62	0.08	0.008	<0.005	0.007	0.02

RESULTS AND DISCUSSION

The effects of oxide thickness on the room temperature mechanical behavior of <123> and <001> oriented single crystals of NiAl cut from material A are illustrated in Figures 2 and 3, respectively. For both orientations but more pronouncedly in the <123> orientation, there are substantial effects on mechanical behavior as a result of varying film thickness. The flow stress decreases initially and then subsequently increases as a function of film thickness. For <123> oriented crystals the flow stress of the film coated material eventually reaches a level nearly equal to that of the uncoated material at thicknesses of greater than approximately 520 nm. Correspondingly, strain to fracture first increases and then decreases as a function of film thickness. For both orientations, a film thickness of approximately 350 nm produces an optimum softening effect (minimum in flow stress and maximum in increased plasticity). The soft orientation <123> crystals coated with 350 nm thick films exhibit an average film-induced reduction in flow stress of approximately 20% ($\sim$ 110 MPa) and as much as a four-fold increase in true strain to fracture, to nearly 0.30 in some specimens. The hard orientation <001> crystals which deform by slip of <111> dislocations exhibit film-induced reductions in flow stress as much as 60% ($\sim$ 850 MPa) and similar four-fold increases in strain to fracture, to true strains of nearly 0.40, as well as exhibiting dramatic changes in work hardening behavior. This large difference in softening behavior for these two orientations is expected on the basis of the previous analysis of dislocation mobilities. The large intrinsic mobility difference between <111> edge and screw dislocations in the hard <001> oriented crystals should permit a much larger film-induced softening effect than observed in <123> soft orientation crystals for which <001> edge and <001> screw dislocation mobilities are not significantly different.

The majority of material A was used in initial experiments to explore the possible existence of surface film softening in β-NiAl and to determine optimum conditions for its manifestation in mechanical testing [1-3,10]. Consequently, an insufficient number of samples was available for purposes of examining the temperature dependence of the flow stress for either film-coated <123> or <001> oriented crystals of material A. Only room temperature mechanical behavior for this material was obtained, examples of which are illustrated in Figures 2 and 3. As a result, the temperature dependence of surface film effects in both <123> and <001> oriented crystals were investigated by obtaining the new material designated B. The oxidation conditions which produced maximum surface film softening in material A (1 hour at 1000°C to produce an oxide thickness of $\sim$350 nm) were employed in the temperature-dependence investigations.

Results for <123> and <001> oriented crystals of material B are given in Figures 4 and 5, respectively. Crystals of both orientations were tested at 210, 298, and 370°K. In both orientations the uncoated crystals exhibit similar flow stresses but significantly larger strains to fracture than the comparable uncoated crystals of material A shown in Figures 2 and 3. Furthermore, material A in the uncoated condition exhibits simple parabolic work hardening behavior in both <123> and <001> oriented crystals, whereas material B exhibits a more complicated stress-strain behavior. Also, material B of <001> orientation never exhibited deformation by kink band formation which is often observed at room temperature and is the dominant mechanism of deformation at temperatures just above room temperature for NiAl of this orientation [10,13,14]. In addition, uncoated <001> crystals of material B exhibited an unusual and very large strain softening behavior at low temperatures, e.g. T = 210°K. It is unlikely that the minor composition differences between these two materials can entirely explain these observed behaviors. Ongoing investigation into this problem by analytical electron microscopy is concentrating

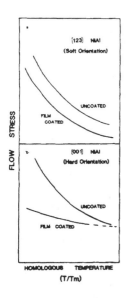

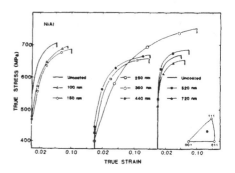

Figure 2 The effect of oxide film thickness on the mechanical behavior of <123> oriented single crystal NiAl (material A) tested in compression at room temperature. Materials were oxidized at 1000°C for various times to produce the thicknesses indicated.

Figure 1. Schematic figures of anticipated effects of surface oxide films on the temperature dependence of the flow stress of NiAl: (a) <123> oriented NiAl, (b) <001> oriented NiAl.

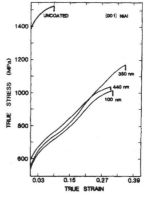

Figure 3. The effect of oxide film thickness on the mechanical behavior of <001> oriented single crystal NiAl (material A) tested in compression at room temperature. Oxidation temperature is 1000°C.

on differences in dendritic structure, subgrain boundaries, microsegregation and other microstructural features which could be responsible for
these differences in behavior between the two heats of material and the
atypical behavior of material B in general.

Inspite of these large differences in behavior between the as-received
materials A and B, the data in Figure 4 for <123> crystals of material
B illustrate that at all temperatures the film coated crystals exhibit
reduced flow stresses relative to their uncoated counterpart crystals
and that the reduction in flow stress is approximately the same (40 MPa)
at all temperatures. This result, illustrated in the inset of Figure
4, is the anticipated result based on dislocation dynamics for <123>
crystals portrayed in Figure 1(a). However, the amount of softening
is somewhat less that that observed in Figure 2 for material A. It should
also be noted that surface films do not significantly alter the already
large strains to fracture observed for uncoated <123> oriented crystals
of material B.

The data in Figure 5 for <001> crystals of material B represent
an unanticipated result. There are essentially no effects of the surface
oxide film on the mechanical behavior of this orientation of material.
The flow stresses of the <001> crystals are largely unchanged by the
presence of the 350 nm film, as are the strains to fracture. This result
when first obtained was totally unexpected and is in apparent contradiction
to the results for <001> oriented crystals of material A in Figure 3.
Also, the results of Figure 5 do not agree with the predicted temperature
dependence proposed in Figure 1(b), which is based on a large intrinsic
mobility difference of <111> edge and <111> screw dislocations in this
orientation.

The unanticipated results of Figure 5 and the smaller extent of
surface film softening observed for <123> crystals of material B compared
to that observed for <123> crystals of material A can be explained in
terms of film adherence. Previous investigations of surface film softening
of bcc metals have mainly involved effects of anodically deposited oxides
on the Group VB and VIB metals Nb, Ta, Mo, and W [3]. Anodic oxides
on these metals are amorphous, adherent to the substrate, and compliant
or plastic enough to deform in concert with the stressed substrate [15].
All of the previously mentioned oxide-bcc metal substrate systems have
exhibited very large film softening effects, much like that portrayed
schematically in Figure 1(b) and experimentally like that for <123> or
<001> NiAl crystals of material A in Figures 2 and 3. In similar fashion
to the anodic films formed on the bcc metals, the delta-Al_2O_3 or theta-
Al_2O_3 films utilized to obtain the large film softening effects in material
A in Figures 2 and 3 have been shown to exhibit remarkable adherence,
even after substantial plastic strains at room temperature [10]. In
contrast, thermal oxide films formed at 1200°C (alpha-Al_2O_3) on material
A have essentially no softening effect on the NiAl substrate [10]. These
oxide films exhibit extensive cracking and spalling on application of
even small strains, thus relieving the compatibility stresses that would
normally develop at the interface during compatible deformation and reducing the effectiveness of the interface region as a source for mobile
dislocations.

We have noted that the thermal oxides grown at 1000°C on material
A exhibit excellent adherence during low temperature plastic deformation
[10], whereas the same oxides formed on material B do not [11]. Some
typical examples of this behavior are presented in Figure 6. In Figure
6(a), it may be noted that the oxide remains predominantly intact and
adherent on <123> crystals of material A after 10% strain. The oxide
does not appreciably spall until very large strains, approaching 20-30%.
In contrast, the same oxide on <123> crystals of material B begins to
spall substantially after only a few percent strain. Figure 6(b) illustrates substantial spalling and minimal adherence of the oxide after

478

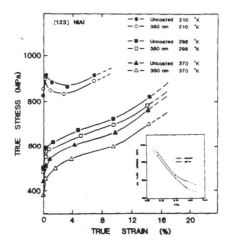

Figure 4. The temperature dependence of surface film effects on mechanical behavior of <123> oriented single crystal NiAl (material B) tested in compression. Materials were oxidized at 1000°C for 1 hour to produce a 350 nm oxide coating.

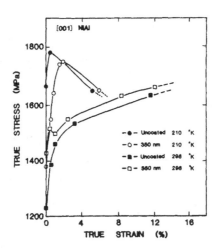

Figure 5. The temperature dependence of surface film effects on mechanical behavior of <001> oriented single crystal NiAl (material B) tested in compression. Materials were oxidized at 1000°C for 1 hour to produce a 350 nm thick oxide coating.

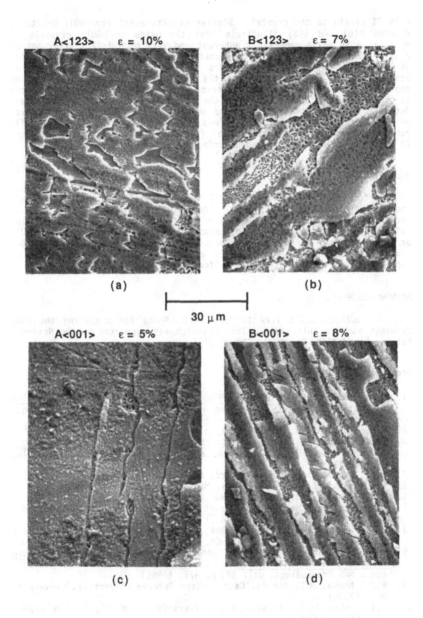

Figure 6. Scanning electron micrographs of the surface oxide films on deformed specimens of materials A and B. Oxidation conditions were 1000°C for 1 hour forming a 350 nm thick delta-Al_2O_3 or twinned theta-Al_2O_3 surface film. (a) <123> crystal, material A, ε=10%; (b) <123> crystal, material B, ε=7%; (c) <001> crystal, material A, ε=5%; (d) <001> crystal, material B, ε=8%.

480

only 7% strain to the crystal. Similar results exist for <001> crystals. Figures 6(c) and 6(d) illustrate that the oxide on <001> crystals of material A is again remarkably adherent and very much intact after appreciable plastic deformation of the substrate, whereas cracking and spalling of the oxide on material B occurs at very small strains and leaves essentially no adherent oxide on the crystal surface after strains on the order of a few percent. The reason for these qualitative differences in oxide adherence can be traced to the zirconium contents of the material. Material A was intentionally doped with additions of Zr for improved oxide adherence while material B was not. These observations on film adherence reaffirm the importance of film-substrate constraint mechanisms to the observation of appreciable surface film softening effects.

SUMMARY

Surface film softening of β-NiAl is a strong function of temperature and crystallographic orientation. The results are best interpreted in terms of intrinsic mobilities and core structures of edge and screw dislocations for crystallographic orientations which dictate either <001> or <111> operative Burgers vectors. Furthermore, adequate adherence characteristics of the surface film are critical to the observation of appreciable surface film softening effects.

ACKNOWLEDGEMENTS

The authors would like to thank A.D. Noebe for preparing the line drawings used in this paper. This research was supported by the National Science Foundation, Grant No. DMR-8506705.

REFERENCES

1. R.D. Noebe and R. Gibala, High Temperature Ordered Intermetallic Alloys, C.C. Koch, C.T. Liu and N.S. Stoloff, eds., MRS Proceedings, Vol. 39, p. 319, (1985).
2. R.D. Noebe and R. Gibala, Scripta Met. 20, 1635 (1986).
3. R.D. Noebe and R. Gibala, Structure and Deformation of Boundaries, K. Subramanian and M.A. Imam, eds., p. 89, TMS-AIME, Warrendale, PA (1986).
4. G.E. Ruddle and H.G.F. Wilsdorf, Appl. Phys. Lett. 12, 271 (1968).
5. V.K. Sethi and R. Gibala, Phil. Mag. 37, 419 (1978).
6. J.D. Eshelby, Phil. Mag. 40, 903 (1949).
7. M. Yamaguchi, Mechanical Properties of BCC Metals, M. Meshii, ed., p. 31, TMS-AIME, Warrendale, PA (1982).
8. J.K. Doychak, M.S. Thesis (1984), Ph.D. Thesis (1986), Case Western Reserve University, Cleveland, OH.
9. J.K. Doychak, T.E. Mitchell and J.L. Smialek, High Temperature Ordered Intermetallic Alloys, C.C. Koch, C.T. Liu and N.S. Stoloff, eds., MRS Proceedings, vol. 39, p. 475, (1985).
10. R.D. Noebe, M.S. Thesis, Case Western Reserve University, Cleveland, OH., (1986).
11. J.T. Kim, Ph.D. Thesis, The University of Michigan, Ann Arbor, MI., (in progress).
12. T.E. Mitchell, private communication.
13. R.T. Pascoe and C.W.E. Newey, Phys. Stat. Sol. 29, 357 (1968).
14. H.L. Fraser, M.H. Loretto, R.E. Smallman and R.J. Wasilewski, Phil. Mag. 28, 639 (1973).
15. V.K. Sethi, R. Gibala and A.H. Heuer, Amer. Cer. Soc. Bull. 57, 308 (1978).

RAPIDLY SOLIDIFIED BINARY TiAl ALLOYS

S. C. HUANG, E. L. HALL and M. F. X. GIGLIOTTI
General Electric Research and Development Center, Schenectady, New York
12301

ABSTRACT

Melt spinning has been carried out on binary TiAl alloys at three
Ti/Al ratios. Antiphase domains were observed in one ribbon specimen, but
no significant disordering was induced by the rapid solidification as indicated by X-ray and electron diffraction analyses. Bending tests of both
the ribbons and the consolidated counterparts showed a decrease in ductility with increasing Al concentration. This compositional effect can be
correlated with the TiAl tetragonality (the c/a ratio) as well as the grain
structure.

INTRODUCTION

The lightweight compound TiAl has the potential for high temperature
structural applications, particularly if its low temperature ductility can
be improved [1,2]. The deformation behavior of TiAl is related to its
crystal structure [3-9], which is based on a face-centered-tetragonal lattice cell ordered in such a way that Ti and Al atoms occupy alternating
(002) planes [10]. The formation of this compound involves a peritectic
reaction and has a phase field ranging from ~50 to ~65 at.% Al [11,12].
The present study investigated the effect of rapid solidification processing on the intermetallic as a function of phase composition. Microstructure, c/a ratio and the degree of order were examined by analytical
electron microscopy and X-ray diffractometry. The bending behavior of
individual melt-spun ribbons and of material consolidated from these ribbons was also examined. The results indicate that while a ductility-
tetragonality correlation can be found, the TiAl ductility is also influenced strongly by microstructure.

EXPERIMENTAL

Three compositions were studied: $Ti_{40}Al_{60}$ and $Ti_{48}Al_{52}$ are single-
phase, and $Ti_{52}Al_{48}$ is two-phase (TiAl+Ti_3Al). Each alloy was made into an
ingot and processed into ribbon form by the melt spinning technique [13].
The melt spinning system involved non-consumable remelting of the ingot in
a cold hearth and bottom ejection of the molten metal onto a spinning
wheel. The ribbons obtained were generally 30 to 50 micrometers in thickness. The ribbon microstructure was studied by optical microscopy and by
analytical electron microscopy (AEM). AEM was carried out using a Hitachi
H-600 microscope which was equipped with an energy dispersive X-ray spectrometer for microchemical analysis. To determine the lattice parameter,
the wheel side surfaces of the ribbons were subjected to X-ray diffraction
in a Siemens D500 diffractometer. The ductility of the ribbons was determined by bending tests, using the formula $\epsilon_b = t/(D+t)$, where t is the ribbon thickness and D is the bend fracture diameter.

Consolidation of the ribbon was carried out by cold compaction, hot
isostatic pressing (HIP) and, subsequently, extrusion. After heat treatment at 1300°C for 2 hours, specimens (1.5 x 3 x 25 mm) were machined from
the consolidated product for 4-point bending tests at room temperature.
The fracture surfaces were studied by scanning electron microscopy.

RESULTS

(A) Ribbon Microstructure

The as cast structures of $Ti_{52}Al_{48}$ and $Ti_{48}Al_{52}$ were two-phase. The as cast structure of $Ti_{60}Al_{40}$ was single-phase. Figure 1a is an optical micrograph of the $Ti_{52}Al_{48}$ ribbon. The ribbon surface that made contact with the melt-spinning wheel surface is near the bottom side of the figure. The structure indicates the presence of multiple phases, although the minor phases are not clearly discernible. No grain structure or evidence of directional growth from the bottom to the top surface of the ribbon can be seen. The microstructure of the $Ti_{48}Al_{52}$ ribbon is similar to Figure 1a, while the $Ti_{40}Al_{60}$ ribbon developed a totally different microstructure as seen in Figure 1b. This ribbon is essentially single-phase, with boundaries which seem to delineate certain solidification features. A few cracks in the plane of the ribbon can be seen.

A transmission electron micrograph obtained from a $Ti_{52}Al_{48}$ ribbon specimen is shown in Figure 2a. The microstructure consists of Ti_3Al particles in a TiAl matrix. The matrix contains numerous faults, mostly twins with a (101) twin plane. The Ti_3Al particles range from 0.1 to 1 micrometers in size and have both rounded and planar surfaces. They are essentially free of internal substructure. Energy dispersive X-ray spectroscopy revealed only a small variation in chemistry existing between the two phases. The Ti_3Al phase contains 45 at.% Al, whereas the TiAl phase contains 46-49 at.% Al. The TEM microstructure of the $Ti_{48}Al_{52}$ ribbon is similar to Figure 2a, but the Ti_3Al particles are fewer and smaller (~0.2 μm). A micrograph of the melt-spun $Ti_{40}Al_{60}$ is shown in Figure 2b, which confirms the single-phase nature of this ribbon. Areas containing antiphase domains, twins, stacking faults and dislocations can be seen, along with featureless areas which seem to develop by recrystallization during the secondary cooling after ribbon solidification.

(B) Ribbon Lattice Parameter

The TiAl lattice constants determined on the three melt-spun ribbons are plotted in Figure 3. With an increasing Al content, the c parameter increases, the a parameter decreases, and the c/a ratio increases. The measured lattice parameter-composition correlations are in agreement with Duwez and Taylor's results [10], which were obtained on equilibrated samples.

(C) Ribbon Bend Tests

Five bending tests were conducted on each TiAl ribbon. The average results on strain to fracture determined by the ribbon bending testing are

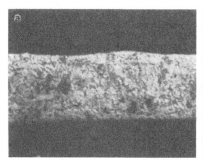

Figure 1. Optical micrographs of melt-spun (a) $Ti_{52}Al_{48}$ ribbon and (b) $Ti_{40}Al_{60}$ ribbon.

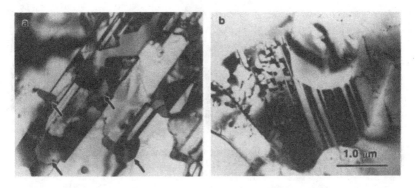

Figure 2. Transmission electron micrographs of (a) $Ti_{52}Al_{48}$ ribbon and (b) $Ti_{40}Al_{60}$ ribbon.

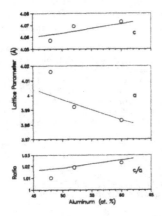

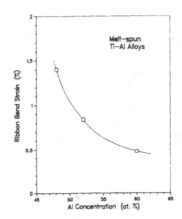

Figure 3. Lattice parameters and axial ratios of melt spun TiAl ribbons as a function of Al concentration.

Figure 4. Bending fracture strain determined on melt spun TiAl ribbons as a function of Al concentration.

shown in Figure 4. As seen, the bending ductilities are relatively low (0.49-1.41%), but show a definite trend of decreasing with increasing Al concentration. The fracture surface of the $Ti_{52}Al_{48}$ ribbon is shown in Figure 5a, which indicates a cleavage fracture. However, several of the surface features seem to reflect the ribbon microstructure. Figure 5b shows the fracture surface of the $Ti_{40}Al_{60}$ ribbon. A more planar failure is seen in this sample.

(D) Ribbon Consolidation and 4-Point Bend Tests

During the consolidation processing, the $Ti_{40}Al_{60}$ alloy experienced difficulties in machining, and no material suitable for bending tests was

484

Figure 5. Scanning electron micrographs of the bend-fracture surfaces of (a) the $Ti_{52}Al_{48}$ ribbon and the (b) $Ti_{40}Al_{60}$ ribbon.

obtained. The other two alloys were successfully consolidated, heat treated (1300°C/2 hours and subsequently 1000°C/2 hours), and machined into bending bars. The heat treatment yielded a uniform, two-phase structure for alloy $Ti_{52}Al_{48}$, having a fine grain size of ~10 micrometers. The minor phase is expected to be either Ti_3Al or transformed β-Ti. The $Ti_{48}Al_{52}$ alloy developed a single-phase TiAl structure consisting of equiaxed grains of ~50 micrometers.

The load-crosshead displacement curves obtained from the 4-point bend testing are shown in Figure 6. Both curves exhibit a deflection from the linear proportionality, indicating plastic deformation. Alloy $Ti_{52}Al_{48}$ appears to be not only stronger, but is also more ductile than $Ti_{48}Al_{52}$. The fractured surfaces of the two specimens are compared in Figure 7. Transgranular cleavage appears to be the predominating mode of failure for both alloys. However, intergranular fracture and secondary cracking can be seen, particularly in the $Ti_{48}Al_{52}$ alloy which has a grain size 5-10 times that of the $Ti_{52}Al_{48}$ alloy.

DISCUSSION

Uniform microstructures were achieved in $Ti_{52}Al_{48}$ and $Ti_{48}Al_{52}$ ribbons. No evidence of directional solidification through the ribbon thick-

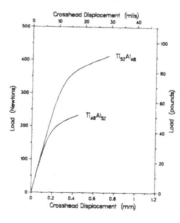

Figure 6. Load-displacement curves from 4-point bending tests of the consolidated TiAl alloys.

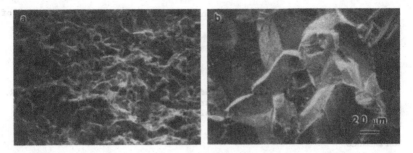

Figure 7. SEM fractography on the consolidated specimens of (a) $Ti_{52}Al_{48}$ and (b) $Ti_{40}Al_{60}$.

ness was observed. This was either the effect of a solid-state transformation which wiped out the solidification-induced microstructure, or a result of the rapid solidification conditions which reduced the degree of segregation. The formation of a second phase was, however, not suppressed. Particles of Ti_3Al have been found which were only 2-4 at.% lower in Al than the nominal alloy compositions. The Ti_3Al phase was thus highly supersaturated with Al when compared with the equilibrium solubility limit of ~37 at.%.

A single-phase TiAl was obtained in the $Ti_{40}Al_{60}$ ribbon. The structure consisted of intragranular faults such as antiphase domains, indicating some extent of disordering. However, the domains were as large as 300 nanometers. Further, our X-ray measurements showed little differences in the c/a ratio compared with data obtained from equilibrated specimens. It might therefore be concluded that no significant disordering in the TiAl lattice was induced by the present rapid solidification process.

The ribbon bending ductility decreased with increasing Al concentration. The antiphase domains in the $Ti_{40}Al_{60}$ ribbon did not appear to affect the bending behavior. A similar compositional effect was observed in the 4-point bending tests of the consolidated material. It has previously been postulated [2] that a composition near the stoichiometric TiAl might be most ductile since it showed a minimum in hardness [12,14]. This hypothesis seems to be supported by compression tests [14] as well as by the present bending tests.

The observed compositional effect might be related to the TiAl tetragonality [15]. With a c/a ratio closer to unity at a lower Al concentration, the ductility might be improved, perhaps due to more isotropic micro-deformation. On the other hand, the fractography indicated a strong involvement of the grain boundary in causing the failure of the alloys. Primary and secondary fracture along grain boundaries has been observed. It is therefore not surprising that the $Ti_{52}Al_{48}$ alloy which had the smaller grain size exhibited more ductility than the $Ti_{48}Al_{52}$ alloy.

CONCLUSIONS

Rapid solidification of $Ti_{52}Al_{48}$, $Ti_{48}Al_{52}$ and $Ti_{40}Al_{60}$ has been carried out using the melt spinning technique. Fine second-phase particles of supersaturated Ti_3Al were found in the low-Al ribbons, while a single-phase structure of TiAl containing antiphase domains was found in the high-Al ribbon. X-ray measurements showed no significant disordering resulting from the rapid solidification conditions. Bending tests of both the ribbons and the consolidated materials showed a decrease in ductility with Al concentration. This compositional effect can be related to the TiAl tetragonality (the c/a ratio) which also increases with the Al content. Additionally, the improved ductility of the hypostoichiometric alloy was

486

attributed to its fine grain size, which was stabilized by the presence of a second phase.

ACKNOWLEDGMENTS

The authors gratefully acknowledge L. C. Perocchi, R. P. Laforce, R. J. Zabala, C. P. Palmer and D. W. Marsh for their technical assistance. They also thank L. A. Johnson for helpful suggestions.

REFERENCES

[1] H.A. Lipsitt, in High-Temperature Ordered Intermetallic Alloys, Materials Research Society Symposia Proceedings, Vol. 39, ed. C.C. Koch, C.T. Liu and N.S. Stoloff, published by Materials Research Society, Pittsburgh, Pennsylvania, 1985, pp. 351-364.
[2] J.B. McAndrew and H.D. Kessler, J. Metals, 8 (1956), p. 1348.
[3] M.J. Marcinkowskli, N. Brown, and R.M. Fischer, Acta Metall., 9 (1961), p. 129.
[4] B.A. Greenberg, phys. stat. sol., 42 (1970), p. 459.
[5] B.A. Greenberg, phys. stat. sol., B55 (1973), p. 59.
[6] D. Shechtman, M.J. Blackburn, and H.A. Lipsitt, Metall. Trans., 5 (1974), p. 1373.
[7] H.A. Lipsitt, D. Shechtman, and R.E. Schafrik, Metall. Trans., 6A (1975), p. 1991.
[8] T. Kawabata, T. Kanai, and O. Izumi, Acta Metall., 33 (1985), p. 1355.
[9] G. Hug, A. Loiseau, and A. Lasalmonie, Phil. Mag., 54A (1986) p. 47.
[10] P. Duwez and J.L. Taylor, Trans. AIME, 196 (1952), p. 70.
[11] H.R. Ogden, D.J. Maykuth, W.L. Finlay, and R.I. Jaffee, Trans. AIME, 195 (1951), p. 1150.
[12] E.S. Bumps, H.D. Kessler, and M. Hansen, Trans. AIME, 196 (1952), p. 609.
[13] R.G. Rowe and R.A. Amato, in Rapidly Solidified Materials, Proc. ASM Conf., Oct. 1986, Orlando, FL, F.H. Froes, publ. ASM, Metals Park, OH, 1986.
[14] T. Tsujimoto, K. Hashimoto, M. Nobuki, and Hiroo Suga, Trans. Japan Inst. Metals, 27 (1986), p. 341.
[15] S.V. Spragins, J.R. Myers, and R.K. Saxer, Nature, 207 (1965), p. 183.

PROCESSING TECHNOLOGY FOR NICKEL ALUMINIDES*

VINOD K. SIKKA
Metals and Ceramics Division, Oak Ridge National Laboratory, P.O. Box X,
Oak Ridge, Tennessee 37831

ABSTRACT

Ductile ordered intermetallic alloys of nickel aluminum or nickel
aluminum chromium have been developed at the Oak Ridge National Laboratory
(ORNL) by optimized additions of boron. These alloys show excellent
elevated temperature mechanical properties and corrosion properties.
However, in order for the alloys to find use in various applications, they
should be fabricable by either the well established or innovative processing
technologies. The present paper will discuss the details of fabrication
technology being pursued at ORNL. The processes being investigated include
powder consolidation by extrusion, powder consolidation by capping,
isothermal forging of powder compacted material, twin-roller casting to thin
sheet followed by cold-rolling, direct casting rod from liquid, extrusion of
billets made by argon-induction melting and electroslag remelting processes,
injection molding of powders, and hot isostatic pressing of powders.
Relative merits of each process will be discussed. Mechanical properties
data on products made by various processes will also be presented and
compared.

INTRODUCTION

The intermetallic compound Ni_3Al has been ductilized [1-6] and its
mechanical and corrosion properties are being determined [7-8]. Based on
the properties, several applications have been identified for these intermetallic compounds. However, the major limit to the commercial use of these
compounds is the development of their processing technology. Significant
progress in this area has been made during the last year at ORNL, and it is
the purpose of this paper to present the development status of various processing technologies. Mechanical properties data on products made by
various processes will also be compared.

PROCESSING TECHNOLOGIES

The processing technology effort is underway on three alloys, Table I.
The IC50 alloy is ordered structure γ'-phase at room temperature and to its
melting point. The IC218 and IC221 are primarily ordered structure
(γ'-phase) with small amounts of disordered structure (γ-phase) at room temperature. The fraction of disordered structure in these alloys increases at
high temperature (>1000°C). The processing technologies explored in the
present research are on all of the above alloys and include powder processing and conventional melting and casting technology. Each of the alloys
is described below:

*Research sponsored by the U.S. Department of Energy, Office of
Conservation and Renewable Energy, Energy Conversion and Utilization
Technologies (ECUT) Program, under contract DE-AC05-84OR21400 with Martin
Marietta Energy Systems, Inc.

Table I. Composition of three nickel aluminide
intermetallics being fabricated at ORNL

Element	Weight percent		
	IC50	IC218	IC221
Al	11.3	8.5	8.5
Cr	--	7.8	7.8
Zr	0.6	0.8	1.7
B	0.02	0.02	0.02
Ni	Bal.	Bal.	Bal.

Powder Processing

Powders of nickel aluminides of all three alloy compositions can be prepared by techniques used for conventional alloys and superalloys. They have already been commercially prepared by Homogeneous Metals Inc. and Universal Cyclops. Powders have been consolidated by extrusion of loose powder at 1100°C to a reduction ratio of 8:1. The extrusion process for direct extrusion of powder has thus far been used for billet sizes of up to 8-in. diam. It is believed that this consolidation process can be scaled up to commercial sizes. Powders have also been compacted by consolidation under atmospheric pressure (CAP). The CAP process is done in a glass container and can be used to make the shape that can be formed in glass mold. The CAP processing has been used to compact billets of up to 8 in. in diameter. As compared to extrusion method, CAP compacts contain more porosity (noninterconnected). The CAP billets can be further consolidated by extrusion at 1100°C. Note however that the billet should be enclosed in a stainless steel can. Complex shapes made by the CAP process can be further consolidated by the hot isostatic pressing (HIP) process. The HIP process can also be used for direct consolidation of powder. A HIP temperature of 1120°C for three hours and a pressure of 30,000 psi was used for consolidation of 8-in.-diam billets of IC221. HIP produced 100% densification with a very fine grain size of 10 to 15 μm. HIP billets can be canned and extruded to smaller sizes. The 8-in.-diam HIP billet of IC221 was enclosed in a stainless steel can and extruded at 1100°C to a reduction ratio of 5:1. Note that the same product could have been obtained by direct extrusion of the loose powder and thereby could have eliminated the HIP step.

Isostatic isothermal forging, an extension of HIP, is the latest method used for consolidation of powders. This method uses an incompressible fluid to transmit pressures of 120,000 psi at temperatures up to 1400°C for times of 1 to 5 s. HIP uses a compressible fluid to apply the pressures. The process produces 100% dense material and retains the size of the starting powder because of minimum time at elevated temperature. This process permits the fabrication of complex shapes and the possibility of varying grain size in the same component by using the powders of different sizes.

Plasma spraying, spray forming, and injection molding are the other methods for consolidation of powders. Plasma and injection molding have been tried in a limited way at ORNL. Plasma forming has been used for nickel aluminide by Chang [8]. Spray forming for nickel aluminide is being explored at Drexel University. Explosive forming, another possibility for powder parts, is being studied by Wright and Flinn [9].

Isothermal Forging

The fine-grained (approximately 10 µm) material, obtained either by
powder extrusion or cast extrusion, show superplastic behavior at temperatures >950°C, Fig. 1. In the superplastic temperature range, total
elongation values are further increased by lowering the strain rate. The
superplastic behavior can be used for isothermal forging of fine-grained
material into near-net shape. The forging of a prototypic turbine disk, a
complex shape, Fig. 2, was performed at 1100°C and at a strain rate of
0.5 per minute. The strain rate used in these experiments was the same as
used in a full-scale commercial isothermal forging press.

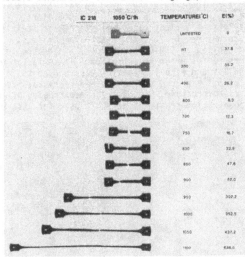

Fig. 1. Photograph showing tensile specimens of cast and extruded fine
grained (~10 µm) IC218. Tests were conducted at a strain rate of 0.05/min
in air. Super plasticity occurred in the temperature range of 950 to
1100°C.

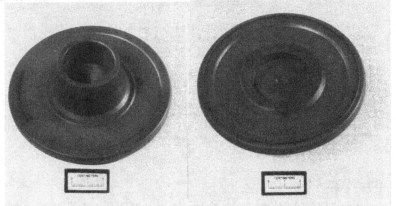

Fig. 2. Prototype turbine disks produced by isothermal forging at
1100°C and a strain rate of 0.5/min from powder extruded bar stock of IC218.

MELTING, CASTING, AND PROCESSING

Nickel aluminides (all three alloys in Table I) have been melted in an air-induction furnace with some argon flowing on the top. The recovery of aluminum and boron in these melts was over 95%. Alloys have also been produced by vacuum-induction melting (VIM), electroslag remelting (ESR), and vacuum-arc double-electrode remelting (VADER).

Billets

All melting techniques have been used to cast billets. Electroslag remelting has been used to cast up to 8-in.-diam billets.

The cast billets of various sizes were canned in stainless steel containers to extrude 3-, 4-, 5 1/2-, and 9-in.-diam billets. The canned billets were homogenized at 900°C for 16 hours and heated to 1100°C for one-half hour per inch radius prior to extrusion. The K-factor without lubrication was 43 to 45 Tsi and with lubrication was reduced to 28 to 30 Tsi. A reduction ratio of 5:1 produced a fine grained recrystallized structure from a large grained cast structure, Fig. 3. The reextrusion billet did not require the homogenization treatment at 900°C used for cast billets.

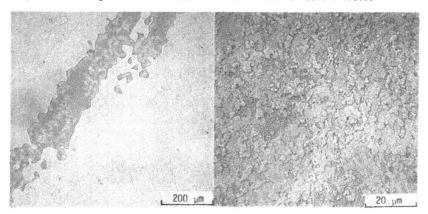

Fig. 3. Photomicrographs showing the generation of a fine grained structure in IC218 by extrusion of 8-in.-diam electroslag billet to a reduction ratio of 5:1. Extrusion was conducted at 1100°C.

Although extrusion of all three alloys was possible, the subsequent hot rolling and hot forging were not consistently successful. The IC50 composition was not hot rollable but the chromium containing alloys IC218 and IC221 were hot rollable with some success. The IC218 composition was also hot forgeable. Work is currently underway to optimize the hot rolling or hot forging conditions for all three alloys. The scale-up to large ingots (approximately 16-in. diam) and their extrusion, forging, and rolling are yet to be demonstrated.

Tube Hollows

Air-induction-melted heats of 500 lb each of IC50 and IC218 have been centrifugally cast into 5-in.-OD × 1-in.-wall × 120-in.-long tubes. The tube hollows showed excellent OD surface, but ID surface needed removal of material to some depth to clean the surface. Once the surfaces are cleaned, tube hollows will be cold Pilgered to various sizes with intermediate anneals at 1050 to 1100°C.

Sheet

The best method for thin sheet (approximately 0.8-mm thick) fabrication of aluminides has been found to be the direct casting of sheet using the twin-roller casting process. This process allows the direct casting of liquid into 1- to 2-mm-thick sheet. This sheet is cold rolled, annealed at 1050°C, cleaned, cold finished, and finally annealed at 1050°C. This process produces sheets of good surface quality and equiaxed grain structure across the thickness. Mechanical properties of this sheet will be discussed in the mechanical properties section.

Bar and Rods

Bar and rods of various sizes can be directly cast from liquid. This is done by withdrawing a small quantity of liquid in a nozzle of desired size on a starting stub. The solidified metal on the stub is pulled out and a new batch of liquid is withdrawn against the freshly solidified stub. This process is repeated until the bar can be withdrawn on a continuous basis. This process has been tried for nickel aluminides, but the first bar stock obtained was of somewhat different analysis than desired. The same process is currently being repeated at Engineering Steel Casting Ltd., Orange, California. Bar obtained by this process can be cold swaged or cold rolled, with intermediate annealing, to a final size.

Mechanical Properties

Grain size was observed to be the most significant controlling factor in the mechanical properties of nickel aluminide fabricated by various processes, Figs. 4 and 5. For example, the powder product (a very fine grain material) showed very high strength at <600°C and lower strength at >600°C. The same product also showed very high ductility for the entire temperature range RT to 1100°C. In fact, the fine grain (<10 µm) material showed superplasticity at temperatures >950°C, see Fig. 1. The coarser grain material, such as castings, showed very low tensile strength at room temperature and very high tensile strength at temperatures >600°C. Creep strength also showed a very strong grain size dependence. The cast material showed up to five times the creep strengh of that of fine grain powder product. This strong effect of grain size is related to the diffusional creep range used for the testing of these alloys.

SUMMARY AND CONCLUSIONS

It has been demonstrated that nickel aluminides can be successfully processed by powder metallurgy. The specific processes that were demonstrated include extrusion of the loose powder, CAP, HIP, and isostatic isothermal forging. Several other techniques of powder consolidation such as spray forming, injection molding, and plasma deposition are underway both at the Oak Ridge National Laboratory (ORNL) and other research centers. Significant progress has also been made in the melting, casting, and processing of nickel aluminides. Specifically the aluminides have been successfully melted using air-induction melting, VIM, ESR, and VADER. The billets made by various melting processes have been extruded both on a small

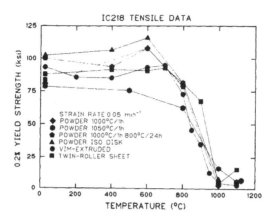

Fig. 4. Yield strength of IC218 produced by various processes.

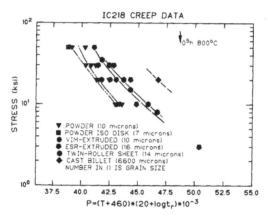

Fig. 5. Larson-Miller parameter plot showing the effect of processing method on creep rupture of IC218 over a temperature range of 649 to 927°C.

scale at ORNL and commercially at Amax Special Metals Corporation, Coldwater, Michigan. It should be noted that in all of the extrusions a stainless steel can was required to get a good extrusion. We are currently developing techniques or conditions under which either the can will not be required or we may be able to use a carbon steel can which would reduce the cost of the can significantly. The hot forging and hot rolling of the nickel aluminide alloys are still not possible although work is underway to make that successful. All three alloys under investigation at ORNL can be cold worked with intermediate anneals. This process has been demonstrated to produce sheet from twin-roller cast sheet products and also on pieces of material that have been cut from powder billets or extruded billets. Work is underway to demonstrate that centrifugally cast tube hollows of nickel aluminides can possibly be cold Pilgered with intermediate anneals to produce seamless tubing of these alloys. The product produced by various processing techniques was machined into specimens for tensile and creep testing. The most important factor causing the difference in properties between various processing is the grain size differences that one gets. For the finer grain sizes, as in the case with powder metallurgy, one gets higher tensile strength at temperature <600°C and the lower strength and higher ductility or even superplasticity at temperatures >600°C. In the case of creep, the powder product or fine grained product produced by casting gives the lowest creep strength, and as the grain size increases, the creep strength increases significantly. It is believed that the creep testing conducted in the present study has been in the temperature range of diffusional creep where minimum creep rate is inversely proportional to the square of the grain size. This implies that the larger the grain size, slower is the creep rate, and that is what was observed in our investigations. Work is currently underway to explore new near-net-shape technologies, near-net-shape fabrication technologies for nickel aluminides, and also to optimize the conditions for hot forging and hot rolling of these alloys. We are very optimistic that in the next one to two years these alloys will be commercially processed by various steel and specialty alloy companies.

REFERENCES

1. K. Aoki and O. Izumi, Nippon Kinzoku Gakkaishi, 43, 1190 (1979).
2. C. T. Liu and C. C. Koch, "Technical Aspects of Critical Materials Used by the Steel Industry, Vol. IIB," National Bureau of Standards, NBSIR83-2679-2, p. 42-1 (1983).
3. C. T. Liu, C. L. White, C. C. Koch, and E. H. Lee, Proc. Symp. High Temp. Materials Chemistry II, The Electrochemical Society, 1983.
4. C. T. Liu, C. L. White, and J. A. Horton, Acta. Metall., 33, 213—19 (1985).
5. A. I. Taub, S. C. Huang, and K. M. Chang, Metall. Trans. A, 15A, 399 (1984).
6. S. C. Huang, A. I. Taub, and K. M. Chang, Acta. Metall., 32, 1703—07 (1984).
7. C. T. Liu and V. K. Sikka, J. Metals, 19 (May 1986).
8. K. M. Chang, S. C. Huang, and A. I. Taub, "Rapidly Solidified Ni$_3$Al-B-Based Intermetallics," GE Report No. 86CRD202 (October 1986).
9. R. N. Wright and J. E. Flinn, "Consolidation and Properties of Rapidly Solidified Nickel Aluminide Powders," presented in session on Processing and Properties of Intermetallics at the Materials Week '86, Lake Buena Vista, Florida, October 2—4, 1986 (unpublished).

ROLLING ANISOTROPY OF Ni₃Al SINGLE CRYSTALS

KATSUYA WATANABE* AND MASAAKI FUKUCHI**
* Hokkaido University, Dept. of Metallurgical Engineering, Sapporo 060, Japan
** Hokkaido Polytechnic College, Otaru 047-02, Japan

ABSTRACT

The rolling anisotropy of Ni₃Al single crystals was studied. A single crystal sheet in the (011) plane showed remarkable anisotropy. Rolling the sheet in the [100] direction was simple but was almost impossible in the [0$\bar{1}$1] direction. Substantial anisotropy was not observed in the (111) and (001) sheets. The texture of the rolled (011) and (111) sheets were {011}<0$\bar{1}$1>. It is concluded that the rolling anisotropy of single crystal sheets is determined by the presence of active slip system related to compressive strain normal to the sheet plane, and tensile strain parallel to the rolling direction.

INTRODUCTION

It has been established that cast polycrystalline Ni₃Al specimens are extremely brittle, while polycrystalline specimens prepared from single crystals by strain annealing have a fairly good plasticity[1]. This suggests that the crystal orientation is an important factor controlling the plasticity of the compound. In this study, rolling anisotropy of Ni₃Al single crystal sheets was studied by examining the process of formation of rolling texture.

EXPERIMENTAL PROCEDURE

Disc-shaped single crystal sheets 1 mm thick and 10 mm in diameter were cut from Ni₃Al single crystals containing 0.5 wt % B to have sheet planes of (011), (111) and (001). After eliminating thickness differences, the sheets were annealed to remove residual stress and, after confirming the single crystal state, the sheets were cold rolled in several crystallographic directions. Each sheet was rolled in specific direction. Pole figures were prepared for the texture analysis.

RESULTS AND DISCUSSIONS

Figure 1 shows the relation between the amount of rolling reduction and the rolling direction for the (011), (111), and (001) single crystal sheets. The (011) sheet showed a remarkable anisotropy. The (011) sheet was easily rolled in the [100] direction up to more than 80 % reduction, but it was almost impossible to roll the sheet in the [0$\bar{1}$1] direction. Substantial anisotropy was not observed for the (111) and (001) sheets, but the 70 % rolling reduction was smaller than when rolling the (011) sheet in

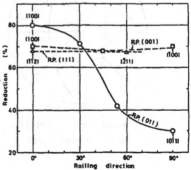

Fig. 1 Relation between rolling direction and rolling reduction.

the [100] direction.

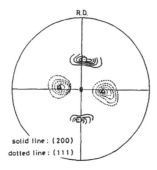

R.D.

solid line : (200)
dotted line : (111)

Fig. 2   (200) and (111) pole figures for
the (011)[100] rolling.

The pole figure of the rolling texture formed in the (011) sheet rolled
in the [100] direction is shown in Fig. 2. In this figure the solid line
corresponds to the (200) pole and the dotted line is for the (111) pole. The
figure shows that the rolling texture is (011)[0Ī1]. Rolling the (011) sheet
in the [0Ī1] direction resulted in a very low reduction value because the
initial crystallographic orientation (011)[0Ī1] is the same as the above
rolling texture and no rotation of plane and direction occurs. This same
texture is formed in the (111) sheet irrespective of the rolling direction.
Rolling the (001) sheet in the [100] direction, although the pole figure
about the same as figure 2 is obtained, the central point of the pole figure
is not the (011) and also the rolling direction is not the [011]. In the
case of the (001) sheets rolling, some other pole figures are obtained
depending on the rolling direction.

Figures 3 and 4 are stereographic projections which illustrate the
crystal rotation when rolling the (011) and (111) sheets. When the axis
perpendicular to the (011) sheet is represented by the point 011 in the unit

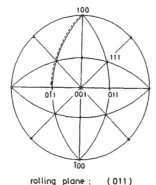

rolling plane :   (011)
rolling direction : |100|
                    |0Ī1|

Fig. 3   Stereographic projection
illustrating crystal ro-
tation due to (011) rolling.

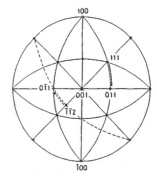

rolling plane :   (111)
rolling direction : |Ī Ī 2|

Fig. 4   Stereographic projection
illustrating crystal ro-
tation due to (111) rolling.

triangle in Fig. 3, vectors within the sheet plane are expressed by the great circle 90° apart from the point 011. This great circle is the 100-0Ī1-Ī00 arc in the figure. The crystallographic slip direction must lie on this arc, since the (011) plane does not rotate by rolling the (011) sheet. The rolling strain can be relieved only by rotation in the slip direction on the great circle, and in this case Ni₃Al can be rolled most satisfactorily in the [100] direction because it has the largest rotation angle, as shown by the dotted arrow in the figure. Fig. 4 shows the crystal rotation for the (111) sheet rolling. Point 111 in the figure, the perpendicular of the sheet plane, will rotate to point 011 as shown by the solid arrow and the slip direction will also rotate from point Ī12 to point 0Ī1 as shown by the dotted arrow. With (001) sheet rolling, the (011)[0Ī1] rolling texture was not observed. This can be understood by noting that the initial axis has several possible directions of crystal rotation and the rolling textures mentioned above are obtained. The texture observed in the (001) sheet seems to be a stage halfway to the final (011)[0Ī1] texture.

The rolling process can be simulated as follows. Two kinds of strain are formed in the sheet, one is a compressive strain, normal to the rolling plane, and the other is a tensile strain, parallel to the rolling direction. The slip systems which contribute to the rolling are the {111}<Ī01>, and the number of the slip systems to be activated for the compressive strain is four for the (011) sheet, six for the (111) sheet, and eight for the (001) sheet. When these slip systems have a sufficient Schmid factor value for the tensile axis parallel to the rolling direction, the sheet can be rolled satisfactorily. For the three sheet planes, the variations in Schmid factor of the slip systems, with the rolling direction, are shown in Figs. 5 to 7. When the

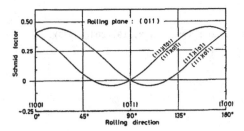

Fig. 5 Relation between rolling direction and Schmid factor for (011) sheet.

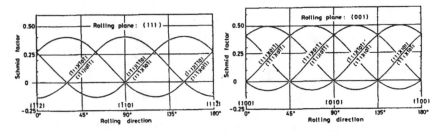

Fig. 6 Relation between rolling direction and Schmid factor for (111) sheet.

Fig. 7 Relation between rolling direction and Schmid factor for (001) sheet.

sheet plane and rolling direction are (011) and [01̄1], as in Fig. 5, none of the four slip systems will be activated because the value of the Schmid factor is zero. In (011)[100] rolling, all four slip systems have a sufficient Schmid factor value and the sheet can be rolled easily with a high value of rolling reduction. When (111) and (001) sheets are rolled, no strong anisotropy is observed because there is at least one slip system which has a reasonable Schmid factor value for any rolling direction as can be seen in Figs. 6 and 7. In these cases, however, the value of the rolling reduction, 70 %, is not as large as the value for (011)[100] rolling. The number of slip system activated can be one of the cause of the difference of the value of rolling reduction. In the (111) and (001) sheets a larger number of slip system than in the (011) sheet can be activated and it is supposed that a heavy mutual interaction between the slip systems has occured in these two sheets.

CONCLUSIONS

The (011) Ni_3Al single crystal sheet shows a remarkable anisotropy under rolling. It was easy to roll the sheet in the [100] direction but almost impossible in the [01̄1] direction. Anisotropy was not substantially observed for the (111) and (001) sheets. The texture formed by rolling the (011) and (111) sheets is (011)[01̄1]. When there are slip systems that contribute to both the compressive strain normal to the rolling plane and the tensile strain in the rolling direction, single crystal sheets can be rolled. In such a case, the sheet can be rolled more satisfactorily with a lower number of slip systems which have a larger Schmid factor for both strain axes.

REFERENCES

1. A. Aoki and O. Izumi, Trans. JIM, 19, 203 (1978).

ROOM TEMPERATURE TENSILE DUCTILITY IN POWDER
PROCESSED B2 FeAl ALLOYS

M. A. Crimp*, K. M. Vedula*, and D. J. Gaydosh**
*Department of Metallurgy and Materials Science, Case Western Reserve
University, Cleveland, Ohio 44106
**NASA Lewis Research Center, Cleveland, Ohio 44135

ABSTRACT

It has been shown that it is possible to obtain significant room
temperature tensile ductility in FeAl alloys using iron-rich deviations from
stoichiometry. A comparison of the room temperature tensile and compressive
behaviors of Fe-50at% Al and Fe-40at% Al shows that FeAl is brittle at higher
Al contents because it fractures along grain boundaries before general
yielding. Lower aluminium contents reduce the yield stress substantially and
hence some ductility is observed before fracture.

Addition of boron results in measurable improvements in ductility of
Fe-40at% Al and is accompanied by an increase in transgranular tearing on the
fracture surface, suggesting a grain boundary strengthening mechanism.
Increasing the cooling rate following annealing at 1273 K results in a large
increase in the yield strength and a corresponding decrease in ductility.

INTRODUCTION

The near equiatomic intermetallic aluminides have been receiving
considerable interest as potential structural materials for moderate to high
temperature applications [1]. One of the aluminides of interest is FeAl which
has a B2 (CsCl) crystal structure.

FeAl, like many other intermetallic alloys suffers from severe room
temperature brittleness in polycrystalline form, although single crystals
display considerable ductility with <111> {110} slip [2,3]. This brittleness
results, even though the observed <111> {110} slip should provide enough slip
systems to satisfy von-Mises' criterion.

Two possible explanations are suggested for the polycrystalline
brittleness of intermetallic alloys. Weak grain boundaries may fail before
sufficient slip systems can be activated or, limited cross-slip may result in
stress concentrations from dislocation pile-ups at grain boundaries,
resulting in brittle failure along grain boundaries [4].

Small additions of boron have been used to increase the ductility of
polycrystalline Ni$_3$Al [5]. The boron was shown to change the fracture from
intergranular to transgranular accompanied by a dramatic increase in
ductility. This transition has been attributed to a strengthening of atomic
bonds at grain boundaries by boron segregation [6]. Recently, it has been
shown that the beneficial effect of boron on the mechanical behavior of Ni$_3$Al
is dependent on cooling rate [7]. Slow cooling has been show to duplicate
the above effect, while water quenching from 1323K results in intergranular
failure due to a large decrease in the amount of boron segregation.

Results of a preliminary study in this laboratory [8] on the effect of
boron additions to near stoichiometric FeAl showed a change in fracture mode
from intergranular to transgranular, indicating a strengthening of grain
boundaries by boron addition in this alloy as well. No significant ductility
at room temperature accompanied this fracture mode change.

Gaydosh and Crimp [9] have shown that for melt spun ribbon, strain to
failure in a room temperature bend test was less than 1% for Fe-50%Al, but
was greater than 7% strain without failure for the iron-rich, Fe-40%Al
composition.

Although the effect of alloy stoichiometry on hardness [10] and bend properties [9] have been reported in the literature, its relationship to the tensile deformation of FeAl is not clear. Hence, the aim of this study is to compare the room temperature deformation properties of polycrystalline Fe-40at%Al and Fe-50at%Al. In addition, the effects of boron additions (0 to 0.2 wt.%) and several cooling rate treatments from 1273 K on the tensile behavior of these alloys have been examined.

MATERIALS AND PROCEDURE

Materials used in this study were processed by hot extrusion of prealloyed Fe-40at.%Al and Fe-50at.%Al elementally blended with boron additions of 0, 0.05, 0.1 and 0.2 wt.%. The initial starting powders had the following chemical analyses in wt%.:

	Al	N	O	Co	Cr	Mn	Ta	Fe
Fe-40Al	24.4	0.004	0.061	0.035	0.16	0.10	0.30	Bal
Fe-50Al	30.5	0.002	0.062	-	0.02	0.01	-	Bal

Chemical analysis after extrusion found no noticeable change in composition or loss of boron. A 2-step extrusion process was conducted at NASA Lewis Research Center. In the first step, 76 mm diameter mild steel cans filled with powder blends were extruded at 1250 K to a 8:1 area reduction ratio. 100 mm lengths of these extrusions were re-extruded in 50 mm diameter containers at 1073 K using an area reduction ratio of 6:1. 50 mm sections were cut and the can was removed by grinding. Heat treatments were performed at 1273 K for 24 hrs. in flowing argon followed by a number of cooling treatments as designated below.
a) slow cooling: 50°K steps/30 min. to 773K, then furnace cooled to 300K
b) air quenching: quickly removed from furnace and cooled in moving air
c) oil quenching: quickly removed from furnace and quenched in oil at 300K
d) water quenching: quickly removed from furnace and quenched in water
Tensile bars were prepared by centerless grinding, with a gage length of 30 mm and a gage diameter of 3 mm. Cylindrical compression test specimens, 5 mm dia. and 10 mm length, were also prepared.
Grain sizes were measured using standard linear intercept methods on optical microstructures. An etchant of 100 parts H_2O, 20 parts HNO_3, 3 parts HF, and 2 part HCl was effective for observing grain boundaries.
Room temperature tensile and compression tests were performed at a strain rate of approximately $3 \times 10^{-3} sec^{-1}$. The resulting tensile fracture surfaces were examined using scanning electron microscopy.

RESULTS AND DISCUSSION

Microstructures

Optical microscopy revealed equiaxed grain structures for all the extruded specimens, indicating that recrystallization has occurred. Typical microstructures are presented elsewhere in these proceedings [11]. The specimens containing boron revealed second phase precipitates within the grains which are suspected to be borides. Such particles are observed even at the lowest boron level (0.05 wt%) suggesting that the solubility of the fairly large boron atom in the matrix is less than 500 wppm.
The as-extruded structures had grain sizes of approximately 10um with changes in stoichiometry and boron addition having little effect. The heat treatment and different cooling rates increased the grain size slightly to approximately 12 um.

Effect of Stoichiometry

While Fe-50Al fails in a brittle intergranular manner in tension at room temperature, Fe-40Al displays some yielding and plastic elongation before intergranular failure. However, in compression, both Fe-40Al and Fe-50Al yield and display substantial ductility. The stress strain behavior of these alloys is displayed in Fig. 1. The major difference is that, while both alloys fail in tension at approximately the same stress level, Fe-40Al yields at a much lower stress than Fe-50Al in compression.

The compressive ductility of Fe-50Al is evidence that the required number of slip systems (presumably <111> type) can be activated for grain boundary compatability in polycrystals provided the stress can be raised high enough. In tension, fracture along grain boundaries occurs at a stress level far below the stress required for general yielding. In the case of Fe-40Al, since the stress required for general yielding is much lower, plastic deformation is observed before intergranular failure. The critical issue, therefore, is the competition between the stress for general yielding and the stress for brittle intergranular failure, which explains the onset of significant tensile ductility in the off-stoichiometric Fe-40Al compared with stoichiometric Fe-50Al.

The dramatic decrease in compressive yield strength with iron rich deviations in stoichiometry is consistent with the decrease in hardness found by Westbrook [10], but the reasons for this behavior are not clear. This behavior is particularly intriguing, since, in the other B2 aluminides, NiAl and CoAl, deviations on both sides of stoichiometry have been shown to result in defect hardening. The observation of <111> slip at room temperature in FeAl in contrast to <100> slip in NiAl and CoAl may have a bearing on this difference. Passage of a a/2<111> dislocation results in an antiphase boundary (APB) and hence dislocations must move in pairs. The antiphase boundary energy of B2 FeAl has been shown to increase with increasing aluminum content from 27 at% to 36 at% [12,13]. If this trend is extrapolated to higher aluminum contents, the APB energy for the a/2<111> superdislocations in Fe-40Al will be substantially lower than that for Fe-50Al. While a decrease in APB energy allows a wider separation of the super dislocation partials, it should be noted that, unlike stacking faults, the burgers vectors of the partials lie in the cross slip plane, and constriction of the partials is not necessary for cross slip to occur. Thus, a decrease in the APB energy may allow the dislocations to move more easily. However, the relationships between APB energy and yield strength in FeAl alloys are not well understood at this time.

Ability to cross-slip is another consideration. If deviations from stoichiometry allow <111> dislocations to cross slip from {110} planes onto {112} or {123} planes, slip across grain boundaries may be easier, resulting in a lower yield stress. A detailed explanation for the lower yield stress of Fe-40at% Al is being sought by an ongoing study of single crystal deformation which will be the subject of another paper.

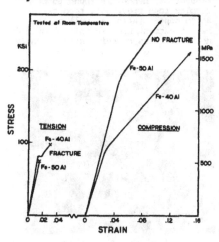

Fig. 1. Effect of stoichiometry on the stress-strain behavior of B2 Fe-50Al and Fe-40Al.

502

Effect of Boron

Addition of small amounts of boron increased the ductility of
as-extruded Fe-40Al to nearly 6%, while keeping the yield strength about the
same. The different levels of boron from 0.05 to 0.2 wt.% show no differences
in strength and ductility in the as-extruded condition. The effect of boron
on the tensile behavior of Fe-50Al is not as significant, with the alloy
continuing to display brittle failure. These results are displayed in Fig. 2.

The increase in tensile ductility of Fe-40Al with boron addition is
believed to be due to grain boundary strengthening and is reflected in the
increased amount of transgranular tearing with boron addition as shown in
Fig. 3. Improved grain boundary cohesion delays fracture beyond yielding to
the point where work hardening of the matrix leads to transgranular tearing.

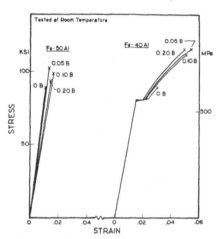

Fig. 2. Effect of boron on the
tensile behavoir of FeAl alloys.

The effect of boron on the
ductility of extruded Fe-40Al did not
vary with boron concentration between
0.05 wt% and 0.20 wt%, suggesting that
the boron effect on ductility and
fracture requires only a small
concentration of boron. Since the
solubility limit of boron appears to
have been reached, higher boron levels
will only result in additional
precipitation and will not increase
the boron concentration at the grain
boundaries.

The effects of boron in Fe-40Al
have not been reproduced in Fe-50Al.
The fracture mode remains
intergranular in all the alloys of
Fe-50Al tested. This is in contrast to
the results of the study reported
earlier [8]. The behavior of
Fe-50at%Al, hence, is found to be more
complex than previously reported.
Boron concentration as well as the
heat treatment used in this study did
not affect its behavior significantly.

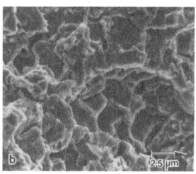

Fig. 3. SEM Micrographs displaying fracture surfaces of a) Fe-40Al
and b) Fe-40Al+0.05wt.%B.

Effect of Cooling Rate

The effect of cooling rate after a 1273 K–24hr heat treatment has a dramatic effect on the tensile behavior of Fe–40Al as shown in Fig 4. It is seen that as the cooling rate is increased, a transition from ductile to brittle behavior occurs. This is related to a large increase in the yield strength with quenching. Additionally, as the cooling rate is increased, the effect of boron on the fracture surfaces is enhanced. Figure 5 shows SEM fractographs of oil quenched Fe–40Al and Fe–40Al+0.05wt%B. Fe–40Al is seen to continue to display intergranular fracture, while the boron containing alloy now displays complete brittle cleavage.

The results of the cooling rate effect on Fe–40Al are not well understood at this time. The increased amount of cleavage fracture with increased cooling rate in samples containing boron is contrary to what has been reported in Ni$_3$Al [7]. This may be due to the considerable amount of metallic contaminants in the material used in this study. These contaminants may be inhibiting the boron effect at slow cooling rates due to grain boundary segregation. At higher cooling rates, the contaminants may be quenched in solution, allowing the boron to contribute more to grain boundary strengthening.

The increase in yield strength with increasing cooling rate is similar to results reported for Fe–Co–2%V [14], another B2 alloy. The explanation for this is not entirely clear. However, an increase in strength due to increased quench temperature has been reported for β–brass [15]. This strengthening has been attributed to vacancies generated during the order–disorder transformation. A similar mechanism may be occurring in Fe–40Al as large concentrations of thermal vacancies are retained in B2 ordered alloys [16] in alloys quenched from high temperatures.

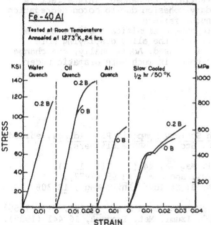

Fig. 4. Effect of cooling rate on the tensile behavior of Fe–40Al alloys.

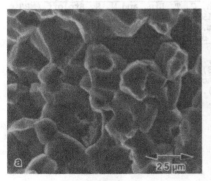
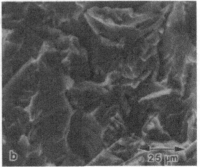

Fig. 5. SEM micrographs displaying fracture surfaces of oil quenched a) Fe–40Al and b) Fe–40Al+0.05wt.%B.

It was not possible to study the effect of cooling rate on the tensile behavior of Fe-50Al due to the fact the quenching resulted in considerable microcracking. However, the effect of cooling rate on the observed fracture surfaces was similar to that observed in Fe-40Al, with the boron resulting in increased cleavage fracture as the cooling rate increased.

CONCLUSIONS

It has been shown that the B2 FeAl alloy can have significant tensile ductility at room temperature if the composition is kept at about Fe-40at.%Al. The reason for this ductility is believed to be due to a low yield strength, allowing some yielding before fracture. Fe-50Al has a much higher yield strength and hence breaks at weak grain boundaries before plastic deformation can begin.

Addition of boron can cause a significant improvement in ductility of Fe-40Al while the yield strength is essentially unchanged. This effect is presumably through improved grain boundary cohesion due to boron segregation and is evidenced by increased transgranular failure.

The cooling rate has been shown to have a significant effect on the ductility and yield strength of Fe-40Al, with the alloy displaying more brittle behavior as the cooling rate is increased. Additionally, the change in fracture mode due to the addition of boron is much more dramatic under conditions of high cooling rate.

REFERENCES

1. J. R. Stephens, Materials Research Society Symposia Proceedings, edited by C. C. Koch, C. T. Liu, and N. S. Stoloff (MRS Publishers, Pittsburgh) 39 381 (1985).
2. Y. Umakoshi and M. Yamaguchi, Phil. Mag. A, 44 (3), 711 (1981).
3. T. Yamagata and H. Yoshida, Mat. Sci. and Eng., 12 95 (1973).
4. T. L. Johnson, R. G. Davies, and N. S. Stoloff, Phil. Mag., 12 305 (1965).
5. C. T. Liu, C. L. White, and J. A. Horton, Acta Met., 33 (2), 213 (1985).
6. T. Ogura, S. Hanada, Masumoto and O. Izumi, Met. Trans.A, 16 441 (1985).
7. A. Choudhury, C. L. White, and C. R. Brooks, Submitted to Scripta Met.
8. M. A. Crimp, and K. Vedula, Mat. Sci. and Eng., 78 193 (1986).
9. D. J. Gaydosh and M. A. Crimp, Materials Research Society Symposia Proceedings, edited by C. C. Koch, C. T. Liu, and N. S. Stoloff (MRS Publishers, Pittsburgh) 39 429 (1985).
10. J. H. Westbrook, J. Electro. Chem. Soc., 103 54 (1956).
11. K. M. Vedula, 1986 MRS Symposium on High-Temperature Ordered Intermetallic Alloys.
12. R. C. Crawford, and I. L. F. Ray, Phil. Mag., 35 (3) 549 (1977).
13. R. C. Crawford, Phil. Mag., 35 (3) 567 (1977).
14. N. S. Stoloff and R. G. Davies, Acta Met., 12 473 (1964).
15. N. Brown, Acta Met., 7 210 (1959).
16. A. Fourdeux and P. Lesbats, Phil. Mag. A, 45 81 (1982).

PART VI

Late Paper Acceptance

HIGH TEMPERATURE MECHANICAL BEHAVIOR OF SOME ADVANCED Ni$_3$Al

S. E. HSU, N. N. HSU, C. H. TONG, C. Y. MA AND S. Y. LEE
Chung Shan Institute of Science and Technology
P. O. Box 1-26-4, Lungtan, Taiwan 32500, Republic of China

ABSTRACT

High temperature mechanical properties of various Zr and Cr strengthened single phase Ni$_3$Al are investigated, with emphasis on the ability of each element to elevate Tp, the temperature corresponding to the peak yield strength. It is observed that Zr is a very effective strengthener, more so below Tp than above it, while a combination of Cr and Zr is capable of shifting Tp to a higher temperature. The combination results in an effective improvement of the rupture strength of Ni$_3$Al. The strengthening mechanisms of each element will be discussed in this paper.

INTRODUCTION

There has been tremendous interest in Ni$_3$Al development in recent years. The anomalous temperature dependence of the yield strength, the inherent good fatigue strength [1] and oxidation resistance, and a lower density compared to nickel base superalloys make Ni$_3$Al a potential alloy for high temperature applications. Among various properties which are important to engineering materials, the high temperature creep strength [2] and ductility [3] are major concerns in current stage of Ni$_3$Al development.

The temperature dependence of yield strength has been studied extensively in binary and ternary Ni$_3$Al with such ternary elements as Ta, Nb, Zr, and Hf, [4,5] etc. Among them, Hf and Zr were found to have very high strengthening effect. However, very few creep data has been published so far. The fact that the creep strength of a 0.5 at% Hf strengthened Ni$_3$Al [6] was moderate when compared to wrought superalloys indicates that development of stronger alloys out of Ni$_3$Al is highly desirable.

In the present work, high temperature mechanical properties, especially the rupture strength, of Zr and Cr strengthened Ni$_3$Al are investigated. Since the mechanical properties of Ni$_3$Al depend on the phases presented in the alloys, only those alloys which have a single L1$_2$ phase, or at least very close to it, will be examined. The attention will be further limited to those aluminides which have enough formability to be considered as wrought alloys.

EXPERIMENTAL

99.9% pure Ni and Al were vacuum induction melted along with high purity Ni-Zr, Ni-Cr and Ni-B master alloys. The compositions of the specimens are listed in Table I, 500 ppm boron was added in each case. Ingots were sliced and cold rolled repeatedly into strips of around 1mm in thickness. Intermediate annealing was carried out from 1000°C to 1150°C in air for 1 to 2 hours.

After the TMT, samples were taken for metallographic examinations and chemical analyses. Plate type tensile test specimens with gauge dimensions of 1mm × 3mm × 15mm were then machined from the strips. Room and elevated temperature tensile tests were conducted in air using a MTS machine. The strain rate was set at 10^{-4}s^{-1}. Stress rupture tests were conducted at 760°C using a Satec machine. The fracture surfaces from mechanical tests were examined in a Philips 500 SEM with an EDX system.

Table I. Chemical Composition of Alloys Tested (at.%)

Specimen No.	Design Composition	Specimen No.	Design Composition
A-1	$Ni_{77}Al_{23}$	C-1	$Ni_{76}Al_{17}Zr_1Cr_6$
A-2	$Ni_{77}Al_{22.5}Zr_{0.5}$	C-2	$Ni_{75}Al_{18}Zr_1Cr_6$
A-3	$Ni_{77}Al_{22}Zr_1$	C-3	$Ni_{74.5}Al_{18.5}Zr_1Cr_6$
A-4	$Ni_{77}Al_{21.5}Zr_{1.5}$	C-4	$Ni_{73}Al_{20}Zr_1Cr_6$
A-5	$Ni_{77}Al_{21}Zr_2$	D-1	$Ni_{74}Al_{16.5}Zr_{1.5}Cr_8$
B-1	$Ni_{77}Al_{17.5}Zr_{1.5}Cr_4$	D-2	$Ni_{73.5}Al_{17}Zr_{1.5}Cr_8$
B-2	$Ni_{76.5}Al_{18}Zr_{1.5}Cr_4$	D-3	$Ni_{72.5}Al_{18}Zr_{1.5}Cr_8$
B-3	$Ni_{75.5}Al_{19}Zr_{1.5}Cr_4$		

RESULTS AND DISCUSSION

Metallography

Typical microstructures, before and after the TMT, are shown in Figure 1 where binary and alloyed Ni_3Al are compared. The as-cast structure is typified by an ordered γ' phase interpenetrated by arrays of dendrites which are composed of the $\gamma + \gamma'$ eutectic structures. The dendrites increase in volume with an increase in degree of departure from stoichiometry. Addition of Zr and Cr notably refines the as-cast dendritic structure. From Figure 1, it is clear that the TMT has effectively broken up the dendrites which would otherwise take very long to homogenize. The TMT also has resulted in a fully recrystallized equiaxed structure in every specimen.

A combination of optical and scanning electron microscopy and X-ray diffraction was used to detect the phases presented in each specimen. The results showed that there was no apparent shift in the Ni-rich γ' phase boundary upon Zr additions if Zr was taken as a substitute for Al. When Cr was added together with Zr, the phases revealed by metallographs were deviated only slightly from the phase diagram obtained by Taylor's [7] for the Ni-Al-Cr ternary system. Figure 2 is a reproduction of Taylor's phase diagram on which our alloys are marked. Closed symbols in Figure 2 represent alloys with a single γ' phase structure, while open symbols represent those with the $\gamma' + \gamma$ or the $\gamma' + \beta$ two phases structures.

At 6 at% Cr level, the single-phased C-4 specimen had only such a limited formability that the TMT schedule had to be modified to make sound specimens. As the Cr content was further increased to 8 at%, the TMT became even more difficult. No single phase material was obtainable for the 8 at% Cr specimens over the aluminum content studied. Specimen D-3 which was closest to the single phase region contained around 1% β phase by volume.

Mechanical Properties

The tensile properties of ternary $Ni_3(Al,Zr)$ specimens with various amount of Zr are plotted in Figure 3. It is noted that Zr is an effective

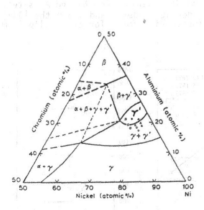

Fig. 1 Optical micrographs showing dendritic or grain structures of (a)-(c): as-cast $Ni_{77}Al_{23}$, $Ni_{77}Al_{22}Zr_1$ and $Ni_{75.5}Al_{19}Zr_{1.5}Cr_4$; (d)-(f): TMTed $Ni_{77}Al_{23}$, $Ni_{77}Al_{22}Zr_1$ and $Ni_{75.5}Al_{19}Zr_{1.5}Cr_4$, respectively.

strengthener. It increases the yield strength of Ni_3Al in the entire tested temperature range. The ductility of $Ni_3(Al,Zr)$ follows the same trend as that of binary Ni_3Al. When Zr is added beyond its solubility, ternary $Ni_3(Al,Zr)$ becomes brittle even at room temperature.

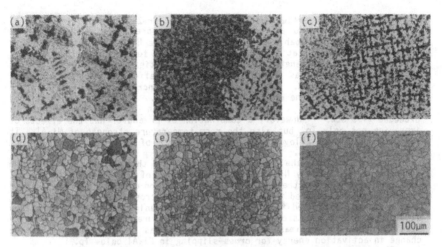

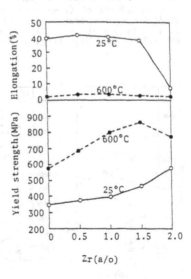

Fig. 2 A reproduction of Taylor's diagram upon which the alloys under study are marked, open symbols are alloys with two-phase structure, while closed symbols are those with single γ' phase structure.

Fig. 3 Yield strength and elonga-tion of $Ni_{77}(Al,Zr)_{23}$ as a function of Zr content.

510

The tensile properties of quarternary Ni-Al-Cr-Zr are shown in Figure 4. in which alloys with variant compositions are compared. It is observed from Figure 4 that Cr is capable of increasing both the high temperature strength and ductility of Ni₃Al at the expense of some room temperature ductility.

When the yield strengths of alloys under investigation are plotted against the test temperature, as shown in Figure 5, the effect of each strengthener can be closely compared. The addition of Zr has increased the yield strength of binary Ni₃Al over the entire temperature range. The strengthening effect of Zr is saturated at a temperature greater than Tp when its content has exceeded 1.5 at%. However, a combination of Zr and Cr has the effects of improving not only Tp, but also the room temperature strength of Ni₃Al. It has resulted in an alloy with a yield strength of over 680 MPa from room temperature up to 800°C.

The results of the present work indicate that Zr is an effective strengthener in ternary Ni₃(Al,Zr) on both sides of Tp. Mishima et al. [5] have studied the strengthening effects of transition elements on Ni₃Al. Their results were interpreted by a thermal activated cross-slip mechanism which, they believed, accounted for most of the strengthening effects of transition elements below Tp, as shown schematically in Figure 6(a). Present results seem to agree to Mishimas' assessment that Zr has caused a large negative change in activation energy for cross-slipping in Ni₃Al below Tp.

The same mechanism can not be accounted for the strengthening effect of Zr on Ni₃Al above Tp, because in this temperature regime, the deformation is believed to be mainly caused by {100} cubic slips. Elements which promote cross-slip to {100} planes will not cause an increase in strength above Tp, as indicated in Figure 6(a). On the contrary, they usually make the subsequent slips on {100}, a lower energy process compared to those which find the cross-slip, difficult. Therefore, the strengthening effect of Zr at temperatures higher than Tp can not be explained by the cross-slip process. It is possibly due to increasing Peierls' stress for dislocations movement.

The effect of Cr on mechanical properties of Ni₃Al has not been well studied. Cr has been considered to partition evenly among Ni and Al sites. It is likely that Cr would reduce the ordering tendency and lower the {111} anti-phase aboundary energy of Ni₃Al. The driving force for cross-slip from

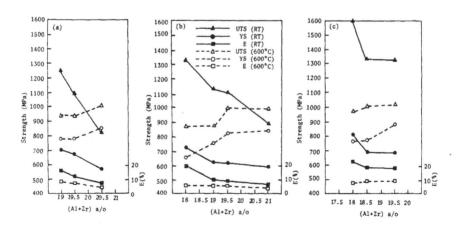

Fig. 4 Tensile properties of nI-Al-Zr-Cr quarternary alloys as a function of (Al+Zr) content, (a) 4 at% Cr, (b) 6 at% Cr, (c) 8 at% Cr.

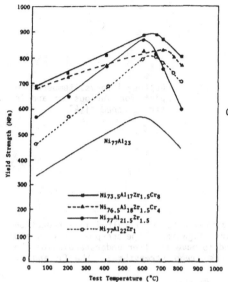

Fig. 5 Yield strength of binary or
alloyed Ni₃Al as a function
of temperature.

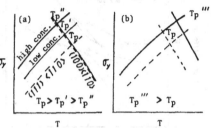

Fig. 6 Schematic drawing showing the
general trend of the yield
strength as a function of
temperature for alloyed Ni₃Al,
(a) assuming that solutes
would promote cross-slips of
dislocations to {100} plane,
(b) that actually observed in
Zr or (Zr+Cr) strengthened
Ni₃Al.

{111} to {100} would be reduced accordingly.   Therefore, it seems that Cr
would reduce the anomalous yield behavior of Ni₃Al. This effect was not
observed in our experiment.   Instead, it was found that Cr, in combination
with Zr, made a great improvement in Ni₃Al's strength over the entire tested
temperatures.   In the meantime, the anomalous yield behavior was perserved.

Stress Rupture Properties

The 760°C rupture strengths of some Ni-Al-Zr-Cr are plotted in Figure 7.
Only the 4 at% and the 6 at% Cr containing specimens with the highest equivalent Al content are single phase materials.   They also have longer rupture
lives.   These two aluminides have their 760°C, 100-hour rupture strengths of
over 500 MPa, which are much higher than that of the binary Ni₃Al's.

When specimens with the same Al + Zr content are compared, the rupture
strength increases with Cr content.   However, it is to be emphasized that the
aluminum content is an important factor in rupture strength comparison.   The
rupture lives of the specimens containing 4 at% or 6 at% Cr were found to
increase with aluminum content.   The stoichiometric effect on creep properties
of Ni₃Al is an important issue and will not be dealt in details in this work.

However, it is interesting to note that there is an inversion in strengths
when Cr containing specimens are evaluated for their tensile or creep properties.   Single-phased γ' specimens have longer lives than the dual-phased γ' +
β ones during creep tests, even though the tensile tests indicated the other
way, the phenomenon is certainly a strain rate effect.   It is possible that
the initial grain sizes of the dual phase specimens are smaller.   However,
there have also been evidences of extensive recrystallization along grain
boundaries in 760°C, 400 MPa rupture test specimens.   The initial grain size
would affect only the transient behavior of the creep process.   Therefore, in
addition to the grain size effect, it is also possible that a stable γ phase

512

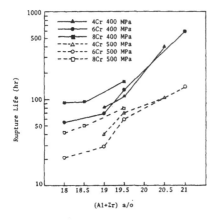

Fig. 7 Rupture life vs. temperature
plot for various Zr and Cr
strengthened Ni₃Al.

has caused an increase in recrystallization rate and resulted in a reduction
in the stress rupture life. Further investigation of the creep mechanisms and
the deformed microstructures are needed to have a better understanding of the
discrepancy between the tensile and the creep strengths of the Cr and Zr
strengthened Ni₃Al.

SUMMARY AND CONCLUSIONS

Near Ll_2 single-phased wrought nickel aluminides were investigated in
this work to search for a material with better mechanical strengths, especially
the rupture strength. It is found that Zr is effective in improving the high
temperature strengths of Ni₃Al, more so at temperatures below Tp than above
it. The strengthening effect of Zr saturates at a level of 1.5 at%.
When added along with Zr, Cr is capable of further improving the strengths
of Ni₃Al up to 800°C. High temperature ductility of Ni₃Al is also improved by
Cr and Zr. It is also found that the rupture strength of Ni₃Al is a strong
function of aluminum content. Ni-19Al-6Cr-1.5Zr alloy has a 760°C, 400 MPa
rupture life of over 600 hours, longer than that of Ni-18Al-8Cr-1.5Zr alloy
which, on the other hand, has a higher tensile strength at 760°C.
The strengthening effect of Zr and Cr can not be fully interpreted by
any single mechanism proposed so far. Creep data and deformed microstructures
are certainly needed in order to have a clear picture of the whole deformation
processes under creep.

REFERENCES

1. N. S. Stoloff and A. K. Kuruvilla, Private Communications, October 1984.
2. "Strucutre Uses for Ductile Ordered Alloys," NMAB-419, National Academy
   Press, Washington, D.C. (1981), 74.
3. C. T. Liu, C. L. White and E. H. Lee, Scr. Metall., 19(1985), 1247.
4. T. Takasugi, O. Izumi and N. Masahashi, Acta Metall., 33(1985), 1259.
5. Y. Mishima, S. Ochiai, M. Yodogawa and T. Suzuki, Trans. Japan Inst.
   Metals, 27(1986), 41.
6. J. H. Schneible, G. F. Petersen and C. T. Liu, J. Mater. Res., 1(1)(1986),
   68.
7. A. Taylor and R. W. Floyd, J. Inst. Metals, 81(1953)451.

Author Index

Subject Index